Q of the Earth: Global, Regional, and Laboratory Studies

Edited by
Brian J. Mitchell
Barbara Romanowicz

1999

Springer Basel AG

Reprint from Pageoph
(PAGEOPH), Volume 153 (1998), Nos. 2/3/4

The Editors:

Brian J. Mitchell
Department of Earth and Atmospheric Sciences
Saint Louis University
St. Louis, Missouri 63103
USA

Barbara Romanowicz
Department of Geology and Geophysics
University of California, Berkeley
Berkeley, California 94720
USA

A CIP catalogue record for this book is available from the Library of Congress, Washington D.C., USA

Deutsche Bibliothek Cataloging-in-Publication Data

Q of the earth : global, regional, and laboratory studies /
ed. by Brian J. Mitchell ; Barbara Romanowicz. - Basel ; Boston ; Berlin :
Birkhäuser 1999
   (Pageoph topical volumes)
   ISBN 978-3-7643-6049-8   ISBN 978-3-0348-8711-3 (eBook)
   DOI 10.1007/978-3-0348-8711-3

© 1999 Springer Basel AG
Originally published by Birkhäuser Verlag in 1999
Printed on acid-free paper produced from chlorine-free pulp

ISBN 978-3-7643-6049-8

9 8 7 6 5 4 3 2 1

Contents

Pure appl. geophys. 153 (1998) 235–236
0033–4553/98/040235–02 $ 1.50 + 0.20/0

Introduction

In terms of its possible contribution to our understanding of the earth, few areas of seismological study have posed as many difficulties as that of seismic wave attenuation, or Q. Many surface-wave seismologists have measured amplitudes at two seismic stations along the same great-circle path from an earthquake and found, even after correcting for instrumental factors, that the amplitude of a particular phase is larger at the more distant station. Similarly, body-wave seismologists have long been aware of the severe effect that velocity gradients and low-velocity zones can have on wave amplitudes. Awareness that such effects can completely swamp those of anelasticity has, undoubtedly, deterred many investigators from pursuing research on seismic-wave attenuation.

In recent years, however, improved instrument deployment and methodologies, as well as the availability of digital data from well-calibrated instruments have spurred progress in obtaining reliable measurements of Q in the earth. In addition, improved laboratory measurements of Q in the seismic frequency band, improved theoretical models, and increasing knowledge of factors that contribute to Q have generated increased interest in this area.

Despite the difficulties inherent in the study of Q, it is a compelling field of endeavor. For a practical point of view, knowledge of Q is important for magnitude determination; for predicting ground motion and using that knowledge to design buildings, bridges, dams and other structures; and for monitoring compliance with nuclear test ban treaties. Knowledge of Q is also important in determining aspects of earth structure that are not easily amenable to study using only seismic velocities. It has long been known that Q is sensitive to temperature, movement of solid-state defects, partial melt, and fluid content to a much greater degree than are seismic velocities. For these reasons, reliable knowledge of the distribution of Q in the earth should provide new insights on its internal structure and evolution.

We therefore think that it is timely to compile a special volume on Q that addresses pertinent issues from a variety of perspectives. Part I of this volume emphasizes global and mantle studies of Q. It includes papers pertaining to a dislocation model for seismic wave attenuation and creep (Karato); the state of global Q tomography (Romanowicz); on teleseismic body waves, both observational results (Der, Morozov *et al.*) and methodology (Bhattacharyya, de Lorenzo); and on subduction zones (Sato *et al.*, Flanagan and Wiens).

Part II emphasizes the crust and includes studies that cover a broad range of scales and distances. These include papers on laboratory studies of crustal rock (Lu and Jackson); surface waves (Cong and Mitchell) and *Lg* coda (Cong and Mitchell, de Souza and Mitchell, Baqer and Mitchell, Mitchell *et al.*) over distances of 100 to 1000 or more km; on regional body waves (Sarker and Albers, Tselentis, Ugalde *et al.*, Gupta); on refraction and reflection surveys (Matheny and Nowack), and on bore hole experiments (Abercrombie, Yoshimoto *et al.*). This diversity of papers illustrates the wealth of questions that can be probed in the field of seismic wave attenuation, and of the oft-required need for novel experimental and theoretical approaches.

We hope that this volume will generate an enhanced appreciation of the difficulties inherent in the study of Q, of the diverse approaches that are required for making progress, and of the promise that Q studies hold for obtaining new and intriguing information about the earth. It is an exciting and multi-faceted discipline that will require the diligent efforts of researchers in observational and theoretical aspects of both seismology and material properties. We believe that observations from global observatories, as well as from controlled experiments on the earth's surface, in bore holes, and laboratory samples, as well as evolving theoretical insights, will continue to contribute to our understanding of the Q distribution in the earth and the factors that control that distribution. Moreover, because mantle Q is intimately related to creep and crustal Q appears to be strongly affected by tectonic and orogenic activity, knowledge of its distribution promises to provide new knowledge regarding mantle flow, plate movement, and crustal deformation in the earth.

Brian J. Mitchell
Department of Earth and Atmospheric Sciences
Saint Louis University
St. Louis, Missouri 63103
USA

and

Barbara Romanowicz
Department of Geology and Geophysics
University of California, Berkeley
Berkeley, California
USA

Part I
Global and Mantle Studies

Pure appl. geophys. 153 (1998) 239–256
0033–4553/98/040239–18 $ 1.50 + 0.20/0

❙ Pure and Applied Geophysics

A Dislocation Model of Seismic Wave Attenuation and Micro-creep in the Earth: Harold Jeffreys and the Rheology of the Solid Earth

SHUN-ICHIRO KARATO[1]

Abstract—A microphysical model of seismic wave attenuation is developed to provide a physical basis to interpret temperature and frequency dependence of seismic wave attenuation. The model is based on the dynamics of dislocation motion in minerals with a high Peierls stress. It is proposed that most of seismic wave attenuation occurs through the migration of geometrical kinks (*micro-glide*) and/or nucleation/migration of an isolated pair of kinks (Bordoni peak), whereas the long-term plastic deformation involves the continuing nucleation and migration of kinks (*macro-glide*). Kink migration is much easier than kink nucleation, and this provides a natural explanation for the vast difference in dislocation mobility between seismic and geological time scales. The frequency and temperature dependences of attenuation depend on the geometry and dynamics of dislocation motion both of which affect the distribution of relaxation times. The distribution of relaxation times is largely controlled by the distribution in distance between pinning points of dislocations, L, and the observed frequency dependence of Q, $Q \propto \omega^\alpha$, is shown to require a distribution function of $P(L) \propto L^{-m}$ with $m = 4 - 2\alpha$. The activation energy of Q^{-1} in minerals with a high Peierls stress corresponds to that for kink nucleation and is similar to that of long-term creep. The observed large lateral variation in Q^{-1} strongly suggests that the Q^{-1} in the mantle is frequency dependent. Micro-deformation with high dislocation mobility will (temporarily) cease when all the geometrical kinks are exhausted. For a typical dislocation density of $\sim 10^8$ m^{-2}, transient creep with small viscosity related to seismic wave attenuation will persist up to the strain of $\sim 10^{-6}$, thus even a small strain ($\sim 10^{-6} - 10^{-4}$) process such as post-glacial rebound is only marginally affected by this type of anelastic relaxation. At longer time scales continuing nucleation of kinks becomes important and enables indefinitely large strain, steady-state creep, causing viscous behavior.

Key words: Seismic wave attenuation, dislocations, geometrical kinks, transient creep, Peierls stress, Bordoni peak, Maxwell time.

Introduction

Understanding the physical mechanisms of anelasticity is important in a number of geophysical problems. For example, interpretation of radial and lateral variation of seismic wave attenuation (e.g., WIDMER *et al.*, 1991; MITCHELL, 1995;

---

[1] University of Minnesota, Department of Geology and Geophysics, Minneapolis, MN 55455, U.S.A. E-mail: karato@maroon.to.umn.edu, Fax: 612-625-3819

ROMANOWICZ, 1994; DUREK and EKSTRÖM, 1996) in terms of geodynamic parameters (such as temperature) requires a physically sound model for the temperature and frequency dependence of attenuation. Interpretation of the distribution of seismic wave velocity also needs to consider the effects of attenuation and resultant velocity dispersion (e.g., KANAMORI and ANDERSON, 1977; KARATO, 1993). Although some microscopic models of dislocation mechanisms have been proposed, for example by GUEGUEN and MERCIER (1973) and MINSTER and ANDERSON (1981), they either fail to explain some of the key parameters such as the activation energy or are inadequate in explaining the origin of the wide distribution of relaxation times implied by the observed frequency dependence. For example, both of these authors assumed that glide motion of dislocations is significantly easier than climb motion and that the former controls seismic wave attenuation and the latter long-term creep. Such a notion is introduced from metallurgy but is not justified for silicate minerals for several reasons. First, such a model would imply that the activation energy of Q^{-1} should be significantly smaller than that of creep. Although such seems to be the case for metals (e.g., FANTOZZI et al., 1982) and MgO (GETTING et al., 1997), the activation energy of Q^{-1} is similar to that of creep in olivine (for a review see KARATO and SPETZLER, 1990). Second, the notion of easy dislocation glide is not supported for silicates where glide is difficult because of high Peierls stresses. For example, GOETZE (1978) showed that creep in olivine can be explained by a glide-controlled model under certain conditions.

Another potentially important application of seismic wave attenuation is the inference of earth's long-term rheology. Because both seismic wave attenuation and long-term rheology involve viscous motion of crystalline defects, some link between them may exist. For example, if the earth behaves like a Maxwell body, then Q^{-1} is directly linked to viscosity by $Q^{-1} = M/\omega\eta$ where M: elastic modulus, η: viscosity and ω: frequency. However, the earth's mantle does not behave like a Maxwell body, but the frequency dependence of Q^{-1} is $Q^{-1} \propto \omega^{-\alpha}$ with $0 < \alpha < 1$ in the seismic frequency range (ANDERSON and MINSTER, 1979; SIPKIN and JORDAN, 1979; ANDERSON and GIVEN, 1982; ULUG and BERCKHEMER, 1984). Such behavior corresponds to transient creep, namely $\varepsilon \sim t^{\alpha}$ and $\dot{\varepsilon} \sim \eta^{-1} \sim t^{\alpha-1}$ where ε: strain, $\dot{\varepsilon}$: strain-rate and t: time. Therefore, if such a frequency dependence is extrapolated to much lower frequencies (i.e., longer time scales), one would predict a significantly larger viscosity for long-term deformation than for shorter term deformation. Such an attempt was made by JEFFREYS (1958, 1976) and ANDERSON and MINSTER (1979) to infer the viscosity relevant to mantle convection and to post-glacial rebound, respectively. JEFFREYS (1958, 1976), in particular, predicted very high viscosities at geological time scales and rejected the convection hypothesis on that ground. Although we now recognize that the earth's mantle can plastically deform at geological time scales by solid state creep, the relationship between long-term creep and seismic wave attenuation has been unclear, and therefore there has not been any clear theoretical basis on which to go beyond JEFFREYS (KARATO and SPETZLER, 1990). A physically sound model to link Q^{-1} with long-term rheology is clearly needed.

Any model of seismic wave attenuation must be consistent with seismological and experimental observations of internal friction and with the microscopic physics of minerals. Both experimental observations on olivine and MgO and theoretical models suggest that mechanisms involving dislocation motion are most likely to be responsible for seismic wave attenuation (GUEGUEN et al., 1989; KARATO and SPETZLER, 1990; JACKSON et al., 1992; GETTING et al., 1997), although grain-boundary mechanisms may be important for fine-grained materials (TAN et al., 1997; LAKKI et al., 1998). Experimental evidence for dislocation mechanisms is well established, however theoretical background of nonelastic relaxation has not been well-understood. One of the most difficult issues is to explain the vast difference in dislocation mobility required to cause attenuation at seismic frequencies and that responsible for long-term deformation. A related issue is to explain observed rather weak frequency dependence of Q over a wide frequency range. Furthermore, the observed difference in activation energies of Q^{-1} in olivine and in MgO must be explained by a microscopic model.

The purpose of this paper is thus to develop a microscopic model of seismic wave attenuation which is consistent with the properties of dislocations in silicates. In particular, I will emphasize the consequence of a high Peierls stress, which gives rise to an important distinction between the migration of geometrical kinks (WÜTHRICH, 1975; SEEGER and WÜTHRICH, 1976) and the nucleation of a pair of kinks (the Bordoni peak; ENGELKE, 1969a,b; FANTOZZI et al., 1982). It is shown that this model provides a clear physical picture for the frequency dependence and activation energy of Q^{-1}, and also allows us to estimate the limit of extrapolation of nonelastic properties obtained from Q^{-1} to longer-term phenomena.

Theory

1. General Considerations

Dislocations can cause a wide range of nonelastic behavior of solids (for review see FANTOZZI et al., 1982; HIRTH and LOTHE, 1982; KARATO and SPETZLER, 1990). At short time scales, micro-motion of dislocations between pinning points causes anelastic behavior. Pinning can occur by various mechanisms including the presence of jogs on screw dislocations and nodes caused by mutual interaction of dislocations and the interaction of dislocations with impurities or vacancies (e.g., HIRTH and LOTHE, 1982). Evidence of some of them are shown in Figure 1. At longer time scales continuous motion of dislocations becomes possible and viscoelastic behavior occurs. A simple phenomenological model to describe these

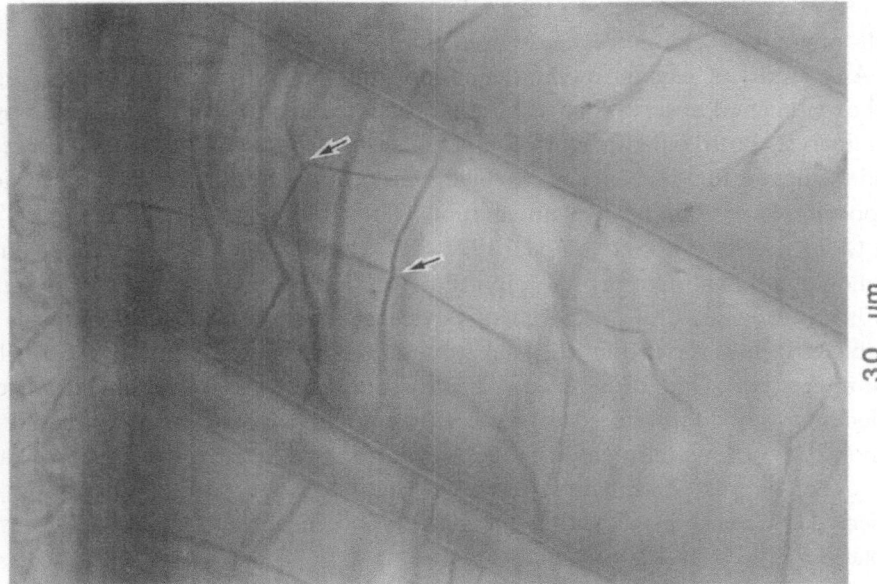

(b)

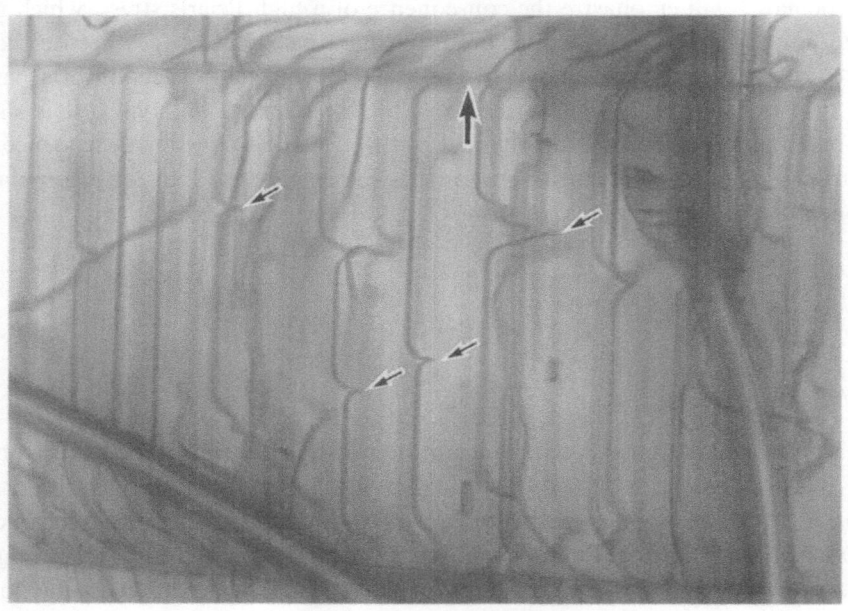

(a)

30 µm

time-dependent behaviors is the Burgers body which contains two relaxation times corresponding to anelastic (τ_a) and viscoelastic (τ_v) behavior ($\tau_a < {} < \tau_v$). The attenuation ($Q^{-1}(\omega)$) and creep function ($\psi(t)$) of such a body are given by (for small attenuation),

$$Q^{-1}(\omega) = \Delta \frac{\omega \tau_a}{1 + (\omega \tau_a)^2} + \frac{1}{\omega \tau_v} \tag{1}$$

and

$$\psi(t) = \frac{1}{M} \left\{ 1 + \Delta \left[1 - \exp\left(-\frac{t}{\tau_a} \right) \right] + \frac{t}{\tau_v} \right\} \tag{2}$$

respectively where M is the elastic modulus and Δ is the relaxation strength (e.g., NOWICK and BERRY, 1972). The transition from anelastic to viscoelastic behavior occurs at $t = \Delta \tau_v = \Delta \eta / M$ where η is the viscosity corresponding to viscoelastic behavior.

2. Seismic Wave Attenuation

The frequency dependence of Q predicted by such a simple model is either $Q \propto \omega$ or $Q \propto \omega^{-1}$ except for a narrow frequency range near $\omega \sim \tau_a^{-1}$ (equation (1)). However, both experimental and seismological observations show much weaker dependence of Q on frequency, $Q^{-1} \propto \omega^{-\alpha}$, with $\alpha = 0.1 - 0.3$, over a wide range of frequencies ($1 - 10^{-3}$ Hz) (e.g., KARATO and SPETZLER, 1990) suggesting a wide distribution of relaxation times, particularly for anelasticity (τ_a).

Figure 1 shows dislocations in olivine. Three points must be noted. First, dislocations are often nearly straight along certain crystallographic orientations, indicating a high Peierls stress. Second, however, dislocations are not always straight, but some curvature is common. The curvature is formed by the mutual interaction of dislocations, and a number of kinks occur on these dislocations. Third, in many cases evidence of pinning can be seen, presumably due to the presence of jogs and/or nodes (Figs. 1a,b) or of impurities. The first observation (a high Peierls stress) indicates that dislocation glide in olivine (and probably in many other silicates) is difficult because of the difficulty in *nucleating* new kinks. However, once kinks are present through ongoing or pre-existing plastic deformation (*geometrical kinks*), these portions of dislocations can move easily because the resistance for kink *migration* due to the Peierls potential of the second kind

Figure 1
Dislocations in olivine in a spinel lherzolite xenolith from Mt. Erebus, Antarctica. (a) Nearly straight screw dislocations organized between (100) subboundaries (shown by a large arrow on the right) indicating a control by a high Peierls potential barrier. All of the screw dislocations are pinned at subboundaries. Cusps (shown by small arrows) suggest pinning due to jogs. (b) Curved dislocations showing geometrical kinks (and jogs) and some nodes (shown by small arrows) caused by dislocation interactions.

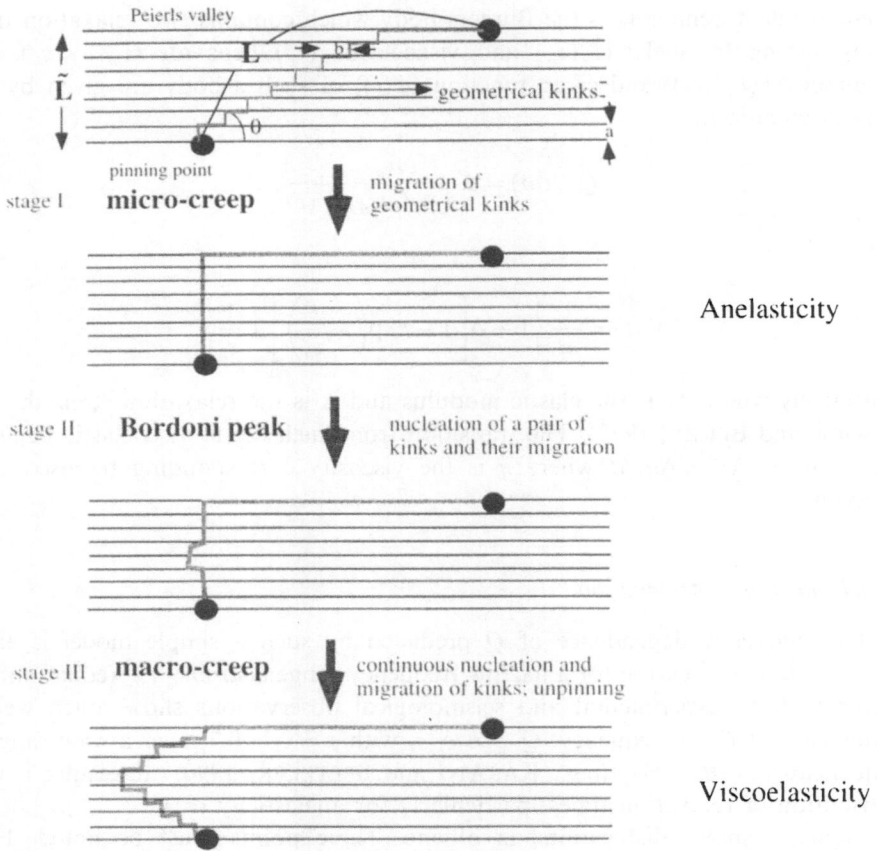

Figure 2

A schematic diagram illustrating the mechanisms of dislocation motion. At stage I (short time and/or low temperatures), only pre-existing geometrical kinks migrate. When all the geometrical kinks have moved, dislocation motion will be terminated until a nucleation of a new pair of kinks occurs (stage II). At stage III (long time and/or high temperatures), continuous nucleation of a pair of kinks and their migration occurs, allowing indefinite deformation.

(BRAISFORD, 1961; WÜTHRICH, 1975; SEEGER and WÜTHRICH, 1976) is not as large as that for kink *nucleation*. Migration of geometrical kinks occurs through an array of Peierls potential wells of the second kind. Therefore each step of migration over one potential well occurs in the anelastic fashion (ENGELKE, 1969a,b). After all the geometrical kinks have moved, the segment of dislocation will have no geometrical kinks (Fig. 2). After this stage, dislocation motion and resultant nonelastic deformation is possible only by nucleation and migration of kinks. Nucleation and migration of new kinks may occur before all the pre-existing geometrical kinks are exhausted. At a certain range of temperature and stress nucleation of a pair of kinks occurs as an isolated event and the process temporarily terminates when migrating kinks reach pinning points, leading to anelastic behavior (the Bordoni peak; e.g., FANTOZZI et al., 1982). However, at higher temperatures and/or lower frequencies the next nucleation event occurs before the previously formed kinks reach pinning points and the dislocation line progressively bows out to cause large-scale deformation, leading to viscous behavior. Unpinning of dislocation may also help the continuous motion of dislocations.

Thus the nature of dislocation motion can be classified into three stages (Fig. 2): (i) stage I: migration of geometrical kinks over the Peierls potential of the second kind. (ii) stage II: isolated nucleation and migration of a pair of kinks. (iii) stage III: continuous nucleation and migration of kinks. The first and the second stage cause anelastic relaxation. The relaxation time τ and relaxation strength Δ due to dislocation motion are given by (FANTOZZI et al., 1982),

$$\tau_I = \left(\frac{L \cos \theta}{b}\right)^2 v^{-1} \exp\left(\frac{E_M}{kT}\right) \tag{3}$$

$$\tau_{II} = v^{-1}\left[\exp\left(\frac{E_N}{kT}\right) + \left(\frac{L \sin \theta}{b}\right)^2 \exp\left(\frac{E_M}{kT}\right)\right] \tag{4}$$

and

$$\Delta_I = \frac{L^3 \cos^2 \theta \, \sin \theta \cdot ab^2}{\pi^4 kT} M\rho_{\text{seg}} \tag{5}$$

$$\Delta_{II} = \rho L^2 \sin^2 \theta \tag{6}$$

where $\tau_{I,II}$ and $\Delta_{I,II}$ are the relaxation times and relation strengths of anelastic relaxation corresponding to stage I and II, respectively and L: distance between pinning points, θ: angle between a dislocation line and the Peierls potential valley, v: frequency of vibration of dislocation segment, E_M: activation energy for kink migration, E_N: activation energy for kink nucleation, b: the length of the Burgers vector, a: the spacing of the Peierls valley, ρ_{seg}: volume density of the dislocation segments, M: elastic modulus, ρ: dislocation density (SEEGER and WÜTHRICH, 1976; FANTOZZI et al., 1982).

Figure 3 illustrates the variation of τ_I and τ_{II} with temperature. τ_{II} involves two activation energies, and therefore the semi-log plot of relaxation time vs. inverse temperature curve has a bend. At low temperatures the nucleation of kinks becomes so difficult that the first term (nucleation time) in equation (4) dominates. Under these conditions τ_{II} has relatively narrow distribution and the activation energy of τ_{II} is nearly equal to that of kink nucleation, E_N. At higher temperatures, however, kink nucleation becomes easy and hence the effects of kink migration (the second term in equation (4)) becomes important. The distribution of τ_{II} is then broad, comparable to that of τ_I. Under these conditions the temperature dependence of τ_I and τ_{II} is similar and the activation energy of relaxation time is nearly equal to that of kink migration, E_M. At yet higher temperatures dislocation motion in stage III will occur which involves successive kink nucleation and/or unpinning. Under these conditions little energy is stored in dislocations and the nonelastic behavior is viscous.

Thus, one can identify four regimes. In regime A (low temperatures, $T < T_1$), τ_{II} is dominated by the nucleation of kinks and is significantly larger than all possible values of τ_I. The distribution function of relaxation times in this regime has a relatively sharp peak, corresponding to τ_{II} separated from a broad peak corre-

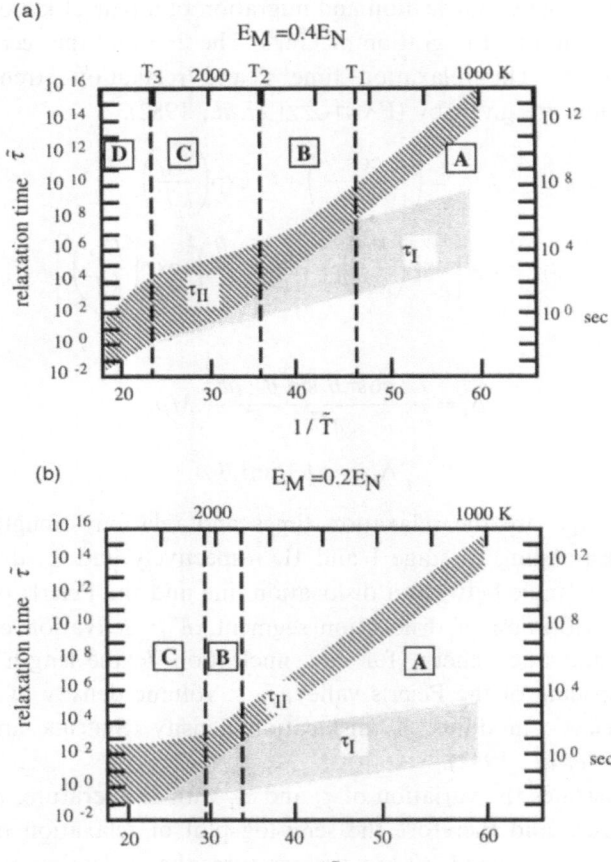

Figure 3

Variation of relaxation times with temperature. Normalized relaxation time, $\tilde{\tau} \equiv \tau v/(L_M/b)^2$ (where τ is either τ_I or τ_{II}) and normalized temperature, $\tilde{T} \equiv RT/E_N$, are used in this plot. $v = 10^{12}$ s^{-1}, $L_M = 10^{-4}$ m, $L_m = 10^{-6}$ m, $b = 0.5$ nm, $E_N = 500$ kJ/mol and $E_M = 0.4E_N$ are assumed in Figure 3(a) and $E_M = 0.2E_N$ for Figure 3(b). The relaxation time for kink nucleation is assumed to have a distribution in two orders of magnitude, corresponding to a variation of activation energy of ~ 60 kJ/mol. Four (three) regimes can be recognized in Figure 3(a) (Figure 3(b)) regarding the relative magnitude of τ_I and τ_{II} (see text for details). Regime B is more distinct in Figure 3(a) (for $E_M = 0.4E_N$) than in Figure 3(b) (for $E_M = 0.2E_N$).

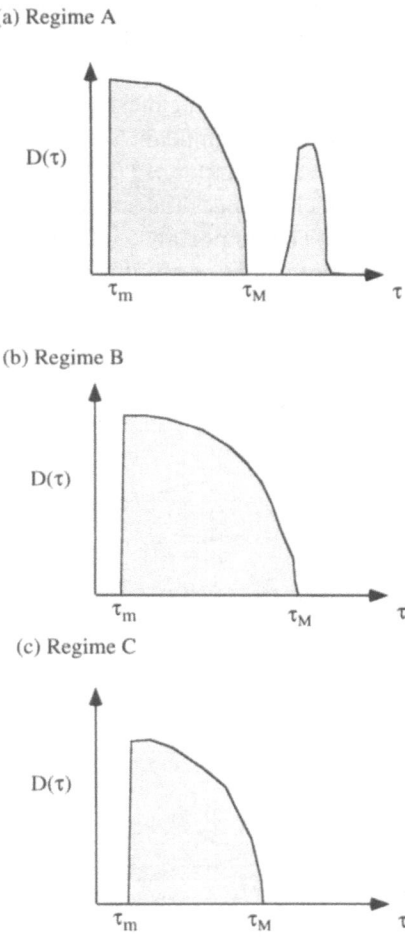

Figure 4
Schematic diagrams showing distribution of relaxation times, $D(\tau)$, for regime A (a), B (b) and C (c).

sponding to τ_I. The maximum (minimum) of τ_I corresponds to the longest (short-est) pinning point distance L (Fig. 4a). In regime C (high temperatures but lower than the temperatures above which viscoelastic flow occurs, $T_3 > T > T_2$), both τ_I and τ_{II} are dominated by kink migration. The spectrum of distribution of character-istic time is continuous (Fig. 4b). Similar to the regime A, the maximum (minimum) of τ_I corresponds to the longest (shortest) L. Between these two regimes there is an intermediate regime in which broad peaks of τ_{II} overlap with those of τ_I. The spectrum of relaxation time is continuous. However, the maximum of relaxation time in this regime is largely controlled by kink nucleation and the activation energy of the maximum relaxation time approximates that of kink nucleation.

The significance of this intermediate regime depends on the distribution of characteristic times for nucleation of kinks. In the simplest model of the Bordoni peak, the characteristic time is determined by the Peierls stress and has a sharp delta function distribution. In this case, regime B will be very narrow. However, most experimental studies evidence a significant distribution of relaxation time for the Bordoni peak (e.g., NOWICK and BERRY, 1972). This is considered to be due to the effects of internal stress, which changes the activation energy of kink nucleation (NOWICK and BERRY, 1972). The importance of this intermediate regime also depends on the ratio of activation energies of kink nucleation and migration. If the activation energy of kink nucleation is substantially larger than that of kink migration, this regime is not important. In contrast, if the difference in activation

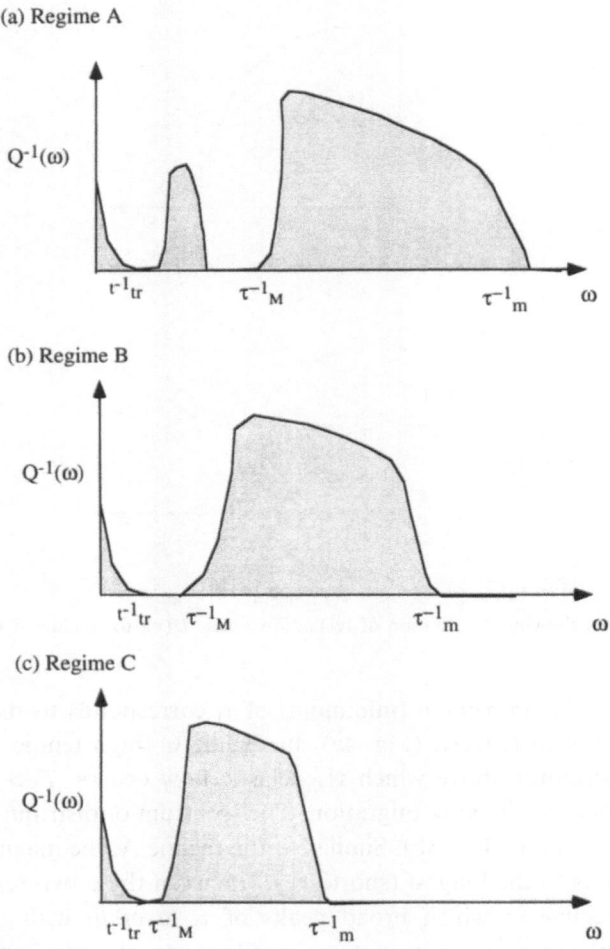

Figure 5
Frequency dependence of Q corresponding to various distribution functions shown in Figure 4.

energies is small, the intermediate regime can be significant (Fig. 3). The difference in activation energies depends critically on the value of the Peierls stress which depends on the nature of bonding and crystal structure. On the one hand, for a crystal with a low Peierls stress (such as fcc metals or ionic crystals with NaCl (e.g., MgO) or CsCl structures), this difference is large and the intermediate stage is of minor importance. On the other hand, for a crystal with strong covalent bonding and/or large unit cells (e.g., olivine, garnet), the Peierls stress is high and the intermediate stage is important. Therefore, the intermediate regime is likely to be important in silicate minerals.

Finally, at very high temperatures (regime D) one has a transition to viscoelastic behavior. This is caused by transition from a single nucleation event to continuous multiple nucleation events and/or unpinning of dislocations (stages II to III).

In each regime where anelastic relaxation occurs (regimes A, B and C), actual frequency and temperature dependence of Q^{-1} is determined by the distribution of characteristic times, viz.,

$$Q^{-1}(\omega, T) = \int_0^\infty \Delta \frac{\omega\tau}{1 + \omega^2\tau^2} D(\tau) \, d\tau \tag{7}$$

where $D(\tau) \, d\tau$ is the number density of characteristic times that occur between τ and $\tau + d\tau$. The distribution function is normalized such that,

$$\int_0^\infty D(\tau) \, d\tau = 1 \tag{8}$$

Q^{-1} in the regime D depends on frequency as,

$$Q^{-1}(\omega, T) = 1/\tau_v\omega = M/\omega\eta. \tag{9}$$

The frequency dependence of Q^{-1} in the anelastic regimes (regimes A, B and C) corresponding to various distribution functions of relaxation times (Fig. 4) is schematically shown in Figure 5. In the low frequency range where the frequencies are near the isolated peak corresponding to τ_{II}^{-1}, there is a Debye-type absorption peak (with some width). As the frequency approaches τ_M^{-1} (τ_M: the maximum of the characteristic time in the absorption band), absorption band starts as $Q^{-1} \propto \omega$. With further increase in frequency ($\tau_M^{-1} < \omega < \tau_m^{-1}$, τ_m: the minimum of relaxation time in the absorption band), the frequency dependence of Q^{-1} becomes weak. At the high end of frequency ($\omega > \tau_m^{-1}$), one has $Q^{-1} \propto \omega^{-1}$.

Let us consider the frequency and temperature dependence in the absorption band ($\tau_M^{-1} < \omega < \tau_m^{-1}$). From equation (7), one derives

$$Q^{-1}(\omega, T) = \omega^{-1} \int_0^\infty \tilde{D}(x/\omega) \frac{x}{1 + x^2} \, dx \tag{10}$$

where $x = \omega\tau$ and $\tilde{D}(x/\omega) \equiv \tilde{D}(\tau) = \Delta(\tau)D(\tau)$. Therefore the power-law relation, $Q^{-1}(\omega, T) \propto \omega^{-\alpha}$, implies,

$$\tilde{D}(\tau) \propto \tau^{\alpha-1}. \tag{11}$$

Such a distribution function can be interpreted in terms of a distribution of pinning point distance L, $P(L)$ defined by,

$$D(\tau)\, d\tau \equiv P(L)\, dL. \tag{12}$$

Here I made a simplifying assumption that the distribution of relaxation times is characterized by the distribution of pinning point distance L alone. Effects of the distribution of angle θ are not explicitly considered. Therefore the pinning point distance in this model should be understood to be $L \langle \cos \theta \rangle$ or $L \langle \sin \theta \rangle$. Using the relations (3) and (5) (I assume that the most important source of distribution of relaxation time is that for migration of geometrical kinks), I have

$$P(L) \propto L^{-m} \quad \text{with} \quad m = 4 - 2\alpha. \tag{13}$$

For a range of $\alpha = 0.1-0.3$ obtained in laboratory experiments, one has m = 3.4–3.8.

Now from the normalization relation (8), it can be shown that $\tilde{D}(\tau) \sim \tau_M^{-\alpha} \tau^{\alpha-1}$ for $0 < \alpha < 1$ and $\tilde{D}(\tau) \sim 1/\tau \, \log(\tau_M/\tau_m)$ for $\alpha = 0$. Therefore, it follows that,

$$Q^{-1}(\omega, T) \propto (\omega \tau_M)^{-\alpha} \quad \text{for} \quad 0 < \alpha < 1 \tag{14}$$

and

$$Q^{-1}(\omega, T) \propto 1/\log(\tau_M/\tau_m) \quad \text{for} \quad \alpha = 0 \tag{15}$$

hence,

$$Q^{-1}(\omega, T) \propto \omega^{-\alpha} \exp\left(-\frac{\alpha E^*}{RT}\right) \quad \text{for} \quad 0 < \alpha < 1 \tag{16}$$

$$Q^{-1}(\omega, T) \propto \frac{1}{[\log(\tau_{Mo}/\tau_{mo}) + (E_M^* - E_m^*)/RT]} \quad \text{for} \quad \alpha = 0 \tag{17}$$

where E^* is the activation energy of the maximum relaxation time, $\tau_{M,mo}$ and $E_{M,m}^*$ are the pre-exponential term and activation energy for $\tau_{M,m}$, respectively. Therefore the activation energy for Q^{-1} in the absorption band for $0 < \alpha < 1$ is,

$$E^* = E_N \quad \text{in regime B} \tag{18}$$

$$E^* = E_M \quad \text{in regimes A and C.} \tag{19}$$

Note that the activation energy in regimes A and C is the activation energy for migration of kinks or that of impurity drag (the latter will be the case for a material with a low Peierls stress) and is significantly smaller than that of long-term creep. In contrast, the activation energy of Q^{-1} in regime B is that for kink nucleation and is close to that of long-term creep. The temperature dependence of Q^{-1} for $\alpha = 0$ is considerably weaker than that for $0 < \alpha < 1$.

3. Micro-creep

Dislocation motion thus analyzed also causes micro-creep (e.g., MCMAHON, 1968; FANTOZZI et al., 1982). Using the creep function defined by (2), it can be shown that strain associated with power-law Q^{-1} ($Q^{-1} \sim \omega^{-\alpha}$) is given by $\varepsilon = \sigma / M [1 + A(t/\tau_M)^\alpha]$ for $\tau_m < t < \tau_M$ where $A = Q_0^{-1}$ (Q_0^{-1}: Q^{-1} at $\omega = \tau_M^{-1}$) and $\dot{\varepsilon} = \sigma / M \alpha A t^{\alpha - 1} \tau_M^{-\alpha}$, similar to the modified Lomnitz's law proposed by JEFFREYS (1958) (MINSTER and ANDERSON, 1981). Therefore the strain rate in this regime will decrease with time, leading to an infinite viscosity if the creep constitutive relation is extrapolated to $t \to \infty$ (JEFFREYS, 1958, 1976). An application of relation $\varepsilon = \sigma / M [1 + A(t/\tau_M)^\alpha]$ would imply that the maximum strain associated with this regime is $\varepsilon_c = \sigma / M (1 + A) \sim \sigma / M$ and therefore the maximum strain caused by this micro-creep is about the same order of the elastic strain. Note that the magnitude of strain associated with the post-glacial rebound, in which the earth behaves like a viscous body, is on the order of ~ 1–10 times elastic strain corresponding to the load due to the melting of ice sheets (KARATO, 1998). Thus the lower limit of frequency (the upper limit of time scale) to which micro-creep responsible for Q^{-1} occurs, is close to the frequency (the time scale) at which deformation is viscous (or viscoelastic).

Now, let us move precisely evaluate this critical strain below which micro-creep occurs. Recall that the transient strain in this regime is caused by the motion of geometrical kinks. Thus the maximum strain associated must correspond to the strain caused by the migration of pre-existing geometrical kinks. Once all geometrical kinks have been exhausted, no further deformation could occur until kink nucleation occurs. Strain caused by the motion of geometrical kinks can be calculated from the area swept by a kink, $L^2 \sin \theta \cos \theta / 2$. Consequently, the maximum strain, ε_c, caused by the motion of geometrical kinks is given by,

$$\varepsilon_c = b \rho_{\text{seg}} L^2 \sin \theta \cos \theta / 2. \tag{20}$$

Now, we assume that the mean spacing of pinning points is the average spacing of dislocations. This is a case for pinning due to jogs or nodes formed by dislocation interactions (see Fig. 1). In this case, $\rho \sim L \rho_{\text{seg}}$ and $\rho \sim L^{-2}$. Thus $\varepsilon_c = b \rho^{1/2} \sin \theta \cos \theta / 2$. For a typical tectonic stress level in convecting mantle ($\sigma \sim \eta \dot{\varepsilon} \sim 10^{21}$ (Pa s) $\times 10^{-15}$ (s^{-1}) ~ 1 MPa), one has $\rho \sim 10^8$ m^{-2} and $L \sim 10^{-4}$ m and hence $\varepsilon_c \sim 10^{-6}$ where I assumed b $= 0.5$ nm and sin θ cos $\theta / 2 \sim 0.1$. One notes that this is comparable to or less than the magnitude of strain associated with the post-glacial rebound, $10^{-6} - 10^{-4}$ (KARATO, 1998).

Therefore it is concluded that the effects of transient creep behavior caused by the migration of geometrical kinks are only marginally important for the flow associated with the post-glacial rebound. In fact, the extrapolation of attenuation data to the time scale of the post-glacial rebound using $\dot{\varepsilon} \propto t^{\alpha - 1} \tau_M^{-\alpha}$ predicts viscosities of $\sim 10^{19}$ and 10^{21} Pa s for the upper and the lower mantles, respectively

(ANDERSON and MINSTER, 1979), which are smaller than but near the viscosities estimated from the analysis of the post-glacial rebound and associated geophysical observations indicating $\eta \sim 10^{20}$ Pa s for the upper mantle and $\eta \sim 10^{22}$ Pa s for the lower mantle, respectively (e.g., NAKADA and LAMBECK, 1989; MITROVICA and FORTE, 1997).

Above the critical strain or the critical time scale, migration of geometrical kinks will no longer occur. Consequently the nature of deformation will change to stage III involving continuous nucleation and migration of kinks causing macro-creep. The transition time scale, t_{tr}, or strain, ε_{tr}, from stage II to stage III deformation can be estimated by comparing strains caused by stage II ($\varepsilon = \sigma/M [1 + A(t/\tau_M)^\alpha]$) and stage III deformation ($\varepsilon = \sigma/\eta\ t$), thus,

$$\frac{t_{tr}}{\tau_v} = Q_0^{-1/(1-\alpha)} \left(\frac{\tau_v}{\tau_M}\right)^{\alpha/(1-\alpha)} \tag{21}$$

or

$$\frac{\varepsilon_{tr}}{\varepsilon_e} = Q_0^{-1/(1-\alpha)} \left(\frac{\tau_v}{\tau_M}\right)^{\alpha/(1-\alpha)} \tag{22}$$

where $\tau_v = \eta/M$ (the Maxwell time) and ε_e is the elastic strain. For $Q_o \sim 50$, $\tau_M \sim 10^6$ sec, $\tau_v \sim 10^{11}$ sec and $\alpha = 0-0.2$, I get $t_{tr} \sim 10^8-10^9$ sec or $\varepsilon_{tr} \sim 10^{-6}-10^{-5}$ where elastic strain is calculated for a tectonic stress of ~ 1 MPa. The transition time to viscoelastic behavior is longer than the maximum relaxation time of anelastic behavior but shorter than the Maxwell time.

Summary and Discussions

The present model provides a theoretical framework to interpret experimental and seismological observations of attenuation of seismic waves caused by dislocation motion in minerals. The model emphasizes the role of migration of geometrical kinks as well as nucleation of kinks and demonstrates that the frequency dependence of $Q^{-1} \propto \omega^{-\alpha}$ can be related to the geometry of dislocations such that the distribution of pinning point distance is given by $P(L) \propto L^{-m}$ with $m = 4-2\alpha$. The activation energy of Q^{-1} is equal to that of kink nucleation for materials with a high Peierls stress, whereas it is equal to that of kink migration (including impurity effects) for materials with a low Peierls stress. A wide distribution of relaxation times ranging from body wave to free oscillation time scales and yet a high activation energy can be explained naturally as a result of the motion of geometrical kinks in crystals with a high Peierls stress.

Some features of the present theory can be compared with available experimental data on single crystals of olivine (GUEGUEN et al., 1989), coarse-grained polycrystalline dunite (JACKSON et al., 1992) and single crystals of MgO (GETTING *et al.*, 1997). In both olivine and MgO the frequency dependence is $Q^{-1} \propto \omega^{-\alpha}$ with $\alpha = 0.1-0.3$. In the present model this implies that the pinning point distance L distributes as $P(L) \propto L^{-m}$ with $m = 3.4-3.8$. Experimental or theoretical studies on the statistics of dislocation geometry will be useful to test this prediction.

The activation energy of Q^{-1} for olivine for dislocation mechanism is $\sim 400-450$ kJ/mol (GUEGUEN *et al.*, 1989; JACKSON *et al.*, 1992) whereas the activation energy of Q^{-1} in MgO is ~ 230 kJ/mol (GETTING *et al.*, 1997). The activation energy of Q^{-1} in olivine is similar to that of creep and of micro-hardness which presumably corresponds to that of nucleation of a pair of kinks (GOETZE, 1978). The activation energy of Q^{-1} in MgO is, in contrast, significantly smaller than those of creep (~ 460 kJ/mol; FROST and ASHBY, 1982). Such a contrast can be understood by the model developed in this paper. The high activation energy in olivine is likely to be due to attenuation in regime B, due presumably to the high Peierls stress. MgO is known to have a considerably smaller Peierls stress than olivine (e.g., FROST and ASHBY, 1982) and attenuation is likely to be due solely to the motion of geometrical kinks resulting in a low activation energy.

The present model of micro-creep predicts that fast deformation due to migration of geometrical kinks should terminate at a critical strain given by (20). Exact evaluation of the critical strain is difficult owing to the uncertainties in the mechanisms of pinning. I used a simple model in which pinning results from dislocation interactions, but other mechanisms such as pinning deriving from impurity atoms could also operate. When micro-creep due to migration of geometrical kinks is completed and continuous deformation involving successive nucleation and migration of kinks occurs, then the creep behavior will change from $\varepsilon = \sigma/M \left[1 + A(t/\tau_M)^\alpha\right]$ to $\varepsilon = \dot{\varepsilon}_S t$ where $\dot{\varepsilon}_S$ is steady-state creep rate. The present theory predicts that the transition between the two behaviors will occur at a time scale between τ_M (the maximum relaxation time) and τ_v (the Maxwell time). Therefore extrapolation of transient creep behavior inferred from seismic wave attenuation to geological time scale exceeding τ_v as was made by JEFFREYS (1958, 1976) is not justified by the microscopic physics of defect motion in minerals. Such a transition in creep behavior could be detected by small strain creep experiments and will be useful in understanding the relationship between various deformation processes in the earth with different time scales.

The frequency and the temperature dependencies of Q^{-1} are mutually related. From equations (16) and (17) I obtain,

$$\frac{\Delta Q^{-1}}{Q^{-1}} = \frac{\alpha E^*}{RT} \frac{\Delta T}{T} \quad \text{for} \quad 0 < \alpha < 1 \tag{23}$$

and

$$\frac{\Delta Q^{-1}}{Q^{-1}} = \frac{\Delta T}{T} \quad \text{for} \quad \alpha = 0 \tag{24}$$

if the distribution of relaxation time is through the distribution of activation energy, or

$$\frac{\Delta Q^{-1}}{Q^{-1}} = 0 \quad \text{for} \quad \alpha = 0 \tag{25}$$

if the distribution of relaxation time is through the distribution of pre-exponential term. Note that the relative variation in Q^{-1} for frequency-independent Q^{-1} is independent of materials parameters and therefore it should be independent of depth. On the other hand, the relative variation in Q^{-1} for frequency-dependent Q^{-1} is proportional to activation energy (enthalpy) and hence increases with depth.

The difference in temperature variation in Q^{-1} between the two cases is substantial. For example for $\Delta T/T = 20\%$ ($\Delta T = 400$ K for $T = 2000$ K), $\Delta Q^{-1}/Q^{-1} = 0$–20% for $\alpha = 0$ but $\Delta Q^{-1}/Q^{-1} = 50\%$ (100%) for $\alpha = 0.1$ ($\alpha = 0.2$) and $E^* = 400$ kJ/mol. The lateral variation in temperature in the upper mantle can be inferred from the lateral variation in seismic wave velocities (e.g., MONTAGNER and TANIMOTO, 1991; SU et al., 1994) using the temperature dependence of seismic wave velocities (KARATO, 1993). The results show $\Delta T/T = 20$–30%. The lateral variation in temperature in the deep upper mantle can also be inferred from the undulation of the topography of the "660 km" discontinuity, giving $\Delta T/T \sim 20\%$ (I assume that the amplitude of undulation is ~ 30 km (SHEARER and MASTERS, 1992) and the Clapeyron slope corresponding to the phase transformation at 660 km is -3 MPa/K (ITO and TAKAHASHI, 1989)). ROMANOWICZ (1994) inferred lateral variation in Q^{-1} assuming that Q^{-1} is independent of frequency. She inferred that the lateral variation in Q^{-1} is $\sim 50\%$ in the deep upper mantle. Similar studies on the shallow upper mantle revealed a lateral variation of Q^{-1} of ~ 50–100% (NAKANISHI, 1978; DZIEWONSKI and STEIM, 1983; ROMANOWICZ, 1990). In contrast to the lateral variation in seismic wave velocities which decreases significantly with depth (e.g., MONTAGNER and TANIMOTO, 1991; SU et al., 1994), lateral variation in Q^{-1} increases with depth in the depth range to ~ 300–400 km. This is consistent with a mineral physics based model, if Q^{-1} is frequency dependent, in which case the lateral variation in Q^{-1} should increase with depth due to the pressure dependence of activation energy (see equation (23)). However, some of the details of such an analysis may need revision. In particular, the inferred large lateral variation in Q^{-1} and a depth variation in lateral variation are difficult to reconcile with the frequency-independent Q^{-1}, which predicts a much smaller lateral variation and no depth dependence in lateral variation in Q^{-1}. A physically more sensible analysis of lateral variation in Q^{-1} should include the frequency dependence of Q^{-1} which is well documented in laboratory studies. Similarly, STACEY's (1995) argument against the importance of the anelastic effects on seismic wave velocities, is based on the frequency-independent Q^{-1} model which is not justified.

Finally, it should be pointed out that Q^{-1} in many minerals (particularly olivine) may also depend on water fugacity (or content). Some preliminary results were already reported by JACKSON *et al.* (1992) who found enhanced attenuation at higher water content. The recent results of deformation experiments showed a significant weakening effect of water in olivine on deformation at moderate temperatures (CHEN *et al.*, 1998), suggesting an enhancement of kink nucleation by water. Therefore, τ_M should decrease and hence attenuation should increase with water fugacity. Some consequences of the effects of water on anelasticity have been discussed by KARATO and JUNG (1998).

Acknowledgments

This work is partly supported by a grant from NSF. I thank Barbara Romanowicz for discussions on seismological constraints on Q. I thank David Kohlstedt and Ian Jackson for critical reading of the manuscript.

REFERENCES

ANDERSON, D. L., and GIVEN, J. W. (1982), *Absorption Band Q Model for the Earth*, J. Geophys. Res. *87*, 3893–3904.

ANDERSON, D. L., and MINSTER, J. B. (1979), *The Frequency Dependence of Q in the Earth and Implications for Mantle Rheology and Chandler Wobble*, Geophys. J. R. Astr. Soc. *58*, 431–440.

BRAISFORD, A. D. (1961), *Abrupt-kink Model of Dislocation Motion*, Phys. Rev. *122*, 778–786.

CHEN, J., INOUE, T., WEIDNER, D. J., WU, Y., and VAUGHAN, M. T. (1998), *Strength and Water Weakening of Mantle Minerals, Olivine, Wadsleyite and Ringwoodite*, Geophys. Res. Lett. *25*, 575–578.

DUREK, J. J., and EKSTRÖM, G. (1996), *A Radial Model of Anelasticity Consistent with Long-period Surface-wave Attenuation*, Bull. Seismol. Soc. Am. *86*, 144–158.

DZIEWONSKI, A. M., and STEIM, J. (1983), *Dispersion and Attenuation of Mantle Waves through Waveform Inversion*, Geophys. J. R. Astr. Soc. *70*, 503–527.

ENGELKE, H. (1969a), *Ein Diffusionsmodell zur Behandlung der Doppelkinkrelaxation in kubisch-flächenzentrierten Metallen*, Phys. Stat. Sol. *36*, 231–244.

ENGELKE, H. (1969b), *Die Deutung sekundärer Eigenschaften des Bordoni-Maximums in einen Mehrmuldenmodell der Doppelkinkrelaxation*, Phys. Stat. Sol. *36*, 245–259.

FANTOZZI, G., ESNOUF, C., BENOIT, W., and RITCHIE, I. G. (1982), *Internal Friction and Microdeformation due to the Intrinsic Properties of Dislocations: The Bordoni Peak*, Prog. Mater. Sci. *27*, 311–451.

FROST, H. J., and ASHBY, M. F., *Deformation Mechanism Maps* (Maps, Pergamon Press, Oxford), (1982) 167 pp.

GETTING, I. C., DUTTON, S. J., BURNLEY, P. C., KARATO, S., and SPETZLER, H. A. (1997), *Shear Attenuation and Dispersion in MgO*, Phys. Earth Planet. Inter. *99*, 249–257.

GOETZE, C. (1978), *The Mechanisms of Creep in Olivine*, Philos. Trans. Roy. Soc. London *A288*, 99–119.

GUEGUEN, Y., and MERCIER, J. C. (1973), *High Attenuation and Low Velocity Zone*, Phys. Earth Planet. Inter. *7*, 39–46.

GUEGUEN, Y., DAROT, M., MAZOT, P., and WOIRGARD, J. (1989), Q^{-1} *of Forsterite Single Crystals*, Phys. Earth Planet. Inter. *55*, 254–258.

HIRTH, J. P., and LOTHE, J., *Theory of Dislocations* (Krieger Publishing Company, Malabar, Florida 1982) 857 pp.

ITO, E., and TAKAHASHI, E. (1989), *Postspinel Transformations in the System Mg_2SiO_4-Fe_2SiO_4 and Some Geophysical Implications*, J. Geophys. Res. *94*, 10,637–10,646.

JACKSON, I., PATERSON, M. S., and FITZGERALD, J. D. (1992), *Seismic Wave Dispersion and Attenuation in Åheim Dunite: An Experimental Study*, Geophys. J. Int., *108*, 517–534.

JEFFREYS, H. (1958), *A Modification of Lomnitz's Law of Creep in Rocks*, Geophys. J. R. Astr. Soc. 7, 92–95.

JEFFREYS, H., *The Earth*, 6th ed. (Cambridge University Press, Cambridge 1976) 574 pp.

KANAMORI, H., and ANDERSON, D. L. (1977), *Importance of Physical Dispersion in Surface Wave and Free Oscillation Problems: Review*, Rev. Geophys. Space Phys. *15*, 105–112.

KARATO, S. (1993), *Importance of Anelastic Relaxation in the Interpretation of Seismic Tomography*, Geophys. Res. Lett. *20*, 1623–1626.

KARATO, S. (1998), *Micro-physics of post-glacial rebound*. In *Ice Age Geodynamics: A New Perspective* (ed. P. Wu) (Trans Tech. Pub., Zürich) pp. 351–364.

KARATO, S., and SPETZLER, H. A. (1990), *Defect Microdynamics in Minerals and Solid-state Mechanisms of Seismic Wave Attenuation and Velocity Dispersion in the Mantle*, Rev. Geophys. *28*, 399–421.

KARATO, S., and JUNG, H. (1998), *Water, Partial Melting and the Origin of Seismic Low Velocity and High Attenuation Zone*, Earth Planet. Sci. Lett. *157*, 193–207.

LAKKI, A., SCHALLER, R. CARRY, C., and BENOIT, W. (1998), *High Temperature Anelastic and Viscoplastic Deformation of Fine-grained MgO-doped Al_2O_3*, Acta Mater. *46*, 689–700.

MCMAHON, C. J., Jr. (1968), *Microplastic Behavior in Iron*, Adv. Material. Sci. *2*, 121–140.

MINSTER, J. B., and ANDERSON, D. L. (1981), *A Model of Dislocation Controlled Rheology for the Mantle*, Philos. Trans. Roy. Soc. London *A299*, 319–356.

MITCHELL, B. J. (1995), *Anelastic Structure and Evolution of the Continental Crust and Upper Mantle from Seismic Surface Wave Attenuation*, Rev. Geophys. *33*, 441–462.

MITROVICA, J. X., and FORTE, A. M. (1997), *Radial Profile of Mantle Viscosity: Results from the Joint Inversion of Convection and Postglacial Rebound Observations*, J. Geophys. Res. *102*, 2751–2769.

MONTAGNER, J-P., and TANIMOTO, T. (1991), *Global Upper Mantle Tomography of Seismic Velocities and Anisotropies*, J. Geophys. Res. *96*, 11,051–11,071.

NAKADA, M., and LAMBECK, K. (1989), *Late Pleistocene and Holocene Sea-level Changes in the Australian Region and Mantle Viscosity*, Geophys. J. Int. *96*, 497–517.

NAKANISHI, I. (1978), *Regional Differences in the Phase Velocity and the Quality Factor Q of the Mantle Rayleigh Waves*, Science *200*, 1379–1381.

NOWICK, A. S., and BERRY, B. S., *Anelastic Relaxation in Crystalline Solids* (Academic Press, New York 1972) 677 pp.

ROMANOWICZ, B. (1990), *The Upper Mantle Degree 2: Constraints and Inferences from Global Mantle Wave Attenuation Measurements*, J. Geophys. Res. *95*, 11,051–11,071.

ROMANOWICZ, B. (1994), *Anelastic Tomography: A New Perspective on the Upper Mantle Thermal Structure*, Earth Planet. Sci. Lett. *128*, 113–121.

SEEGER, A., and WÜTHRICH, C. (1976), *Dislocation Relaxation in Body-centered Cubic Metals*, Nuovo Cimento *33B*, 38–75.

SHEARER, P. M., and MASTERS, T. G. (1992), *Global Mapping of Topography on the 660 km Discontinuity*, Nature *355*, 791–796.

SIPKIN, S. A., and JORDAN, T. H. (1979), *Frequency Dependence of Q_{ScS}*, Bull. Seismol. Soc. Am. *69*, 1055–1079.

STACEY, F. D. (1995), *Theory of Thermal and Elastic Properties of the Lower Mantle and Core*, Phys. Earth Planet. Inter. *89*, 219–245.

SU, W-J., WOODWARD, R. L., and DZIEWONSKI, A. M. (1994), *Degree-12 Model of Shear-velocity Heterogeneity in the Mantle*, J. Geophys. Res. *99*, 6945–6980.

TAN, B. H., JACKSON, I., and FITZGERALD, J. D. (1997), *Shear Wave Dispersion and Attenuation in Fine-grained Synthetic Olivine Aggregates: Preliminary Results*, Geophys. Res. Lett. *24*, 1055–1058.

ULUG, A., and BERCKHEMER, H. (1984), *Frequency Dependence of Q for Seismic Body Waves in the Earth's Mantle*, J. Geophys. *56*, 9–19.

WIDMER, R., MASTERS, G., and GILBERT, F. (1991), *Spherically Symmetric Attenuation within the Earth from Normal Mode Data*, Geophys. J. Int. *104*, 541–553.

WÜTHRICH, C. (1975), *On the Internal Friction in b.c.c. Metals due to the Motion of Geometrical Kinks*, Scripta Metal. *9*, 641–646.

(Received February 26, 1998, revised May 13, 1998, accepted May 20, 1998)

Pure appl. geophys. 153 (1998) 257–272
0033–4553/98/040257–16 $ 1.50 + 0.20/0

Pure and Applied Geophysics

Attenuation Tomography of the Earth's Mantle: A Review of Current Status

BARBARA ROMANOWICZ[1]

Abstract—Resolving the lateral variations of attenuation in the deep mantle by tomographic methods holds potential for constraining its thermal structure and dynamics. It is a challenging subject which has been addressed by only a few studies until now. We here review the main motivations behind pursuing this challenge, the difficult issues involved in separating effects of anelastic attenuation from scattering and focusing due to propagation in 3-D elastic structure and finally discuss the current status of global attenuation tomography.

Key words: Tomography, attenuation, earth's mantle.

Introduction

Global anelastic tomography is a difficult subject that can bring important constraints on the thermal structure of the mantle and therefore its dynamics, in complement to those provided by elastic tomography. Global elastic tomography has made great strides since the pioneering first studies of the last two decades (DZIEWONSKI *et al.*, 1977; WOODHOUSE and DZIEWONSKI, 1984; NATAF *et al.*, 1986). It is currently possible to resolve whole mantle structure at degree 12 with reasonable agreement between different studies (SU *et al.*, 1994; LI and ROMANO-WICZ, 1996; MASTERS *et al.*, 1996) and upper mantle structure with even greater detail, with lateral resolution reaching on the order of 500–1000 km (MONTAGNER and TANIMOTO, 1991; EKSTRÖM *et al.*, 1997; LASKE and MASTERS, 1996; TRAMPERT and WOODHOUSE, 1996). Such a resolution is also progressively attainable in the lower mantle, but only in the vicinity of subduction zones (VASCO *et al.*, 1995; GRAND *et al.*, 1997; VAN DER HILST *et al.*, 1997).

In contrast, anelastic tomography has been lagging somewhat behind. One-dimensional profiles of the variations with depth of the quality factor Q have now reached a high level of consensus (e.g., ROMANOWICZ, 1994a; DUREK and

[1] Seismological Laboratory and Department of Geology and Geophysics, University of California at Berkeley, Berkeley, CA 94720, U.S.A. E-mail: barbara@seismo.berkeley.edu, Tel.: (510) 643 56 90

EKSTRÖM, 1996; BHATTACHARYYA *et al.*, 1996) with generally only minor departures from the reference model PREM (DZIEWONSKI and ANDERSON, 1981). Degree 2 structure seems to be reasonably well constrained in the upper mantle from both normal mode and surface wave data (ROMANOWICZ, 1990; SUDA *et al.*, 1991; DUREK *et al.*, 1993), however very few studies have been completed that explore higher degrees. In fact only two groups to date have looked at both even and odd heterogeneity in the upper mantle at shorter wavelengths (ROMANOWICZ, 1994a, 1995, 1997; BHATTACHARYYA *et al.*, 1996), and only one published study addresses whole mantle 3-D global attenuation, using body-wave data (BHATTACHARYYA, 1996). In the upper mantle, agreement between body wave and surface wave results is variable. Emerging on-going studies may help resolve these disagreements (e.g., REID and WOODHOUSE, 1997).

In what follows, we briefly discuss the significance of global anelastic tomography, the technical reasons for which it is lagging behind elastic tomography and finally review its current status.

Anelastic Tomography: Goals and Issues

The existence of large lateral variations of Q in the crust (see MITCHELL, 1995 for a review) and the upper mantle, is well documented from various regional studies, either using surface waves (e.g., MITCHELL, 1975; CANAS and MITCHELL, 1978, 1981; NAKANISHI, 1979a; BUSSY *et al.*, 1993) or body waves, such as multiple *ScS* phases (NAKANISHI, 1979b; SIPKIN and JORDAN, 1980; LAY and WALLACE, 1983; CHAN and DER, 1988; SIPKIN and REVENAUGH, 1994) or multiple *S* and depth phases (SHEEHAN and SOLOMON, 1992; FLANAGAN and WIENS, 1990, 1994; DING and GRAND, 1993). These lateral variations can be an order of magnitude larger than observed lateral variations in velocity, commonly exceeding 50–100%.

There are two main reasons that make resolving 3-D anelastic structure in the deep mantle a worthwhile goal. First, the quality factor Q of the earth is considerably more sensitive to temperature than elastic velocity, as shown by laboratory and theoretical studies (e.g., MINSTER and ANDERSON, 1981; BERKHEMER *et al.*, 1982; GUEGUEN *et al.*, 1989; KARATO and SPETZLER, 1990; JACKSON *et al.*, 1992) and this sensitivity differs from that of elastic velocity. As argued by ROMANOWICZ (1994b), the nonlinear (Arrhenius law) sensitivity of Q to temperature implies that, in principle, attenuation tomography should be able to resolve hot regions (high attenuation) better than elastic tomography. In addition, elastic velocity can be affected to a large degree by compositional variations (e.g., YAN *et al.*, 1989; BINA and SILVER, 1997), so, ultimately, mapping regions with various degrees of agreement between velocity and attenuation distributions should help us constrain the distribution of chemical versus thermal heterogeneity in the mantle. At shallow depths, as discussed by MITCHELL (1995), and perhaps also near the core-mantle boundary, the effect of fluid inclusions and partial melt on attenuation also requires consideration.

The other main reason pertains to the dispersion effect of Q on elastic velocities (FUTTERMAN, 1962). For example, LIU et al. (1976) and KANAMORI and ANDERSON (1977) demonstrated that the baseline shift between velocity models of the mantle obtained from low frequency free oscillations and surface waves on the one hand, and short-period body waves on the other, could be accounted for by introducing the dispersive effect of Q, assuming an absorption band model with high frequency cut-off in the vicinity of 1 sec, as implied by body-wave observations (e.g., SIPKIN and JORDAN, 1979). When lateral variations of Q are present, the dispersion corrections also vary laterally. This has implications not only for the comparison of elastic models obtained in different seismic frequency bands, but also for geodynamic studies that jointly utilize seismic tomographic models and geoid data to infer the viscosity structure of the mantle (e.g., HAGER et al., 1985; FORTE et al., 1996). Indeed, the conversion factor between velocity and density is a crucial parameter in these studies, and, in the presence of 3-D Q structure, it should vary both with position and with the dominant frequency of the elastic models considered. KARATO (1993) illustrated the importance of anelasticity correction for the radial profile of dv/dr (the density derivative of velocity) and ROMANOWICZ (1990) showed that the misalignment in phase between the degree two pattern of fundamental mode-free oscillation frequency shifts, and that of the geoid, could be explained by the effects of lateral variations in anelastic dispersion.

Seismic Measurements of Attenuation in the Mantle

To measure attenuation in the deep mantle, one can use either low frequency surface wave and free-oscillation data, or deep turning body-wave data. Low frequency techniques typically employ either a travelling wave or a standing wave, normal mode formalism. The travelling wave formalism is well adapted to fundamental mode surface waves to periods of about 250 s, that are well isolated on the seismograms and sample the first 400 km of the upper mantle. Typically, in this approach, the amplitude spectrum $A_i(\omega)$ of each wave packet i is computed at frequency ω after appropriate windowing and tapering and an attenuation coefficient η is defined such that:

$$A_i(\omega) = A_o(\omega) \exp(-\eta_i(\omega)X_i) \tag{1}$$

where X_i is the epicentral distance in (km), and $A_o(\omega)$ represents the amplitude at the source.

The standing wave approach allows us to sample deeper into the mantle and to measure Q not only along the fundamental mode branch. In this approach, two types of methods are commonly used. One relies on the fitting of resonance peaks to spectra of individually observed free oscillations. The fitting is done either on the complex spectra (e.g., MASTERS and GILBERT, 1983) or on the amplitude spectra (e.g., DUREK and EKSTRÖM, 1997) of long time windows (at least 12 hours or more, depending on the Q of the mode). DUREK and EKSTRÖM (1997) have shown that the differences between the two approaches are not significant. The other type of method relies on measuring the amplitude decay with time of individual normal modes, by computing the spectrum successively shifted in time and measuring the slope of the resulting amplitude/time curve (e.g., ROULT, 1975; SAILOR and DZIEWONSKI, 1978). When applied carefully, both modal approaches yield similar results. However, significant discrepancies (on the order of 15%) exist between fundamental mode measurements of Q using the travelling wave and the standing wave approach, in their common frequency domain of application, that are still currently the subject of controversy (e.g., DUREK and EKSTRÖM, 1997).

The measurement of attenuation of body wave is generally a differential measurement and involves two "related" phases (such as S and SS or successive multiple ScS phases) to help minimize effects of the source and near-source and near-receiver structure. These measurements can also be done either in the frequency domain, by looking at the spectral ratios of two related phases, and measuring the slope of the spectrum as a function of frequency, or in the time domain, by computing a transfer function from the wave form of the first phase to the second one (e.g., BHATTACHARYYA, 1996).

While it is easily recognized that accurately resolving 3-D anelastic structure of the mantle could be very useful to further our understanding of mantle dynamics, progress has to date been slow because of the inherent difficulty of measuring attenuation and especially its lateral variations. Indeed, the amplitude of seismic waves travelling through the earth is affected not only by anelastic attenuation but also by focusing and scattering effects due to propagation in a 3-D elastic medium. The latter can be as large or larger than anelastic effects and depend strongly on the short wavelength details of the elastic structure, which are at present not very well constrained. Indeed, as shown by WOODHOUSE and WONG (1986) in the framework of ray theory and by ROMANOWICZ (1987) and PARK (1987) in the framework of asymptotic normal mode theory, to first order, the focusing terms due to elastic structure depend on the transverse gradients of velocity along the propagation path.

Asymptotically, if A_l is the amplitude of a normal mode of angular order l, then the perturbation due to focusing takes the form:

$$\delta A_l = (1 + \delta F_l) \tag{2}$$

where δF_l, which represents the focusing/scattering term, has the form (ROMANO-WICZ, 1987):

$$\delta F_l = \frac{-a\Delta}{2Uk}\tilde{D} \tag{3}$$

here a is the earth's radius, Δ epicentral distance, U group velocity and $k = l + 0.5$. $\tilde{D}$ is the minor arc average of the transverse derivative term D, which, in turn, can be expressed in terms of local coordinates (θ, ϕ) on the surface of the sphere as:

$$D = \frac{\sin(\phi - \Delta)}{\sin(\Delta)} [\partial_\theta^2 \delta\omega_k^0(\theta, \phi) \sin \phi - \partial_\phi \delta\omega_k^0(\theta, \phi) \cos \phi]. \tag{4}$$

In equation (4) $\delta\omega_k^0$ is the "local frequency" (JORDAN, 1978), which, to zeroth order, represents the integrated effect of structure beneath the local point (θ, ϕ) on the surface of the earth.

For a velocity model described by an expansion in spherical harmonics with coefficients (Cst, Sst), transverse gradients depend on terms of the form $s^2 Cst, s^2 Sst$, and are therefore sensitive to large values of s, in other words small wavelengths.

If the elastic structure of the earth were perfectly known, one could first correct for its effects using either linear theory (Born approximation), as described above, or, preferably, a more complete formalism including multiple scattering effects (e.g., LOGNONNÉ and ROMANOWICZ, 1990; FRIEDERICH, 1997; GELLER and HARA, 1993). We expect that in the near future, global 3-D models will become reliable enough at short wavelengths so that, combined with the increase in computer power, such corrections will become feasible and accurate.

Until now however, indirect methods of dealing with focusing and scattering have generally been used. ROMANOWICZ (1990) and DUREK et al. (1993) exploited the fact that, for surface waves and in the case of linear theory, focusing and anelastic effects could be separated by combining measurements over several consecutive wavetrains for a single recording, because focusing terms change sign with the direction of propagation whereas attenuation terms are always additive. This is for example how elastic focusing can be visually detected in actual recordings, when successive wavetrains present alternating high and low amplitudes (e.g., LAY and KANAMORI, 1985). Figure 1 shows an example of vertical component recordings for the M 7.5 Chile earthquake of 03/03/1985, observed at Geoscope stations and compared with predictions calculated for the PREM model (DZIEWONSKI and ANDERSON, 1981). Several records (stations SSB, KIP, WFM) exhibit anomalously high amplitudes for later arriving trains. In order to remove focusing effects, at least 4 consecutive surface wavetrains are needed. As pointed out by ROMANOWICZ (1994a), the drawback of this approach is that, the longer the travel path, the more the waves are affected by 3-D elastic structure, and the harder it is to account for that in an approximate, linear fashion. Moreover, such a technique is only applicable to surface waves. An alternative approach, favored by ROMANOWICZ (1995) for surface waves and BHATTACHARYYA et al. (1996) for body waves, is to reject data that are strongly affected by focusing, after visual inspection. In the case of body waves, BHATTACHARYYA et al. (1996) used a technique in which the attenuation operator t^* is inferred from the slope of the variation with frequency of SS to S amplitude ratios. Data for which a smooth variation of this ratio with frequency cannot be obtained are rejected. In the case of surface waves, ROMANOWICZ (1994a) devised a technique that allows to 1) keep only data for first arriving trains that have travelled the shortest paths (e.g., R1 and R2 for Rayleigh waves), and 2) among those data, reject those that do not present a smooth variation of attenuation coefficient with period. This approach appears to be successful, provided very strict rejection criteria are applied, which limits the coverage of the earth that can be achieved and therefore the spatial resolution of three-dimensional structure.

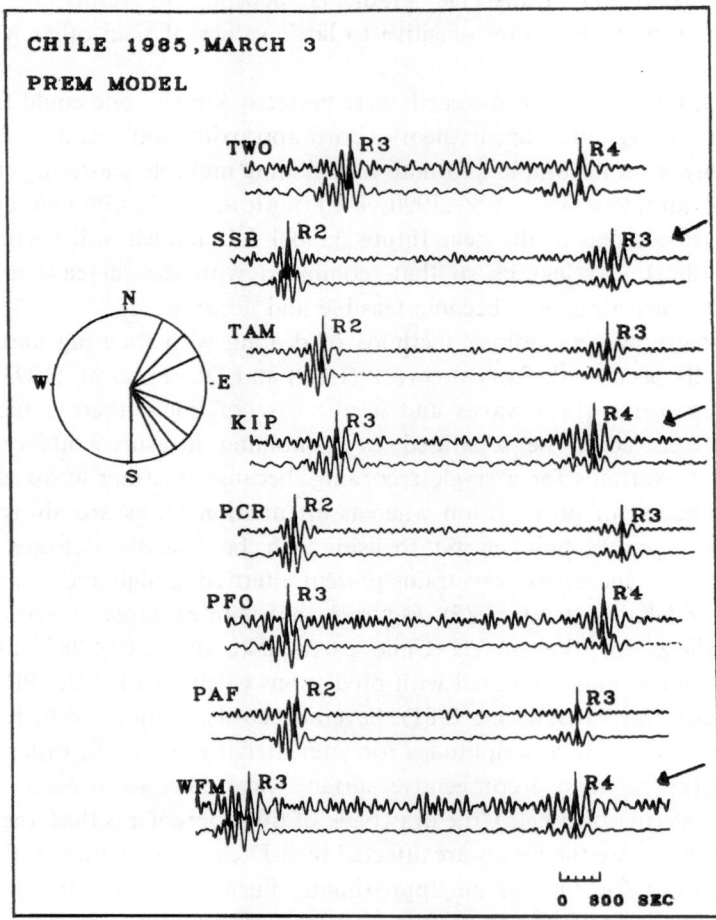

Figure 1

Example of vertical component records for the Chile earthquake of 03/03/85 at Geoscope stations (top traces) compared to PREM synthetics (bottom traces). Stations indicated by an arrow exhibit anomalously large amplitudes of R3 or R4 trains compared to preceding trains, indicating the presence of focusing effects.

An additional concern is that potentially major uncertainties in the amplitude at the source (scalar moment) must be allowed for. In the case of body waves, this is dealt with by considering spectral ratios between phases for which source take-off angles are not very different. In the case of surface waves, the method designed by ROMANOWICZ (1994a) involves computing the scalar moment bias by comparing attenuation coefficients measured on first arriving trains with those measured using three consecutive trains. In the latter case, the source effect is cancelled out, however the attenuation measurement is generally less accurate due to the increased influence of elastic structure over the longer R3 (or G3) path. To determine the source correction factor, an interactive graphic procedure was designed, which involves the superposition of the attenuation curves obtained both ways.

The comparison of maps of lateral variations of attenuation coefficient of Rayleigh waves at different periods obtained using the former approach, in which four consecutive wavetrains are required (ROMANOWICZ, 1990) and the latter approach, in which only first arriving trains (R1 and R2 for Rayleigh waves) are used (ROMANOWICZ, 1994a), indicates good agreement at long wavelengths.

The published models to date generally rely on amplitude measurements in the frequency domain. This is suitable for isolated phases, such as fundamental mode surface waves or specific body-wave phases such as S, SS and ScS. More recently, we have started to explore the possibility of using time-domain wave-form information to invert for anelastic structure (ROMANOWICZ et al., 1996; ROMANOWICZ, 1997), which should allow us to extend attenuation measurements to a larger portion of the seismogram, and therefore increase sampling of the deep mantle. In the case of surface waves, a time-domain method based on the comparison of observed and synthetic wave forms would also help resolve the problem of contamination of fundamental modes by higher mode energy, inherent in frequency domain methods, and, more generally, the problem of length of time window considered to compute the spectrum, and the dependence of amplitudes on chosen tapers. Such an approach requires the ability to accurately account for 3-D elastic structure. We will discuss it briefly in a later section.

Existing Global Models and Stable Features

As mentioned, only a few models of attenuation have been published, because of the difficulties involved in obtaining reliable measurements of attenuation. The status of attenuation tomography today is comparable to that of elastic tomography in the early 1980s, with stable long wavelength features emerging, although still little quantitative information at scales beyond degrees 5–6 of spherical harmonic expansion of lateral heterogeneity. Moreover, most studies as yet assume that lateral variations in Q are confined to the upper mantle.

Even though the degree 2 pattern in Q in the upper mantle is less prominent than that of elastic velocities in studies of free oscillations and surface waves (SMITH and MASTERS, 1989), it appears to be reasonably well constrained (ROMANOWICZ et al., 1987; ROULT et al., 1990; ROMANOWICZ, 1990; SUDA et al., 1991; DUREK et al., 1993), with significant correlation between the results of surface-wave and body-wave studies (BHATTACHARYYA et al., 1996). This structure is also well correlated with degree 2 in elastic velocities, and in particular manifests a similar shift in phase towards the west as one goes from the uppermost mantle (first 300 km) into the transition zone (400–600 km), which is indicative of the predominantly thermal nature of heterogeneity at these long wavelengths, (e.g., MONTAGNER and ROMANOWICZ, 1992). Even at such long wavelengths, there is however some disagreement concerning the depth distribution of attenuation. DUREK et al. (1993) argued that the lateral variations in Q are primarily confined to the depth range corresponding to the seismic low velocity zone (80–220 km), whereas more recent studies indicate significant heterogeneity persisting into the transition zone (ROMANOWICZ, 1995; BHATTACHARYYA et al., 1996).

At shorter wavelenghts there are currently only two published global models that consider both even and odd terms of lateral heterogeneity in Q: one using surface waves (ROMANOWICZ, 1995) and the other body waves (BHATTACHARYYA et al., 1996). Comparison of these models evidences agreement in some regions of the world, with low attenuation in Eurasia, western Australia and the Himalayas (BHATTACHARYYA et al., 1996; Fig. 2) and high attenuation in the mid-Pacific, eastern Australia and China. In general, models obtained to date at the global scale concur with regional scale models on the existence of a correlation of lateral variations in Q with tectonic province, in the first 250 km of the upper mantle, with high Q under shields and low Q under oceans, particularly so under young oceans. Surface wave derived models (ROMANOWICZ, 1990, 1995; DUREK et al., 1993) tend to indicate high attenuation under young oceans (as confirmed by regional studies, e.g., DING and GRAND, 1993). ROMANOWICZ (1994b) demonstrated the correlation of Q with the age of the sea floor in the first 250 km of the upper mantle, both in the Pacific and in the Atlantic Oceans (Fig. 3). In this depth range, the correlation of Q structure with heat flow is also observed. At greater depths in the South Pacific the shift to the west of the high Q maximum observed at degree 2 also persists at shorter wavelengths, perturbing the correlation with age of the sea floor. In this depth range, the correlation of Q structure with hotspot distribution appears to be significant (ROMANOWICZ, 1994b, 1995). The disagreement between the predictions of model QR19 (ROMANOWICZ, 1995) and the cap-averaged t^* measurements of BHATTACHARYYA et al. (1996) under the young ocean in the South Pacific may come in part from the different way in which SS waves average structure over depth beneath their bounce point as compared to surface waves, which have inherently greater depth resolution.

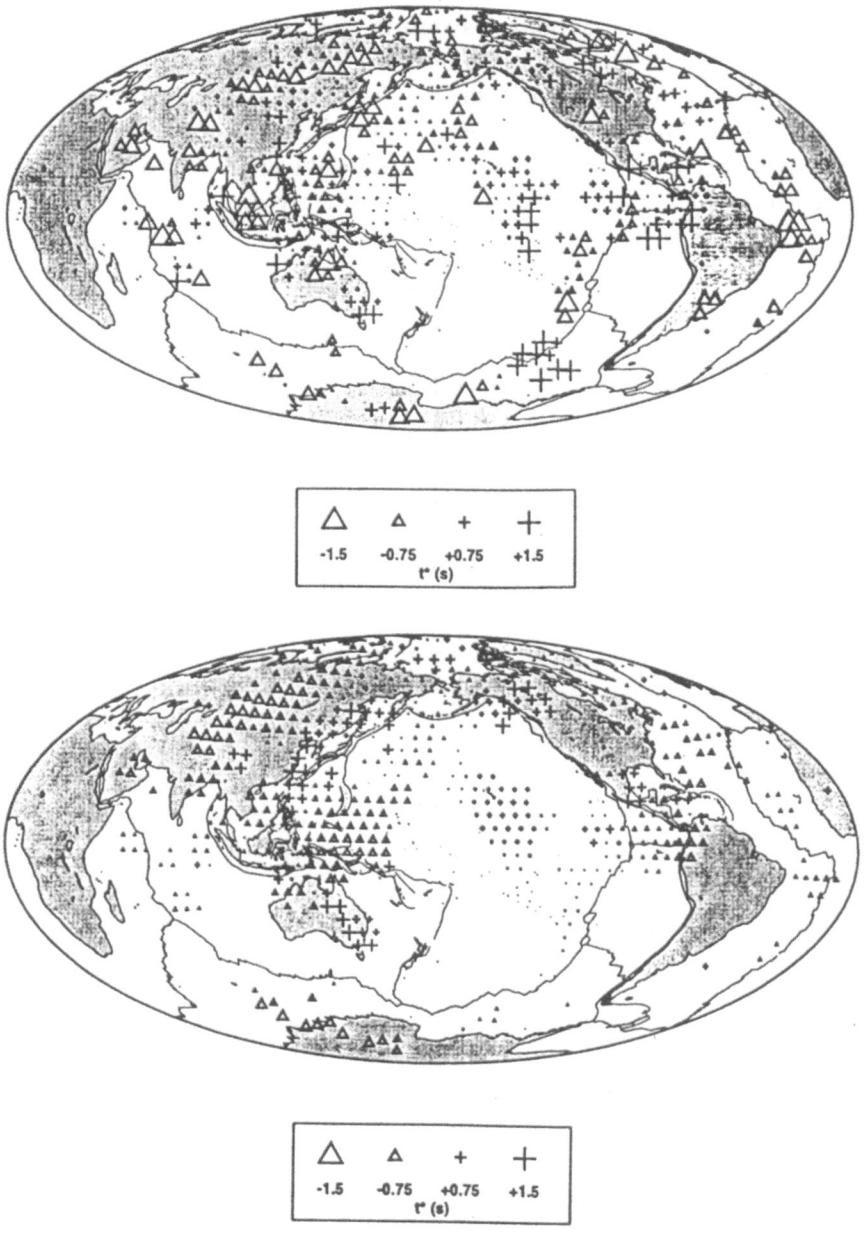

Figure 2

Top: Cap averaged residual t^* values obtained from SS/S spectral amplitude ratios by BHAT-TACHARYYA *et al.* (1996). Bottom: For comparison, average t^* values predicted by model QR19 (ROMANOWICZ, 1995) plotted at 5° caps. In both cases, average values have been subtracted before plotting. From BHATTACHARYYA *et al.* (1996).

In an on-going study, we are investigating the retrieval of global 3-D mantle Q structure using surface and body wave-form data (ROMANOWICZ *et al.*, 1996; ROMANOWICZ, 1997). This approach follows the general framework of global wave-form inversion for elastic structure developed by LI and ROMANOWICZ (1995, 1997) and proceeds in an iterative manner: in the first step, a spherically symmetric attenuation model is assumed, and wave-form data are inverted for 3-D elastic structure. In the second step, the derived 3-D elastic structure is used as the starting

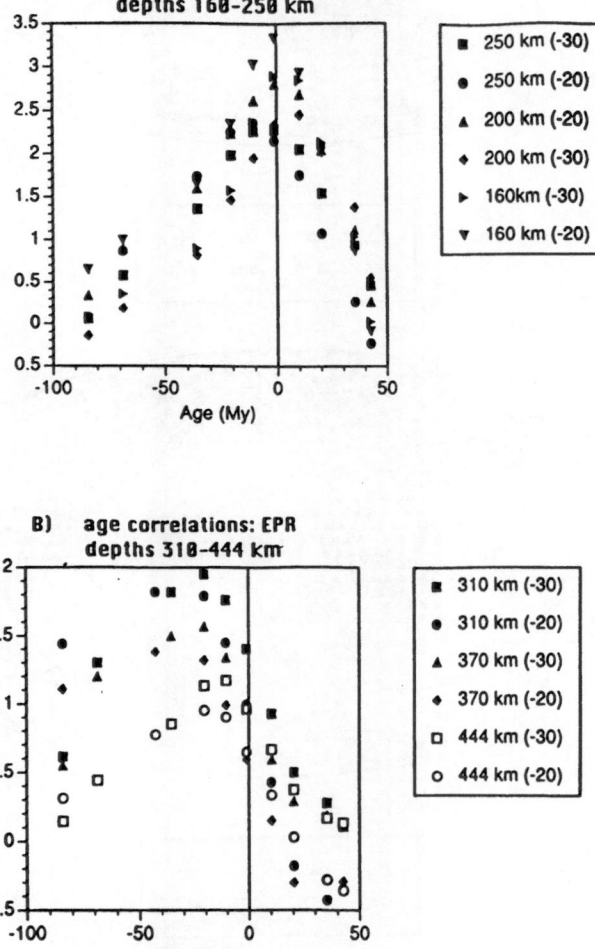

Figure 3
Dependence on age of the sea floor of the southern Pacific Ocean of the anomaly in Q in the upper mantle in different depth ranges, according to model QR19 (ROMANOWICZ, 1995). Top: 0–25 km; Bottom: 250–450 km. The symmetry around the ridge is broken in the deeper range, where lowest Q is found in the central Pacific. From ROMANOWICZ (1994b).

model for an inversion for 3-D anelastic structure. In this second step, focusing terms due to the elastic structure can be incorporated in the forward computation of seismograms. Preliminary results indicate that this approach is promising and results in models that are at least qualitatively compatible with those obtained earlier using a different dataset and a spectral approach. (Fig. 4).

Whether any significant lateral variations in Q exist in the lower mantle is currently an open question. Regional studies based on multiple *ScS* data cannot discriminate between lateral variations spread over the entire mantle or concentrated in the upper mantle. It may be that lateral variations in temperature in the lower mantle are mostly confined to narrow upwellings or plumes and their detection must await a significant increase in our ability to resolve small-scale lateral variations in anelastic structure. Conversely, it is expected that stronger lateral variations in Q exist in D'', because of the boundary layer nature of this portion of the mantle (e.g., LOPER and LAY, 1995). Our preliminary whole mantle wave-form inversion results for *SH* waves (ROMANOWICZ, 1997) indicate a stable pattern of degree two in the lowermost mantle, correlated with that which is well constrained in elastic tomographic models (Fig. 5), with, as in the upper mantle, high attenuation corresponding to low velocity. This would confirm the thermal origin of the velocity lows observed in the central Pacific and under Africa, that have often been interpreted as associated with rising thermal plumes (e.g, STACEY and LOPER, 1983). The existence of melt inclusions (e.g., WILLIAMS and GARNERO, 1996) could also contribute to correlated patterns of velocity and attenuation.

The preliminary pattern obtained up to degree 4 is highly correlated with a global map of *P* velocities in D'' obtained by WYSESSION (1996) from the study of travel times of *P*-diffracted waves. High attenuation in D'' in the Pacific Ocean and beneath Africa has also been suggested by BHATTACHARYYA (1996) in a whole mantle inversion in which the bulk of the lower mantle is assumed to have laterally homogeneous Q.

Conclusions

The study of lateral variations of Q in the deep mantle is still in its infancy. Quantitative models are few and their reliability is difficult to assess, since many discrepancies exist between different studies. Some qualitative features of the models can nevertheless be considered as well established. There is general agreement that lateral variations of Q are strongest in the crust and the uppermost mantle, where they are correlated with tectonic features and elastic velocities, and that some heterogeneity persists into the transition zone, at least at the longest wavelengths, indicating a strong thermal component to the low degree elastic structure in this depth range. In this depth range, the long wavelength distribution of Q manifests a correlation with that of hot spots. Lateral variations of Q in the

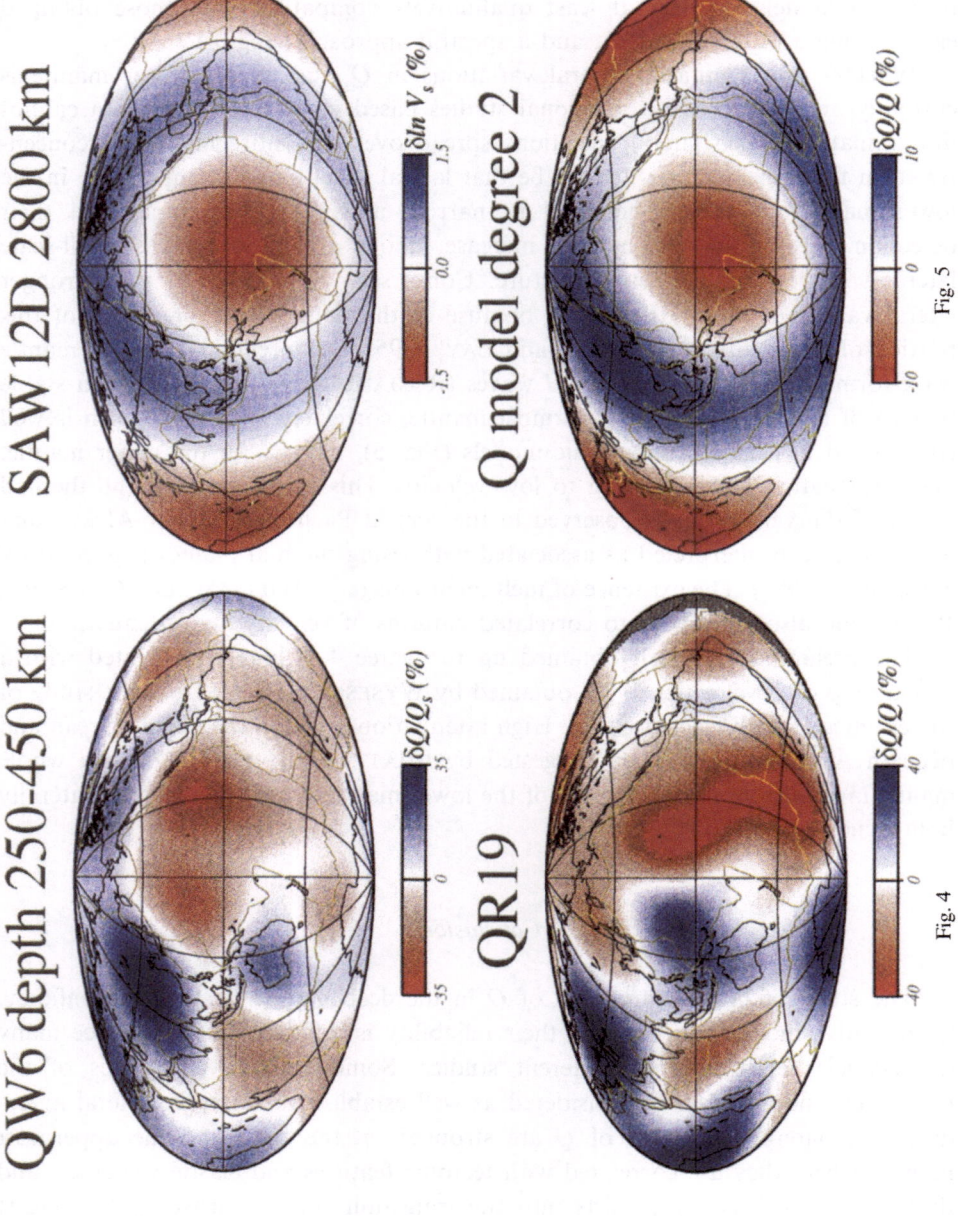

Fig. 5

Fig. 4

lower mantle are much less well constrained, although there seem to be indications of correlation of anelastic and elastic structures in the lowermost mantle, at least at very long wavelengths.

Our ability to further constrain Q models quantitatively relies strongly on how successful we will be, in the near future, in modeling the effects of elastic structure on seismic wave amplitudes (both for low frequency surface waves and shorter period body waves). This is contingent on the construction of global elastic 3-D models with well constrained small-scale features, likely beyond degree 30, which are beginning to be available in subduction zone regions (VAN DER HILST et al., 1997; GRAND et al., 1997) but also on the incorporation of more exact theoretical formalism in the inversion (e.g., ROMANOWICZ, 1987; LOGNONNÉ and CLEVEDE, 1996).

Acknowledgements

This work was partially supported by NSF Grant EAR9706380. It is Berkeley Seismological Laboratory Contribution NR 9809.

REFERENCES

BERKHEMER, H., KAMPFMANN, W., AULBACH, E., and SCHMELING, H. (1982), *Shear Modulus and Q of Forsterite and Dunite near Partial Melting from Forced Oscillation Experiments*, Phys. Earth Planet. Inter. *29*, 30–41.

BHATTACHARYYA, J. (1996), *Three-Dimensional Anelastic Structure of Earth*, Ph.D. Thesis, UC San Diego.

BHATTACHARYYA, J. G., MASTERS, G., and SHEARER, P. (1996), *Global Lateral Variations of Shear Attenuation in the Upper Mantle*, J. Geophys. Res. *101*, 22,273–22,290.

BINA, C. R., and SILVER, P. G. (1997), *Bulk Sound Travel Times and Implications for Mantle Composition and Outer Core Heterogeneity*, Geophys. Res. Lett. *24*, 499–502.

BUSSY, M. J., MONTAGNER, J. P., and ROMANOWICZ, B. (1993), *Tomographic Study of Upper Mantle Attenuation in the Pacific Ocean*, Geophys. Res. Lett. *20*, 663–666.

CANAS, J. A., and MITCHELL, B. J. (1978), *Lateral Variation of Surface-wave Anelastic Attenuation across the Pacific*, Bull. Seismol. Soc. Am. *68*, 1637–1650.

CANAS, J. A., and MITCHELL, B. J. (1981), *Rayleigh Wave Attenuation and its Variation Across the Atlantic Ocean*, Geophys. J. R. Astron. Soc. *67*, 159–176.

CHAN, W. W., and DER, Z. A. (1988), *Attenuation of Multiple ScS in Various Parts of the World*, Geophys J. Int. *92*, 303–314.

Figure 4
Comparison of maps of lateral variations in Q in the upper mantle obtained for (top) a preliminary model based on wave-form data inversion (ROMANOWICZ, 1997) and (bottom) model QR19 (RO-MANOWICZ, 1995).

Figure 5
Comparison of degree 2 maps in D'' for S velocity (top) as in model SAW12D (LI and ROMANOWICZ, 1996) and for Q (bottom), from a preliminary inversion of wave-form data (ROMANOWICZ, 1997).

CLEVEDE, E., and LOGNONNÉ, P. (1996), *Frechét Derivatives of Coupled Seismograms with Respect to an Anelastic Rotating Earth*, Geophys. J. Int. *124*, 456–482.

DING, C.-Y., and GRAND, S. P. (1993), *Upper Mantle Q Structure beneath the East Pacific Rise*, J. Geophys. Res. *98*, 1973–1985.

DUREK, J. J., and EKSTRÖM, G. (1996), *A Model of Radial Anelasticity Consistent with Long-period Surface Wave Attenuation*, Bull. Seismol. Soc. Am. *86*, 144–158.

DUREK, J. J., and EKSTRÖM, G. (1997), *Investigating Discrepancies among Measurements of Traveling and Standing Wave Attenuation*, J. Geophys. Res. *102*, 24,529–24,544.

DUREK, J. J., RITZWOLLER, M. H., and WOODHOUSE, J. H. (1993), *Constraining Upper Mantle Anelasticity Using Surface-wave Amplitudes*, Geophys. J. Int. *114*, 249–272.

DZIEWONSKI, A. M., HAGER, B. H., and O'CONNELL, R. J. (1977), *Large-scale Heterogeneities in the Lower Mantle*, J. Geophys. Res. *82*, 239–255.

DZIEWONSKI, A. M. (1984), *Mapping the Lower Mantle: Determination of Lateral Heterogeneity in P Velocity up to Degree and Order 6*, J. Geophys. Res. *89*, 5929–5952.

DZIEWONSKI, A. M., and ANDERSON, D. L. (1981), *Preliminary Reference Earth Model*, Phys. Earth Planet. Int. *25*, 297–356.

EKSTRÖM, G. J., TROMP, J., and LARSON, E. (1997), *Measurements and Global Models of Surface Wave Propagation*, J. Geophys. Res. *102*, 8137–8157.

FLANAGAN, M. P., and WIENS, D. A. (1990), *Attenuation Structure beneath the Lau Backarc Spreading Center from Teleseismic S Phases*, Geophys. Res. Lett. *17*, 2117–2120.

FLANAGAN, M. P., and WIENS, D. A. (1994), *Radial Upper Mantle Structure of Inactive Backarc Basins from Differential Shear-wave Measurements*, J. Geophys. Res. *99*, 15,469–15,485.

FUTTERMAN, W. I. (1962), *Dispersive Body Waves*, J. Geophys. Res. *67*, 5279–5291.

FORTE, A. M., DZIEWONSKI, A. M., and WOODWARD, R. L., *Aspherical structure of the mantle, tectonic plate motions, nonhydrostatic geoid, and topography of the core-mantle boundary*. In *Dynamics of the Earth's Deep Interior and Earth Rotation* (eds. J.-L. Le Mouel, D. E. Smylie, and T. Herring) (AGU, Washington DC 1996) pp. 135–166.

FRIEDERICH, W. (1997), *Propagation of Seismic Shear and Surface Waves in a Laterally Heterogeneous Mantle by Multiple Forward Scattering*, Geophys. J. Int., in press.

GELLER, R. J., and HARA, T. (1993), *Two Efficient Algorithms for Iterative Linearized Inversion of Seismic Waveform Data*, Geophys. J. Int. *115*, 699–720.

GRAND, S., VAN DER HILST, R., and WIDIYANTORO, S. (1997), *Global Seismic Tomography: A Snapshot of Convection in the East*, GSA Today 7(4), 1–7.

GUEGUEN, Y., DAROT, M., MAZOT, P., and WOIRGARD, J. (1989), *Q-1 of Forsterite Single Crystals*, Phys. Earth Planet. Inter. *55*, 254–258.

HAGER, B., CLAYTON, R. W., RICHARDS, M. A., COMER, R. P., and DZIEWONSKI, A. M. (1985), *Lower Mantle Heterogeneity, Dynamic Topography and the Geoid*, Nature *313*, 541–545.

JACKSON, I., PATERSON, M. S., and FITZGERALD, J. D. (1992), *Seismic Wave Dispersion and Attenuation in Aheim Dunite: An Experimental Study*, Geophys. J. Int. *108*, 517–534.

JORDAN, T. H. (1978), *A Procedure for Estimating Lateral Variations from Low Frequency Eigenspectra Data*, Geophys. J. R. Astr. Soc. *52*, 441–455.

KANAMORI, H., and ANDERSON, D. L. (1977), *Importance of Physical Dispersion in Surface-wave and Free-oscillation Problems: Review*, Rev. Geophys. Space Phys. *15*, 105–112.

KARATO, S. I., and SPETZLER, H. A. (1990), *Effect of Microdynamics in Minerals and Solid-state Mechanisms of Seismic Wave Attenuation and Velocity Dispersion in the Mantle*, Rev. Geophys. *28*, 399–421.

KARATO, S. I. (1993), *Importance of Anelasticity in the Interpretation of Seismic Tomography*, Geophys. Res. Lett. *20*, 1623–1626.

LASKE, G., and MASTERS, G. (1996), *Constraints on Global Phase Velocity Maps from Long-period Polarization Data*, J. Geophys. Res. *101*, 16,059–16,075.

LAY, T., and WALLACE, T. (1983), *Multiple ScS Travel Times and Attenuation beneath Mexico and Central America*, Geophys. Res. Lett. *10*, 301–304.

LAY, T., and KANAMORI, H. (1985), *Geometric Effects of Global Lateral Heterogeneity on Long-period Surface Wave Propagation*, J. Geophys. Res. *90*, 605–621.

LI, X.-D., and ROMANOWICZ, B. (1995), *Comparison of Global Wave-form Inversions with and without Considering Cross-branch Modal Coupling*, Geophys. J. Int. *121*, 695–709.

LI, X.-D., and ROMANOWICZ, B. (1996), *Global Mantle Model Developed Using Nonlinear Asymptotic Coupling Theory*, J. Geophys. Res. *101*, 22,245–22,272.

LIU, H. P., ANDERSON, D. L., and KANAMORI, H. (1976), *Velocity Disperson due to Anelasticity: Implication for Seismology and Mantle Composition*, Geophys. J. R. Astron. Soc. *47*, 41–58.

LOGNONNÉ, P., and ROMANOWICZ, B. (1990), *Fully Coupled Earth Vibrations: The Spectral Method*, Geophys. J. Int. *102*, 365–395.

LOPER, D. E., and LAY, T. (1995), *The Core Mantle Boundary Region*, J. Geophys. Res. *100*, 6397–6421.

MASTERS, G., and GILBERT, F. (1983), *Attenuation in the Earth at Low Frequencies*, Phil. Trans. R. Soc. London, Ser. A *308*, 479–522.

MASTERS, G., JOHNSON, S., LASKE, G., and BOLTON, H. (1996), *A Shear Velocity Model of the Mantle*, Philos. Trans. R. Soc. London A *354*, 1385–1411.

MINSTER, J. B., and ANDERSON, D. L. (1981), *A Model of Dislocation Controlled Rheology for the Mantle*, Philos. Trans. R. Soc. London A. *299*, 319–356.

MITCHELL, B. J. (1975), *Regional Rayleigh Wave Attenuation in North America*, J. Geophys. Res. *80*, 4904–4916.

MITCHELL, B. J. (1995), *Anelastic Structure and Evolution of the Continental Crust and Upper Mantle from Seismic Surface Wave Attenuation*, Rev. Geophys. Space Phys. *33*, 441–462.

MONTAGNER, J. P., and TANIMOTO, T. (1991), *Global Upper Mantle Tomography of Seismic Velocities and Anisotropies*, J. Geophys. Res. *96*, 20,337–20,351.

MONTAGNER, J. P., and ROMANOWICZ, B. (1992), *Degree 2,4,6 Inferred from Seismic Tomography*, Geophys. Res. Lett. *20*, 631–634.

NAKANISHI, I. (1979a), *Phase Velocity and Q of Mantle Rayleigh Waves*, Geophys J. R. Astron. Soc. *58*, 35–59.

NAKANISHI, I. (1979b), *Attenuation of Multiple ScS Waves Beneath the Japanese Arc*, Phys. Earth Planet. Inter. *19*, 337–347.

NATAF, H. C., NAKANISHI, I., and ANDERSON, D. L. (1986), *Measurement of Mantle Wave Velocities and Inversion for Lateral Heterogeneity and Anisotropy, 3, Inversion*, J. Geophys. Res. *91*, 7261–7307.

PARK, J. (1987), *Asymptotic Coupled Mode Expressions for Multiplet Amplitude Anomalies and Frequency Shifts on an Aspherical Earth*, Geophys. J. R. Astr. Soc. *90*, 129–164.

REID, F., and WOODHOUSE, J. H. (1997), *Measurement of Differential Travel Times and t* and the Structure of the Upper Mantle*, EOS Trans AGU, F485.

ROMANOWICZ, B. (1987), *Multiple-multiplet Coupling due to Lateral Heterogeneity: Asymptotic Effects on the Amplitude and Frequency of the Earth's Normal Modes*, Geophys. J. R. Astr. Soc. *90*, 75–100.

ROMANOWICZ, B. (1990), *The Upper Mantle Degree 2: Constraints and Inferences from Global Mantle Wave Attenuation Measurements*, J. Geophys. Res. *95*, 11,051–11,071.

ROMANOWICZ, B. (1994a), *On the Measurement of Anelastic Attenuation Using Amplitudes of Low Frequency Surface Waves*, Phys. Earth Planet. Int. *84*, 179–191.

ROMANOWICZ, B. (1994b), *Anelastic Tomography: A New Perspective on Upper-Mantle Thermal Structure*, Earth. Planet. Sci. Lett. *128*, 113–121.

ROMANOWICZ, B. (1995), *A Global Tomographic Model of Shear Attenuation in the Upper Mantle*, J. Geophys. Res. *100*, 12,375–12,394.

ROMANOWICZ, B., MEGNIN, C., and LI, X.-D. (1996), *Wave-form Inversion for Global Mantle Elastic and Anelastic Structure*, EOS Trans AGU, 77, 482.

ROMANOWICZ, B. (1997), *3-D Models of Elastic and Anelastic Structure in the Mantle*, EOS Trans AGU, 78, F466.

ROMANOWICZ, B., ROULT, G., and KOHL, T. (1987), *The Upper Mantle Degree Two Pattern: Constraints from Geoscope Fundamental Spheroidal Model Eigenfrequency and Attenuation Measurements*, Geophys. Res. Lett. *14*, 1219–1222.

ROULT, G. (1975), *Attenuation of Seismic Waves of Very Low Frequency*, Phys. Earth Planet. Inter. *10*, 159–166.

ROULT, G., ROMANOWICZ, B., and MONTAGNER, J. P. (1990), *3-D Upper Mantle Shear Velocity and Attenuation from Fundamental Mode-free Oscillation Data*, Geophys. J. Int. *101*, 61–80.

SAILOR, R. V., and DZIEWONSKI, A. M. (1978), *Measurement and Interpretation of Normal Mode Attenuation*, Geophys. J. Astron. Soc. *53*, 559–581.

SHEEHAN, A., and SOLOMON, S. C. (1992), *Differential Shear-wave Attenuation and its Lateral Variation in the North Atlantic Region*, J. Geophys. Res. *97*, 15,339–15,350.

SIPKIN, S. A., and JORDAN, T. H. (1979), *Frequency Dependence of QScS*, Bull. Seismol. Soc. Am. *69*, 1055–1079.

SIPKIN, S. A., and JORDAN, T. H. (1980), *Regional Variations of QScS*, Bull. Seismol. Soc. Am. *70*, 1071–1102.

SIPKIN, S. A., REVENAUGH, J. (1994), *Regional Variation of Attenuation and Travel Times in China from Analysis of Multiple ScS Phases*, J. Geophys. Res. *99*, 2687–2699.

SMITH, M. F., and MASTERS, G. (1989), *Aspherical Structure Constraints from Free Oscillation Frequency and Attenuation Measurements*, J. Geophys. Res. *94*, 1953–1976.

STACEY, F. D., and LOPER, D. E. (1983), *The Thermal Boundary Layer Interpretation of D and its Role as a Plume Source*, Phys. Earth Planet. Inter. *33*, 45–55.

SU, W.-J., WOODWARD, R., and DZIEWONSKI, A. M. (1994), *Degree 12 Model of Shear Velocity Heterogeneity in the Mantle*, J. Geophys. Res. *99*, 6945–6980.

SUDA, N., SHIBATA, N., and FUKAO, Y. (1991), *Degree-2 Pattern of Attenuation Structure in the Upper Mantle from Apparent Complex Frequency Measurements of Fundamental Spheroidal Models*, Geophys. Res. Lett. *18*, 1119–1122.

TRAMPERT, J., and WOODHOUSE, J. H. (1986), *High Resolution Global Phase Velocity Distributions*, Geophys. Res. Lett. *23*, 21–24.

VAN DER HILST, R. D., WIDIYANTORO, S., and ENGDAHL, E. R. (1997), *Evidence for Deep Mantle Circulation from Global Tomography*, Nature *386*, 578–584.

VASCO, D., JOHNSON, L. R., and PULLIAM, R. J. (1995), *Lateral Variations in Mantle Velocity Structure and Discontinuities Determined from P, PP, S, SS and SS-SdS Travel Time Residuals*, J. Geophys. Res. *100*, 24,037–24,059.

WIDMER, R., MASTERS, G., and GILBERT, F. (1991), *Spherically Symmetric Attenuation within the Earth from Normal Mode Data*, Geophys. J. Int. *104*, 541–553.

WILLIAMS, Q., and GARNERO, E. J. (1996), *Seismic Evidence for Partial Melt at the Base of the Earth Mantle*, Science *273*, 1528–1530.

WOODHOUSE, J. H., and DZIEWONSKI, A. M. (1984), *Mapping the Upper Mantle: Three-dimensional Modeling of the Earth Structure by Inversion of Seismic Wave Forms*, J. Geophys. Res. *89*, 5953–5986.

WOODHOUSE, J. H., and WONG, Y. (1986), *Amplitude, Phase and Path Anomalies of Mantle Waves*, Geophys. J. R. Astron. Soc. *87*, 753–773.

WYSESSION, M. E. (1996), *Large-scale Structure at the Core-mantle Boundary from Core-diffracted Waves*, Nature *382*, 244–248.

ZHANG, Y. S., and TANIMOTO, T. (1993), *High-resolution Global Upper Mantle Structure and Plate Tectonics*, J. Geophys. Res. *98*, 9793–9823.

(Received February 2, 1998, revised July 30, 1998, accepted August 21, 1998)

To access this journal online:
http://www.birkhauser.ch

Pure appl. geophys. 153 (1998) 273–310
0033–4553/98/040273–38 $ 1.50 + 0.20/0

❙ Pure and Applied Geophysics

High-frequency *P*- and *S*-wave Attenuation in the Earth

ZOLTAN A. DER[1]

Abstract—Investigations of the spectral characteristics of teleseismic body waves revealed that the spectral falloff rate between 1 Hz and 10 Hz is primarily controlled by anelastic attenuation along the path. In addition, the amount of high-frequency energy in teleseismic body waves is far above the level expected on the basis of Q estimates at low frequencies, thus leading to the idea of frequency dependence in Q. Q variations in the earth's mantle can be investigated by mapping out the variations of high frequency (4–10 Hz) energy relative to the low frequency (1–3 Hz) energy in teleseismic P waves, and similar ratios at lower frequencies in teleseismic S waves. Because of the extreme sensitivity of spectral content of short-period body waves to Q variations, large uncertainties in other factors affecting spectral content can be tolerated in such studies. With the increasing number and density of broadband seismic stations recording at high sampling rates, tomographic studies of Q at high frequencies become possible.

Key words: Attenuation, body waves, seismic, high frequency.

1. *Historical Overview*

The measurement of seismic anelastic attenuation in the earth is a difficult undertaking. Anelastic losses must be estimated from changes in the amplitudes of seismic waves after corrections have been made for other known factors that affect them. The observed changes in the seismic wave amplitudes and spectral contents may be generated by numerous causes besides anelastic attenuation. Unfortunately, these other factors are mostly unfamiliar to us. The problem of attenuation losses, similar to other problems in seismology, seem to be more tractable when signals of very low frequencies are analyzed. Consequently, many of the early estimates of the Q structure of the earth derive from studies of the free oscillations of the earth and surface waves that repeatedly circled the earth. Q estimates for individual lines in the earth free oscillation spectra were obtained from the temporal decay rates of individual spectral lines and the sharpness of resonance peaks.

Moving towards higher frequencies, methods were developed to estimate the average Q structure of the earth from spectral amplitudes of long-period surface waves (ANDERSON and ARCHAMBEAU, 1964; ANDERSON *et al.*, 1965). The result-

---

[1] ENSCO Inc., Springfield, VA, U.S.A.

ing Q-structure estimates showed that most of the anelastic losses in the earth occurred in the upper mantle roughly corresponding to the depth of the low velocity zone. The picture that emerged from these studies is that of an earth with high anelastic losses and frequency-independent Q structures. Regardless of the exceptional progress made even at these relatively low frequencies, the influence of other factors could not be entirely eliminated. Lateral heterogeneities in the earth alter the amplitudes of free oscillations by leakage of the energy from one mode to the other by scattering. Surface waves cross major structural boundaries causing energy losses, multipathing, mode conversions, and they undergo focusing effects. These factors make the study of Q from surface waves more difficult.

Simultaneously with these studies, there was a steady increase of the body wave attenuation analyses for specific regions. Many of these studies were often qualitative in nature and provided relatively few numerical Q estimates. Most of them simply noted patterns of P- and S-wave paths of high and low amplitudes and changes in the visually detectable frequency content, and interpreted them in terms of Q changes in the mantle and crust. These studies led to dramatic progress in our understanding of the internal dynamics of the earth, the new global tectonics. They also showed that the Q in the upper mantle is highly variable, it has low Q regions under backarc basins, mid-ocean ridges and tectonically active regions (OLIVER and ISACKS, 1967; BARAZANGI et al., 1975). Q values, on the other hand, were shown to be higher under old oceanic crust and shield areas. The scope of such studies of Q was limited by the limited frequency response of the photographically recording World Wide Seismic Network (WWSSN) stations, although the very existence of this network made these studies possible.

During the early 1970s progress in digital computer technology made the routine numerical simulation of seismograms possible. Such simulation studies, mostly aimed at understanding the mechanics of earthquake sources and seismic path structures, advanced our understanding of the finer features of long-period seismograms and source-path interactions. Most of the data thus analyzed initially still came from the long-period WWSSN network with only occasional inclusion of short-period seismograms. The inclusion of short-period seismograms in such studies is made more difficult by the fact that waveforms of short-period body waves are hard to model because the detailed structural information at the scale of their wavelengths, needed for accurate modeling, is generally not available. Moreover, because of the source-path tradeoffs most Q values derived from waveform modeling studies in the short-period band should be viewed with some skepticism. Studies of waveforms, amplitudes and spectra of short-period body waves at seismic arrays revealed that these vary considerably and apparently randomly over short distances, thus requiring, in their analyses, a quite different philosophy and a theoretical framework of propagation of seismic waves through random media.

It gradually became apparent that frequency independent Q models of the earth, incorporating the low Q values in the upper mantle from long-period studies, are inappropriate. Many studies indicated that the Q values in the short-period band are much lower than in the long-period band (e.g., FRASIER and FILSON, 1972; DOUGLAS et al., 1972). These apparently contradictory results, alongside some similar, sporadic results from the 1960s, peacefully coexisted in the literature during these years. ARCHAMBEAU et al. (1969), using recording systems with better frequency response at high frequencies (the VELA Long Range Seismic Measurement, LRSM, station network), found that their data required a frequency dependent Q. SIPKIN and JORDAN (1979) and LUNDQUIST and CORMIER (1980) came to the same conclusions using WWSSN data. Support for the ideas of frequency-dependency of Q were also provided by theoretical studies of anelastic losses (e.g., MINSTER and ANDERSON, 1973; MINSTER, 1980; LUNDQUIST and CORMIER, 1980; ANDERSON and GIVEN, 1982).

The early 1970s have also produced several studies of the Q structure under North America. In the long-period band SOLOMON and TOKSÖZ (1970) found wide areal differences in Q under this region. These results were further supported by inversion of surface wave data for Q structures (LEE and SOLOMON, 1975, 1979). In the short-period band several studies of short-period attenuation in this region, using digitized analog seismic recordings, showed major variations in mantle Q (DER et al., 1975; DER and MCELFRESH, 1976, 1977).

During the 1980s there was rapid progress in the understanding of short-period wave attenuation. Substantial research was fueled by the need of the U.S. Department of Defense (DoD) for accurate estimates of nuclear explosion yields by analyses of seismic *P*-wave amplitudes. Consequently, most of the work was concentrated on differences in mantle Q under areas where test sites were located. The ensuing debates resolved many of the apparent discrepancies among the results of various groups, and confirmed the existence of a magnitude "bias" between two of the major test sites: the Nevada test site and the Kazakh test site (LAY and HELMBERGER, 1981; DER et al., 1982a; TAYLOR et al., 1986; DOUGLAS, 1987, 1991; DOUGLAS and MARSHALL, 1996).

There have been several later studies of the Q structure in North America (LAY and HELMBERGER, 1981; DER et al., 1982a; TAYLOR et al., 1986) confirming the large variations in Q beneath the continent. A study by BACHE et al. (1985) showed that the mantle under most of Kazakh and Russia has a high Q. A worldwide compilation of t_p^* estimates, orginally byproducts of discrimination studies, showed that large variations in mantle Q relevant to the short-period band exist worldwide (DER et al., 1982b).

Several studies were performed to delineate the ranges of frequency-dependent models compatible with the available broadband data beneath North America (DER et al., 1982a; LAY and HELMBERGER, 1981; DER and LEES, 1985) and Eurasia (DER et al., 1986; LEES et al., 1986). A worldwide study of frequency dependence (CHAN and DER, 1988) complements the regional variations of Q_{ScS} by SIPKIN and JORDAN (1979) and shows that Q in the short-period band varies in the same sense as the variations of Q_{ScS} worldwide. Other studies support this finding (e.g., LAY and WALLACE, 1983; NAKANISHI, 1979). During the late 1980s and the 1990s there was a decrease of research activity in Q studies. Part of the reason is that the interest in seismic nuclear monitoring has shifted to regional distances and that many have felt that the quality of the then available data was inadequate for achieving rapid progress in this area of research. The latter was indeed a major difficulty in many of the studies quoted above. Often it was necessary to analyze photographic records with poor time resolution and dynamic range and resort to crude measurements, such as those of rise times, in order to put limits on Q. Such crude methods can be entirely avoided with modern digital data.

The recent decade has seen the rapid expansion of digital data networks, both for multiple use and regional networks for earthquake studies. Substantial amounts of data can be accessed through the Internet. With the increasing availability of high-quality high-frequency digital seismic data and the improved understanding of the velocity structure of the earth from tomographic studies, the time is ripe for tomographic mapping of the Q in the earth at high frequencies. At the same time, the results from velocity tomography made the classification of the regions of the earth into the old stereotypes obsolete. We now know that many of the high Q zones in the upper mantle may be remains of old slabs, suggesting that the three-dimensional Q structure of the mantle may follow similar patterns (GRAND et al., 1991; VAN DER HILST et al., 1997; WIDIYANTORO and ENGDAHL, 1997; VAN DER LEE and NOLET, 1997).

This review paper attempts to summarize our knowledge of short-period seismic wave attenuation, as seen through the experience of the author. This subject comprises a large volume of literature, one cannot do justice to it in this short paper. Consequently, some relevant work of considerable merit may not be quoted for the sake of brevity. Moreover, I exclude numerous studies of attenuation at regional distances where the attenuation mechanism is probably a combination of anelastic Q and scattering, and mention long-period surface wave attenuation studies in passing only.

2. Teleseismic Energy Loss in the Short-period Band

The well known standard formula for anelastic attenuation

$$A \sim \exp\left(-\pi f \frac{t}{Q}\right)$$

reveals the extreme sensitivity of seismic amplitudes to Q at high frequencies. In this formula t is the travel time, f is frequency and Q is the dimensionless quality factor the reciprocal of which is the fractional energy loss during one cycle (KNOPOFF, 1964). The quantity $t^* = t/Q$ is a commonly used parameter in attenuation studies. Because of the concentration of attenuation losses in the upper mantle low velocity zone of the earth, t^* is a weak function of distance at teleseismic distances. Ignoring the distance dependence, a t^* value of 1 sec. for *P* waves is commonly used in simulations of long-period seismograms. Such a value would reduce the amplitude of the 4 Hz component of a body wave by four orders of magnitude relative to the 1 Hz component, which causes such component for most events to fall below noise levels and thus renders it unobservable by most known seismic recording systems. This would rule out observation of frequencies as low as 4 Hz in the seismic data altogether. On the other hand, as we shall show below, *P* waves can be shown to contain energy at teleseismic distances to frequencies above 6 Hz.

Various attempts have been made to reconcile the data with the low constant *Q* values accepted in low frequency modeling and explain the apparent presence of such high frequencies in the data. Alternative explanations suggested and discarded included spectral leakage due to the data windows used in spectral analyses, instrument nonlinearities, noise, crustal amplification, variations in source spectra, etc. Extreme models of crustal amplification or surface reflection for explosions can be concocted to shape the spectrum such that the high frequencies would be suppressed. Such crustal models must be quite peculiar and thus they are not credible as explanations for the absence of high frequency energy for the wide variety of regions where *P* wave *Q* was studied. Likewise, cancellation of energy by surface reflection (*pP*) is also unlikely. This author has only seen two examples of nuclear explosions where the *pP* nulls were evident, after studying hundreds of events. In the real world enough high-frequency energy would escape the source region due to scattering by the heterogeneities in the earth or the roughness of the free surface such that it would be still far above the levels allowable by the extremely low *Q* models derived from low frequency data. It is apparent, therefore, that discrepancies of orders of magnitude in spectral amplitudes cannot be consistently explained by any combination of those suggested alternative explanations. Simply stated, the value of the high frequency *Q* along a path through the earth is thus tied to the presence or absence of high frequency energy (in the band 4–10 Hz) observable in *P* waves along that path. Because of the extreme sensitivity of high frequency energy levels to *Q*, all other conceivable factors play only a secondary role. This statement may appear to be too simplistic, nonetheless it is supported by considerable observational evidence.

Most of the work regarding attenuation of body waves in the earth is consistent with the notion that the anelastic energy losses are associated with losses in shear deformation. Although some studies indicated a contribution from losses in compression (TAYLOR *et al.*, 1986) these claims are contradicted by other studies. In any case, all studies agree with the assumption that anelastic losses occur *mostly* in shear deformation. Another issue is the contribution of scattering to attenuation that requires further studies. The shear loss mechanism of attenuation translates into the approximate relationship $t_s^* = 4\, t_p^*$ when the typical mantle velocities for P and S waves are considered. The approximate validity of the relationship between shear wave and P-wave attenuation makes it possible to merge and jointly interpret P- and S-wave atenuation measurements (DER *et al.*, 1986a; LEES *et al.*, 1986) and thus make use of the greater sensitivity of S waves to build attenuation models (DER *et al.*, 1986b).

It appears that frequency dependent Q is inherent in the commonly accepted mechanism of attenuation in the earth's mantle. Anelastic media can be modeled with the following mathematical form of the general stress-strain relationship

$$\sigma + \tau_\sigma \dot{\sigma} = M_R(\varepsilon + \tau_\varepsilon \dot{\varepsilon})$$

where τ_σ and τ_ε are relaxation times for stress and strain and M_R is an elastic modulus and the dot denotes a time derivative. Thus both the stresses and strains are also dependent on the rates of stress and strain. This relation assumes a dynamic loss as the material goes through a stress and strain cycle. The stress-strain relationship above gives rise to a frequency dependent Q of the form

$$\frac{1}{Q} = \frac{\omega(\tau_\varepsilon - \tau_\sigma)}{1 + \omega^2 \tau_\varepsilon \tau_\sigma}$$

where $\omega = 2\pi f$. Relaxation times are typically dependent on the temperature in the fashion

$$\tau = \tau_0 \exp\left(\frac{(E^* + PV^*)}{RT}\right)$$

and E^*, V^* and τ_0 are material constants, P is pressure and T is temperature, R is the gas constant. Thus knowing the pressures and the material constants, the attenuation measurements will provide information with regard to the temperatures in the earth. Most anomalously high attenuation zones for high frequency seismic waves in the earth seem to be associated with high temperatures, conductivities and heat flow. High temperatures thus will give rise to higher attenuation (SOLOMON, 1972). Superposing simple absorption band models such as that described above one can construct more general frequency versus Q relationships. A model of a broad absorption band often used has the form (e.g., LUNDQUIST and CORMIER, 1980)

$$|F(\omega)| = \exp\left[-\frac{\omega t_m^*}{\pi} \arctan \frac{\omega(\tau_1 - \tau_2)}{1 - \omega^2 \tau_1 \tau_2}\right]$$

where t_m^*, τ_1 and τ_2 are appropriately chosen constants with the dimension of time. In the time domain the attenuation-related dispersion associated with such Q models can be described in terms of a minimum phase filter with the appropriate amplitude response (DER and LEES, 1985; CHOY and CORMIER, 1986). Given the strong likelihood of the frequency dependence of Q in the earth, Q estimates in limited frequency bands cannot be extrapolated to other bands.

If Q is frequency dependent, appropriate ways of measuring it must be applied. In the case of constant Q the simple method of fitting linear functions to spectral ratios in a linear frequency vs. the logarithm of the spectral ratio can be used. The slope of such line will be inversely proportional to the value of Q. If Q is frequency dependent, the averaged slope of such line has little meaning by itself. On the left side of the following equation we define apparent t^*, $\bar{t}^*$. This quantity can be related to the actual value of $t^*(f)$ by the following formula,

$$\bar{t}^* = t^* + f\left(\frac{dt^*}{df}\right) = \frac{-1}{\pi}\frac{d(\ln A)}{df},$$

where f is frequency and A is amplitude. The formula states that in the frequency-dependent case a spectral ratio used to measure t^* will be a biased; the bias consisting of the gradient term $f(dt^*/df)$. The question is how much error is caused by Q estimation by this term. As we shall show below this error is slight, because according to most evidence Q (or t^*) changes only slowly with frequency.

If t^* varies slowly with frequency this formula could be used to derive t^* by solving this differential equation although there will still be an unknown integration constant. Absolute values of t^* can be obtained from some types of measurements that include the decay rates of multiple *ScS* and free oscillations of the earth and can be used, in principle, to estimate the unknown integration constant.

3. Estimation Methodology Issues

Most studies of the mechanisms of individual earthquakes and earth structures on a larger scale employ waveform modeling as a tool. This popular methodology attempts to reproduce the observed seismograms by incorporating the earth structures along the paths involved, assuming the source mechanisms and time functions and Q (or t^*) values. All these parameters are adjusted until a good visual or RMS fit between the observed waveforms and the synthetics is achieved. Despite the fact that in most studies typically t^* values near 1 sec. are used for P waves and $t^* = 4$ sec. for S waves, and these seem to be unacceptable based on spectral evidence, no adjustments are usually made to these t^* values. The primary reason is that even large t^* values do not affect the low frequency waveforms being modeled, especially those of the relatively narrow-band short-period seismograms from the old WWSSN system. Relatively small adjustments to the source time functions and structures could always be made to improve the fits between the synthetic and observed waveforms. Moreover, good waveform fits can also be obtained by inflexibly using the fixed t^* values mentioned above even though other data indicated sizeable regional variations in the Q values along the teleseismic paths involved. The problems are associated with the low sensitivity of the waveform fitting procedure at low frequencies ($f < 1$ Hz) to Q variations. The predominantly low frequency character of raw teleseismic body wave seismograms is due not only to Q but to the source spectra, and it is relatively easy to match the overall waveform of a teleseismic body wave by matching the low frequency amplitude and phase only, while ignoring the relatively higher frequency energy riding over it. The station-averaged coherences between the synthetic and actual seismograms become lower at higher frequencies and the actual spectral shapes of the two differ also (DER and SHUMWAY, 1988).

Part of the historical controversy between waveform modelers and spectral modelers was based on differences in philosophical outlook. Waveform modeling is inherently a totally deterministic procedure which clearly breaks down at frequencies above 1 Hz. In some waveform modeling work the waveforms are actually low-pass filtered to "improve" the fit. If, instead, high-pass filtering above 1.5 Hz would be applied to teleseismic body wave seismograms, the waveform modeling task would become utterly impossible. At high frequencies the wave amplitudes and waveforms vary considerably over distances of a few kilometers as evidenced by numerous studies at various seismic arrays. Even though the amplitude variations can be large, they do not amount to orders of magnitude. The appropriate models applicable here are random media, a concept alien to the deterministic mindset used in seismic modeling. Thus the attenuation estimates at high frequencies must, instead, be based on statistical and spectral methods rather than on waveform modeling. Waveform fitting is a procedure that is more appropriate and useful for source mechanism studies at low frequencies and for large events. The problems described above are not inevitable in time domain waveform modeling studies. If spectral fitting is performed simultaneously with the time domain waveform fitting and care is taken that key time domain features of the seismograms, such as rise times, also fit, results more consistent with the Q values derived from frequency domain estimation procedures can be obtained (CHOY and CORMIER, 1986).

The observed properties of short-period body waves support the statements made above. As expected for heterogeneous random media, amplitudes of teleseismic body waves at seismic arrays vary considerably over short distances. Many of these variations are systematic for sources at the same location, but vary with azimuth and slowness (CHANG and VON SEGGERN, 1980). Such variations in waveforms can also be effectively modeled by treating the problem statistically, using the theories developed for modeling random media (CAPON and BERTEUSSEN, 1974; AKI et al., 1985). The amplitude variations in such cases may be regarded as products of the focusing and defocusing of the energy related to small-scale heterogeneities in the crust and the mantle. Moreover, there may be strong variations in amplitude levels, although lesser variations in spectral shapes, among various stations are obviously not related to Q but caused by near-surface layering (DER *et al.*, 1980). Thus body wave amplitude variations among some individual sites may tell us less about the Q in the underlying mantle than spectral variations, although *on the average* both are diagnostic.

Analogously to amplitude variations the spectral shapes also vary slightly across seismic arrays. Taking spectral ratios between all pairs of sensors at NORSAR for a set of 10 teleseismic events, the variations in the apparent t_p^* thus obtained have a standard deviation of about 0.06 sec., presumably due to the waveform fluctuations and not actual variations in attenuation. Such variations put a limit on the accuracy of relative t^* measurements. Nevertheless, these statistical fluctuations are smaller than the regional differences in t_p^* found and the more pronounced differences are found in the long-period band. It appears therefore that the magnitudes of the errors in absolute and relative t^* and relative amplitude measurements due to the heterogeneity of the media can be reduced by averaging multiple spectral measurements with different paths.

Frequency dependence introduces another complication into the methodology of attenuation measurements in the short-period band. Theoretically, it should be possible to detect frequency dependence by comparing observed spectra with well-known source spectra. For the frequency-dependent case their ratio should have a noticeable curvature compared to the ratio associated with a constant Q. BACHE (1980) and WALCK (1988) found that certain instrument corrected spectral ratios in the short-period band had such curvature. It appears, however, that the curvature was detected at the steeply descending portion of the short-period instrument response curve that applied in theses cases. It is somewhat doubtful that the actual short-period instrument response at the frequencies, where it falls off rapidly is known well enough to make such a claim. Therefore it is preferable to tie together independent short and long period t^* estimates; each derived from the appropriate instrument or broadband instruments. At this point the question of frequency dependence cannot be decided convincingly based solely on data in the short-period band data (DOUGLAS, 1991; DER and LEES, 1985), nonetheless it is an inescapable conclusion when short- and long-period Q measurements are compared.

Relative t^* measurements between two stations must also depend on spectral shapes more than on amplitude data. In Figure 1 histograms of relative t^* and $\log_{10}$(amplitude) $\sim$ magnitude differences are shown for the station pair OB2NV (at the Nevada test site) and RKON (Red Lake Ontario) for a large number of events. Note that the t^* measurements group tightly while magnitude differences have a large scatter. By regression analyses of a large data set we have found the relation $\Delta m_b \sim 1.35 \Delta t^*$ (DER *et al.*, 1979), and this apparently is not valid for the individual event pair measurements.

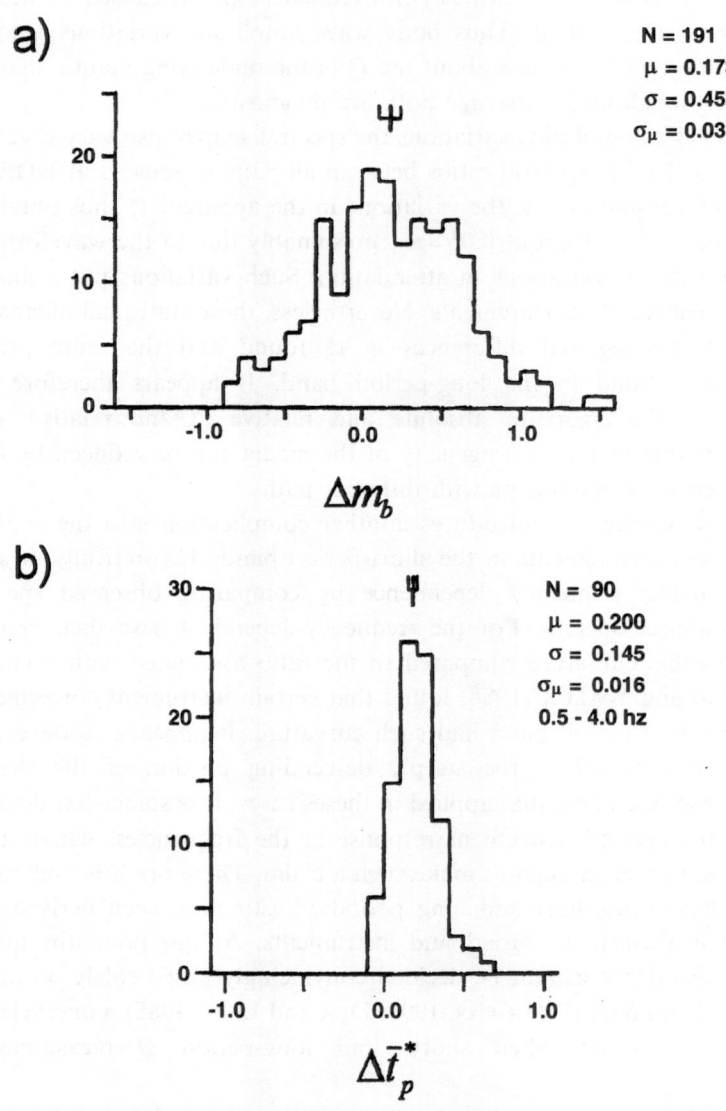

Figure 1

Histograms of the relative P wave t^* differentials and m_b differentials for a suite of common teleseismic events between the stations RKON (Red Lake, Ontario) and the station OB2NV at the Nevada test site. Both stations were located on granite, thus near-surface layering effect is minimized. Note the small scatter in the relative t^* values and the large scatter in relative m_b. Assuming a dominant frequency of 1 Hz, the attenuation could be described in terms of t_p^* differentials as $\Delta m_b = 1.35 \, \Delta t^*$. Part of the scatter in m_b differences may be due to radiation patterns, although most must be due to the lateral heterogeneities under the two stations. This figure should be indicative of the relative importance of spectral t^* and amplitude measurements in attenuation studies.

Table 1

Typical time domain measurements to be fitted to differential t*(f) models (shield-to-shield ~ deep earthquake-to-shield).

S-wave periods (sec.) shield-to-shield	*S*-wave periods (sec.) shield-to-tectonic	Corresponding amplitude ratios shield/tectonic
1.3	2.5	8
1.8	3	5
2.3	3.4	3.5
P-wave period (sec.) shield-to-shield	*P*-wave period (sec.) shield-to-tectonic	
0.7	0.9	2–3

With frequency independent Q the spectral ratio and related measurements can define the absolute Q unambiguously. If Q is frequency dependent then the measurement of $t^*(f)$ becomes less direct. Slopes of spectral ratio estimates, which comprise the bulk of attenuation-related data, are mostly influenced by the apparent t^*, $\bar{t}^*$. Over limited frequency ranges any possible gradual changes in these slopes cannot be detected because of the inherent errors in such measurements. Thus the function $t^*(f)$ must be constructed for each path by piecing together various types of information consisting of spectral ratios between assumed source spectra and absolute Q estimates derived from source independent amplitude decay measurements. Additional constraints may be provided from *S*-wave measurements, making the assumption that most losses occur in shear deformation (DER and LEES, 1985; DER *et al.*, 1986a).

The observed short-period seismic body wave data characteristics from deep earthquakes in South America and in two broad regions of North America (designated as tectonic-*T* and shield-*S*) which must be matched by modeling, are summarized in Table 1. Because the *S* waves observed had varying waveforms in both U.S. regions, and the differential in periods increased with decreased period, this phenomenon by itself is an indication that the change in period was due to attenuation (Table 1). Each pair of models had to account for the observed differential in apparent t^* values, amplitude and period changes for both *P* and *S* waves (DER *et al.*, 1980).

When the differential attenuation between two paths, or average differential attenuation between paths involving two types of tectonic regimes must be estimated, then there will be two functions of frequency $t^*(f)$ that must be estimated. Spectral differences expressed as spectral ratios are not sufficient to constrain any of these. It is possible, however, to constrain the relative spacing of such curves by fitting amplitude, spectral content and time domain characteristics for both *P* and *S* waves over the two types of paths. An example of such a procedure is described by DER and LEES (1985) for constraining the relative differences in $t^*(f)$, involving the passage body waves through the upper mantle under the "tectonic" western U.S. mantle versus the "shield" type central U.S. mantle. As stated above, the available measurements consisted of station-averaged time domain amplitude levels, spectral ratios over the short-period band, mostly from high frequency analog stations (LRSM) and time domain average wave periods from WWSSN network for both P and S waves. The issue was to decide which of three types of pairs of $t^*(f)$, each pair corresponding to a shield-to-shield (S-S) path and for a shield-to-tectonic (S-T) type path, describes the properties of the data observed. Since the paths from deep earthquakes cross the upper mantle only once, the missing leg of the path is considered 'shield' type implicitly. Two of the three pairs of $t^*(f)$ that have been examined converge at low frequencies and have either strong (CS) or weak (CW) convergence above 1 Hz, the third pair prescribed two curves that were quasi-parallel and did not converge at low frequencies. The high t^* curve for all models was arbitrarily made to converge to 0.8 sec., a value that is not critical for the argument regarding the *relative* run of the curves in each pair. These models are shown in Figure 2.

The issue of such models arose because of the possibility that some pairs of *divergent* $t^*(f)$ models could be consistent with both the observed spectral differences between the Kazakh and Nevada test sites and small differences in explosion magnitude estimates between the two test sites, the so-called 'magnitude bias'. In this case the central United States may be considered as an analog of the Kazakh test site. It was also proposed, on geological grounds, that a better analog of the Kazakh test site would be the northeastern United States (Maine). On the other hand, it is clear, based on geophysical observations, that this cannot be true. Studies of teleseismic body-wave spectra indicate that the northeastern United States is underlain by a mantle of moderate attenuation, resulting in the severe loss of high frequency energy in P and S waves (SOLOMON AND TOKSÖZ, 1970; DER et al., 1982a). In contrast, the P waves from Kazakh explosions typically contain a considerable amount of high-frequency energy, indicating a high Q mantle underneath. Thus despite any geological similarities between Kazakh and the Northeastern United States, the attenuative properties of the mantle underlying them are quite different.

Simulations of the consequences of the three types of models assuming various source pulse shapes (a combined result of sources and attenuation crossing the downgoing leg of the paths) are shown in Figures 3a–c. Source pulses of various duration were convolved with the WWSSN instrument response and the causal minimum phase attenuation operators associated with the frequency-dependent t^* models in Figure 2. Despite the limited bandwidth, the three pairs of $t(f)$ models have quite noticeable consequences with regards to differentials in wave amplitudes and periods, associated with the differences in the two types of upper mantle under North America, the upgoing leg of the ray paths. The QP and CS models both

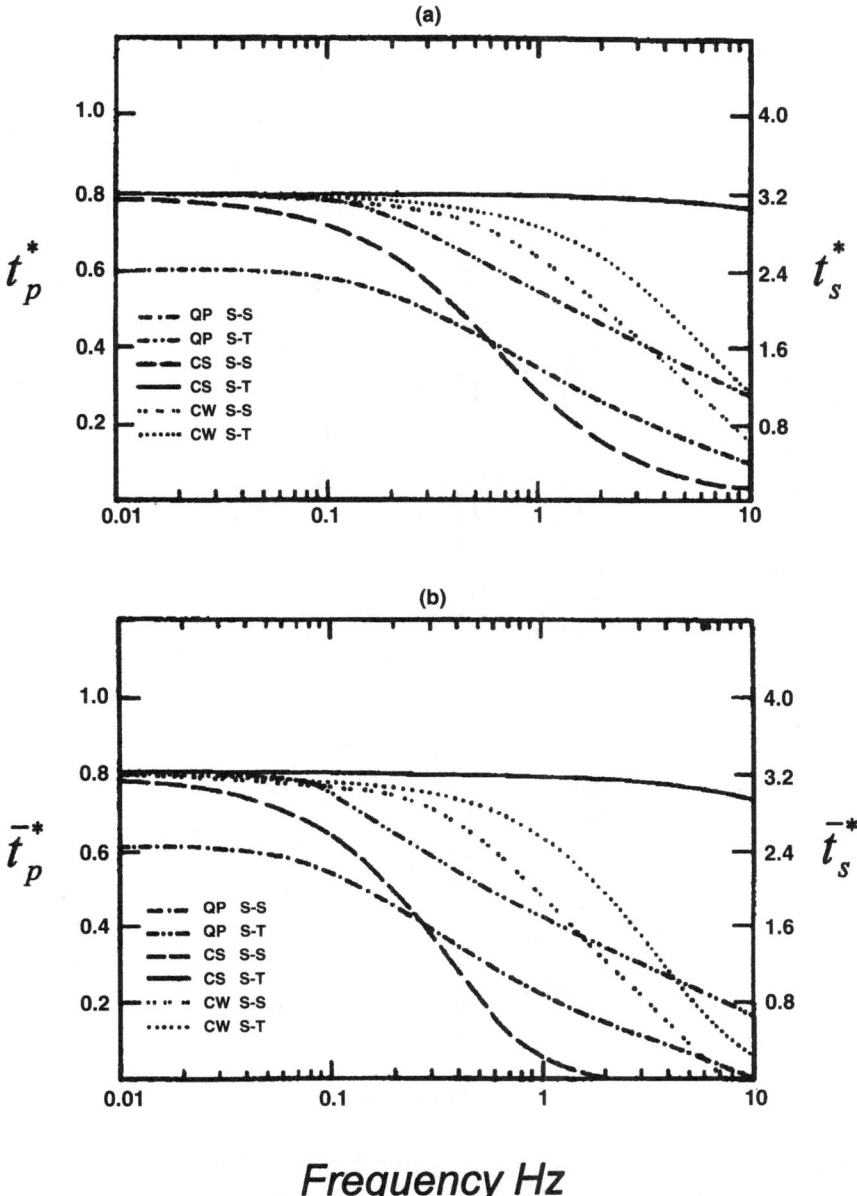

Figure 2
Three pairs of $t^*(f)$ models; a weakly convergent (CW) pair, a strongly convergent (CS) pair and a quasi-parallel (QP) pair. These are to be matched to spectral and waveform data in Table 1 derived from the central and western United States short-period stations.

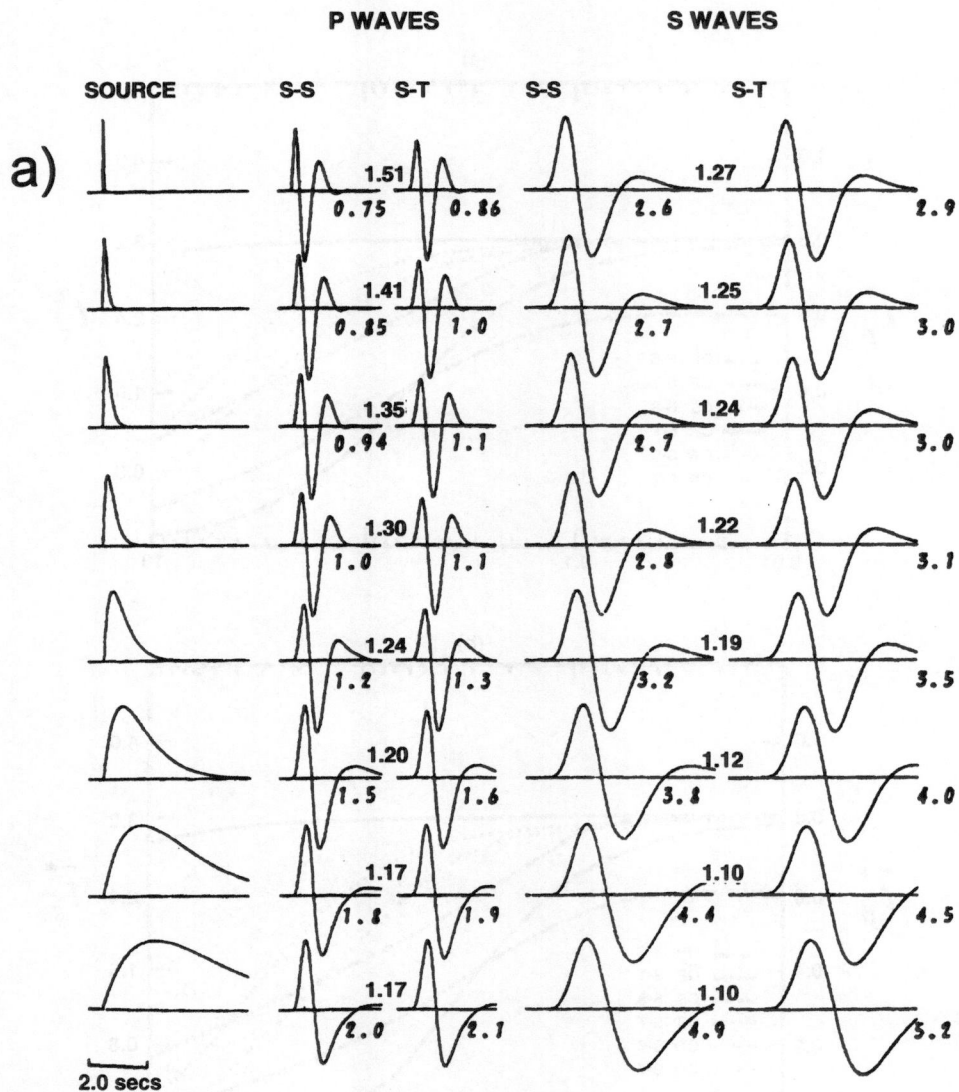

Figure 3(a)

reproduce the observed periods for the waves. Conversely, the observed apparent t^* values, derived from slopes of spectral ratios near 1 Hz, are near 0.2 sec. for P waves and 0.8 sec. for S waves and are correct only for the QP model. The CS model, because of its strong frequency dependence near 1 Hz, overstates these by a factor of 3.5. The CS model also yields amplitude differentials that are too large. The CW model is also totally unacceptable, because it causes excessively small changes in both amplitudes and wave periods.

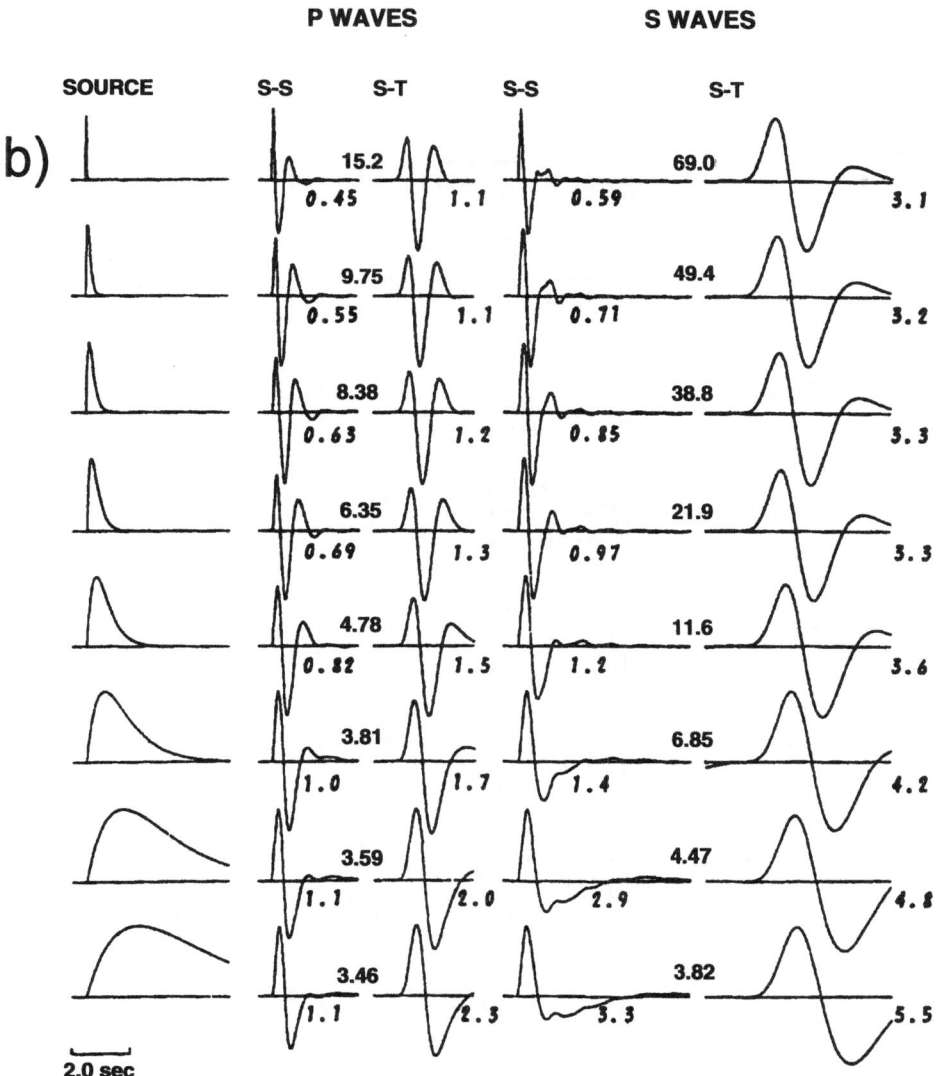

Figure 3(b)

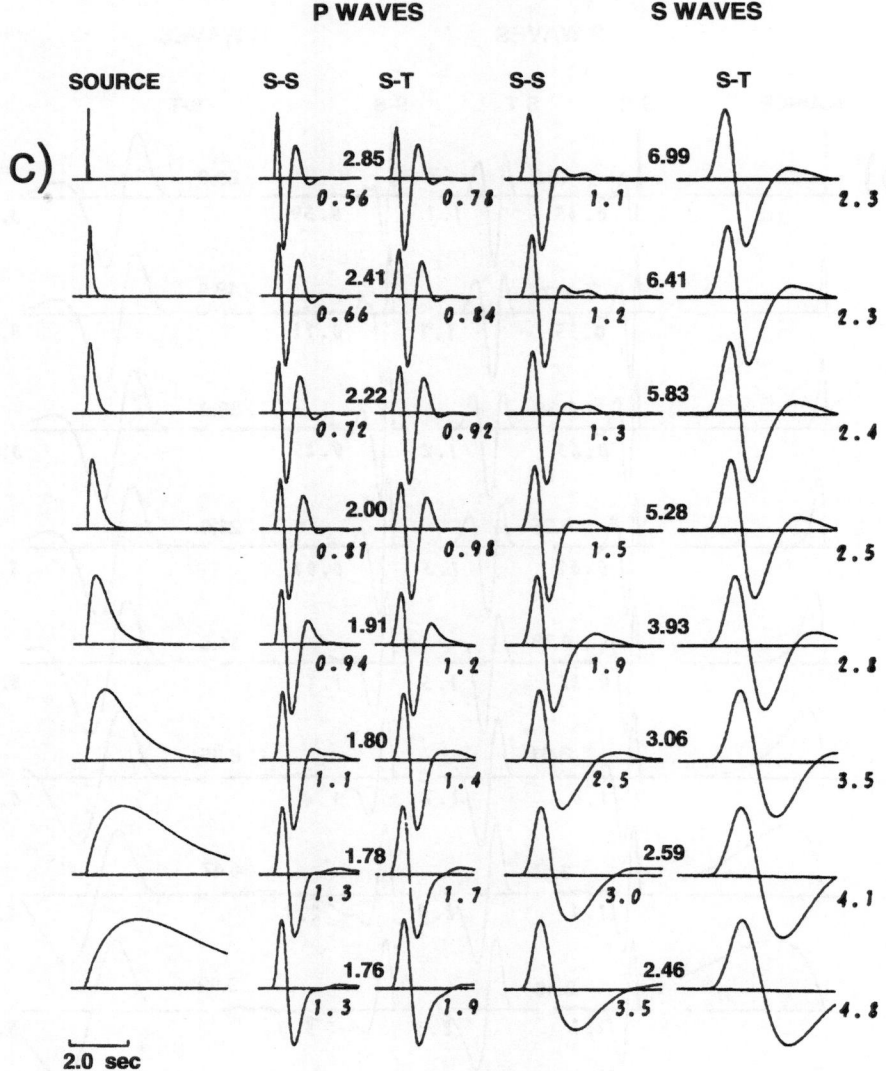

Figure 3(c)

Figures 3 a–c. Synthetic *P* and *S* waveforms corresponding to the shield-to-shield (S-S) and shield-to-tectonic (S-T) paths for the (a) weakly convergent (*CW*), (b) strongly convergent (*CS*), and (c) quasi-parallel (*QP*) pair of $t^*(f)$ models computed by using the short-period WWSSN response. All waveforms were normalized to the same amplitude in the plot in order to better observe them. The source pulses are shown on the left. The peak-to-peak amplitude ratios are shown between each pair of waveforms, the dominant periods are shown below each trace in italics. Only the *QP* pair of frequency dependencies matches the observations described in Table 1.

Although these results do not apply directly to the test sites, they drastically reduce the likelihood of the kind of regional variations that could result in such scenarios. Despite the limitations of the data in the standard WWSSN short-period band (0.5–5 Hz), these three models could be easily differentiated by simple simulations, although these results by themselves do not prove frequency dependence. The common trends in the QP model could be removed and the differentials in amplitude and frequency contents in the data would still be satisfied. This exercise demonstrates, however, that large divergences in regional $t^*(f)$ in the short-period band are not likely.

The question of the 'magnitude bias', the prime motivator for this work, was finally resolved by a direct experiment (Joint Verification Experiment–JVE) which established that nuclear explosions of identical size at the Nevada and Kazakh test sites indeed would result in a difference in magnitude estimates at teleseismic distances and that the size of this difference was in good agreement with that predicted by the QP type models (DOUGLAS and MARSHALL, 1996; SYKES and EKSTRÖM, 1989) or a constant t^* differential. This finding also removes the need for arguments concerning Kazakh test site 'analogs' and models very different from the QP types. Further evidence for the inadmissibility of strongly frequency-dependent models in the short-period band was provided by DOUGLAS (1991), who demonstrated that Yellowknife seismograms of the Alma Ata earthquake deconvolved using such models which resulted in physically improbable displacement histories where forward and reversed displacements followed in quick succession.

In order to measure attenuation differences among various paths for common events, ratios of body-wave spectra associated with two paths have been commonly utilized. If the shape and magnitude of the source spectrum can be estimated theoretically, the spectral ratios between an observed spectrum and some assumed source spectrum can be utilized to estimate the attenuation. Obviously, in either case, the fit cannot extend to the frequency range where the noise approaches the signal in amplitude. Thus t^* for paths with high attenuation cannot be measured very accurately because only a small, low frequency, portion of the spectrum is observable above the background noise. Moreover, in such cases relatively small uncertainties in the assumed source functions, which mostly affect the spectral details at low frequencies but not the asymptotic falloff rates, can cause sizeable variations in the t^* derived. This problem is demonstrated by the debate between DOUGLAS *et al.* (1993) and ZHU *et al.* (1993) which involved t^* estimates for a path between Tuamotu and Yellowknife. The apparently high t^* value for this path is apparently due to high mantle temperatures under the Tuamotu region. Nevertheless, such arguments over difference in t_p^* of less than 0.2 sec. (rather than between 0.5 and 1 sec.!) demonstrate the recent progress made in this field.

Some researchers question the appropriateness of a least-squares fitting procedure to obtain slopes of spectral ratios. Clearly, the distribution of the estimated spectral ratio around the theoretical best fit is not normal. Nevertheless, it is unlikely that major errors in the t^* estimated have resulted from such procedures. The other question asks for the best way to compute the spectrum of body wave transients. In most early studies some tapering window function was used and the fitting was done in the frequency range where the S/N ratio was high. For strongly varying spectra the errors of spectral leakage are of some concern. Multi-taper methods (PERCIVAL and WALDEN, 1993; ZHU et al., 1989) reduce such problems if very short (~ 3 sec.) waveform segments are used in computing spectra. Oppositely, using such short segments was avoided in practically all earlier studies and the general results produced by this approach (ZHU et al., 1989) seem to differ little from those of earlier studies. Nevertheless, multi-taper methods are generally preferable to single taper analyses.

Another approach, running the waveforms through a bank of sharply tuned band-bass filters and measuring the energy in various bands, can also reduce the problems with spectral leakage. It was discovered however, that this method produces results very similar to the windowed spectra if tapers with low leakage, such as the Parzen taper, were used (SHARROCK et al., 1995). Even though some of the earlier methods may have biases, estimated much below 0.1 sec., they were certainly adequate to support the most important conclusions drawn from them. These conclusions involve the absolute sizes of t^* in the short-period band (much less than 1 sec. for P waves), the sizes of relative variations across North America 0.2–0.3 sec.), and the general need for frequency dependence. The main limitations in such studies are the variability and scatter in spectral and amplitude data at high frequencies and the uncertainties in the assumed source spectra, not the methodology in computing spectra.

In view of new tomograpic studies of the earth's velocity and Q structure (ROMANOVICZ et al., 1987; ROMANOVICZ and MONTAGNER, 1990; ROMANOVICZ, 1990), the crude regional subdivisions of the upper mantle into generic tectonic regimes used in past studies appear to be obsolete and require re-evaluation. With the increasing number of broadband digital seismic stations and regional networks, true tomographic studies of Q at high frequencies are now feasible. Due to the extreme sensitivity of the spectra of body waves to Q at high frequencies, the Q effects dominate. Thus one can map out the absorbency of the earth in three-dimensional grids using simple tools of spectral analysis. Whenever applicable, variations due to other factors, such as near-surface layering may be corrected, but such effects are of minimal importance.

In computing the apparent t^* from spectral ratios of body waves and theoretical source spectra errors will occur due to the uncertainties in assumed source spectra. There are several models for source spectra from nuclear explosions. Most assume a ω^{-2} falloff rate of spectral amplitude of the source spectra at high frequencies (VON SEGGERN and BLANDFORD, 1972; MUELLER and MURPHY, 1971). Some assume a ω^{-3} falloff rate at high frequencies adjoining a transitory ω^{-2} rate at lower frequencies (HELMBERGER and HADLEY, 1981). All these have different scaling laws and spectral overshoots. None of these models are well enough defined to rule out others. Typically the reduced displacement potentials (RDPs) for above may fit the data optimally, on occasion, for individual explosions. Consequently there will be unavoidable uncertainties in Q (or t^*) depending on the models. Most seismologists prefer a ω^{-2} falloff rate for earthquakes (e.g., AKI, 1967). A more complex earthquake source spectral model that includes ω^{-1}, ω^{-2} and ω^{-3} falloff rates was proposed by GELLER (1976). Assuming that t^* is derived from spectral ratios at the high frequency end beyond the corner frequencies in the short-period band up to 10 Hz, and the ω^{-2} and ω^{-3} falloff rates translate into a difference of t^* of about 0.1 sec., with the *lower* value associated with the ω^{-3} falloff. In any case despite the uncertainties the t^* are fairly well constrained at high frequencies.

An additional complication factor for nuclear explosion is that anelastic losses seem to occur in the near explosion environment, in the range where linearity was assumed previously (TRULIO, 1978; MINSTER, 1978a,b; DAY and MINSTER, 1984). Such losses were not accounted for by the source models mentioned above. These losses explain some apparent discrepancies between absolute amplitudes of the elastic waves leaving explosion sources at close distances and those observed teleseismically that apparently required large values of t^*.

Despite the fact that digitized old photographic recordings from narrow-band systems are generally unsuitable for spectral analyses, some time domain features of these seismograms can be used to delimit $\bar{t}^*$. Assuming various limiting values of source time functions (such as delta functions, or double delta functions with opposite polarities), it can be shown that dominant periods or rise times measured from the time domain records are limited by the Q along the path (STEWART, 1984). Such arguments were used to put lower limits on Q in several studies (DER *et al.*, 1982a, 1986). The study by CHOY and CORMIER (1986) has shown how such constraints and fitting of spectra can contribute to better Q estimates in time domain studies.

4. Regional Variations of Seismic Body-wave Attenuation at High Frequencies

Large-scale regional variations of body-wave attenuation have been discovered early during the development of plate tectonic theory in areas of subduction near Fiji (OLIVER and ISACKS, 1967) and were found later in other areas as well. Although some values were assigned to Q_p and Q_s, most of these studies relied on qualitative observations, noting the variations in the dominant periods and amplitudes of body waves. Although such variations could be explained by other means for individual events, such as source directivities, the authors of such studies presented convincing evidence that such variations can be explained consistently only by regional variations in Q in the upper mantle. The variations in Q as postulated from these anomalies in wave amplitudes and frequency contents generally coincide with corresponding variations in mantle velocities, low Q values being accompanied with the lowering of both the body-wave and shear-wave velocities by increased temperatures. Later studies which used surface reflections also confirmed the existence of low Q zones under the backarc basins of many subduction zones (BARAZANGI *et al.*, 1975). Manifestations of Q variations in the short-period band are quite obvious in the visible variations of the amplitudes and dominant periods of body waves, especially of S waves. These findings correlate well with other indications of mantle temperature variations such as heat flow and mantle conductivity.

Figure 4 presents examples of 3-component seismograms recorded on the WWSSN short-period system for two stations in the Tonga-Fiji region. The station NIU (top) shows high-frequency waveforms associated with a high Q path, the lower set of seismograms is the same event seen at VUN along a low Q path with considerably lower dominant frequency of waveforms. Despite narrow bandwidth of the WWSSN system, the contrast is quite visible. These are just two examples of a large-scale study of waveform frequency contents and amplitudes that found large lateral variations in the upper mantle Q associated with a subducting slab. Equivalent variations have also been confirmed in other subduction regions.

Another example of visible Q effects is from a study of the upper mantle Q in areas behind island arcs using pP waves from deep earthquakes (BARAZANGI *et al.*, 1975). The top four traces in Figure 5 display teleseismic seismograms where the pP phase crossed high Q upper mantle (slab), and two at the bottom where the pP

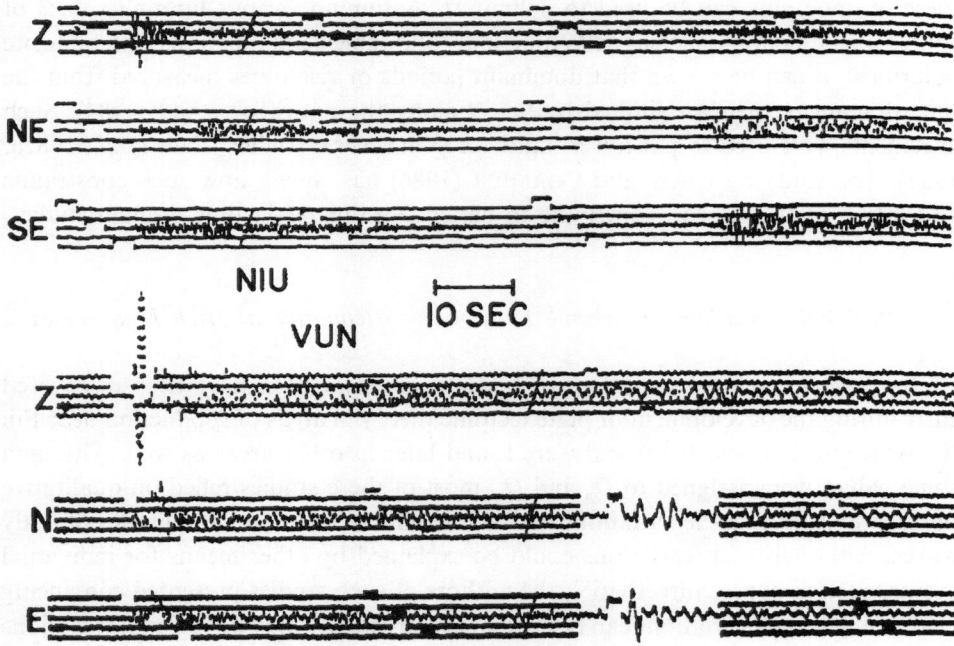

Figure 4
Short-period seismograms recorded at the station NIU, Tonga, involving a high Q path, and at the station VUN, Fiji, along a low Q path from an event of December 10, 1965 at 18.1°S, 179.3°W, depth 624 km. Note the contrast between the high frequency nature of both the P and S phases at NIU as contrasted by the much lower frequencies for the two phases at VUN (after OLIVER and ISACKS, 1967).

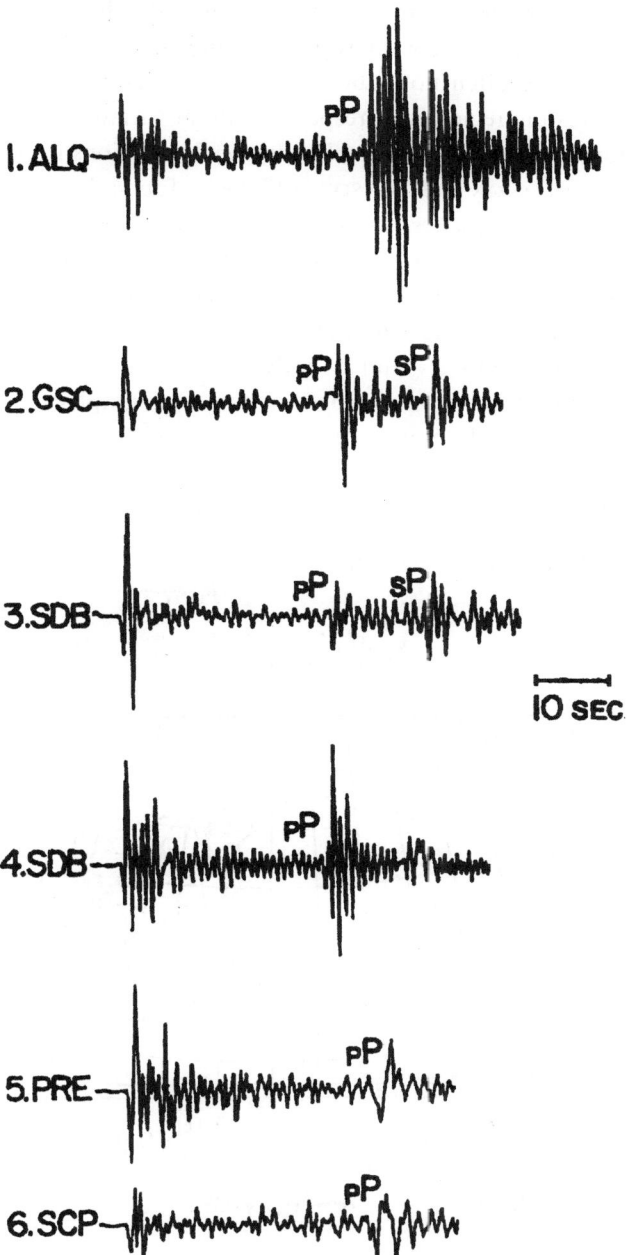

Figure 5

Examples of WWSSN seismograms containing *pP* phases from deep earthquakes that passed through portions of upper mantle with low attenuation (1, 2, 3, 4) and high attenuation (5 and 6). (After BARAZANGI *et al.*, 1975). Note the differences in the frequency content of *pP*.

phase traveled through low Q mantle (backarc basin). The differences in the sizes and frequency contents of these arrivals are quite obvious. Thus Q effects in the short-period band are often not subtle, but may be detected by visual inspection.

Another interesting region with respect to lateral variations in high frequency Q already mentioned is North America. The regional variations in this continent were first outlined by SOLOMON and TOKSÖZ (1970) who found, by the analyses of P and S waves from deep earthquakes recorded over North America, a low Q region west of the Rocky Mountains, high Q areas in central North America and the Pacific Coast and some lower Q under the eastern part of the United States. Even though the manifestations of this variation in Q are subtle at low frequencies, they are quite evident when short-period P- and S-body wave data are analyzed. Firstly the short-period P- and S-wave amplitudes vary in a geographical pattern consistent with the picture provided by SOLOMON and TOKSÖZ (1970), NORTH (1977), BOOTH et al. (1974), and DER et al. (1980). A simple montage of S waveforms from deep earthquakes as observed across North America shows the radical changes in their

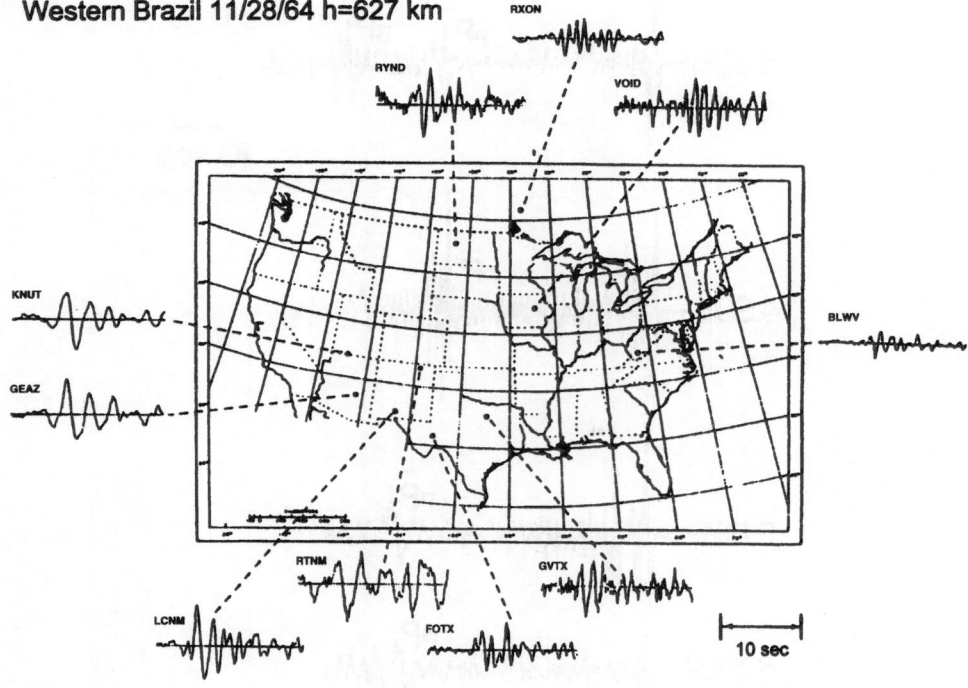

Figure 6

Teleseismic S waveforms from a deep earthquake across the United States. These waveforms are tracings from film and were not corrected for the scalar instrument magnification but the instruments match in the shapes of frequency responses. This figure indicates comparable attenuation-induced variability in frequency contents as those encountered in the studies of backarc basins, high frequency waveforms east of the Rocky Mountains and low frequency waveforms west of it (after DER et al., 1980).

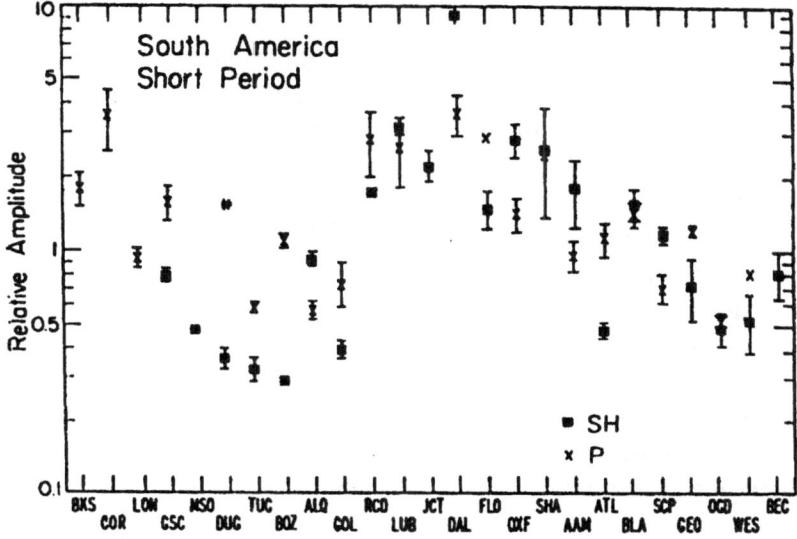

Figure 7

Relative *P*- wave and *SH*-amplitude variations across North America (after LAY and HELMBERGER, 1981) derived from deep earthquake data. The amplitudes were corrected for the appropriate double-couple radiation patterns. The seismic stations are arranged to form a west-to-east cross section. The *P* and *SH* variations track each other, the variations in the latter being only somewhat larger because typically the *P* waves have more high frequency content than the *SH*. Note the low amplitudes in the western mountainous regions (DUG, TUC, ALQ, GOL, BOZ) and the high amplitudes in central North America (RCD, LUB, JCT, DAL, FLOW and OXF) and the West Coast (BKS and COR). Intermediate and low values are found on the East Coast (ATL, BLA, SCP, GEO, OGD and WES).

spectral contents (Fig. 6), the *S* waves observed in the southwestern United States have considerably longer periods than those elsewhere. Amplitudes of both *P* and *SH* waves from deep earthquakes disclose a remarkable pattern (Fig. 7). First of all, the amplitude patterns of *P* and *SH* waves from a later study track each other however, the contrast for *SH* waves is larger (LAY and HELMBERGER, 1981). They also follow the pattern of SOLOMON and TOKSÖZ (1970) and later workers; low amplitudes in the Basin and Range, high amplitudes in central North America and the Pacific Coast. The *P*-wave spectral and amplitude differentials between the Basin and Range and the central United States found in the studies of DER and McELFRESH (1976, 1977) can be explained by two models of upper mantle shown in Figure 8. The western U.S. model (WUS) in this figure is derived by a reinterpretation of the findings of ARCHAMBEAU *et al.* (1969) in which the *Q* losses are reassigned mostly to the western end of their SW-NE profile, which extended from the Nevada Test Site to the north central U.S.

In contrast to the old narrow-band WWSSN data the differences in mantle Q_p can be easily seen when modern new data from the U.S. National Seismic Network are analyzed. We have selected six stations of this network with identical amplitude responses. Two from the central U.S (MIAR–Mt. Ida, AR, AAM–Ann Arbor MI), two from the eastern seaboard (BINY–Binghamton, NY, BLA–Blacksburg, VA) and two from the Basin and Range (ALQ–Albuquerque, NM, TPNV–Topopah, NV). According to previous studies (e.g., SOLOMON and TOKSÖZ, 1970) these three pairs are listed in the order of increasing attenuation. Unlike the P waves from the WWSSN network, the decrease in high frequency contents in P waveforms are immediately visible from top to bottom of Figure 9. Passing a pair of these seismograms for AAM and ALQ through a bank of band-pass filters also evidences the contrast (Fig. 10). Finally, the spectra for eight stations computed by averaging five eigenspectra which utilize the multi-taper method (Fig. 11) clearly show the contrast between the spectra observed in the central United States (Fig. 11a) with those of the stations west of the Rocky Mountains (Fig. 11b). While the former have spectra that decrease gradually and, where the background noise level allows it, remain above the noise

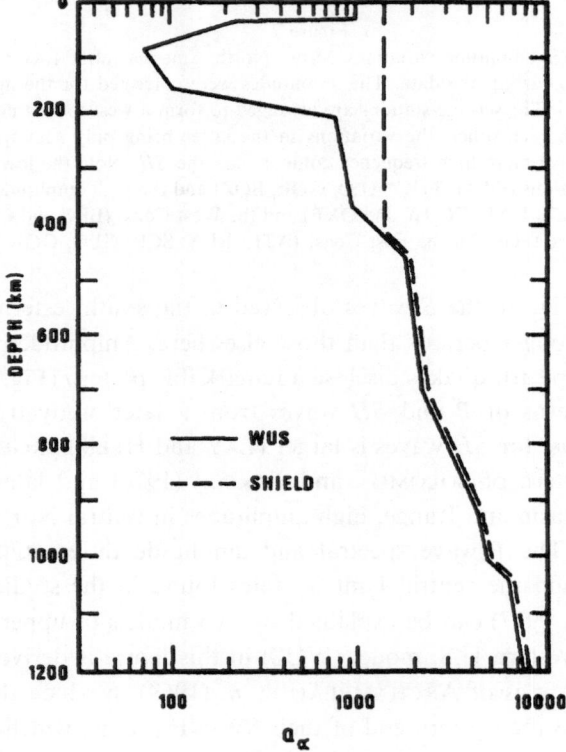

Figure 8
Proposed Q_p models for short-period P waves under the tectonically active western United States (WUS) versus the north central United States (shield). The WUS model is a modified version of the model of ARCHAMBEAU *et al.* (1969).

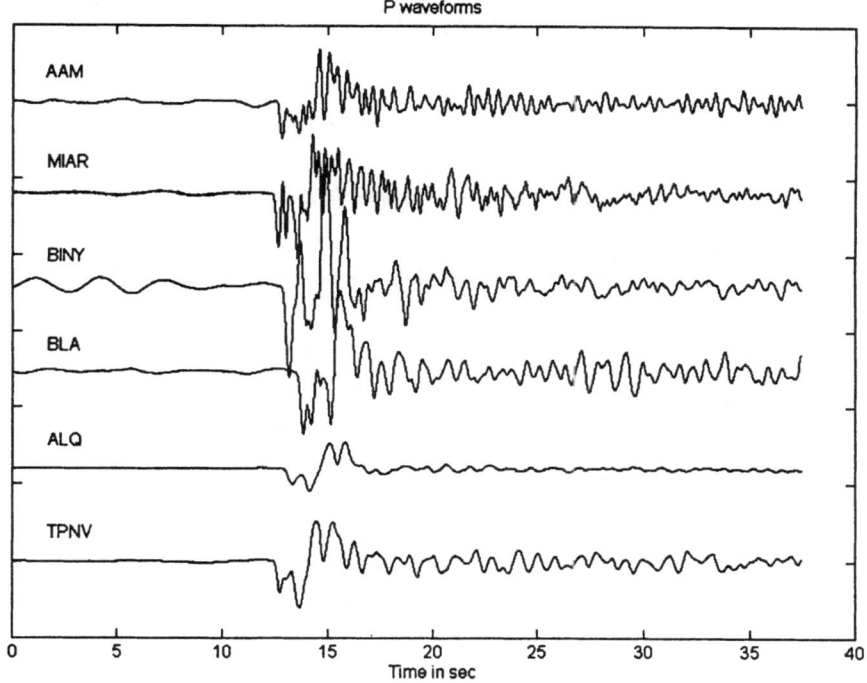

Figure 9

P-wave waveforms from a deep earthquake in Jujuy Province, Argentina (July 20, 1997, 22.69S, 66.02W, depth = 256.5 km) as recorded at stations of the National Seismic Station Network. Since the frequency responses of these stations increase and flatten out towards the higher frequencies, the visible differences in the frequency contents of these waveforms at the stations in the central (left) and western (right) United States are quite dramatic. This phenomenon is considerably harder to see with seismic system responses employed earlier.

at frequencies up to 7–10 Hz, the spectra associated with the latter rapidly decrease with frequency and reach the noise level below 6 Hz. This happens even though the overall signal-to-noise ratio is lower at the western stations. The results are in agreement with the picture of the regional variations from other studies presented above. It is noteworthy that the relative spectral contents of seismograms, the *S* waveforms in several studies quoted above (DER *et al.*, 1980, 1982a; LAY and HELMBERGER, 1981) and the *P*-wave spectral results by DER *et al.* (1980) correlate very well with the upper mantle velocity inversion results which exhibit a pronounced upper mantle low velocity layer in the western part of the continent (VAN DER LEE and NOLET, 1997). The features of their map show very low velocities in the Basin and Range, higher velocities on the eastern seaboard and California; details that correlate well with attenuation measurements. Besides the broad regional features similarities include the sudden transition from Texas to New Mexico (DER and MCELFRESH, 1977) and the higher velocities (lower attenuation) in the Montana and Idaho area.

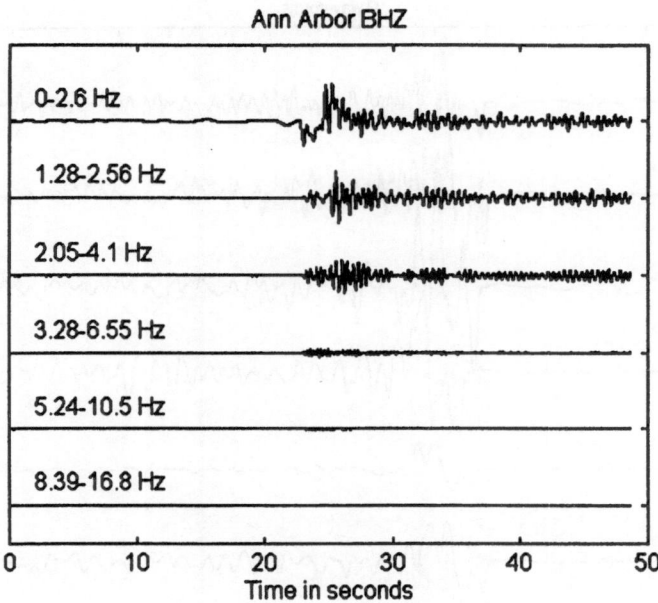

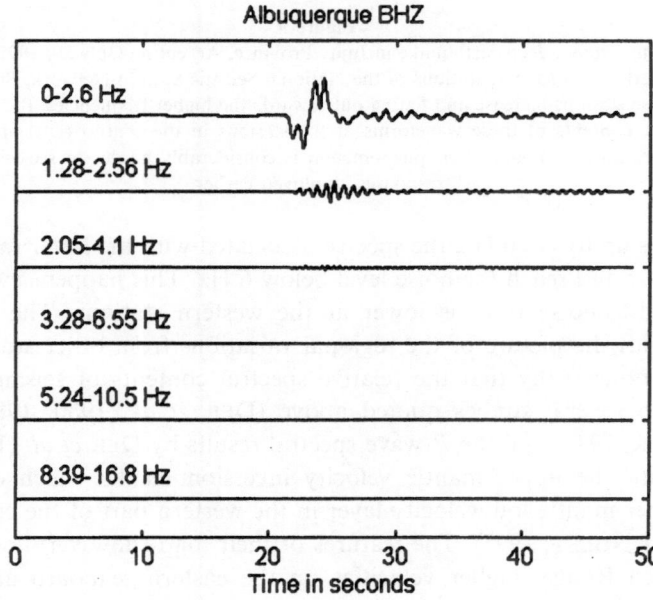

Figure 10
Band-pass filter results for a pair of short-period vertical seismograms for two stations (Ann Arbor, MI and Albuquerque, NM). Note the differences in spectral contents and the diminished high frequency content of the lower example.

The differences in the mantle Q under various former nuclear test sites can be detected *reciprocally*, noting spectral differences at common high Q recording sites. All *P* waveforms from the Nevada test site appear to be deficient in high frequency energy when compared to test sites in high Q areas such as the Kazakh test site as shown in Figure 12. In this figure we compare spectra of Nevada (Yucca Flats and Pahute Mesa) and Kazakh (Degelen and Shagan) nuclear explosions at the common sensor A0 of NORSAR which were arranged in the order of decreasing estimated yields (body-wave magnitudes) from top to bottom. The Kazakh explosion spectra fall off less rapidly with increasing frequency and remain above the noise level to higher frequencies (DER *et al.*, 1985). DOUGLAS (1987) and BACHE *et al.* (1985) reported the corresponding phenomenon at the British array sites. Because of explosion source scaling effects, this effect is seen better at low yields in the figure. The same is true for earthquakes; intraplate earthquakes inside areas underlain by high Q mantle contain more high frequency energy. Another study that confirmed the low Q nature of the mantle under much of western North America consisted of

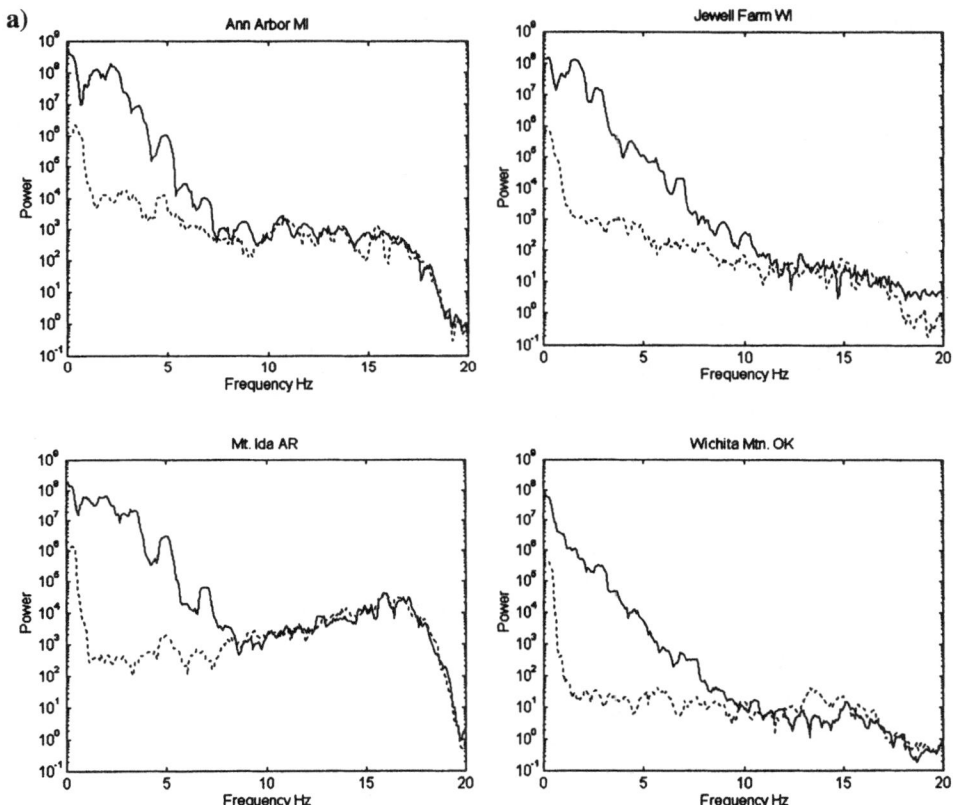

Figure 11(a)

b)

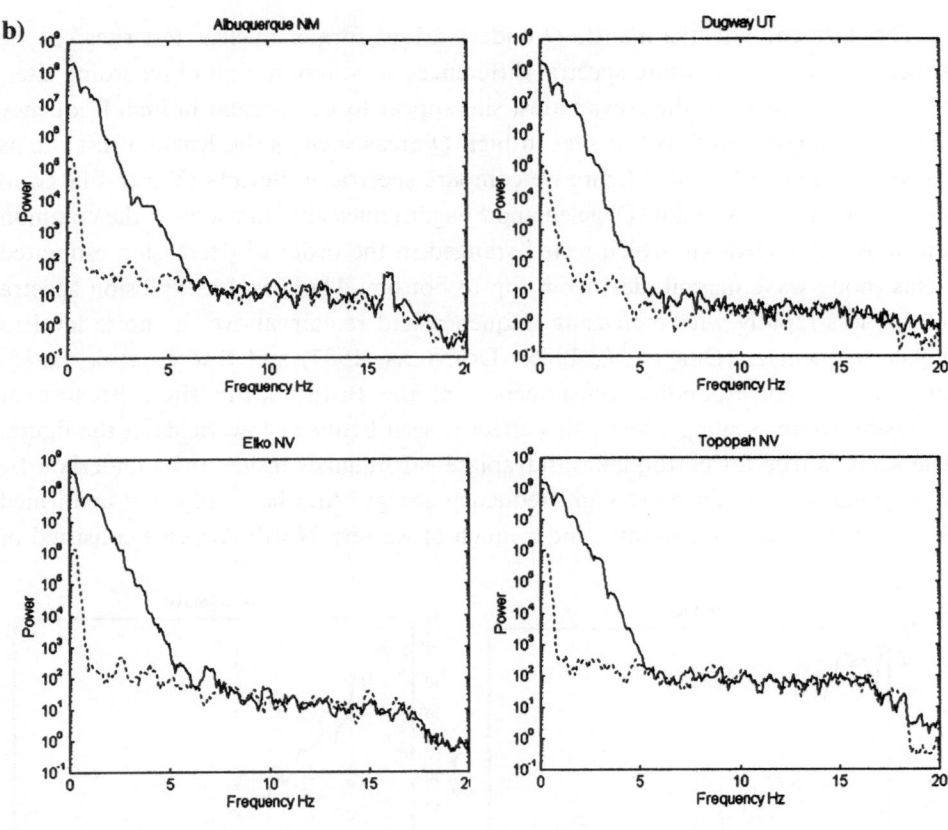

Figure 11(b)

Figure 11a & b. (a) Spectral of *P* waves of the Argentina events as recorded in the central United States, (b) as recorded west of the Rocky Mountains. Dotted lines are noise spectra. These spectra were computed by the multi-taper method averaging the first five eigenspectra and were not corrected for the identical instrument responses. Note that while the spectra fall off gradually and are above the noise in the 6–8 Hz range in the central United States, those in the west fall off sharply and show no significant energy above the noise above 5 Hz.

the spectral analyses of *P* waveforms from the Salmon nuclear explosion in Mississippi (DER and MCELFRESH, 1977). The *P* waves from this explosion contain

Figure 12
Comparison of *P*-wave power spectra of nuclear explosions from the Nevada test sites (Yucca Flats and Pahute Mesa) and the Kazakh test sites (Shagan and Degelen) computed for the A0 array site of the Norwegian Seismic Array (NORSAR). The estimated yields increase from top to bottom. The spectra were not corrected for instrument and were terminated as the *S/N* ratio fell below 2. Note that all spectra from the Kazakh test sites fall off slowly with increasing frequency and indicate the presence of high frequency signal energy reaching 9 Hz. In comparison with these, the spectra of Nevada explosions fall of faster and do not show energy above the noise beyond 3.5 Hz. These spectral differences indicate high mantle *Q* under Kazakh and low mantle *Q* beneath the Nevada test site.

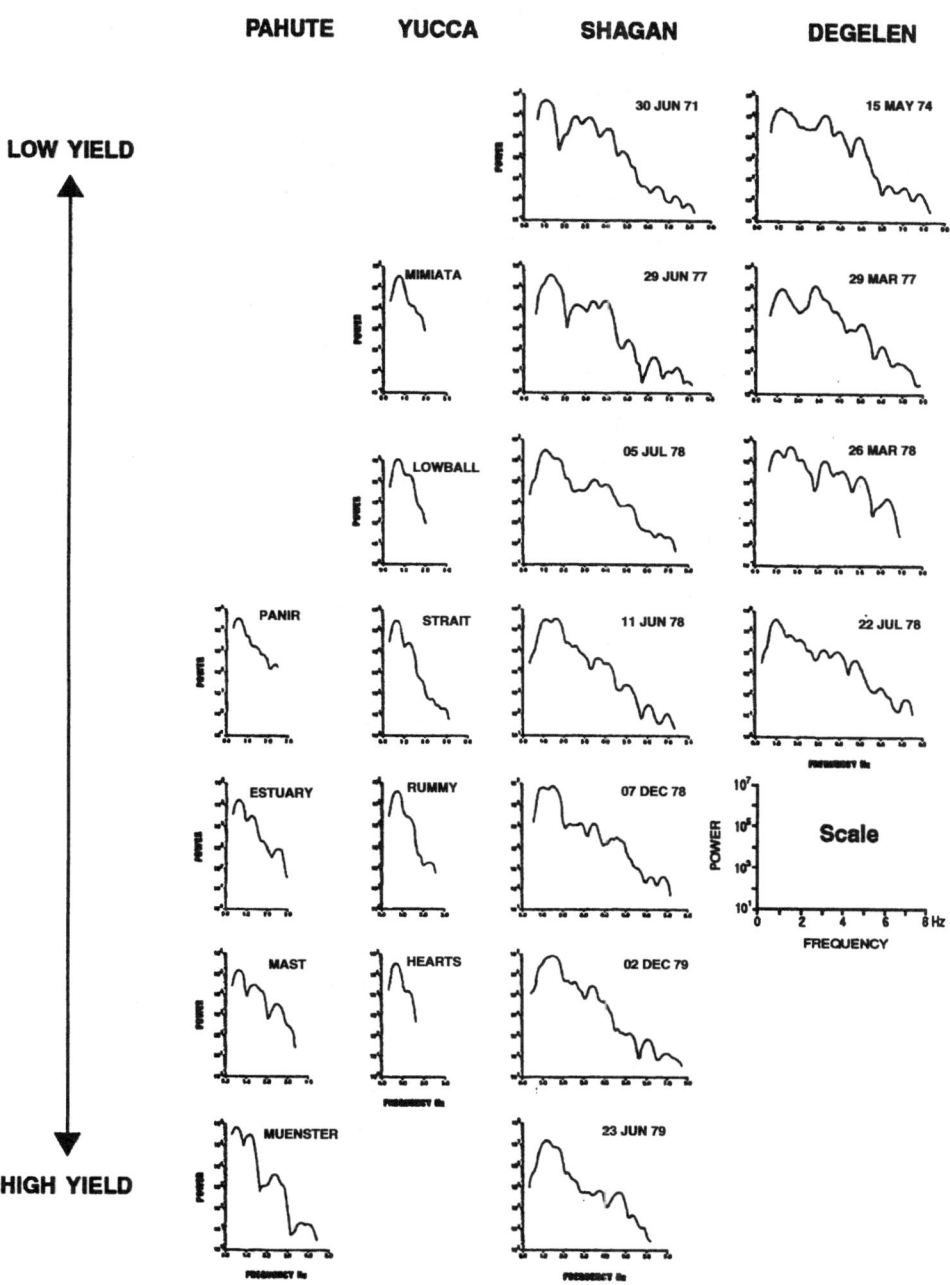

Figure 12

more high frequency energy at the seismic stations located in the eastern United States than those located in the Colorado Plateau and the Basin and Range.

Assembling worldwide t_p^* estimates derived from high frequency falloff rates of spectra from both earthquakes and explosions and sorting them according to generic types of paths (Fig. 13) reveals a large variability in t_p^* (DER *et al.*, 1982b). The means vary from 0.15 sec. to 0.4 sec. and the variance for the mixed and tectonic path categories is larger. This must be a reflection of the heterogeneity of the 'tectonic' population. Even though these t^* values are crude, it is revealing that only a single estimate approximated 1 sec. A similar observation was made by SHARROCK *et al.* (1995) for teleseismic observations of P waves from Pacific subduction zone events. It appears that high frequency attenuation varies considerably worldwide, but most t_p^* values remain in the 0–0.5 sec. range and values close to 1 sec. must be rare.

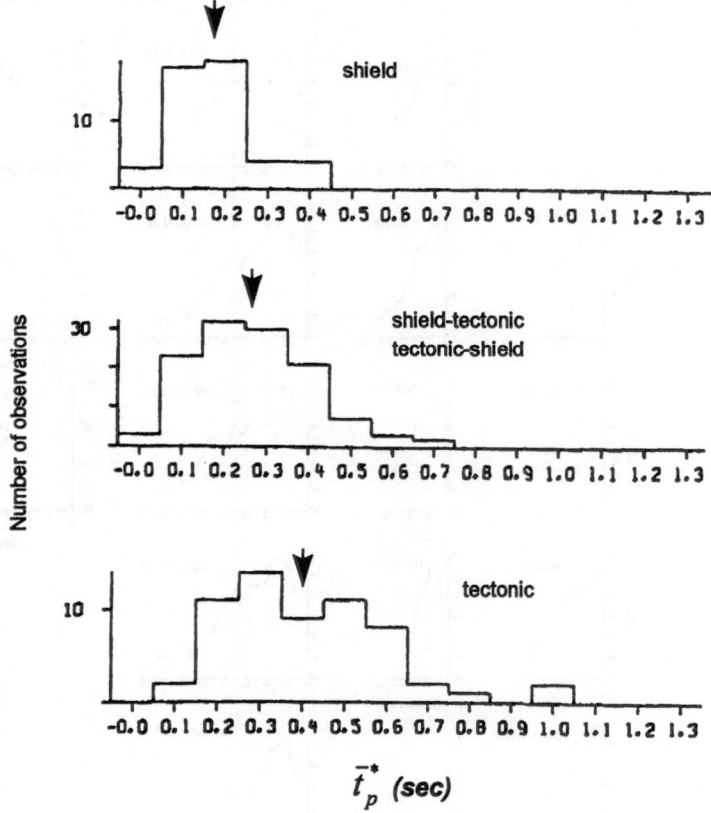

Figure 13

Histograms of worldwide P-wave t^* estimates sorted with respect to generic path types. The mean t_p^* increases from top to bottom (arrows) for each of these populations, indicating that t_p^* varies considerably on the global scale and increases with increasing 'tectonic' component in the propagation path.

5. Frequency Dependence

It has been pointed out by many investigators that it is generally impossible to reconcile the Q values derived from long-period data with the spectral content of teleseismic body waves (SIPKIN and JORDAN, 1979; LUNDQUIST and CORMIER, 1980; DER et al., 1982a; CLEMENTS, 1982). If teleseismic body-wave spectra are corrected for a constant t^* then the high frequency end of the spectrum will increase exponentially in such a way that the result is incompatible with source proposed spectra of either earthquakes or explosions (AKI, 1967; MUELLER and MURPHY, 1971; HELMBERGER and HADLEY, 1981).

Unfortunately, most absolute Q figures derived from long-period data are also subject to unknown errors due to various lateral heterogeneities in the earth. The highly heterogeneous nature of the core mantle boundary and the earth's near surface will have an effect on the multiple *ScS* measurement. Similarly, lateral heterogeneities will cause a loss of energy in the various free oscillation modes through scattering and intermode conversions. If most of the measured losses in such measurements result from anelastic Q then frequency dependence must exist. To derive the actual function $t^*(f)$ one must reconcile both absolute Q estimates and relative measurements of $\bar{t}^*$ for a given region. This was attempted by DER et al. (1986c) for the shield areas of Eurasia and by DER and LEES (1985) for the continental United States roughly subdivided into central and western regions. In both of these studies we attempted to match a variety of frequency domain and time domain characteristics of seismograms for both P and S waves to models of frequency dependence.

The first study fitted a complex Q model to a large set of diverse data. As was typical for the mid-1980s we had only digital array data (NORSAR), some digital Seismic Research Observatory (SRO) data and WWSSN photographic data. For the latter only waveform characteristics such as dominant periods and rise times could be fitted. Hand-digitized WWSSN data were found to be too unreliable for analyses of high frequencies. The array and SRO data provided digital data for multiple *ScS* and body-wave (P and S) surface reflections, a profile consisting of Russian PNE (peaceful nuclear explosions). Added to these were some PP/P spectral ratios from NORSAR for constraining Q_p. Collecting all these we have derived a frequency-dependent Q model summarized in Figure 14. The $t^*(f)$ and $\bar{t}^*(f)$ curves appropriate to these data are shown with the constraints (shaded boxes) imposed by various kind of measurements in Figure 15. It appears, however, that in spite of the uncertainties in these constraints, frequency dependence is required.

The latter study also demonstrated the difficulties in reconciling a variety of observations and measurements to arrive at some satisfactory solution for a $t^*(f)$ for two generalized regions. In the latter study we matched the relative averaged time domain amplitudes, spectral ratios, dominant periods of short period P and S waves from deep earthquakes across the United States to three pairs of frequency-dependent

Eurasian Shield Q(f) Model

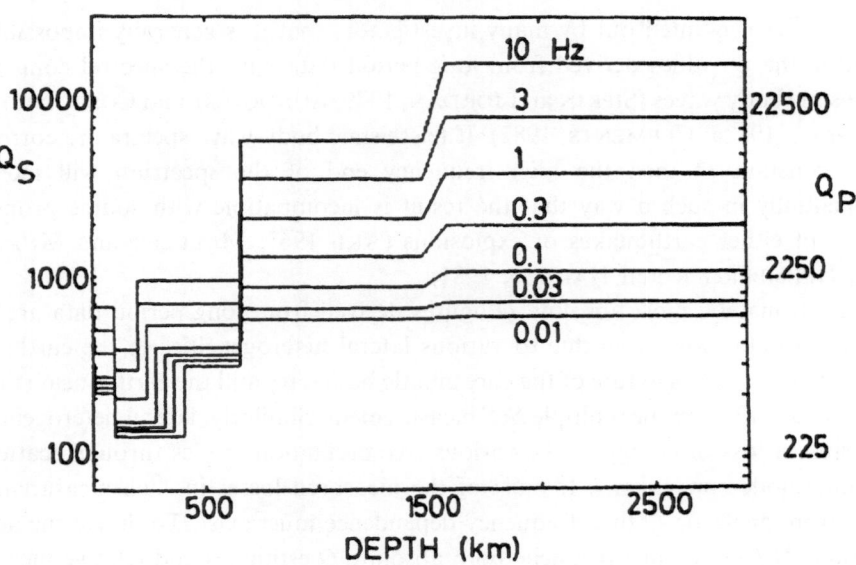

Figure 14

The EURS Q model. Each line is the plot of Q versus depth for a different frequency. This is essentially the same generic kind of model as those proposed by MINSTER and ANDERSON (1973) and LUNDQUIST and CORMIER (1980), but adapted to a high mantle Q 'shield' environment. (After DER et al. 1986b.)

models. The procedures followed and the results are discussed as a demonstrative example in the methodology section above.

Worldwide it was found that the values of low frequency attenuation measurements from multiple ScS and the overall frequency characteristics of long-period body waves as interpreted in terms of Q crossing the upper mantle, varied in the same sense (CHAN and DER, 1988; SIPKIN and JORDAN, 1980; LAY and WALLACE, 1983; NAKANISHI, 1979). Typically relatively low attenuation (t_s^* values in the range 1.5–2.5 for two-way travel through the upper mantle) was found under old continental shields and old oceans, while higher values (t^*s in the range 3.5–5) are characteristic for ocean ridges (SCHLUE, 1981; SHEEHAN and SOLOMON, 1992), tectonically active areas of Eurasia (DER et al., 1986a), Mexico (LAY and WALLACE, 1983). The picture that emerges from worldwide studies defines an earth with large three-dimensional variations in frequency-dependent Q which have been outlined only in some interesting, highly anomalous areas. Much of Eurasia outside of tectonically stable areas is still poorly explored. The forms of frequency dependence are still not understood well.

There has been considerable recent progress in the understanding of the earth's three-dimensional velocity structure recently based on results of tomographic studies. Therefore, several of the old categories such as 'shield' and 'tectonic' lost

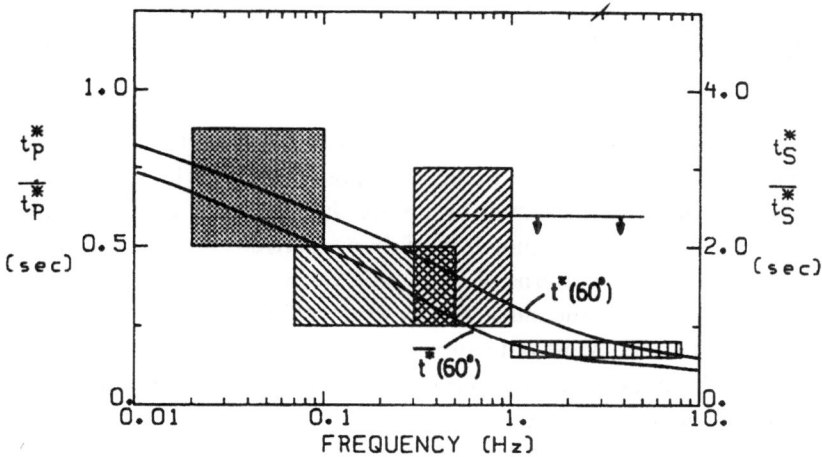

Figure 15

Plot of $t^*(f)$ and $\bar{t}^*(f)$ for direct *S* and *P* at 60° as predicted by the EURS *Q* model. The curves are superposed on constraints imposed by various $\bar{t}^*$ observations (boxes); among them spectral constraints in the short-period band (rightmost box), short-period waveform constraints (downward arrows). In the long-period range these include *S*-*SS*-*SSS* and multiple *ScS* constraints (boxes on the left). Note the differences between the actual and apparent t^*. This is an illustration of how various types of measurements can be fitted together to construct a frequency dependent t^* model (after DER *et al.*, 1986; LEES *et al.*, 1986).

their original meaning as laterally homogeneous provinces of the earth. The emerging picture of the velocity structure of the earth requires the reevaluation and reinterpretation of the *Q* structure as well. With the increasing number of digital seismic stations recording at high sampling rates (see the report by the NATIONAL RESEARCH COUNCIL, 1995) and broad frequency bands, the time has come to explore global variations of *Q* by mapping the absorbency of the upper mantle to high frequency *P* and *S* waves. As can be seen from the discussions above, such mapping can be accomplished by applying quite simple, broadband frequency domain analysis methods to high frequency seismograms. The determination of the three-dimensional *Q* structure of the earth will provide useful constraints on the temperature distribution and thus on the deductions of the internal dynamics of the earth.

6. Conclusions

Due to the spatial variability of high frequency teleseismic *P* and *S* waveforms, and to a lesser degree, of their seismic spectra, deterministic waveform modeling is not an effective method for estimating anelastic attenuation at high frequencies.

The reasons are that spectra of synthetic seismograms generally do not match the spectra of the data at higher frequencies, i.e., waveform fitting becomes increasingly ineffective for higher frequencies.

The issue is, essentially, the absorbency of the lithospheric and upper mantle structures to high frequency seismic body waves. Owing to the extreme sensitivity of the spectral shapes to Q variations, as opposed to other factors, the Q at high frequencies can be easily measured by noting the variations in the spectral shapes along various paths. The regional variations derived from spectral variations correlate well with gross patterns of amplitude variations of both P and S waves at high frequencies, despite the substantially greater scatter in amplitude data occasioned by other factors. Most data support anelastic losses in shear deformation in the mantle and no evidence has been found which favors significant losses in compression at high frequencies.

Q in the earth appears to be frequency dependent, based on the consistent pattern of lower Q estimates from long-period data than for high frequency data for the same general regions. Nevertheless, no evidence has been found for rapid variations of teleseismic t^* values with frequency. The available data can be easily reconciled with slow, gradual variations of Q throughout the observable seismic band. Gross regional variations in t_p^* are in good agreement with t_s^* variations after the latter are divided by the factor of four in both the short- and long-period bands, while the absolute values of these appear higher in the long-period band. This supports the idea of quasi-parallelism of regional $t^*(f)$ curves.

The degree of anelastic attenuation of high frequency body waves varies considerably over the mantle. Large variations have been shown to exist in many areas of the world where subduction takes place. While the downgoing slabs are characterized by low attenuation, the upper mantle Q values under backarc extensional basins are generally low. The mantle attenuation beneath the oceans seems to decrease with the increasing age of the overlying crust, starting with extremely strong attenuation near mid-ocean ridges. There is a large range of variation in the attenuative properties of the mantle under continental North America. The Basin and Range province is characterized by high mantle attenuation, central North America with low attenuation while the eastern seaboard has a moderate degree of mantle attenuation. Many areas of tectonically active parts of Eurasia are only poorly explored in this respect but tend to have lower mantle Q than the 'shield' areas. South America is characterized by pockets of low mantle Q in some areas, such as the Altiplano.

With the increasing number of digital seismic stations recording at high sampling rates and broad frequency bands, there is an opportunity to explore and reevaluate the global variations in the absorbency of the upper mantle to high frequency P and S waves and thus refine the broad picture outlined above. Regional seismic networks and portable seismometer configurations are especially suitable to map small-scale variations in mantle Q. Many of the difficulties encountered in earlier studies due to photographic recordings can be avoided by using broadband data. Such data can be applied to the direct estimation of broadband attenuation operators. The information derived from Q tomography, complementary to velocity tomography, will result in a better understanding of the temperature variations within the earth and the dynamics of the upper mantle.

Acknowledgements

The author is indebted to Alan Douglas and Keith McLaughlin for their reviews which improved the paper. Chris Schneider provided assistance in data acquisition, spectral analyses and the production of the figures. The author thanks the Royal Astronomical Society and the Seismological Society of America for granting permission to reproduce previously published figures in this paper.

REFERENCES

AKI, K. (1967), *Scaling Law of the Seismic Spectrum*, J. Geophys. Res. *72*, 1212–1231.

AKI, K., WU, R., and NOWACK, R. (1985), *Seismic scattering and lithospheric heterogeneity*. In *The VELA Program; A Twenty-five Year Review of Basic Research* (ed. A. U. Kerr) Defense Advanced Research Projects Agency.

AKI, K., and RICHARDS, P. G., *Quantitative Seismology* (W. H. Freeman & Co. 1980)

ANDERSON, D. I., and ARCHAMBEAU, C. B. (1964), *The Anelasticity of the Earth*, J. Geophys. Res. *69*, 2071–2084.

ANDERSON, D. L., BEN-MENAHEM, A., and ARCHAMBEAU, C. B. (1965), *Attenuation of Seismic Energy in the Upper Mantle*, J. Geophys. Res. *70*, 1441–1448.

ANDERSON, D. L., and HART, R. S. (1976), *The Q of the Earth*, J. Geophys. Res. *83*, 5869–5882.

ANDERSON, D. L., and GIVEN, J. W. (1982), *Absorption Band Q Model for the Earth*, J. Geophys. Res. *87*, 3893–3904.

ARCHAMBEAU, C. B., FLINN, E. A., and LAMBERT, D. H. (1969), *Fine Structure of the Upper Mantle*, J. Geophys. Res. *74*, 5825–5865.

BACHE, T. C., MARSHALL, P. D., and BACHE, L. B. (1985), *Q for Teleseismic P Waves from Central Asia*, J. Geophys. Res. *90*, 3575–3587.

BARAZANGI, M., PENNINGTON, W., and ISACKS, B. (1975), *Global Study of Seismic Attenuation in the Upper Mantle behind Island Arcs Using pP Waves*, J. Geophys. Res. *80*, 1079–1092.

BHATTACHARYYA, J., MASTERS, G., and SHEARER, P. (1996), *Global Lateral Variations of Shear-wave Attenuation in the Upper Mantle*, J. Geophys. Res. *101*, 22273–22289.

BOOTH, D. C., MARSHALL, P. D., and YOUNG, J. B. (1974), *Long- and Short-period P-wave Amplitudes in the Range of 0°–114°*, Geophys. J. R. Astr. Soc. *39*, 523–537.

CAPON, J., and BERTEUSSEN, K. A. (1974), *Random Medium Analysis of Crust and Upper Mantle Structure under NORSAR*, Geophys. Res. Lett. *1*, 327–328.

CHAN, W. W., and DER, Z. A., *Attenuation of Multiple ScS in Various Parts of the World*, Geophys. J. *92*, 303–314.

CHANG, A. C., and VON SEGGERN, D. H. (1980), *A Study of Amplitude Anomaly and m_b Bias at LASA Subarrays*, J. Geophys. Res. *85*, 4811–4828.

CHOY, G. L., and CORMIER, V. F. (1986), *Direct Measurement of the Mantle Attenuation Operator from Broadband P and S Waveforms*, J. Geophys. Res. *91*, 7326–7342.

CHUN, K.-Y., ZHU, T., and WEST, G. F. (1991), *Teleseismic Attenuation and Nuclear Source Functions Inferred from Yellowknife Array Data*, J. Geophys. Res. *96*, 12083–13097.

CLEMENTS, J. R. (1982), *Intrinsic Q and Frequency Dependence*, Phys. Earth Planet. Int. *27*, 286–299.

DAY, S. M., and MINSTER, J. B. (1984), *Numerical Simulation of Attenuated Wavefields Using the Pade Approximant Method*, Geophys. J. R. Astr. Soc. *78*, 105–118.

DER, Z. A., MASSE, R. P., and GURSKI (1975), *Regional Attenuation of Short-period P and S Waves in the United States*, Geophys. J. R. Astr. Soc. *40*, 85–106.

DER, Z. A., and MCELFRESH, T. W. (1976), *Short-period P-wave Attenuation along Various Paths in North America as Determined from P-wave Spectra of the SALMON Nuclear Explosion*, Bull. Seismol. Soc. Am. *66*, 1609–1622.

DER, Z. A., and MCELFRESH, T. W. (1977), *The Relationship between Anelastic Attenuation and Regional Amplitude Anomalies of Short-period P Waves in North America*, Bull. Seismol. Soc. Am. *67*, 1303–1317.

DER, Z. A., MCELFRESH, T. W., and MRAZEK, C. P. (1979), Interpretation of Short-period *P*-wave Magnitude Anomalies at Selected LRSM Stations, Bull. Seismol. Soc. Am. *69*, 1149–1160.

DER, Z. A., SMART, E., and CHAPLIN, A. (1980), *Short-period S-wave Attenuation in the United States*, Bull. Seismol. Soc. Am. *70*, 101–125.

DER, Z. A., MCELFRESH, T. W., and O'DONNELL, A. (1982a), *An Investigation of the Regional Variation and Frequency Dependence of the Anelastic Attenuation under the United States in the 0.5–4 Hz Band*, Geophys. J. R. Astr. Soc. (1982), J. Geophys. Res. *87*, 3893–3904, 67–99.

DER, Z. A., RIVERS, W. D., MCELFRESH, T. W., O'DONNELL, A., KLOUDA, P. J., and MARSHALL, M. E. (1982b), *Worldwide Variations in the Attenuative Properties of the Upper Mantle as Determined from Spectral Studies of Short-period Body Waves*, Phys. Earth. Planet. Int. *30*, 12–25.

DER, Z. A., and LEES, A. C. (1985), *Methodologies for Estimating t*$*(f)$ from Short-period Body Waves and Regional Variations of t*$*(f)$ in the United States, Geophys. J. Int. *82*, 125–140.

DER, Z., MCELFRESH, T., WAGNER, R., and BURNETTI, J. (1985), *Spectral Characteristics of P Waves from Nuclear Explosions and Yield Estimation*, Bull. Seismol. Soc. Am. *75*, 379–390.

DER, Z. A., LEES, A. C., CORMIER, V. F., and ANDERSON, L. M. (1986a), *Frequency Dependence of Q in the Mantle Underlying the Shield Areas of Eurasia. Part I: Analyses of Short- and Intermediate-period Data*, Geophys. J. Int. *87*, 1057–1084.

DER, Z. A., LEES, A. C., and CORMIER, V. F. (1986b), *Frequency Dependence of Q in the Mantle Underlying the Shield Areas of Eurasia. Part III: The Q Model*, Geophys. J. Int. *87*, 1103–1112.

DOUGLAS, A., CORBISHLEY, D. J., BLAMEY, C., and MARSHALL, P. D. (1972), *Estimating the Firing Depth of Underground Explosions*, Nature *237*, 26–28.

DOUGLAS, A. (1987), *Differences in Upper Mantle Attenuation between the Nevada and Shagan Test Sites: Can the Effects Be Seen in P-wave Seismograms?* Bull. Seismol. Soc. Am. *77*, 270–276.

DOUGLAS, A. (1992), *Q for Short-period P Waves; Is it Frequency Dependent?* Geophys. J. Int. *108*, 110–124.

DOUGLAS, A., *Broadband estimates if the seismic source functions of Nevada explosions from far-field observations of P waves*. In *Explosion Source Phenomenology* (eds. S. R. Taylor, H. J. Patton and P. G. Richards) (Geophysical Monograph Series, American Geophysical Union 1991).

DOUGLAS, A., and MARSHALL, P. D., *Seismic source size and yield for nuclear explosions*. In *Monitoring a Comprehensive Test Ban Treaty* (eds. E. S. Husebye and A. M. Dainty) (Kluwer Academic Publishers 1996).

DOUGLAS, A., MARSHALL, P. D., and YOUNG, J. B. (1993), Comment on 'Teleseismic *P*-wave attenuation and nuclear explosion source functions inferred from Yellowknife array data' by Kin-Yip Chun, Tianfei Zhu and Gordon F. West. With reply. J. Geophys. Res. *98*, 9921–9927.

FRASIER, C. W., and FILSON, J. (1972), *A Direct Measurement of Earth's Short-period Attenuation along a Teleseismic Ray Path*, J. Geophys. Res. *77*, 3782–3787.

GELLER, R. J. (1976), *Scaling Relations for Earthquake Source Parameters and Magnitudes*, Bull. Seismol. Soc. Am. *66*, 1501–1523.

GRAND, S. P., VAN DER HILST, R., ENGDAHL, R., SPAKMAN, W., and NOLET, G. (1991), *Tomographic Imaging of Subducted Lithosphere below Northwest Pacific Arcs*, Nature *353*, 37–43.

HELFFRICH, G., and STEIN, S. (1993), *Study of the Structure of Slab-mantle Interface Using Reflected and Converted Seismic Waves*, Geophys. J. Int. *115*, 14–40.

HELMBERGER, D. V., and HADLEY, D. M. (1981), *Seismic Source Functions and Attenuation from Local and Teleseismic Observations of the NTS Events JORUM and HANDLEY*, Bull. Seismol. Soc. Am. *71*, 51–67.

JORDAN, T. H. (1975), *The Continental Tectosphere*, Rev. Geophys. Space Phys. *13*, 1–12.

KNOPOFF, L. (1964), *Q*, Reviews of Geophysics *2*, 625–660.

LAY, T., and HELMBERGER, D. V. (1981), *Body-wave Amplitude Patterns and Upper Mantle Attenuation Variations Across North America*, Geophys. J. R. Astr. Soc. *66*, 691–731.

LAY, T., and WALLACE T. C. (1983), *Multiple ScS Travel Times and Attenuation beneath Mexico and Central America*, Geophys. Res. Lett. *10*, 301–304.

LEE, W. B., and SOLOMON, S. C. (1975), *Inversion Schemes for Surface Wave Attenuation and Q in the Crust and the Mantle*, Geophys. J. R. Astr. Soc. *43*, 47–71.

LEE, W. B., and SOLOMON, S. C. (1979), *Simultaneous Inversion of Surface Wave Phase Velocity and Attenuation, Rayleigh and Love Waves over Continental and Oceanic Paths*, Bull. Seismol. Soc. Am. *69*, 65–96.

LEES, A. C., DER, Z. A., CORMIER, V. F., MARSHALL, M. E., and BURNETTI, J. A. (1986), *Frequency Dependence of Q in the Mantle Underlying the Shield Areas of Eurasia. Part II: Analyses of Long-period Data*, Geophys. J. Int. *87*, 1085–1101.

LUNDQUIST, G. M., and CORMIER, V. C. (1980), *Constraints on the Absorption Model of Q*, J. Geophys. Res. *85*, 5244–5256.

MINSTER, J. B., and ANDERSON, D. L. (1973), *A Model of Dislocation Rheology for the Mantle*, Phil. Trans. R. Soc. *299*, 319–356.

MINSTER, J. B. (1978a), *Transient and Impulse Responses of a One-dimensional Linearly Attenuating Medium. Part I; Analytical Results*, Geophys. J. R. Astr. Soc. *52*, 479–502.

MINSTER, J. B. (1978b), *Transient and Impulse Responses of a One-dimensional Linearly Attenuating Medium. Part II; Parametric Study*, Geophys. J. R. Astr. Soc. *52*, 503–524.

MINSTER, J. B., *Anelasticity and attenuation*. In *Physics of the Earth's Interior* (Soc. Italiana di Fisica, Bologna, Italy 1980).

MINSTER, J. B., and Day, S. M. (1986), *Decay of Wavefields Near an Explosive Source Due to High-Strain, Nonlinear Attenuation*, J. Geophys. Res. *91*, 2113–2122.

MUELLER, R. A., and MURPHY, J. R. (1971), *Seismic Characteristics of Underground Nuclear Detonations; Seismic Scaling Law of Underground Detonations*, Bull. Seismol. Soc. Am. *61*, 1975.

NAKANISHI, I. (1979), *Attenuation of Multiple ScS Waves beneath the Japanese Arc*, Phys. Earth. Planet. Int. *19*, 337–347.

NATIONAL RESEARCH COUNCIL, *Seismological Research Requirements for a Comprehensive Test Ban Monitoring System* (National Academy Press, Washington D.C. 1995).

OLIVER, J., and ISACKS, B. (1967), Deep Earthquake Zones, Anomalous Structures in the Upper Mantle and the Lithosphere, J. Geophys. Res. *72*, 4259-4275.

NORTH, R. G. (1977), *Station Magnitude Bias—Its Determination, Causes and Effects*, Lincoln Laboratory, MIT, Technical Note 1977–24.

PERCIVAL, D. B., and WALDEN, A. T., *Spectral Analysis for Physical Applications; Multitaper and Conventional Univariate Techniques* (Cambridge University Press 1993).

ROMANOVICZ, B. (1990), *3-D Upper Mantle Degree 2; Constraints and Inferences from Global Mantle Wave Attenuation Data*, J. Geophys. Res. *95*, 11051–11071.

ROMANOVICZ, B., and MONTAGNER, J. P. (1990), *3-D Upper Mantle Velocity and Attenuation from Fundamental Mode Free Oscillation Data*, Geophys. J. Int. *101*, 61–80.

ROMANOVICZ, B., ROULT, G., and KOHL, T. (1987), *The Upper Mantle Degree Two Pattern; Constraints from GEOSCOPE Fundamental Spheroidal Mode Eigenfrequency and Attenuation Measurements*, Geophys. Res. Lett. *14*, 1219–1222.

SHARROCK, D. S. MAIN, I. G., and DOUGLAS, A. (1995), *Observations of Q from the Northwest Pacific Subduction Zone Recorded at Teleseismic Distances*, Bull. Seismol. Soc. Am. *85*, 237–253.

SHARROCK, D. S., MAIN, I. G., and DOUGLAS, A. (1995), *A Two-layer Attenuation Model for the Upper Mantle at Short Periods*, Geophys. Res. Lett. *22*, 2561–2564.

SCHLUE, J. W. (1981), *Differential Shear-wave Attenuation (δt*) Across the East Pacific Rise*, Geophys. Res. Lett. *8*, 861–864.

SHEEHAN, A. F., and SOLOMON, S. C. (1992), *Differential Shear Wave Attenuation and Its Lateral Variation in the North Atlantic Region*, J. Geophys. Res. *97*, 15339–15350.

SHORE, M. J. (1983), *Short Period P Wave Attenuation in the Middle and Lower Mantle of the Earth*, Bull, Seismol. Soc. Am.

SIPKIN, S. A., and REVENAUGH, J. (1994), *Regional Variation of Attenuation and Travel Time in China from Analysis of Multiple ScS Phases*, J. Geophys. Res. *99*, 2687–2699.

SIPKIN, S. A., and JORDAN, T. H. (1979), *Frequency Dependence of Q_{ScS}*, Bull. Seismol. Soc. Am. *69*, 1069–1079.

SIPKIN, S. A., and JORDAN, T. H. (1980), *Regional Variations of Q_{ScS}*, Bull. Seismol. Soc. Am. *70*, 1071–1102.

SOLOMON, S. C., and TOKSÖZ, M. N. (1970), *Lateral Variations of Attenuation of P and S Waves beneath the United States*, Bull. Seismol. Soc. Am. *60*, 819–838.

SOLOMON, S. C. (1972), *Seismic Wave Attenuation and Partial Melting in the Upper Mantle of North America*, J. Geophys. Res. *77*, 1483–1501.

STEWART, R. C. (1984), *Q and the Rise and Fall of a Seismic Pulse*, Geophys. J. R. Astr. Soc. *76*, 793–805.

SYKES, L. R., and EKSTRÖM, G. (1989), *Comparison of Seismic and Hydrodynamic Yield Determinations from the Soviet Joint Verification Experiment of 1988*, Proc. Natl. Acad. Sci. *86*, 3456–3460.

TAYLOR, S. R., BONNER, B. P., and ZANDT, G. (1986), *Attenuation and Scattering of Broadband P and S Waves Across North America*, J. Geophys. Res. *91*, 7309–7325.

TRULIO, J. G. (1978), *Simple Scaling and Nuclear Monitoring*, Applied Theory Inc. Los Angeles.

VAN DER HILST, R. D., and WIDIYANTORO, S. (1997), *Global Seismic Tomography: A Snapshot of Convention in the Earth*, GSA Today *7*, 1–7.

VAN DER LEE, S., and NOLET, G. (1997), *Upper Mantle S Velocity Structure of North America*, J. Geophys. Res. *B10*, 22815–22838.

VON SEGGERN, D. H., and BLANDFORD, R. R. (1972), *Source Time Functions and Spectra of Underground Nuclear Explosions*, Geophys. J. R. Astr. Soc. *31*, 83–97.

WALCK, M. (1988), *Spectral Estimates of Teleseismic P-wave Attenuation to 15 Hz*, Bull. Seismol. Soc. Am. *78*, 726–740.

WIDIYANTORO, S., and ENGDAHL, E. R. (1997), *Evidence for Deep Mantle Circulation from Global Tomography*, Nature *386*, 578–584.

WILCOCK, W. S., SOLOMON, S. C., PURDY, G. M., and TOOMEY, D. R. (1992), Seismic Attenuation Structure of a Fast-spreading Mid-ocean Ridge, Science *258*, 1470–1474.

ZHU, T., CHUN, K-Y., and WEST, G. F. (1989), *High Frequency P-wave Attenuation Determination Using the Multiple-window Spectral Analysis Method*, Bull. Seismol. Soc. Am. *79*, 1054–1069.

(Received November 18, 1997, revised May 4, 1998, accepted June 1, 1998)

To access this journal online:
http://www.birkhauser.ch

Pure appl. geophys. 153 (1998) 311–343
0033–4553/98/040311–33 $ 1.50 + 0.20/0

❙Pure and Applied Geophysics

2-D Image of Seismic Attenuation beneath the Deep Seismic Sounding Profile "Quartz," Russia

Igor B. Morozov,[1] Elena A. Morozova,[1] Scott B. Smithson[1]
and Leonid N. Solodilov[1]

Abstract—We present a 2-D image of the upper mantle attenuation using nuclear explosion data from the ultra-long refraction/reflection profile "Quartz." Our analysis is based on a modified common spectrum technique followed by least-squares inversion for Q and iterative ray tracing in the velocity structure obtained earlier. The resulting attenuation structure corroborates the earlier model for northern Eurasia, as well as our recent estimate based on the analysis of the long-range P_n phase, and provides significantly more detail than the existing models. The resulting upper mantle attenuation structure is characterised by Q values ranging from 400 to 1800. Down to the depths of 150–190, and probably 400 km, the attenuation increases horizontally in SE direction, away from the Baltic Shield. Our model exhibits strong 2-D, vertical and horizontal attenuation contrasts. A high-attenuation layer in the depth range of 120–150 to 160–180 km can apparently be associated with the presence of a partial melts within the base of the lithosphere.

Key words: Attenuation, two-dimensional, upper mantle, peaceful nuclear explosions.

1. Introduction

Average Q values for the earth are well defined from the measurements of its free oscillations (e.g., Anderson and Given, 1982) in the long-period band. Surface-wave studies allow observations of regional variations of the crustal and mantle attenuation down to about 400 km (e.g., Patton, 1980; Mitchell, 1995). However, short-period, body-wave attenuation measurements still remain very limited (Der *et al.*, 1986a). Global attenuation models are dominated by oceanic areas, whereas mantle attenuation shows a wide range of regional variations, as well as with depth and frequency (e.g., Mitchell, 1995; Der *et al.*, 1986b). Regional attenuation models, both derived from surface-wave and short-period array studies, remain highly averaged at regional scales, and they are typically

---

[1] Department of Geology and Geophysics, University of Wyoming, PO Box 3006, Laramie, WY 82701-3006. Tel.: (307) 766 3363, Fax: (307) 766 6679, E-mail: morozov@uwyo.edu
[2] Centre for Regional Geophysical and Geoecological Research (GEON), Moscow, Russia.

presented in the form of regional 1-D attenuation columns, and thus their consistency and resolution still remain moderate. Also, these models mostly constrain S-wave attenuation and apply (semi-)empirical relations to estimate attenuation factors of P waves (DER *et al.*, 1986b).

DER *et al.* (1986b) and LEES *et al.* (1986) presented a regional frequency-dependent attenuation model for northern Eurasia. The short-period part of the model was derived from the analysis of Soviet PNEs recorded by the NORESS array (DER *et al.*, 1986a). Within the 1–10 Hz frequency range, this model predicts Q_p values in the uppermost mantle between 1000–2000, with a stronger attenuating layer ($Q_p \approx 400–600$) between 100 km and 200–250 km.

Northern Eurasia presents a unique region to study regional variations of the seismic parameters of the upper mantle. A series of ultra-long refraction/reflection profiles using nuclear explosions (PNEs) acquired during the Deep Seismic Sounding (DSS) program in the former USSR have produced an unparalleled source of information about 2-D structure of the mantle down to the depth of 600–800 km. Densely spaced (10–15 km) 3-component instruments deployed along linear 3000–4000 long reversed profiles allow the actual observation of continuous lateral variations of the structure of the mantle (for reviews of the DSS program, see, e.g., RYABOY, 1989, and KOZLOVSKY, 1990). However, to date, the analysis of mantle attenuation using DSS data is limited to five earlier profiles (YEGORKIN and KUN, 1978; YEGORKIN *et al.*, 1981) and to an extrapolation of V_p values using an empirical $Q (V_p)$ relation (YEGORKIN and KUN, 1978; RYABOY, 1989). The analysis procedure employed by YEGORKIN and KUN (1978) and YEGORKIN *et al.* (1981) was based on the evaluation of spectral ratios for selected first arrivals. No attempts to derive a consistent model for *in situ* Q or to use secondary seismic phases were made. Also, apparently no corrections for instrument noise were applied. The resulting values of Q found for various diving waves were associated with the depth of their turning points in the mantle, yielding average values of Q_p increasing from 210–230 below the Moho to 440 at 150 km, then sharply decreasing to 190 and reaching 345 at the depth of 400–450 km (YEGORKIN and KUN, 1978; YEGORKIN *et al.*, 1981). These values are significantly lower than the values reported by DER *et al.* (1986b), and above the 200-km depth they are closer to the results for the westen U.S., in a totally different tectonic environment (DER *et al.*, 1986b). YEGORKIN and KUN (1978) revealed strong (about 70%) regional variations of attenuation. The short-period attenuation obtained from DSS studies was found to be approximately constant with frequency (YEGORKIN and KUN, 1978), in contrast to the findings of DER *et al.* (1986a,b). Partly this frequency independence was explained by the relatively narrow DSS frequency band (RYABOY, 1989); this frequency independence was also used to explain the generally low Q values obtained, which were more consistent with the long- and intermediate-period results (YEGORKIN and KUN, 1978; YEGORKIN, personal communication).

Due to limited sampling of the mantle, the existing regional attenuation model by DER *et al.* (1986b) strongly averages regional characteristics and is still necessarily defined as a one-dimensional depth column. The depth and amplitude of the sharp attenuation constrast near 100 km depth in this model are simplifications and apparently cannot be understood in terms of stable regional characteristics. Indeed, all existing DSS upper mantle velocity models exhibit significant horizontal velocity variations to a depth of at least 150–200 km (MECHIE *et al.*, 1993; EGORKIN *et al.*, 1987; RYBERG *et al.*, 1996; PAVLENKOVA, 1996; PAVLENKOVA *et al.*, 1996), and thus associated variations of Q might also be expected. On the other hand, the sampling density provided by DSS profiles (Fig. 1) greatly exceeds the density of data characterized as "detailed" in the study by DER *et al.* (1986a). Therefore, by using DSS data we should be able to obtain more adequate and detailed attenuation images of the uppermost mantle. A clear indication of this potential is immediately seen from a high-pass filtered PNE record (Fig. 1, bottom), showing that the refractions and reflections penetrating deeper than about 150 km depth return to the surface significantly depleted of high frequencies, which indicates a sharp increase of attenuation near this depth (MOROZOV *et al.*, 1998).

Our principal objective in this paper is to present a joint velocity/attenuation image of the uppermost mantle along one of the most studied DSS profiles— "Quartz" (Fig. 2), acquired in 1984–1987. Also, we will demonstrate how DSS PNE recordings lead to images of seismic velocity and attenuation structures within the top 400–500 km that allow observations of the detailed structure of the uppermost mantle that cannot be resolved by other techniques.

To date, the data from the profile "Quartz" have been studied extensively by several groups. Detailed crustal models were developed using ray-tracing (EGORKIN and MIKHALTSEV, 1990; MOROZOVA *et al.*, 1997) and refraction/reflection tomography (SCHUELLER *et al.*, 1997). For the uppermost mantle, 1-D velocity profiles were published by MECHIE *et al.* (1993), RYBERG *et al.* (1996) and MOROZOVA *et al.* (1997) derived 2-D velocity structures (Fig. 3). After applying high-pass filtering to "Quartz" records, RYBERG *et al.* (1995) observed the teleseismic P_n phase propagating within the upper approximately 80 km of the mantle, and proposed that this phase is due to the strong scattering within the mantle layer beneath the Moho. This concept was further supported by 1-D modelling by TITTGEMEYER *et al.* (1996). Nevertheless, recently we interpreted the teleseismic P_n as multiple P_n refractions, or whispering-gallery modes, and demonstrated how this phase can be used to constrain a vertical attenuation contrast within the mantle (MOROZOV *et al.*, 1998). Thus, although a significant controversy in the results of the above authors still exists, the crustal and upper mantle velocity models developed by different researchers generally agree and may be considered as well established. In this paper, we present the first study of spectral characteristics of "Quartz" records and include seismic attenuation information in our 2-D model of the uppermost mantle structure.

Our approach to attenuation measurement and inversion described in Section 2 comprises two major steps. First, we obtain apparent attenuation parameters, t^*, using a technique related to the common spectrum method (HALDERMAN and DAVIS, 1991). Unlike the traditional spectral ratio methods (e.g., BÅTH, 1974;

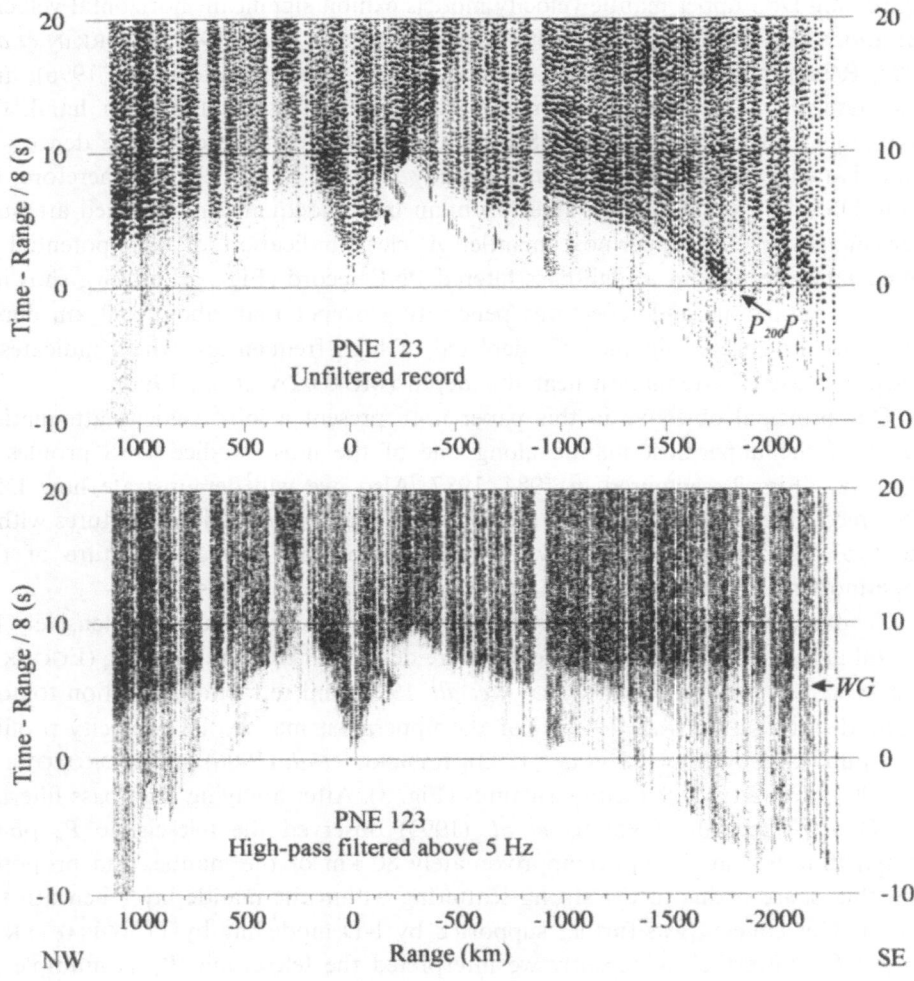

Figure 1

Vertical-component record from the northern PNE 123 of the DSS profile "Quartz" (see the map in Fig. 2). *Top*: unfiltered section, *bottom*: the same section after high-pass filtering above 5 Hz. Note dramatic change in the frequency content of the phases penetrating deeper than 150 km: the reflection from the top of the asthenospheric low-velocity zone at approximately 200 km depth ($P_{200}P$) nearly disappears from the filtered records which in turn are dominated by the long-range phase that we interpreted as whispering-gallery phase (labelled WG; MOROZOV et al., 1998). The continuity of the coverage of the uppermost mantle allows a detailed estimation of its seismic velocity and attenuation structure.

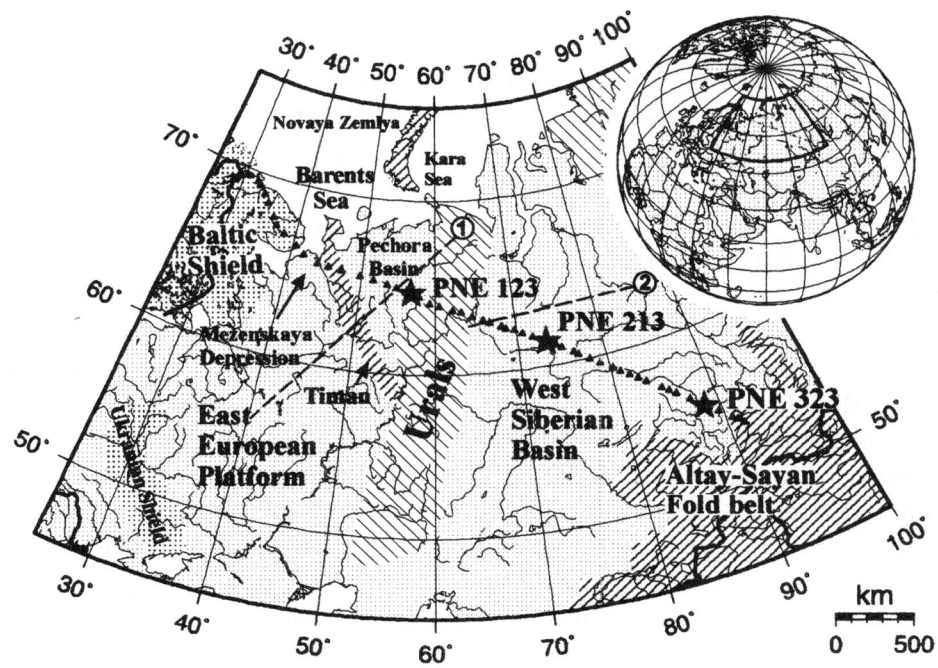

Figure 2

Map of the western part of the former USSR showing profile "Quartz." Circles indicate the locations of the nuclear explosions (PNEs) recorded by the profile. Numbered dashed lines indicate two of the profiles crossing "Quartz" and analysed previously: 1) by YEGORKIN and KUN (1978) and 2) by YEGORKIN et al. (1981).

MATHENEY and NOWACK, 1995), rise time techniques (GLADWIN and STACEY, 1974) and wave-form matching methods (MATHENEY and NOWACK, 1995), our method does not require an identification of a "reference" event or a source signature. Instead, source power spectrum and geometric spreading and coupling factors are estimated simultaneously with the attenuation parameters, and all data records are treated in a uniform manner. Most importantly, such an approach allows examination of the resolution and variance of the apparent attenuation estimates and of the trade-off between the values of t^*, of geometric spreading (and coupling) parameters, and of the shape of the effective source power spectrum.

Apparent attenuation profiles measured for all 3 PNEs of the profile "Quartz" are then used in Section 2.2 to invert for the *in situ* attenuation structure of the earth's mantle along the profile. By using a least-squares inversion technique together with iterative forward modelling, we incorporate attenuation values in our 2-D velocity structure (Fig. 3) derived recently using all 3 PNEs and 48 chemical explosions along the profile (MOROZOVA et al., 1997; 1998).

2. Measurements of Attenuation

Anelastic properties of the earth can be estimated from seismic amplitude decay (BRAILE, 1977; BRZOSTOWSKI and MCMECHAN, 1992; GRAD and LUOSTO, 1994), rise times (GLADWIN and STACEY, 1974) pulse broadening (WRIGHT and HOY, 1981), spectral ratio methods (e.g., BÅTH, 1974; CARPENTER and STANFORD, 1985), common spectrum method (HALDERMAN and DAVIS, 1991; GAO, 1997), wavelet modelling (JANNSEN *et al.*, 1985; MATHENEY and NOWACK, 1995; MATHENEY *et al.*, 1997), and analytical signal methods (TANER *et al.*, 1979). JANNSEN *et al.* (1985), BADRI and MOONEY (1987), TONN (1989, 1991), and MATHENEY and NOWACK (1995) compared different techniques, finding that each of these methods work well in favourable situations while some of them have significant limitations. Thus, amplitude-decay techiques strongly depend on the correct evaluation of the geometric spreading effects; pulse broadening and time rise approaches are often source-dependent (BLAIR and SPATHS, 1984). On the contrary, spectral ratio methods and related techniques employing instantaneous wavelet attributes are less affected by focusing and source signature. A refined form of the spectral ratio technique, the common spectrum method (HALDERMAN and DAVIS, 1991) that also has the advantage of a symmetric treatment of all records and does not require selection of a "reference" station is used in this study.

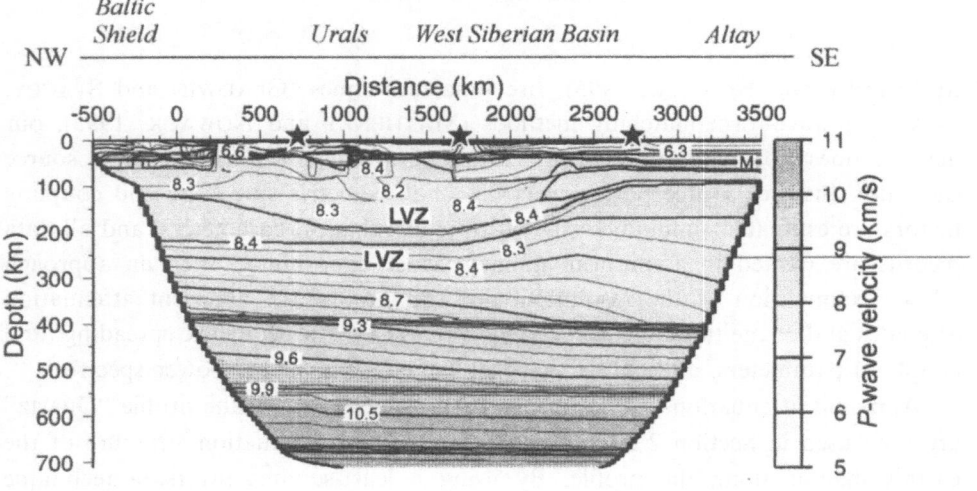

Figure 3
2-D *P*-wave velocity structure within the uppermost mantle obtained by travel-time inversion of the full "Quartz" data set (MOROZOVA *et al.*, 1997, 1998).

In the common spectrum method, the determination of seismic attenuation usually incorporates two steps. First, the attenuation parameter t^* is estimated from the spectral contents of the data records. This parameter is related to the *in situ* attenuation:

$$t^* = \int \frac{ds}{vQ},$$ (1)

where s is the ray path, v is the seismic velocity, and Q is the quality factor of the medium.

The evaluation of t^* can be carried out without reference to a specific subsurface structure, using the frequency dependence of the signal amplitude decay:

$$P_i(f) \cong S(f)G_i \exp(-2\pi t_i^* f),$$ (2)

where $P_i(f)$ is the spectral power of the signal recorded by recording channel i at the frequency f, $S(f)$ is the spectral power of the source, and G_i stands for the geometrical spreading factor describing focusing and defocusing effects, and also accounting for the frequency-independent variations in the geophone coupling. The frequency independence of the amplitude parameters G_i is the strongest (yet necessary) limitation of this approach, based on the ray theoretical approximation ignoring near-source tuning and frequency-dependent wave propagation effects.

In all methods based on modelling of the power spectrum (2), a fundamental ambiguity is present, due to the invariance of the equation (2) under the following simultaneous transformation of the source spectrum and t^* values:

$$\begin{cases} S(f) \to a \, \exp(2\pi bf)S(f), \\ G_i \to \dfrac{G_i}{a}, \\ t_i^* \to t_i^* - b, \end{cases}$$

where a and b are arbitrary constants. This ambiguity must be eliminated before the inversion of Q values using the relation (1); this problem is in a certain sense related to the difficulty of choosing the reference signal in the spectral ratio methods. On the other hand, after resolution of this difficulty, equation (2) leads to the estimates of true (effective) attenuation factors t^*, rather than of the apparent $\overline{t}^*$ characterising the slope of the amplitude spectrum calculated after source and receiver corrections (cf., DER *et al.*, 1986a):

$$\overline{t}^* = \frac{1}{2\pi} \frac{d(\ln P_{\text{corrected}})}{df} = t^* + f\frac{dt^*}{df}.$$ (3)

Once t^* values are determined, the intrinsic attenuation Q is obtained by ray tracing and solution of equation (1) which becomes a linear system if the velocity structure is determined independent of the attenuation. Although the apparent attenuation can be included into the inversion on equal grounds with the travel times, in a manner similar to MATHENEY and NOWACK (1997), who used amplitudes to constrain small-scale velocity variations, this complication does not appear to be justified in our case, since the error inherent in t^* measurements significantly exceeds travel-time uncertainties, and therefore the Q structure cannot be resolved to the same degree of detail as the velocity structure. For this reason, it appears that the only way to obtain the attenuation structure is to start from the velocity model that is constrained by all available travel time and amplitude information, and to assign Q values to its structural units. Moreover, the knowledge of the velocity structure, phase association, and ray sampling provide important guidance during the process of t^* measurements, and thus the resulting attenuation model is strongly tied up to the velocity structure (Fig. 3).

In the following sections, after an examination of the available seismic ray coverage, we discuss the two steps of attenuation analysis separately.

2.1. PNE Ray Coverage

The 2-D model of the upper mantle P-wave velocity structure along the profile "Quartz" (Fig. 3) was obtained using iterative travel-time inversion applying the ray-tracing program RAYINVR by ZELT and SMITH (1992). Our system for seismic data analysis (MOROZOV and SMITHSON, 1997) includes tools capable of tracing rays within several types of velocity models, including layered velocity-interface structures used by RAYINVR program, and providing a "seamless" transition from the iterative ray tracing performed by MOROZOVA *et al.* (1997, 1998) to our attenuation inversion described below.

The parameterisations used in t^* measurements and in the inversion for attenuation must be chosen carefully in order to account for the characteristic features of the models and at the same time to avoid overparameterisation. We prefer this explicit association of attenuation parameters with the velocity structure (Fig. 3) to damping schemes typically used to stabilise solutions of underdetermined systems (e.g., MENKE, 1989). The generally accepted damping approaches (e.g., a depth- and structure-independent smoothness constraint) are also equivalent to inclusion of additional *a priori* information which may not be sufficiently justified but dictated only by the considerations of processing simplicity (cf., SCHUELLER *et al.*, 1997).

In this work, instead of overparameterising our inversion, we chose to employ specially chosen parameterisation schemes, making use of the ray coverage of the subsurface estimated from the ray tracing. This knowledge of ray coverage controls both our t^* and Q parameterisations. Since we intend to use only the first arrivals in our analysis, we start with a selection of the rays which represent the first arrivals and provide as uniform a depth sampling as possible (Fig. 4). This ray distribution

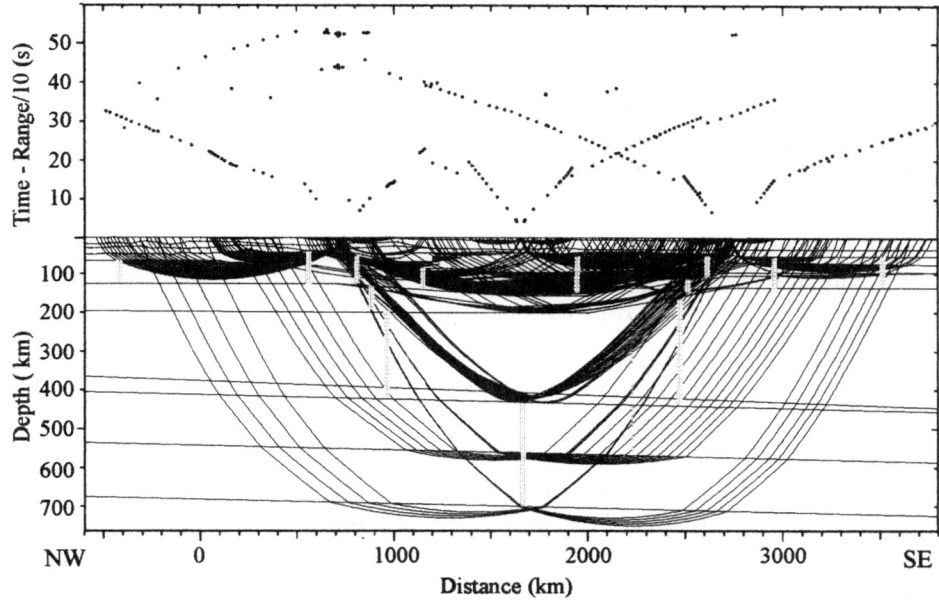

Figure 4

Ray coverage from all 3 PNEs used in our inversion. Note that the depth sampling is "model-driven": the rays are shot from their turning points, allowing a controlled, more uniform coverage of the subsurface. Above the model calculated travel times reduced using the reduction velocity of 10 km/sec are shown. Thick grey bars in the velocity model indicate the positions of the nodes at which the values of $1/Q$ are specified within different layers. See the text for more details.

was obtained by tracing rays which have their turning points spaced at regular depth intervals and by rejecting the branches that result in second arrivals. Such ray-tracing style is free from the difficulties of two-point ray tracing from the surface and results in a comparatively uniform, representative, and economical ray coverage of the subsurface.

As Figure 4 shows, the ray coverage of the deeper layers is highly nonuniform. Only narrow depth regions of 20–40 km near the 410-km, and 30–60 km beneath the 560-km and 660-km discontinuities are imaged by subhorizontal ray segments, and the layers below 200 km are traversed by more steeply dipping rays. Such ray coverage explains the contrast between the abundance of detail observed in the uppermost mantle structure above 200 km, and the relatively simple 2-D structure below this level in our velocity model (Fig. 3). Indeed, the structure below 200 km shown in Figure 3 was derived from the analysis of the first arrivals plus the reflections from these prominent mantle discontinuities; however, due to the nature of the mantle velocity profile, most of the refracted arrivals are shadowed by the deeper arrivals arriving earlier (Fig. 4). For this reason, the velocity structures beyond the regions sampled by subhorizontal rays, as shown in Figure 4, are known in an averaged sense. This consideration equally refers, of course, to the earlier "Quartz" mantle models derived by MECHIE *et al.* (1993) and RYBERG *et al.* (1996), and to other DSS profiles that exhibit sharp velocity contrasts and velocity inversion within the upper mantle (MECHIE *et al.*, 1997).

Another important limitation of the ray-theoretical velocity model displayed in Figure 3 lies in the bounds of its vertical resolution imposed by the wave nature of seismic signals. An estimate of the vertical Fresnel zone radius at the turning point of a refracted ray might be obtained using dynamic ray tracing (e.g., CERVENY *et al.*, 1984); however, we can use a rough estimate based on the total length L of the ray and the dominant wavelength λ (corresponding to the radius of the first vertical-incidence reflection Fresnel zone at the depth of $L/2$; see SHERIFF, 1989):

$$R_F \approx \sqrt{\lambda \frac{L}{4}}. \qquad (4)$$

For a wave turning at the depth of 410 km and holding the dominant frequency of 1 Hz, $\lambda \approx 9$ km and $L \approx 2200$ km, and therefore R_F equals about 70 km, exceeding the thickness of the regions below the mantle discontinuities covered by subhorizontal rays shown in Figure 4. This estimate demonstrates once again that the attenuation structure below 200 km inferred from these refractions should be limited to relatively simple thick layers like those shown in Figure 3.

In the following analysis, we employ the knowledge of the ray coverage shown in Figure 4 in two ways:

1) During the measurements of t^*, we split the offset ranges into segments corresponding to the onsets of different phases and associate the t^* parameters with the phases rather than with the recording channels. This allows us to reduce the number of measured parameters and to increase the stability of the estimates.

2) During the inversion for Q values, we calculate t^* values along the rays shown in Figure 4 and calculate Q that allows us to match the observed t^* *(offset)* curves. We restrict our inversion to vertically constant Q within the velocity model layers, with varying numbers of Q nodes in the horizontal direction (Fig. 4).

2.2 Attenuation Factors t^*

Although the equation (2) can be readily linearised by taking logarithms, in the common spectrum method by HALDERMAN and DAVIS (1991) it is solved directly by using a nonlinear least-squares inversion, or by using a Bayesian inversion (GAO, 1997). This significant complication is caused by the desire to avoid the increased weighting of the logarithms of small amplitudes within the spectral holes (HALDERMAN and DAVIS, 1991). On the contrary, due to its correspondence to the exponential nature of energy decay, taking the logarithms assures much greater clarity of the approach, computational efficiency, and independence from the details of the nonlinear inversion algorithms. Also it is advantageous in the sense that it increases the weighting of the lower amplitudes at the high end of the spectrum that are due to the attenuation. On the other hand, the problem of transformation of variables in the equation (2) is quite different from the choice of the weighting of various terms required for its numerical solution, and this problem needs to be examined separately. Note that since HALDERMAN and DAVIS (1991) use amplitude spectra in their attenuation measurements, their transformation of the spectral power $A_i(f) = \sqrt{P_i(f)}$ stretches the low-amplitude part of the spectrum in a way similar to the effect of taking logarithms, and the choice of Euclidean norm in the spectral amplitude space during the nonlinear least-squares inversion is not dictated by physical considerations.

The optimal approach to an approximate solution of (2) would tend to suppress the spectral holes while targeting the exponential decay of the spectrum. In the "common spectral balancing method" that we propose below, we achieve these two goals by explicit linearisation of the relation (2), followed by amplitude-dependent weighting and by an iterative "balancing" of the spectra.

2.2.1 Common spectrum method

By rewriting the equation (2) in the logarithmic form:

$$\ln P_i(f) \cong \ln S(f) + \ln G_i - 2\pi t_i^* f, \tag{5}$$

we obtain a system of linear equations with respect to the logarithms of spectral density of the source $\ln S(f)$ and of the geometric factors $\ln G_i$, and of the attenuation parameters t_i^*.

Before the relation (5) can be employed for the inversion, three significant factors must be evaluated: 1) time windowing of the arrival used for the measurements, 2) the influence of seismic noise, 3) the appropriate (generally channel-dependent) frequency band.

Careful windowing of the first arrivals allows a decrease of the influence of interfering secondary phases and an increase of the signal-to-noise ratio. We perform the windowing by visually inspecting three-component trace amplitudes obtained by applying the method described by MOROZOV and SMITHSON (1996). Figure 5 illustrates the resulting first break amplitudes from the PNE 323 (Fig. 2) and the first-arrival time windows employed in our spectral analysis. The spectra are obtained by using Hanning windows and are smoothed within the frequency ranges Δf which are related to the time durations T of the windowed signals by the uncertainty relation:

$$\Delta f = \frac{1}{T}. \tag{6}$$

In far-offset recordings which utilise portable instruments, seismic noise may cause a significant bias in the value of t^* if the measurement is carried out within too wide a frequency band (MATHENEY and NOWACK, 1995). To correct the values of the signal spectral power $P_i(f)$ for the effects of seismic noise, we subtract the spectral power of the pre-shot noise from the spectral power of recorded signal:

$$P_i(f) \approx P_i^{\text{recorded}}(f) - N_i(f), \qquad (7)$$

where $N_i(f)$ is the noise spectral power measured in the channel i before the first arrival. Relation (7) is based on the assumption that noise is uncorrelated with the signal, and can be utilised for noise-adaptive filtering of seismic records (ANDERSON and MCMECHAN, 1988).

Measurements of the pre-shot seismic noise also provide a means to find the offset-dependent frequency band within which the relation (5) is valid. Following MATHENEY and NOWACK (1995), we restrict the application of equation (5) to the frequency range where $P_i^{\text{recorded}}(f)$ exceeds the noise level $N_i(f)$ by 5 dB. The frequency band used in our attenuation measurements is shown in Figure 6.

Having determined the appropriate frequency ranges, for each PNE we discretise equation (5) on a regular frequency grid and apply weights w_k to the data points, transforming it into a quadratic optimisation problem:

$$\sum_k w_k \left(\ln P_k - \sum_m A_{km} u_m \right)^2 \to \min, \qquad (8)$$

where the index in the data space $k \equiv \{i, f\}$ represents a combination of the recording channel number i and a sampled frequency value f, and the model space index m comprises the parameterisation of the entire model:

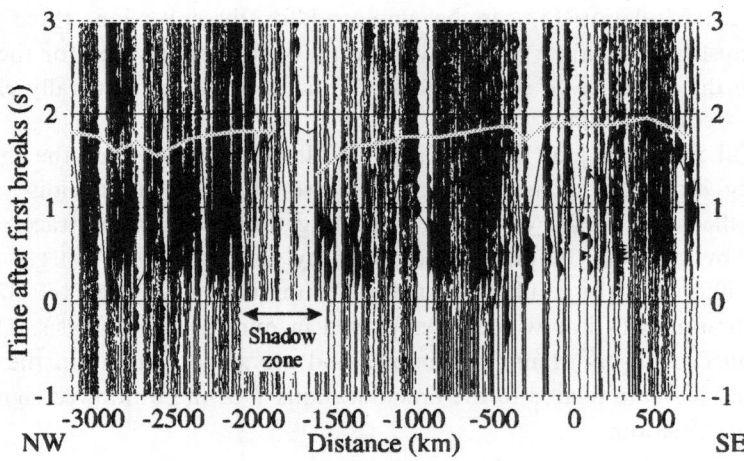

Figure 5
Three-component vector amplitudes of the first arrivals from the southern "Quartz" PNE 323 aligned at the picked first break onset times. Variable-length time windows of about 1.8 sec length around the first arrivals were used for spectral estimates.

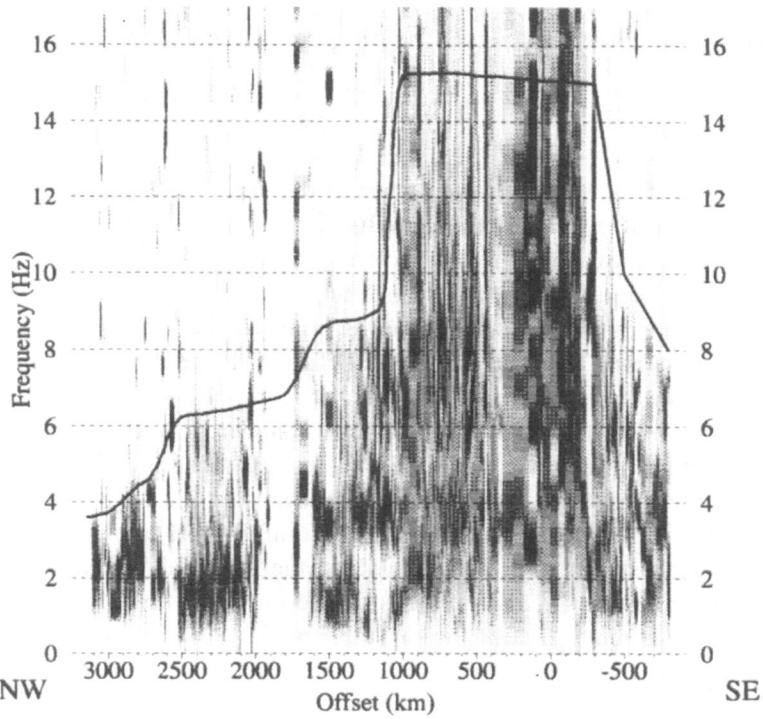

Figure 6

Record-dependent frequency band used in our attenuation measurements. We plot the smoothed spectral power of the recorded first arrivals after the noise subtraction, as given in equation (7), in grey scale, and pick the upper frequency bound. The frequency band of the signal is about 0.5–15 Hz within 0–1000 km of offset, thereafter it quickly drops down to about 0.5–8 Hz and decreases to 0.5–4 Hz at 3000 km.

$$\mathbf{u} = (\ln S_1, \ldots, \ln S_{N_f} | \ln G_1, \ldots, \ln G_{N_G} | t_1^*, \ldots, t_{N_t}^*)^T. \tag{9}$$

In equation (9), N_G is the number of parameters sampling the receiver locations; N_t is the number of parameters controlling the t^* (*offset*) dependence, and N_f is the total number of frequencies. S_i is the source amplitude at the corresponding frequencies, G_i and t_i^* are the parameters sampling the corresponding functions in equation (2) using the linear interpolation described in the Appendix. Functions G and t^* (equation (2)) do not require sampling at the same points, nor do grid numbers N_G and N_t have to be the same. In our study, we took N_G equal to the number of recording stations and tried two different parameterisations for t^*, as described below. The frequency sampling was chosen at a regular 0.5 Hz interval (Fig. 6).

The weights w_k in (8) may depend on the corresponding data point, and in particular, on the observed spectral power. For example, after setting:

$$w_k = \frac{(\ln P)'}{(\sqrt{P'})} \propto \sqrt{P},$$

(P is the value of the spectral power density) the linearised least-squares minimisation of the form (8) would produce results close to those obtained by a nonlinear iterative inversion used by HALDERMAN and DAVIS (1991).

To suppress the effects of the spectral bursts and holes, we reduce the weights w_k for the data points with spectral power that are significantly lower or higher than the average for a given record (Fig. 7). The reference cut-off levels for the low and high spectral density values are estimated from the histograms of the spectra. Coupled with the amplitude-dependent weights shown in Figure 7, trace-dependent weights reflecting the overall quality of data records also may be applied; however such weights were not used in the present study.

Due to the invariance of the original equation (5) with respect to the transformation (3), a constraint must be imposed on the solution. We employ an explicit constraint included in the discretisation of the model described in the Appendix, after which the minimisation of the form (8) is carried out by applying a standard least-squares method (e.g., MENKE, 1989), resulting in desired t^*, ln S, and ln P values:

$$\mathbf{u} = \mathbf{G}^{-gT}\mathbf{d} \equiv (\mathbf{A}^T\mathbf{W}\mathbf{A})^{-1}\mathbf{A}^T\mathbf{W}\mathbf{d}, \tag{10}$$

where $\mathbf{d}$ is the data vector of ln P values, $\mathbf{A}$ is the matrix defined in the relation (8), $\mathbf{u}$ is the model vector from (9), $\mathbf{G}^{-gT}$ denotes the generalised inverse, and $\mathbf{W} = \mathrm{diag}(w_1, w_2 \ldots)$ is the matrix of the weights applied to the data points in (8).

The application of spectral power-dependent weights (Fig. 7) which reduce the contributions from low amplitudes in the objective function (8), may have the

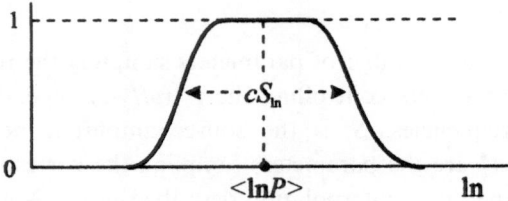

Figure 7

The dependence of the weight of the data point in the objective function (8) on the logarithm of the spectral power density. The shape of the function is controlled by the average value of the spectral density $\langle \ln P \rangle$ and its standard deviation $S_{\ln P}$ that are estimated from the data after a removal of the contributions of the lowest amplitudes. The parameter c is the scaling of the width of the weight function; we chose $c = 3$ in our analysis.

undesirable effect of biasing of the t^* estimates toward higher values. This problem is equally present in the method of HALDERMAN and DAVIS (1991), by which this power-dependent weighting is achieved through a coordinate transformation as discussed above. Fortunately, the structure of our basic equation (5) and our "adaptive" weighting technique (Fig. 7) provide an elegant way to remove this bias and simultaneously examine the stability of the solution.

2.2.2. Common spectrum balancing

In order to make the inversion independent of the nonlinear weighting, we iteratively "balance" the spectrum of each receiver channel using the results of the previous iteration, and repeat the least-squares inversion (10). Thus, having obtained the values of the parameters $S = S^n(f)$, $G = G^n$, and $t^* = t^{*n}$ after n-th iteration of the inversion, the power spectra in the equation (5) can be balanced by compensating the estimated systematic frequency dependence in both its sides:

$$\ln P_i(f) - \ln S^n(f) + 2\pi t_i^{*n} f = \Delta \ln S(f) + \ln G_i - 2\pi \Delta t_i^* f, \tag{11}$$

where $\Delta \ln S(f) = \ln S^{n+1}(f) - \ln S^n(f)$ and $\Delta t_i^* = t_i^{*n+1} - t_i^{*n}$ are the corrections of the corresponding parameters. If our common-spectrum approximation (2) is valid, and the solution approaches correct values of $\ln S$, $\ln G$, and t^*, the balanced spectra are close to a constant within the utilised frequency band. Once the source spectrum and the spectral slope due to the attenuation have been removed in the left-hand side of equation (11), the outliers suppressed by the weighting (Fig. 7) are mostly due to amplitude variations that cannot be accounted for by our regression model (2), and t^* values are less subject to the bias caused by the amplitude-dependent weighting.

By solving equations (11) applying the method described in the previous section, and iterating this procedure until the values of $\ln S$, $\ln G$, and t^* become stationary, we obtain the final solution. In our case, only three to five spectral-balancing iterations were necessary.

2.2.3. Resulting t^* values

The resulting attenuation factors t^* measured from the records of all three PNEs are presented in Figure 8, and the estimated source spectra of all three PNEs are shown in Figure 9. In the calculation of the source spectra, only the arrivals of the waves turning above the asthenospheric LVZ, corresponding to the ranges to 1600–1800 km were used (Fig. 4). At larger ranges, where the signal is less consistent (Fig. 8), the source spectra in equation (5) were fixed, although the spectral balancing technique was still employed during the inversion of the t^* values and receiver terms.

At the first step, t^* values were parameterised by the corresponding record numbers, resulting in recorded signals being treated independently. To further compensate the various sources of instability of the measurement, four inversions were carried out using 20% uniformly distributed random variations of the fre-

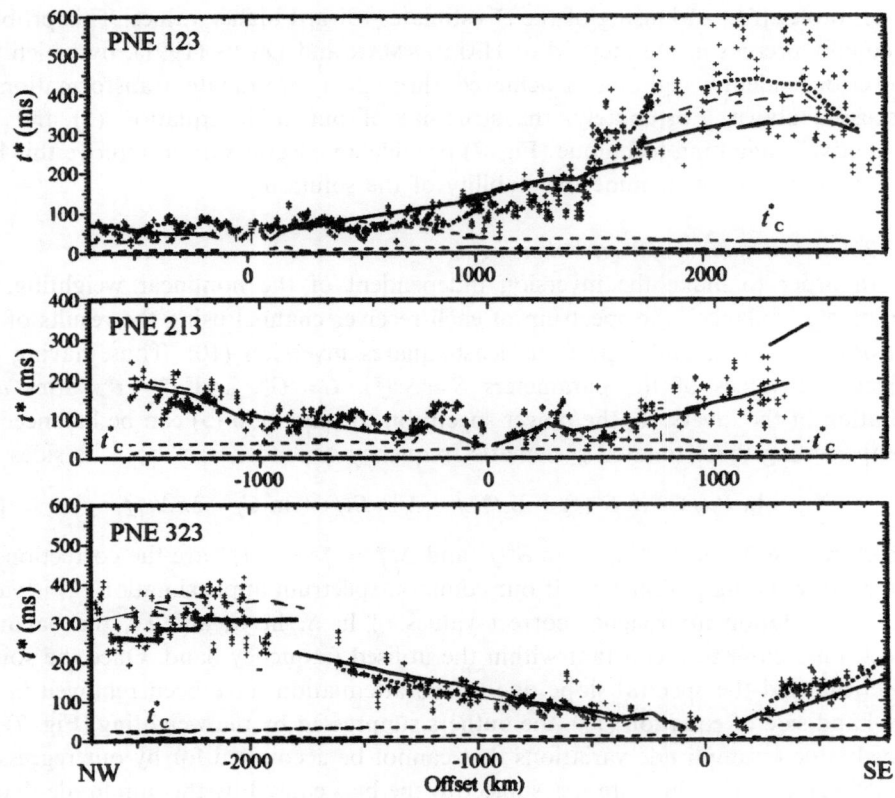

Figure 8

Measured t^* values for the PNEs vs. receiver offset. Crosses represent individual t^* measurements; the dotted line segments show the smoothed t^* (offset) dependence obtained by a more sparse t^* parameterisation within several offset ranges selected, based on the examination of the ray coverage (Fig. 4). This dependence is used in further inversion for *in situ* attenuation. These plots also display the t^* curves calculated for our two alternative attenuation models (thick and dashed lines) and the contribution to the t^* values due to our crustal velocity-attenuation model (dashed lines labelled t^*_C). The two t^*_C curves correspond to the effects of the sediment cover (*bottom*) and of the entire crust (*top*).
See the following sections for discussion.

quency band (TONN, 1991). All resulting estimates of t^* were plotted in the form of a scatter graph (Fig. 8).

The t^* values shown in Figure 8 exhibit significant scatter that is due to the effects of near-receiver conditions, variations of crustal Q, and noise and is not related to the variations of Q within the mantle. In order to smooth these values, we employ the same common-spectrum balancing method, selecting only a few (3–6) t^* parameterisation points within several offset ranges that correspond to the onsets of the waves turning within different depth ranges (Fig. 4). Between these points, t^* values are linearly interpolated, as described in the Appendix. As a result, we arrive at the averaged t^* curves (Fig. 8) that are not produced by simple filtering of the scattered individual t^* values, but are averaged, based on a physical assumption of the continuity of the t^* (*offset*) dependence within the same wave group.

In order to account for the unconstrained additive constant in t^* due to the fundamental ambiguity of the constant-spectrum method (3), constant shifts (accompanied by a corresponding correction of the source spectra) were applied to all t^* values from every PNE, to ensure that modelled t^* matches the observed values at near offsets.

2.2.4. t^*-model covariance

In conjunction with the use of the more natural logarithmic parameterisation of the spectral power and its computational efficiency, the least-squares method is importantly advantageous by providing a straightforward means to examine the quality of the solution. With our parameterisation, the inversion problem is purely overdetermined (e.g., MENKE, 1989), and thus the least-squares inversion requires no damping, and its resolution matrix is unity: $\mathbf{R} = \mathbf{I}$. The model covariance matrix corresponding to the solution (10) is expressed through the covariance of the logarithms of the spectral power values [cov $\mathbf{d}$]:

$$[\text{cov } \mathbf{u}] = \mathbf{G}^{-g}[\text{cov } \mathbf{d}]\mathbf{G}^{-gT} = (\mathbf{A}^T\mathbf{W}\mathbf{A})^{-1}\mathbf{A}^T\mathbf{W}[\text{cov } \mathbf{d}]\mathbf{W}^T\mathbf{A}(\mathbf{A}^T\mathbf{W}^T\mathbf{A})^{-1}. \quad (12)$$

For our estimates we assume that the power density values are uncorrelated and have the same variance $(\sigma^2_{\ln P})_i = 1/2b_i T_i$, where b_i are the channel-dependent bandwidths employed in the measurements, and T_i are the time windows used in the measurements of the spectra (WHITE, 1992). The data covariance matrix is

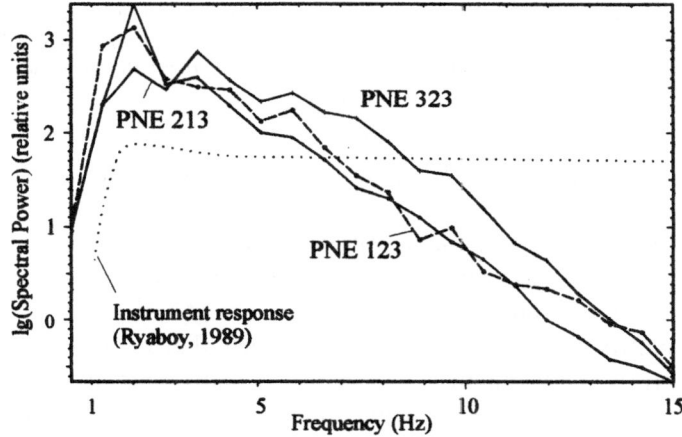

Figure 9

Estimated source spectra from the PNEs, obtained using our common-spectrum balancing method. Note the higher-frequency content of the signal from the southern PNE 323. Dotted line shows the instrument response of the "Taiga" recording stations (RYABOY, 1989) that were deployed throughout the offset ranges used in this study.

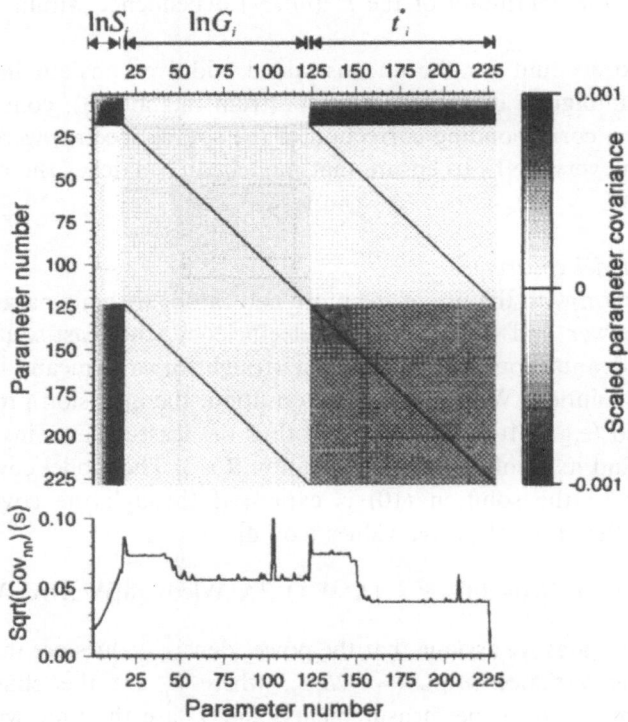

Figure 10

Model covariance matrix of t^* measurements (12). The rows and columns of the matrix correspond to the common-spectrum model parameters (9), as labelled above the plot. The first 123 rows and columns are multiplied by the factor 0.2 in order to equalise the display which is normalised to show the covariance of t^* values. Note the significant correlation between the values of t^*, the corresponding ln G, and spectral power values, indicated by darker areas. *Bottom*: the diagonal of the covariance matrix.

therefore $[\text{cov } \mathbf{d}] = \text{diag}\ (1/2b_1 T_1, 1/2b_2 T_2, \ldots)$, and the matrix (12) can be evaluated during the inversion.

Figure 10 presents the model covariance matrix which corresponds to the last step of our inversion, using the offset range 0–1500 km. Since we used smoothing of the spectra based on the relation (6), $b_i T_i \approx \Delta f_i T_i = 1$. The diagonal of the covariance matrix (Fig. 10, bottom), demonstrates that the estimated uncertainties of t^* values (rows and columns 124–226 of the matrix) are about 40 ms, although a significant covariance between these values and the corresponding ln G and ln S values is present (Fig. 10, top). Also the t^* uncertainties at the offsets beyond 1000 km (columns 123–152) are larger, about 70–80 ms, because of the narrower useable frequency band (Fig. 6; cf. WHITE, 1992). An outlier at the parameter number 210 (t^*) and 105 (ln G) in the bottom of Figure 10 corresponds to a trace with significantly lower S/N ratio.

2.3. Interval Attenuation

Equation (1) relates the measured values of the attenuation factors t^* to the intrinsic attenuation Q of the subsurface. The scatter of the values of t^*, however, is quite large (Fig. 8), and their variances are significant (Fig. 10), therefore we cannot expect that the measured apparent attenuation can contribute substantially in constraining the velocity-interface structure. As has been observed (WHITE, 1992), fine-scale parameterisations of the Q model do not lead to correct results and produce underconstrained inversion problems, unambiguous treatment of which remains highly problematic (SCHUELLER *et al.*, 1997). Therefore, we use a simple parameterisation of the upper mantle Q-structure by assigning values of Q to the layers of its velocity-interface structure developed earlier by using seismic travel-time tomography (SCHUELLER *et al.*, 1997) and iterative ray tracing (MOROZOVA *et al.*, 1997, 1998).

2.3.1. Model parameterisation

To achieve an adequate and robust parameterisation of the Q structure, we split the layers of the model into six groups, corresponding to the basic structural units of the model: 1) the crust, the attenuation of which is fixed during the inversion; 2) the uppermost mantle above the LVZ; 3) the top LVZ and the underlying layer that is too thin to be resolved in our Q model; 4) the major LVZ and the gradient layer immediately above the 410-km discontinuity; and 5)-6) two mantle layers distinguished below the 410-km boundary.

Due to limited sampling and significant t^* uncertainties, crustal attenuation cannot be constrained from available PNE data. Therefore, within the crustal layers we set values of $Q = 200$ for the sediments, $Q = 600$ for the rest of the upper crust, and $Q = 1200$ for the lower crust, in this way approximating the averaged attenuation profile in the crust of tectonically stable regions (MITCHELL, 1995). Note, however, that although crustal attenuation makes a significant contribution to t^* values (40–50 ms), beyond the offsets of 300–400 km this contribution is largely reduced to a nearly constant shift of the t^* curves (Fig. 8) that can be removed by using the uncertainty relations (3).

Within each of these mantle layer groups we select a set of 1 to 7 nodes, depending on ray sampling, as is also shown in Figure 4. The values of Q are constant vertically within each layer group, and are interpolated linearly in the horizontal direction between the Q nodes (Fig. 4). In the language of the Finite Element method this parameterisation is expressed as an expansion of $1/Q$ in terms of a set of basis functions $\psi_{l,i}(x, z)$:

$$\frac{1}{Q(x, z)} = \sum_{l,i} q_{l,i}\psi_{li}(x, z) \equiv \sum_{k} q_k\psi_k(x, z), \tag{13}$$

where l is the number of the layer, i is the number of the Q node of the layer, and q is the internal friction parameter. Each basis function $\psi_{l,i}(x, z)$ equals 0 beyond the layer group, and changes linearly between 1 at the location of node i to 0 at the adjacent nodes. As before, k is the cumulative index of the node: $k = \{l, i\}$.

Within the class of the piecewise-linear functions (13), the set of integral equations (1) for all rays becomes an overdetermined system of linear equations

$$\sum_k G_{rk} q_k = t_r^*. \tag{14}$$

In these equations, r is the index of ray traced through the model, t_r^* is the observed attenuation factor along this ray, obtained by interpolation of the appropriate $t^*(x)$ branch shown in the bottom part of Figure 11, and the kernel G_{rk} is the path integral along the ray r:

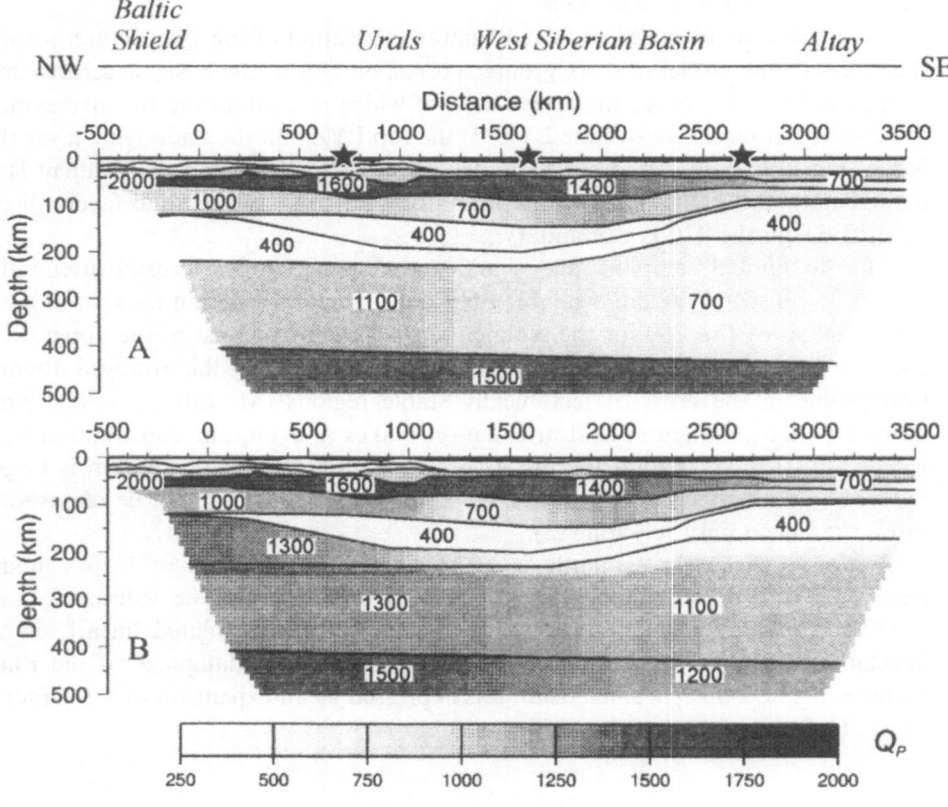

Figure 11
"Quartz" *P*-wave attenuation models: (A) obtained using the linear inverse of the first arrival spectra from all three "Quartz" PNEs; (B) an alternative model obtained from the model *A* by applying additional constraints which use iterative forward ray tracing.

$$G_{rk} = \int_r \psi_k(x, z) \frac{ds}{v(x, z)}. \tag{15}$$

The integrals (15) were evaluated numerically during ray tracing, and the approximate solution of the system (14) was again obtained employing the least-squares method, resulting in the attenuation model A presented in Figure 11.

2.3.2. Q model

The model A (Fig. 11) minimises the RMS misfit associated with the least-squares solution of unweighted equations (14). The residual RMS misfit of t^* in model A is 46 ms within the 1500 km offset range, which is comparable to the scatter and 40 to 70 ms uncertainties of the observed t^* values. Beyond the offsets of 1500 km the RMS t^* misfit is 91 ms, which is also comparable to the increased variance (Fig. 10) and scatter (Fig. 8) of the t^* values. Given this inherent uncertainty of t^* values, and also the limited vertical resolution of inversion using refracted arrivals (equation (4)), the minimisation of the RMS residual does not appear as a strict criterion, and it may be partly traded for a proximity to some preferred *a priori* model, as is often done in tomography (e.g., MENKE, 1989).

Exploiting this ambiguity of t^* inversion, we modified the model A (Fig. 11, top), making it satisfy additional constraints arising from our analysis of "Quartz" PNE data: 1) the attenuation within the LVZ below 190 km is concentrated near its top, and thus the sharp attenuation drop at the depth of 190 km in model A is redistributed more smoothly; 2) better correspondence to the attenuation model of DER et al. (1986b) (see the *Discussion* section); 3) the unconstrained attenuation values within the layer 100–190 km, left of km 1200 of the profile (cf., the ray coverage diagram in Fig. 4) is reduced, allowing efficient propagation of the high-frequency teleseismic P_n to offsets of 3000 km (MOROZOV et al., 1998); and 4) the presence of a high-velocity, high-gradient, and probably, high-Q layer underlying the top LVZ (this layer is indicated by the strong $P_{200}P$ arrival in Figure 1; MOROZOVA et al., 1997). We did not attempt to incorporate these constraints into the least-squares inversion; instead, we used iterative forward modelling to examine the changes in the attenuation structure below 150–190 km where Q is constrained only in an averaged sense.

The resulting alternative model B (Fig. 11, bottom) also provides a satisfactory t^* fit (RMS t^* misfit of 116 ms at the offsets exceeding 1500 km). The models A and B represent two members of a suite of possible Q models that account for the t^* data nearly equally well. Although a complete characterisation of this suite still has not been achieved (one possible way to accomplish it was described by MOROZOVA et al. (1998)), we favour the model B as more consistent with other observations.

Four major features of both obtained mantle attenuation structures (Fig. 11) are also apparent from an examination of the measured t^* distributions shown in the bottom of Fig. 11: 1) the attenuation increases at the level of the top LVZ between 120–150 km depth, corresponding to the arrivals at offsets between 1000–1500 km;

2) below the 410-km discontinuity the attenuation is significantly lower, as indicated by the change in slope of the t^* (*offset*) dependence; 3) variations in the slope of t^* with distance are consistent between the reversed PNEs, demonstrating that these variations are related to the variations in Q between different mantle layers; and 4) the measured t^* distributions are asymmetric, indicating an increase in attenuation within the top of about 150–190 km in the mantle in SE direction.

In the vertical direction our Q model displays low attenuation within the mantle layer immediately below the crust and extending to the depth of the regional 8° discontinuity (Fig. 11; THYBO and PERCHUC, 1997); MOROZOVA *et al.* (1997, 1998) have identified this boundary at the depth of 60–90 km from the "Quartz" data). Within this layer, however, Q values decrease from about 1800 under the Baltic Shield to 700–800 beneath the Altay-Sayan foldbelt (Fig. 11). Below this layer, and to the top of the first LVZ at a depth of 140–150 km in the middle of the profile (Fig. 3), the attenuation is higher, $Q \approx 1000$–1200 (Fig. 11). Within the first LVZ the attenuation increases to $Q \approx 400$–500, which is also consistent with our earlier estimates using the spectral content of the teleseismic P_n phase (MOROZOV *et al.*, 1997). Although the region between 140–150 and 190 km depth is subdivided into two layers with contrasting velocity and reflectivity, we cannot resolve its fine attenuation structure from our data.

Within the prominent LVZ below the depth of 150–190 km, the ray coverage is significantly poorer, and the velocity and attenuation structure is constrained only by the waves from two PNEs which penetrate deeper than about 380 km (Fig. 4). Above this level our model also shows an increase in attenuation in SE direction, with Q values ranging between 400–800 at its top to 1200–1500 deeper than approximately 250 km. Although because of the significant scatter in our t^* measurements this horizontal attenuation gradient is not very confidently established, it is still favoured by the apparent asymmetry of the t^* values obtained from PNEs 123 and 323 at far offsets (Fig. 11, bottom).

Below the depth of the 410-km discontinuity our model approximates the mantle attenuation by a single parameter, $Q \approx 1500$. Although, as we will show in the following section, the trade-off between the Q values within the LVZ and below the 410-km discontinuity is significant, the subhorizontal slope in the t^* values for the rays originating below 400 km (Fig. 11) still indicates a decrease in the attenuation below this depth.

Attempts to evaluate the errors in our attenuation model in a formal manner, by calculating its covariance matrix (12), lead to the result showing a significant trade-off between the $1/Q$ values withing the LVZ and below the 410-km discontinuity, since these regions are penetrated by the same rays (Fig. 4). This trade-off does not allow us to provide simple estimates of the uncertainties of Q values, or give an unambiguous constraint on the attenuation structure below 190 km. Instead, we assess this ambiguity by proposing two representative models A and B (Fig. 11).

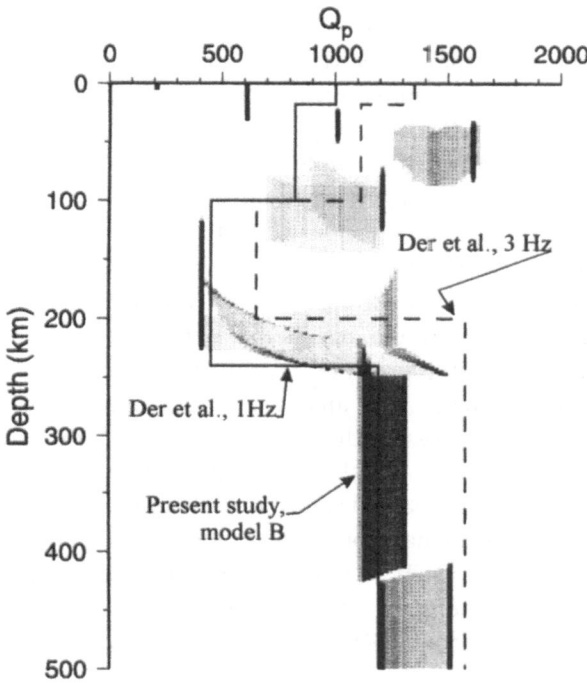

Figure 12

Comparison of the vertical cross section of our Q model with the top part of the frequency-dependent model for the Eurasian Shield by DER *et al.* (1986). Our model B (Fig. 11) is presented as in the form of a histogram; the density of grey shading is proportional to the frequency of occurrences of the corresponding *(depth Q)* values within the middle (500–2500 km) of our 2-*D* model B (Fig. 11).

3. Discussion

Our measurements result in upper mantle attenuation values (Fig. 11) that correspond well to the estimates by DER *et al.* (1986a). Figure 12 shows a vertical cross section across the middle of our attenuation model B (profile distances 1000 to 2000 km in Fig. 11) as compared to the upper 500 km of the attenuation structures at 1 and 3 Hz for the northern part of Eurasia obtained by DER *et al.* (1986b), also using the recordings of Soviet PNEs at NORSAR. Although the attenuation structure given by DER *et al.* (1986b) is significantly more averaged regionally, its correspondence with our attenuation column is notable (Fig. 12). In both models, the major contribution to the upper mantle attenuation comes from the relatively thin layer between the depths of 100 and 200 km, while the attenuation below about 250 km is lower ($Q \approx 1000–1500$). This correspondence is especially remarkable because DER *et al.* (1986b) used a significantly different

P-wave velocity model without the asthenospheric LVZ (KING and CALCAGNILE, 1976). At the same time, our model shows strong lateral variations in *Q* values within the top 150 km of the mantle (Figs. 11 and 13), and also suggests that the observed horizontal attenuation gradient may extend deep into the prominent asthenospheric LVZ (Fig. 11). Note that the somewhat higher attenuation below 200 km revealed in our study (Fig. 12) also apparently corroborates the west-east increase of mantle attenuation suggested in our model B (Fig. 11), since the model by DER *et al.* (1986b) refers to a region west of the profile "Quartz."

Although short-period *Q* values in the lithosphere are known to span a wide range of values and to exhibit strong regional variations (MITCHELL, 1995), a comparison of our model to the results of YEGORKIN and KUN (1978) and YEGORKIN *et al.* (1981) still presents a problem, since the difference is quite significant, while two of the profiles analysed by YEGORKIN and KUN (1978) and YEGORKIN *et al.* (1981) crossed the profile "Quartz" (Fig. 2). Probably the heterogeneity of the uppermost mantle can partly account for the unusually low *Q* values reported by YEGORKIN and KUN (1978) and YEGORKIN *et al.* (1981). However, an apparently systematic bias toward low *Q* in their models suggests that their measurements might have been influenced by the choice of analysis techniques and parameters. We see one of these potential problems of the attenuation measurements by YEGORKIN and KUN (1978) and YEGORKIN *et al.* (1981) in their choice of very short range intervals (70–200 km, or 5–10 data points) for $1/Q$ estimations. Although leading to an apparently detailed mapping of the attenuation values, such an approach may nevertheless result in poorly constrained *Q* models (WHITE, 1992).

Following YEGORKIN and KUN (1978), we examined the correlation between the revealed and attenuation values. By extracting velocity and *Q* values from different depth levels of our model within the region of ray coverage, we obtained a scatter plot showing a positive correlation between *V* and *Q* at greater depth (Fig. 13). In Russian DSS studies a positive correlation was found between the values of V_p and Q_p, with a correlation coefficient of about 0.6; however, significant deviations from this correlation were detected (YEGORKIN and KUN, 1978; RYABOY, 1989). We find that the lateral and vertical variation in $1/Q$ values by far exceed the variations of V_p (Fig. 13), and thus the attempts to associate the values of *Q* with V_p through a predominant $Q(V_p)$ relation throughout the entire depth range may not be justified. However, within the lithospheric mantle, in the depth range 50–190 km, our data suggest a more systematic relation between $1/Q$ and V_p, as shown in Figure 14. The points $(1/Q, V_p)$ in our model appear to be arranged in three clusters (Fig. 14), probably indicating at least three different types of physical conditions of the mantle rocks. For all three clusters the $1/Q - V_p$ correlation is negative, due to the observed increase of the attenuation toward the base of the lithosphere. This result appears to be consistent with the decrease of the mechanical strength of the continental lithosphere between 100–200 km depth (e.g.,

KOHLSTEDT *et al.*, 1995). However, this conclusion is again in conflict with the findings of YEGORKIN and KUN (1978) and of YEGORKIN *et al.* (1981).

A comparatively low attenuation within the upper 80 km of the mantle revealed in our study does not support the model of a strongly scattering mantle proposed by RYBERG *et al.* (1995) and TITTGEMEYER *et al.* (1996), based on their analysis of the same profile. The interpretation by RYBERG *et al.* (1995) and TITTGEMEYER *et al.* (1996) is based solely on the explanation of the relatively high frequency and long incoherent coda of the teleseismic P_n phase, labelled *WG* in Figure 1. These authors, however, disregard mantle attenuation below the level of 150–190 km that accounts equally well for the observed contrast in frequency contents between the teleseismic P_n and deep refracted waves (Fig. 1; MOROZOV *et al.*, 1998). If the uppermost 80 km of the mantle are strongly scattering within the short-period frequency band, it would have a profound effect on the seismic attenuation measured using the same PNE records. The efficient propagation of the high-freqency teleseismic P_n within the upper 80–90 km of the mantle to 3000 km ranges, an equivalent of 1600–2000 wavelengths, (Fig. 1) indicates in itself low effective attenuation rather than strong multiple scattering within this layer.

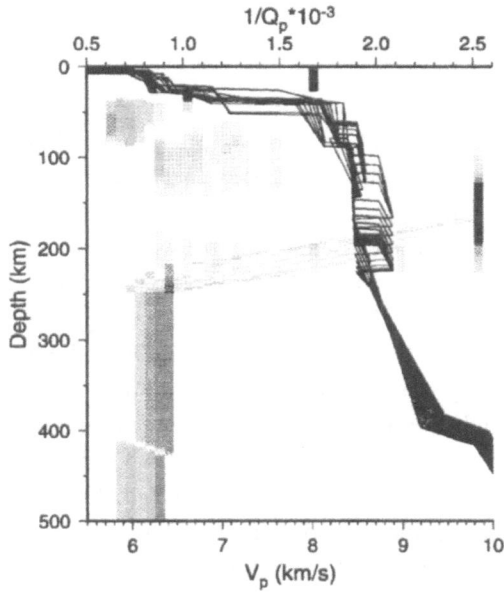

Figure 13

A comparison of the obtained internal friction $1/Q$ and P-wave velocity values along several vertical cross sections between 1000 and 2000 km in our model *B*. The ranges of $1/Q$ values are represented in grey scale, solid lines are vertical velocity profiles. Within the entire depth range the trend of the variations of $1/Q$ with depth and its wide ranges are significantly different from the depth variations V_p.

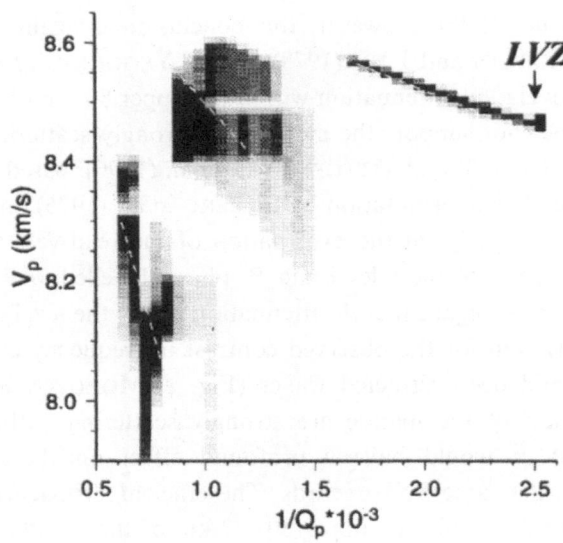

Figure 14

Correlation of the obtained internal friction $1/Q$ and *P*-wave velocities within the lithosphere. Grey-scale levels represent different frequencies of appearance of $(1/Q, V_p)$ values within the top 200 km of our velocity-attenuation model. The $(1/Q, V_p)$ points appear to be separated into three clusters, with negative correlation between $1/Q$ and V_p within each of them indicated with dashed lines. A group of points corresponding to the narrow LVZ between 140–150 km is labelled.

The physical causes of the observed mantle attenuation might be scattering by mantle heterogeneities as well as inelastic energy dissipation. However, the relatively clear onset of the high-frequency energy in the teleseismic P_n phase (Fig. 1) suggests that the increase of attenuation below the level of 150–190 km occurs predominantly due to the increase in intrinsic (inelastic) attenuation. Elastic scattering tends to delay the arrival of high-frequency energy, while intrinsic absorption removes this energy from the wavefield (RICHARDS and MENKE, 1983). Therefore, if significant scattering was present below the depth of about 150 km, we would expect to observe high-frequency diffraction tails originating at the first arrivals at offsets exceeding 1000 km and extending across the gather; such diffractions are not found in the records (Fig. 1; see also MOROZOV *et al.*, 1998). Thus "Quartz" PNE records suggest that the increase of high-frequency energy is associated with the phases propagating within the top 100–110 km of the mantle and is not likely related to the scattering of the deeply refracted waves.

Given the variability of Q in space and frequency, caution should be exercised for implications of the model regarding the physical and chemical state of the mantle. Nevertheless, the attenuation structure revealed in this study plus the observed velocity/interface structure suggest an identification of several major tectonic units in the mantle. We associate the low-attenuation mantle layers down to about 220 km in the middle of the profile with the base of the lithosphere (Fig. 11, bottom). Although the resolution of our Q model is not sufficient to constrain smaller-scale variations of the attenuation, the velocity and reflectivity structure between 150–220 km in the middle of the profile (Fig. 3) suggests that the increased attenuation in the base of the lithosphere may be localised within the narrow LVZ between 130–170 km depth, as suggested in our model B (Fig. 11). This vertically localised, low-velocity, attenuating structure within the base of the lithosphere can probably be associated with the presence of partial melt and fluids, as recently suggested by PERCHUC and THYBO (1996). In "Quartz" records, this layer is identified not only by the gaps in the refracted arrivals, but its top and bottom are also bounded by reflections associated with the 8° and the Lehmann discontinuities, respectively (Fig. 15; MOROZOVA et al., 1997, 1998; THYBO and PERCHUC, 1997). Based on their worldwide compilation of long-range refraction/reflection profiles, THYBO and PERCHUC (1997) suggested such a partially molten layer might be a global feature of "cold," stable continental areas. The thin, high-velocity layer underlying this LVZ imaged by the travel-time inversion between about 180–220 km depth in the middle of the profile (MOROZOVA et al., 1997) is probably more dense and rigid than the asthenospheric mantle below, and its concave shape might indicate a major lithospheric delamination event (Fig. 15). The pronounced horizontal increase of Q in our model toward the shield area and extending to the depths probably close to 400 km (Fig. 11) suggests that the compositional and mechanical heterogeneity within the tectonically involved "tectosphere" (JORDAN,

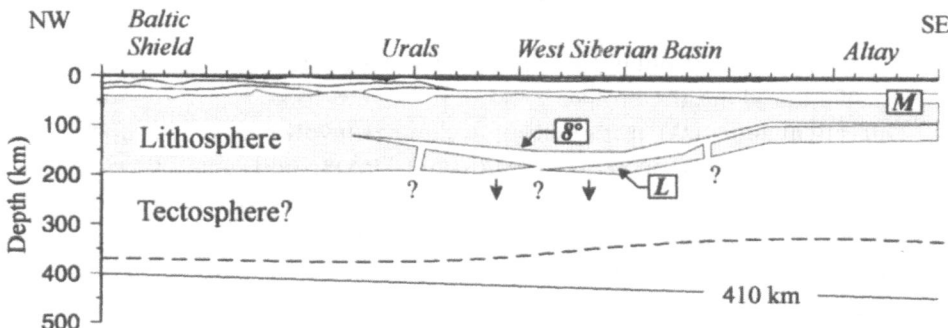

Figure 15

A possible interpretation of the structure of the lower lithosphere (compare to Fig. 11, bottom). The labels indicate the Moho boundary (M), the inferred 8° and Lehmann (L) discontinuities (MOROZOVA et al., 1997, 1998). The presence of the low-Q LVZ between 140 and 170–180 km depth, between the 8° and Lehmann discontinuities (THYBO and PERCHUC, 1997), and a high-velocity underlying layer, suggest possible mechanical instability of the base of the lithosphere under the West Siberian Basin. Given the interpreted thick lithosphere under the Baltic Shield, the depth of the thermal boundary layer (tectosphere) should be near the 400-km discontinuity (cf., JORDAN, 1988).

1988) might also be pronounced within the stable continental mantle. It is remarkable that the low-attenuation sub-Moho mantle extends completely under the East European Platform and the Ural Mountains, and the increase of the attenuation occurs above the region of proposed lithospheric delamination.

Finally, a few limitations of the present study also need to be mentioned. Because of the limited recorded frequency band, we did not attempt to derive a frequency-dependent attenuation model. Velocity inhomogeneities at the scale comparable to the Fresnel zone radius (4) may cause a more complicated wave propagation pattern then modelled in our ray-theoretical approaches and may lead to a systematic bias in the attenuation estimates that is difficult to evaluate at present. However, dense DSS PNE recordings still probably represent the best data that might help tackle this problem. Also, the uncertainty of our attenuation model (as well as of most mantle Q models) is quite large. We assessed this uncertainty by presenting two alternatives; nevertheless, for a comprehensive interpretation it is desirable to have a mechanism that could outline the set of multiple alternatives. Due to the presence of many correlated parameters, to the nonlinear and somewhat interpretative nature of the inversion, this task is not trivial and deserves special study, similar to the multivariate analysis of the "Quartz" velocity model sensitivity carried out by MOROZOVA *et al.* (1998).

4. Conclusions

In this paper, we analysed the upper mantle attenuation in northern Eurasia using the PNE data from the 3850-km Russian DSS profile "Quartz." By applying an improved t^* measurement technique followed by attenuation tomography and iterative ray tracing in the "Quartz" velocity structure obtained earlier (MOROZOVA *et al.*, 1997, 1998), we obtained a model of Q structure to the depth of approximately 500 km along the profile.

Our attenuation (t^*) measurement technique involves a modification of the common-spectrum method (HALDERMAN and DAVIS, 1991) that allows inversion for source spectra as well as for frequency-independent receiver terms and attenuation factors. We extend this method to an adaptive spectrum balancing scheme which combines computational efficiency with a robust treatment of spectral holes and outliers. Another advantage of our t^* measurement approach in the use of a flexible model parameterisation which takes advantage of the ray coverage established by travel-time inversion methods.

The attenuation structure revealed in this study corroborates the earlier model by DER *et al.* (1986b) and provides significantly more detail. Also, this structure supports our earlier estimate based on the analysis of the teleseismic P_n phase (MOROZOV *et al.*, 1998). We demonstrated that the upper mantle structure is characterised by:

1) mantle Q values ranging from 500 to about 2000;
2) general horizontal variations, with attenuation increasing distant from the Baltic Shield within all layers down to 150–190 km, and probably to the 410-km discontinuity;
3) strong vertical contrasts in the attenuation, especially its increase at the level 140–150 km that can apparently be associated with the base of the lithosphere;
4) significant horizontal attenuation contrasts above 190 km depth;
5) lower attenuation within the prominent asthenospheric LVZ between the depths of 190 and 410 km;
6) the analysis of temporal variations of the frequency content of the records suggests that the increase of the attenuation below 150–190 km is predominantly due to the inelastic absorption of seismic energy.

Acknowledgements

The acquisition of the data was carried out by the Centre GEON (the Special Regional Geophysical Expedition at that time) in 1984–87. Digitisation of the records to their full length, and our processing and interpretation of "Quartz" data were sponsored by the U.S. Air Force Office for Scientific Research under Grants F49620-94-1-0134 and F49620-94-A-0134. Comments by R. L. Nowack and by an anonymous reviewer have greatly contributed to the improvement of the manuscript. GMT programs (WESSEL and SMITH, 1995) were used in the preparation of some figures.

Appendix: Parameterisation in the Simultaneous Inversion for Source Power Spectrum, Geometric Divergence, and Attenuation Parameters

In order to build a discrete counterpart of equation (5), we expand the unknown functions ln S, ln G, and t^* in terms of three sets of basis functions:

$$\ln S(f) = \sum_i \psi_i^s(f) s_i,$$

$$\ln G(x) = \sum_i \psi_i^G(x) g_i,$$

$$t^*(x) = \sum_i \psi_i^t(x) t_i. \tag{A1}$$

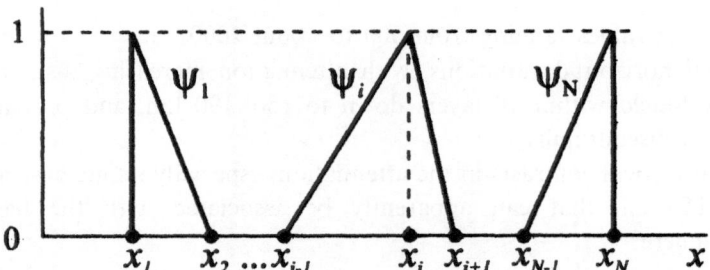

Figure A1

Basis functions $\psi(x)$ used in the discretisation of the model parameters (A1). For the source spectrum the first and the last of these functions are excluded from the parameterisation in order to remove the ambiguity (3).

Here, we abandon for a moment the labelling of the labelling of the values of G and t^* with the channel number, but use a more general "coordinate" x that may correspond to any record specification. The choice of these coordinates, and also the number and the type of the basis functions ψ influence the stability and the resolving power of the inversion, and therefore this choice must employ all available information on the shape of the expected solution.

Since the recording conditions at the stations spaced 10–15 km apart can be considered uncorrelated, the "coordinate" x specifying the geometric spreading and receiver coupling can be associated with the receiver number, and the total number of variables must be equal to the number of records. On the other hand, the values of t^* are associated with seismic phases, rather than with recording channels. Therefore, we can employ the seismic phase information discussed in section 2.1 and build a set of functions $\psi_i^t(x)$ in the definitions (A1) that incorporate this information, reducing the number of variables, increasing the stability of the inversion, and reducing the need for an additional interpretative averaging of t^* values (e.g., such as the averaging employed by MATHENEY *et al.*, 1997). To build such a parameterisation, we use the offset as the argument x of the basis functions $\psi_i^t(x)$ in (A1) and split the range of offsets into segments corresponding to the onsets of different phases, shown in Figure 4. Within each of these regions we specify a set of 2–4 piecewise-linear basis functions illustrated in Figure A1. Thus the smoothing of the t^* (*offset*) dependence is carried out during the inversion, and no smoothing is done across the different phase offset boundaries.

In the parameterisation of $S(f)$, the same set of basis functions is used (Fig. A1). To eliminate the degree of freedom (3) that is not constrained by our common spectrum model (2), we simply require $S(f_{min}) = S(f_{max}) = 0$ at the boundaries of the frequency range used in the inversion. With our choice of basis functions, this constraint is implemented by excluding the first and the last basis functions shown in Figure A1 from the representation space of the model.

In the actual program implementation of the inversion, we take advantage of the flexible input/output and a powerful job description language provided by our seismic processing system (MOROZOV and SMITHSON, 1997) that allowed us to experiment with several user-defined parameterisations of the model described above.

REFERENCES

ANDERSON, D. L., and GIVEN, J. W. (1982), *Absorption Band Q Model of the Earth*, J. Geophys. Res. *87*, 3893–3904.

ANDERSON, R. G., and MCMECHAN, G. A. (1988), *Noise-adaptive Filtering of Seismic Shot Records*, Geophysics *53*, 638–649.

BADRI, M., and MOONEY, H. M. (1987), *Q Measurement from Compressional Seismic Waves in Unconsolidated Sediments*, Geophysics *52*, 772–784.

BÅTH, M., *Spectral Analysis in Geophysics* (Elsevier, Amsterdam 1974).

BLAIR, D. P., and SPATHS, A. T. (1984), *Seismic Source Influence in Pulse Attenuation Studies*, J. Geophys. Res. *89*, 9253–9258.

BRAILE, L., *The Earth's Crust*, AGU Monograph Series *20* (Am. Geophys. Un., Washington, DC 1977).

BRZOSTOWSKI, M. A., and MCMECHAN, G. A. (1992), *3-D Tomographic Imaging of Near-surface Seismic Velocity and Attenuation*, Geophysics *57*, 396–403.

CARPENTER, P. I., and STANFORD, A. R. (1985), *Apparent Q for Upper Crustal Rocks of the Central Rio Grande Rift*, J. Geophys. Res. *90*, 8661–8674.

CERVENY, V., KLIMES, L., and PSENCIK, I. (1984), *Paraxial Ray Approximation in the Computation of Seismic Wavefields in Inhomogeneous Media*, Geophys. J. *79*, 80–104.

DER, Z. A., LEES, A. C., CORMIER V. F., and ANDERSON, L. M. (1986a), *Frequency Dependence of Q in the Mantle Underlying the Shield Areas of Eurasia, Part I: Analyses of Short and Intermediate Period Data*, Geophys. J. R. Astr. Soc. *87*, 1057–1084.

DER, Z. A., LEES, A. C., and CORMIER, V. F. (1986b), *Frequency Dependence of Q in the Mantle Underlying the Shield Areas of Eurasia, Part III: the Q Model*, Geophys. J. R. Astr. Soc. *87*, 1103–1112.

DER, Z. A., MASSE, R. P. and GURSKI (1975), *Regional Attenuation of Short-period P and S Waves in the United States*, Geophys. J. R. Astr. Soc. *40*, 85–106.

DOUGLAS, A., CORBISHLEY, D. J., BLAMEY, C. and MARSHALL, P. D. (1972), *Estimating the Firing Depth of Underground Explosions*, Nature *237*, 26–28.

EGORKIN, A. V., and MIKHALTSEV, A. V., *The results of seismic investigations along geotraverses*. In *Super-deep Continental Drilling and Deep Geophysical Sounding* (eds. Fuchs, K., Kozlovsky, Y. A., Krivtsov, A. I., and Zoback, M. D.) (Springer, Berlin 1990) pp. 111–119.

EGORKIN, A. V., ZYUGANOV, S. K., PAVLENKOVA, N. I., and CHERNISHOV, N. M. (1987), *Results of Lithospheric Studies from Long-range Profiles in Siberia on Profiles in Siberia*, Tectonophysics *140*, 29–47.

GAO, S. (1997), *A Bayesian Non-linear Inversion of Seismic Body-wave Attenuation Factors*, Bull. Seismol. Soc. Am. *87*, 961–970.

GELLER, R. J. (1976), *Scaling Relations for Earthquakes Source Parameters and Magnitudes*, Bull. Seismol. Soc. Am. *66*, 1501–1523.

GLADWIN, M. T., and STACEY, F. D. (1974), *Anelastic Degradation of Acoustic Pulses in Rock*, Phys. Earth Planet. Int. *8*, 332–336.

GRAD, M., and LUOSTO, U. (1994), *Seismic Velocities and Q-factors in the Uppermost Crust beneath the SVEKA Profile in Finland*, Tectonophysics *230*, 1–18.

HALDERMAN, T. P., and DAVIS, P. M. (1991), Q_p *beneath the Rio Grande and East African Rift Zones*, J. Geophys. Res. *96*, 10113–10128.

JANNSEN, D., VOSS, J., and THEILEN, F. (1985), *Comparison of Methods of Determine Q in Shallow Marine Sediments from Vertical Reflection Seismograms*, Geophys. Prospect. *33*, 479–497.

JORDAN T., H. (1988), *Structure and Formation of the Continental Tectosphere*, J. Petrology, Special Lithosphere Issue, 11–37.

KING, D. W., and CALCAGNILE, G. (1976), *P-wave Velocities in the Upper Mantle Beneath Fennoscandia and Western Russia*, Geophys. J. R. Astr. Soc. *46*, 407–432.

KOHLSTEDT, D. L., EVANS, B., and MACKWELL, S. J. (1995), *Strength of the Lithosphere: Constraints Imposed by Laboratory Experiments*, J. Geophys. Res. *100*, 17,587–17,602.

KOZLOVSKY, Y. A., *The USSR integrated program of continental crust investigations and studies of the earths deep structure under the globus project*. In *Super-Deep Continental Drilling and Deep Geophysical Sounding* (eds. Fuchs, K., Kozlovsky, Y. A., Krivtoz, A. I., and Zoback, M. D.) (Springer, Berlin 1990) pp. 90–103.

LEES, A. C., DER, Z. A., and CORMIER, V. F. (1986), *Frequency Dependence of Q in the Mantle Underlying the Shield Areas of Eurasia, Part II: Analysis of Long-period Data*, Geophys. J. R. Astr. Soc. *87*, 1085–1101.

MATHENEY, M. P., and NOWACK, R. L. (1995), *Seismic Attenuation Values Obtained from Instantaneous Frequency Matching and Spectral Ratios*, Geophys. J. Int. *123*, 1–15.

MATHENEY, M. P., NOWACK, R. L., and TRÉHU, A. M. (1997), *Seismic Attribute Inversion for Velocity and Attenuation Structure Using Data from the GLIMPCE Lake Superior Experiment*, J. Geophys. Res. *102*, 9949–9960.

MECHIE, J., A. EGORKIN, V., FUCHS, K., RYBERG, T., SOLODILOV, L., and WENZEL, F. (1993), *P-wave Velocity Structure beneath Northern Eurasia from Long-range Recording along the Profile Quartz*, Phys. Earth Planet Inter. *79*, 269–286.

MECHIE, J., EGORKIN, A. V., SOLODILOV, L., FUCHS, K., LORENZ, F., and WENZEL, F., *Major features of the mantle velocity structure beneath northern Eurasia from long-range seismic recordings of peaceful nuclear explosions*. In Upper Mantle Heterogeneities from Active and Passive Seismology (ed. Fuchs, K.) (Kluwer Academic Publ., Dordrecht 1997) pp. 33–50.

MENKE, W., *Geophyscial Data Analysis: Discrete Inverse Theory* (Academic Press, San Diego 1989) pp. 35–60.

MITCHELL, B. (1995), *Anelastic Structure and Evolution of the Continental Crust and Upper Mantle from Seismic Surface Wave Attenuation*, Rev. Geoph. and Space Phys. *33*, 441–462.

MOROZOV, I. B., and SMITHSON, S. B. (1996), *Instantaneous Polarization Attributes and Directional Filtering*, Geophysics *61*, 872–881.

MOROZOV, I. B., and SMITHSON, S. B. (1997), *A New System for Multicomponent Seisimic Processing*, Computers and Geosciences *23*, 689–696.

MOROZOV, I. B., MOROZOVA, E. A., SMITHSON, S. B., and SOLODILOV, L. N. (1988), *On the Nature of the Teleseismic P_n Phase Observed on the Ultra-long Range Profile "Quartz," Russia*, Bull. Seismol. Soc. Am. *88*, 62–73.

MOROZOVA, E. A., MOROZOV, I. B., and SMITHSON, S. B., *Heterogeneity of the uppermost Eurasian mantle along the DSS profile Quartz, Russia*. In: *Upper Mantle Heterogeneities from Active and Passive Seismology* (Kluwer Academic Publ., Dordrecht 1997) pp. 139–146.

MOROZOVA, E. A., MOROZOV, I. B., SMITHSON S. B., and SOLODILOV, L. N. (1998), *Heterogeneous Structure of the Uppermost Mantle Imaged by the Ultra-long Range Profile "Quartz," Russian Eurasia*, J. Geophys. Res., submitted.

PATTON, H. (1980), *Crust and Upper Mantle Structure of the Eurasian Continent from the Phase Velocity and Q of Surface Waves*, Rev. Geoph. and Space Phys. *18*, 605–625.

PAVLENKOVA, N. I. (1996), *General Features of the Uppermost Mantle Stratification from Long-range Seismic Profiles*, Tectonophysics *264*, 261–278.

PAVLENKOVA, N. I., PAVLENKOVA, G. A., and SOLODILOV, L. N. (1996), *High Velocities in the Uppermost Mantle of the Siberian Craton*, Tectonophysics *262*, 51–65.

PERCHUC, E., and THYBO, H. (1996), *A New Model of Upper Mantle P-wave Velocity below the Baltic Shield: Indication of Partial Melt in the 95 to 160 km Depth Range*, Tectonophysics *253*, 227–245.

RICHARDS, P. G., and MENKE, W. (1983), *The Apparent Attenuation of a Scattering Medium*, Bull. Seismol. Soc. Am. *73*, 1005–1022.

RYABOY, V. (1989), *Upper Mantle Structure Studies by Explosion Seismology in the USSR*, Delphic Associates, 138 pp.

RYBERG, T., WENZEL, F., MECHIE, J., EGORKIN, A., FUCHS, K., and SOLODILOV, L. (1996), *Two-dimensional Velocity Structure beneath Northern Eurasia Derived from the Super Long-range Seismic Profile Quartz*, Bull. Seismol. Soc. Am. *86*, 857–867.

RYBERG, T., FUCHS, K., EGORKIN, V., and SOLODILOV, L. (1995), *Observations of High-frequency Teleseismic P$_n$ on the Long-range Quartz Profile across Northern Eurasia*, J. Geophys. Res. *100*, 18151–18163.

SCHUELLER, W., MOROZOV, I. B., and SMITHSON, S. B. (1997), *Crustal and Uppermost Mantle Velocity Structure of Northern Eurasia along the Profile "Quartz,"* Bull. Seismol. Soc. Am. *87*, 414–426.

SHERIFF, R. E., *Geophysical Methods* (Prentice Hall, 1989) 333 pp.

THYBO, H., and PERCHUC, E. (1997), *The Seismic 8° Discontinuity and Partial Melting in Continental Mantle*, Science *275*, 1626–1629.

TITTGEMEYER, M., WENZEL, F., FUCHS, K., and RYBERG, T. (1996), *Wave Propagation in a Multiple-scattering Upper Mantle—Observations and Modelling*, Geophys. J. Int. *127*, 492–502.

TONN, R. (1989), *Comparison of Seven Methods for the Computation of Q*, Phys. Earth Planet. Int. *55*, 259–268.

TONN, R. (1991), *The Determination of the Seismic Quality Factor Q from VSP Data: A Comparison of Different Computational Methods*, Geophys. Prospect. *39*, 1–27.

WESSEL, P., and SMITH, W. H. F. (1995), *New Version of the Generic Mapping Tools Released*, EOS Trans. Am. Geophys. U. *76*, 329 pp.

WHITE, R. E. (1992), *The Accuracy of Estimating Q from Seismic Data*, Geophysics *57*, 1506–1511.

WRIGHT, C., and HOY, D. (1981), *A Note on Pulse Broadening and Anelastic Attenuation in Near-surface Rocks*, Phys. Earth Planet. Int. *25*, P1–P8.

YEGORKIN, A. V., and KUN, V. V. (1978), *Absorption of Longitudinal Waves in the Earth's Upper Mantle*, Physics of the Solid Earth *14* (4), 262–269.

YEGORKIN, A. V., KUN, V. V., and CHERNYSHEV, N. M. (1981), *Absorption of Longitudinal and Transverse Waves in the Crust and Upper Mantle of the West Siberian Plate and Siberian Platform*, Physics of the Solid Earth *17* (2), 105–115.

ZELT, C. A., and SMITH, R. B. (1992), *Seismic Travel-time Inversion for 2-D Crustal Velocity Structure*, Geophys. J. Int. *108*, 16–34.

(Received November 10, 1997, revised June 2, 1998, June 23, 1998)

 To access this journal online:
http://www.birkhauser.ch

Pure appl. geophys. 153 (1998) 345–375
0033–4553/98/040345–31 $ 1.50 + 0.20/0

▌Pure and Applied Geophysics

Attenuation of Broadband *P* and *S* Waves in Tonga: Observations of Frequency Dependent *Q*

MEGAN P. FLANAGAN[1] and DOUGLAS A. WIENS[2]

Abstract—Teleseismic broadband recordings of intermediate and deep focus earthquakes are used to quantify both compression (Q_α) and shear (Q_β) wave attenuation within the Lau backarc basin. A spectral-ratio method is employed to measure differential attenuation (δt^*) between the depth phases sS, pP, and sP and the direct S and P phases over the frequency band 0.05 and 0.5 Hz. We use a stacking algorithm to combine the spectra of several phase pairs from a single event, having similar azimuth and range, to obtain more robust δt^* measurements; these estimates are then used to compute the average Q above the focal depth. Q_β and Q_α are measured directly from the sS-S and pP-P phase pairs respectively, however, the interpretation of δt^* measured from sP-P requires assumptions about the ratio Q_α/Q_β. We find an empirical ratio of $Q_\alpha/Q_\beta = 1.93$ for this region and use it to compute Q_α and Q_β from the Q_{sP} observations.

We observe lateral and depth variations in both Q_β and Q_α beneath the tectonically active Lau Basin and the geologically older, inactive Lau Ridge and Fiji Plateau. The upper 200 km beneath the Central and Northern Lau Basin show a Q_β of 45–57 and a Q_α of 102–121, and Q appears to increase rapidly with depth. The upper 600 km beneath the Lau backarc basin has a Q_β of 118–138, while over the same depth interval we observe a higher Q_β of 139–161 beneath the Lau Ridge and Fiji Plateau. We also find Q_α of 235–303 beneath the northern Lau Basin and a higher Q_α of 292–316 beneath the Fiji Plateau and the Lau Ridge measured directly from pP-P phase pairs. These geographic trends in the broadband Q measurements correlate with our previous long-period estimates of Q_β in this region, however, the broadband measurements themselves are higher by about a factor of two. These observations suggest substantial frequency dependence of Q in the upper mantle, beginning at frequencies less than 1.0 Hz and consistent with the power-law form: $Q \propto \omega^\alpha$ with α between -0.1 and -0.3.

Key words: Attenuation, Frequency dependence, subduction zone.

Introduction

Investigating the anelastic structure of the uppermost mantle continues to be an important problem in geophysics. In particular, the observation of attenuation

[1] Cecil H. and Ida M. Green Institute of Geophysics and Planetary Physics, Scripps Institution of Oceanography, University of California San Diego, La Jolla, U.S.A. Fax: 619-534-5332, E-mail: megan@mahi.ucsd.edu

[2] Department of Earth and Planetary Sciences, Washington University, St. Louis, MO, U.S.A.

heterogeneity near subduction zones has implications for the dynamics of oceanic lithosphere and asthenosphere. Interactions of seismic waves with such attenuating structures can provide new information about these complex regions to complement the rather well-known elastic structure. Although substantial progress has been made in understanding lateral variations of Q in the upper mantle, there are many unanswered questions. Robust constraints on Q and its regional and frequency dependence, especially in oceanic regions, remain sparse.

Much of our current understanding of the upper mantle is based on inversions for attenuation structure from long-period seismic data (frequency < 1.0 Hz). Improved data coverage and measurement techniques have led to numerous models of attenuation on both global scales (DUREK et al., 1993; ROMANOWICZ, 1995; BHATTACHARYYA et al., 1998) and regional scales (REVENAUGH and JORDAN, 1991; SHEEHAN and SOLOMON, 1992; DING and GRAND, 1993; SIPKIN and REVENAUGH, 1994; FLANAGAN and WIENS, 1994; MITCHELL, 1995). A universal feature of these models is the presence of a low-Q asthenosphere and upper mantle with lateral variations in attenuation which are quite strong.

Intrinsic attenuation also requires velocity dispersion to satisfy causality (FUTTERMAN, 1962) and needs to be quantified to reconcile seismic velocity models of the earth produced from both short-period body waves and long-period surface waves and normal modes. Interpreting Q observations and quantifying dispersion for application to mantle travel times require knowledge of the frequency dependence of Q, which is still not well developed. The absorption band model is commonly used to explain both the absorption and dispersion of seismic waves in the mantle (LIU et al., 1976; ANDERSON and GIVEN, 1982). The behavior of this mechanical model explains discrepancies among long- and short-period seismic observations in that the apparently weaker damping observed for higher frequency waves (e.g., SIPKIN and JORDAN, 1979; LUNDQUIST and CORMIER, 1980) is a rheological property of mantle materials.

Quantitative measurements of intrinsic attenuation in the mantle at higher frequencies (above 1.0 Hz) have been less prevalent due to the difficulty in estimating attenuation from short-period seismograms. However, some work has been done on attenuation in the frequency range 1.0 to 8.0 Hz (DER et al., 1982; BACHE et al., 1986; BOWMAN, 1988) and higher (WALCK, 1988). Absolute values of upper-mantle attenuation from low frequency studies have traditionally been difficult to reconcile with those from high-frequency studies because most investigations comprise seismic waves with fundamentally different travel paths. However, the broadband analyses presented in this paper show much merit in maintaining a large frequency-bandwidth and bridging the gap between the long- and short-period bands commonly considered in global seismology while restricting the study to one geographic region for consistency.

Here, we make Q measurements from broadband teleseismic wave forms and combine these with previous observations at both long and short periods in an attempt to quantify any frequency dependence of attenuation in the upper mantle of this convergent margin. We choose the Tonga-Fiji region not solely because it produces the greatest number of deep earthquakes (thereby supplying a large database), but because of the wealth of attenuation studies completed there over the years with which we can compare our new results. We examine the depth phases *sS*, *pP*, and *sP*, a phase which to our knowledge has not been previously used in studies of backarc basins, and quantify the observations in terms of the decay rate of the spectral ratio of *sS* to *S*, *pP* to *P*, and *sP* to *P* waves. The geometry of the *P*, *pP*, and *sP* phases through the mantle and a representative vertical component seismogram are illustrated in Figure 1; ray paths of *S* and *sS* paths are similar but not shown. This phase-pair approach has been used quite successfully to measure the attenuation in the mantle wedge above the Benioff zone (FLANAGAN and WIENS, 1990, 1994). Note the broadness of the *pP* and *sP* waves relative to the direct *P* wave; this is due to attenuation in the backarc region above the slab. After quantifying the attenuation of both *P* and *S* waves by the quality factors Q_α and

Geometry of Raypaths

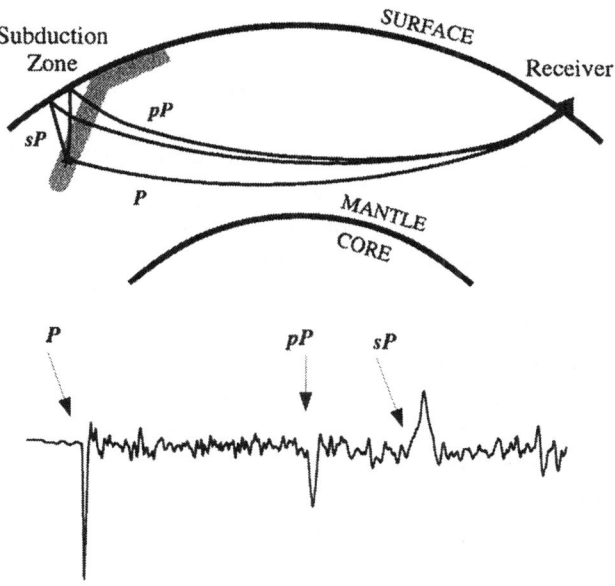

Figure 1

Earth cross section displaying ray paths of *sP*, *pP*, and *P*; similar paths exist for the *sS* and *S* phases. The light gray region indicates the Tonga slab, and the star represents a deep focus earthquake. Ray paths through the lower mantle are similar for each phase pair, constraining the differential attenuation to lie above the earthquake source. The vertical-component displacement seismogram below shows the relative arrival of the aforementioned phases. Care is taken to select broadband records in which the depth phases *pP* and *sP* are well separated from other arrivals.

Table 1

Earthquake source parameters from the CMT catalog

Event	Date	Time (UT)	Latitude (deg.)	Longitude (deg.)	Depth (km)	Moment (Nm)	Strike/Dip/Slip (deg.)
1	89–11–16	08:39:42.3	−17.69	−179.06	531.0	2.3×10^{18}	092/46/027
2	90–05–20	07:32:36.9	−18.09	−175.34	232.0	2.7×10^{18}	260/32/−025
3	91–05–10	13:33:54.2	−16.07	−174.20	135.0	2.6×10^{18}	197/14/−075
4	91–10–18	17:22:55.1	−24.20	−177.61	194.0	5.0×10^{18}	245/61/−011
5	91–11–11	16:15:50.9	−24.05	−177.46	178.0	1.1×10^{18}	235/21/−037
6	92–08–04	06:58:35.8	−21.58	−177.32	278.0	1.2×10^{18}	108/56/−163
7	92–08–30	20:09:06.9	−17.74	−178.77	573.0	4.7×10^{18}	014/49/−138
8	93–03–21	05:04:59.0	−17.97	−178.53	584.0	3.3×10^{18}	066/41/−177
9	93–04–24	09:54:21.2	−17.73	−179.81	600.0	2.7×10^{18}	086/29/−148
10	93–07–09	15:37:55.1	−19.79	−177.54	412.0	1.5×10^{18}	137/10/−174
11	93–10–11	13:07:29.9	−17.80	−178.76	556.0	1.2×10^{18}	027/45/177

Q_β, we combine our broadband observations with a suite of previously determined attenuation observations in the Tonga-Fiji region to investigate the frequency dependence of Q.

Broadband Data

We conduct a survey of intermediate and deep earthquakes in the Tonga subduction zone and select those characterized by simple source mechanisms and good excitation of *P*- and *S*-wave energy. The data are three-component broadband seismograms (BB? channel, 20 Hz sample rate) recorded by the GDSN and the IRIS-IDA networks. We apply the same criteria for these broadband data as described in FLANAGAN and WIENS (1994) for the long-period seismograms: epicentral distance range is 30° to 90° to avoid both upper mantle triplications and core-mantle boundary interactions; focal depth must be greater than 130 km to ensure clear separation of *pP* and *sP* pulses; and slab diffraction effects are minimized by eliminating phases which travel within 20° of the Tonga slab. We impose a minimum moment threshold of 9×10^{17} Nm, higher than the value for the long-period study, as a number of smaller events were examined and discarded due to unacceptably low signal-to-noise ratios (SNR). Likewise, large events (moment $> 6 \times 10^{18}$ Nm) are not used due to complex rupture characteristics which are more evident in the broadband wave forms.

Eleven earthquakes are found which yield suitable high quality seismograms for attenuation analysis. The source parameters for these events, taken from the Harvard Centroid Moment Tensor (CMT) catalog (e.g., DZIEWONSKI and ANDERSON, 1981) and are listed in Table 1. We use both the vertical and tangential component records, obtained by rotating the horizontal components using the CMT epicentral information, and visually examine them for clear depth phases pP, sP, sS and direct P, S phases.

The use of broadband data to study attenuation structure offers considerable advantages over long-period data: most notably, improved lateral and depth resolution. The broadband instruments have a frequency passband with a flat velocity between approximately 0.005 and 2.0 Hz, relevant to this study, and this broad bandwidth is an important aid to the resolution of our techniques. The higher frequencies and hence shorter wavelengths of the broadband data enable better lateral resolution due to the smaller Fresnel zones. Figure 2 demonstrates the resolving power of using broadband observations, especially with regard to rays sampling the oceanic crust of the northern Lau Basin. These are ground displacement records produced by deconvolving the instrument response over the passband 0.01 to 0.1 Hz for the long-period trace and 0.01 to 2.0 Hz for the broadband trace. The pulse-broadening, indicative of intrinsic attenuation, of the pP and sP relative to the direct P wave is more obvious on the broadband seismograms (channel BBZ) than on the long-period records (channel LPZ).

Also note that the crustal precursors seen in the broadband wave forms are more distinct than on the long-period records, and offer a better opportunity to resolve fine crustal layering. Broadband data, because they include higher frequency waves, are more susceptible to scattering effects which may mimic intrinsic attenuation and thus must be taken into account. This becomes important in the attenuation analysis as it has been shown that interference from energy multiply reflected in crustal structure may obscure the depth phase wave forms in which we are interested (SIPKIN and REVENAUGH, 1994). Although this is not the case for the previous SH studies (FLANAGAN and WIENS, 1990, 1994), where thin oceanic crust does not grossly perturb the long-period wave forms, it is more problematic with broadband data and more so for the P waves. As we include higher frequencies we must also consider that multiple reflections within the crust at the bounce point location may produce arrivals with sufficient time separation to distort the depth phase wave forms. This could potentially perturb the Q measurements, thus this elastic effect must be accounted for in order to isolate the anelastic attenuation (discussed below). In Figure 2, the records from Event 10 recorded in China (e.g., SSE) sample the Lau Ridge and a rather strong crustal precursor to sP on the broadband seismogram, although this feature is somewhat muted on the long-period seismogram. Similarly, an sP phase recorded in North America (e.g., PFO) and sampling the northern Lau Basin, shows a small precursor, or change slope at the onset of the main pulse, on the broadband trace which is completely absent in the long-period trace.

Also note the loss of high frequency content of sP in comparison to pP in Figure 2, indicating the strong attenuation of shear energy relative to compressional energy. Using the sP phase is ideal as it retains much of the high frequency content

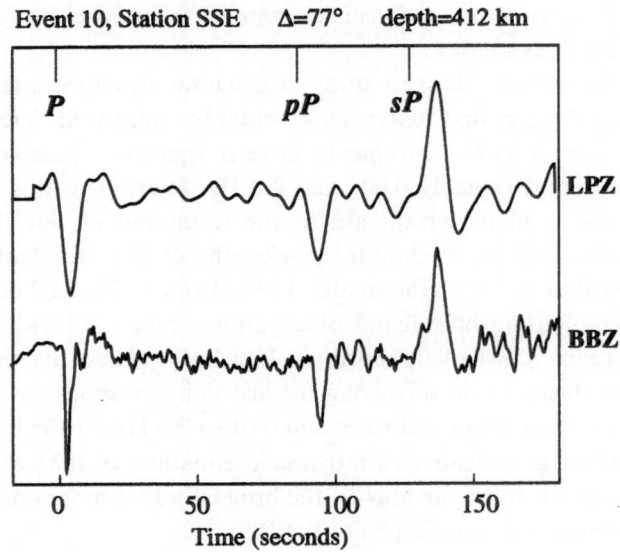

Event 10, Station SSE Δ=77° depth=412 km

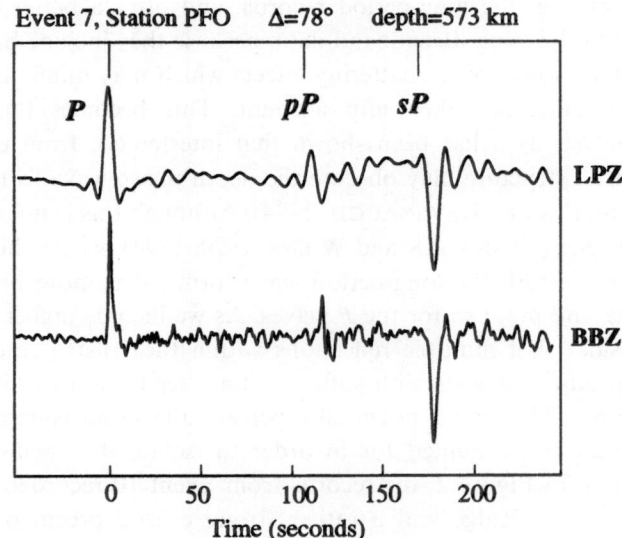

Event 7, Station PFO Δ=78° depth=573 km

Figure 2

Comparison of broadband and long-period vertical-component displacement seismograms recorded at the same station for Event 10 (top) with rays sampling the Lau Ridge and Event 7 (bottom) with rays sampling the northern Lau Basin. Displacement records are obtained by removing the instrument response over the frequency range 0.01 to 2.0 Hz for the broadband and 0.01 to 0.1 Hz for the long-period data. Note the loss of higher frequencies in the depth phases relative to the direct P phases, as well as sP relative to pP, due to anelastic attenuation in the Lau backarc region.

necessary to measure δt^*, and to model Moho precursors yet by its nature does not exhibit as much complexity from the water reverberations as *pP* does. In addition, the use of *P* and *S* phases in concert allows us to better understand the mechanism of attenuation in the upper mantle of backarc regions. Because compressional-wave and shear-wave attenuation vary independently, depending on the properties of the mantle rocks, by measuring both types we can exploit this complementary information to more accurately interpret the geologic nature of the attenuation structures observed.

Wave-form Analysis

We adopt the time and frequency domain methods of δt^* measurement from FLANAGAN and WIENS (1990) to model the crustal velocity and mantle attenuation structure of the Lau backarc basin. Both methods have been applied to data sets of long-period *SH* wave forms with great success. The time domain procedure is a forward modeling approach to determine the best-fitting crustal and attenuation parameters. The frequency domain procedure is a spectral ratio technique which measures the differential attenuation operator (δt^*) between two seismic phases.

To the greatest extent possible, identical data sets are used in both the time and frequency analyses, however, the interference of extraneous phases sometimes precludes this. Due to source-receiver geometry there is often interference of the *PP* phase arriving just after the *sP* phase on the vertical component. Likewise, the *ScS* sometimes interferes with the *sS* phases on the tangential component records. This interference is not as crucial for the analysis of crustal structure because we focus mainly on modeling of precursors to the depth phases; it is, however, problematic for the attenuation measurement. The depth phases must be windowed carefully to avoid contamination from other arrivals, therefore on a record by record basis all phases are visually examined and many were subsequently discarded as being unsuitable for frequency analysis. The result is that the data eligible for the frequency domain δt^* analysis are merely a subset of the entire data set used for the time domain crustal modeling. Our goal is to model the crustal structure as well as possible so it can be included in the attenuation analysis to minimize any error induced in the δt^* measurements due to inadequate compensation for elastic structure. A total of 118 individual seismograms from 11 events are used in our modeling of crustal structure; a subset of these data, 83 seismograms, is used in the attenuation analysis. Complete discussion of the crustal modeling is given in FLANAGAN (1994); we only outline the procedure briefly here as it applies to this study.

Modeling in the Time Domain

Our forward modeling technique differs from other approaches using syntheti-cally generated seismograms as we model differences between the direct (P) and reflected (pP) wave forms in a manner which suppresses source and receiver effects. The difference in travel time between the depth phase and the Moho precursor and the amplitude ratio of the two pulses are the parameters which allow us to determine the crustal thickness and velocity contrast across the Moho. A grid search algorithm is employed to find the best-fitting crustal structure such that it can be subsequently removed from the frequency domain δt^* measurements.

As the broadband data contain higher frequency energy, we found it useful to stack many records together to improve the SNR instead of modeling individual seismograms. Stacking should enhance the coherent Moho underside reflections and subdue random noise, thereby allowing more robust modeling of the crustal precursors. We stack only records from the same event and not from several events because variation in the source-time functions and attenuation is of greater concern for broadband data. Similarly we cannot stack all the records from a single event because the bounce points may be in completely different crustal locations, affect-ing the shape of the depth phase. Subsets of the data from a single event which show consistent crustal precursors may be stacked together under the assumption that the crust has laterally uniform properties over a short distance range (~ 50 km). Clearly, stacking results averaging over the region sampled by the depth phases, however if we restrict the stacks to phases with similar azimuths and with similar bounce points, we minimize this effect. Lack of stations in the Southern Hemisphere and the remote location of the Tonga Subduction Zone results in poor azimuthal coverage of the observations. Usually there were two possible groups of data to be stacked: phases traveling to stations in North America and phases traveling to stations in Asia.

We align the seismograms on the pP, sP or sS phase for a given event and azimuth, normalize the amplitudes to unity then stack them in the time domain; the aligned phases stack coherently and noise stacks incoherently. Stacks are also made for the direct P and S phases separately and are used as the source pulse in the modeling; Figure 3 shows two examples of this stacking procedure. On the left, the results of stacking five seismograms from Event 7 recorded in North America and having bounce points (typically within 0.5° of each other) beneath the northern Lau Basin. These records show a somewhat small Moho precursor to sP due to the thin oceanic crust, nonetheless the good quality of these records and the stacking procedure enhances the $s_m P$ precursor (the underside Moho reflection). On the right, the results of stacking nine seismograms from Event 10 recorded in Asia and having bounce points beneath the Lau Ridge. Note the SNR is greatly improved by stacking the records together, specifically for the pP phase.

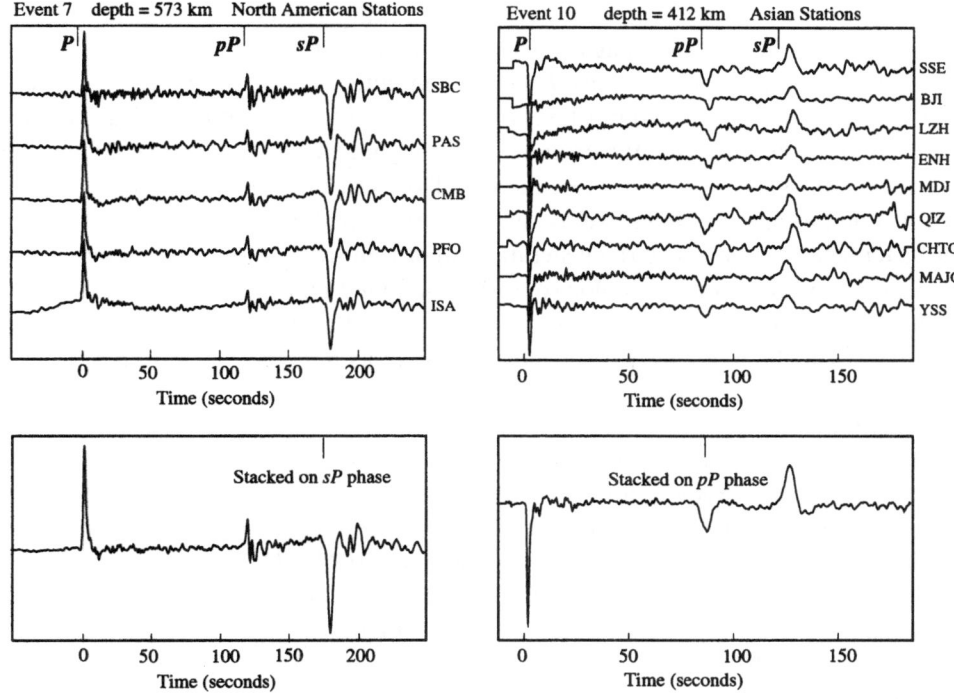

Figure 3

Individual and stacked seismograms from Event 7 (left) and Event 10 (right). On the top, displacement records for each station are aligned and stacked separately on the *sP*, *pP*, and *P* phases. Identifying the s_mS and p_mP precursors in individual traces is sometimes difficult, however, stacking the seismograms enhances the SNR and allows the Moho reflection to be seen more clearly (bottom right).

We use the forward modeling technique described in FLANAGAN and WIENS (1990) which requires time domain convolution. The reflected $pP(t)$ and direct $P(t)$ wave forms are related by: $pP(t) = CP(t)*R(t)*A(t)$. The (stacked) direct pulse $P(t)$ contains the source and receiver functions and the lower mantle path structure while the upgoing phase $pP(t)$ contains the additional information of upper mantle attenuation $A(t)$ and crustal structure at pP bounce point $R(t)$ (*C* is a frequency independent constant related to the source radiation pattern and geometric spreading).

This modeling is performed on both stacked and single station wave form pairs: pP-P; sP-P; and sS-S. The attenuation functions $A(t)$ are computed using a linear causal constant *Q* model (LIU *et al.*, 1976; KJARTANSSON, 1979) where *Q*, and consequently δt^*, is the parameter we vary. As the amplitudes of the wave forms are normalized in the stacking algorithm, the constant *C* is not necessary for the calculation; however, the radiation pattern (taken from the CMT focal mechanisms) is required to determine the polarity of the two phases. We include δt^* as a modeling parameter here, although not as an absolute measure of the differential attenuation; the spectral ratio and spectral stacking techniques described below offer better resolution and thus are used to quantify δt^* for each phase pair.

The crustal transfer functions $R(t)$ are computed using two-dimensional geometric ray theory for horizontal layers since near-critical and post-critical interactions are not significant for the cases we are studying (KROEGER, 1987). Geometric ray theory provides a good representation of the crustal transfer functions and is not as computationally intensive as the reflectivity method (FUCHS and MÜLLER, 1971; KENNETT, 1983). The ray generation algorithm traces through any number of horizontal layers of uniform velocity to the computed amplitude and time delay of each reflected and converted phase until some small amplitude cut-off is reached (i.e., 0.02% of the reference phase). This results in as many as 500 rays for pP and sP and about 100 rays for sS although typically only the first twenty rays are of considerable amplitude.

There are several parameters which affect the modeling of the crustal precursors, and previous studies show that crustal thickness (h), Moho velocity change ($\Delta\alpha$, $\Delta\beta$) and Moho density change ($\Delta\rho$) are more significant than attenuation (δt^*) in time-domain modeling (ZHANG and LAY, 1993). The velocity and density remain constant within each crustal layer, and the sub-Moho and water properties are fixed at: α_m, β_m, $\rho_m = 7.8$ km/s, 4.5 km/s, 3.3 kg/m^3; α_w, β_w, $\rho_w = 1.5$ km/s, 0.0 km/s, 1.0 kg/m^3. In addition, we assume $\beta = \alpha/\sqrt{3}$ and that the density scales with velocity in the crust as $\Delta\rho_c = 0.32\Delta\alpha_c$ such that the only free parameters in the modeling are α, h, and δt^*.

A grid search algorithm is used to determine the crustal thickness, velocity and attenuation providing the best fit to the observed wave forms. This procedure involves iteratively computing crustal functions, convolving them with the direct P phase, and measuring the misfit by computing the L_2 norm (MENKE, 1989). We iterate over a large range of values to start (e.g., $h \approx 5$ to 30 km, 0.5 to 2.5 km for sediments, $\alpha \approx 3.5$ to 7.5 km/s, $Q \approx 40$ to 400), and once rough estimates are reached we further iterate over a smaller range, and/or include a second crustal layer (such as sediments), then loop simultaneously over both layers to achieve the best-fitting parameters. For each phase pair there may be from 200 to 3000 computations, depending on the number of crustal layers and how fine the model increments may be, thus visual inspection is not always practical.

We focus on models producing the lowest L_2 norm and visually check the wave-form fits to exclude spurious results. Figures 4 and 5 display modeling of crustal layers overlain by a water layer taken from well-constrained bathymetric maps if not modeled. The effect of varying crustal thickness is quite pronounced at the onset of the pulse due to the increasing time difference between pP and p_mP for increasing h (Figs. 4b, 5b) but the effect is also noticeable at the end of the pulse as the first crustal reverberation, having opposite polarity, serves to sharpen the tail of the main pulse. The velocity contrast at the Moho, $\Delta\alpha$, largely influences the amplitude of the Moho reflection p_mP or s_mP (Figs. 4c, 5c), and the differential attenuation, δt^*, changes the overall width of the pulse as well as the broadness of the tail end (Figs. 4d, 5d).

There is some coupling between the crustal thickness and velocity for a given δt^*, however there appears to be little trade-off between attenuation and overall crustal structure using this time domain technique. As a test of the method, we computed synthetic seismograms using reflectivity and perform the phase-pair modeling on these "perfect data." Results indicate errors of ± 0.05 to ± 0.2 km/s in velocity and ± 1 to 2 km in crustal thickness, the same range of uncertainties we find from modeling the real data (see FLANAGAN, 1994, for complete details).

Modeling *pP* Waveforms

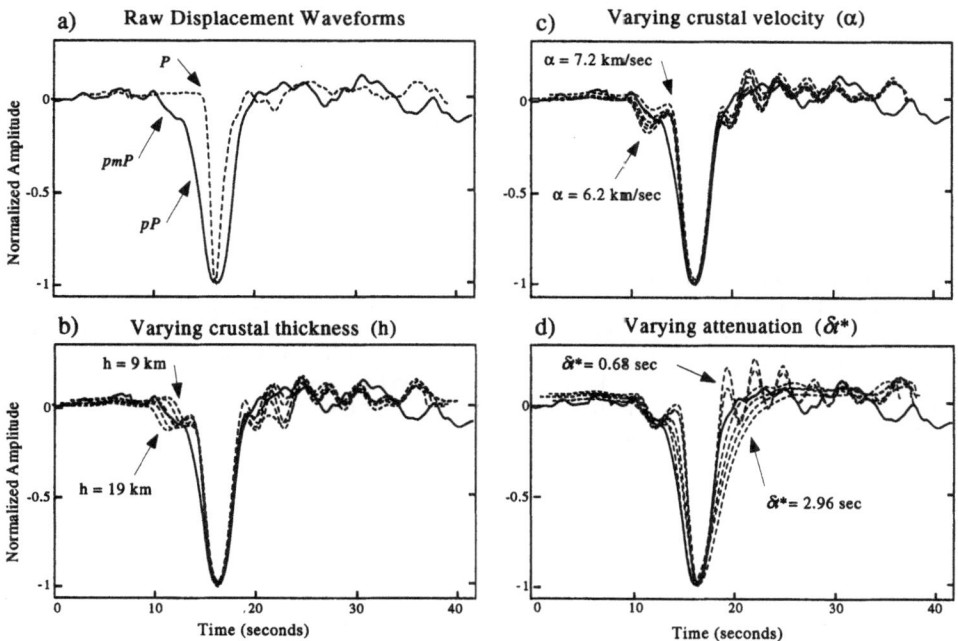

Figure 4

Example of time domain wave-form modeling *pP*; the results of convolving the direct *P* pulse with different crustal transfer functions, and attenuation operators to match the *pP* pulse. Data are stacked Asian stations from Event 10 which bounce on the Lau Ridge. a) Normalized displacement *P* (dashed) and *pP* (solid) wave forms showing the p_mP precursory Moho reflection. Dashed lines are the wave forms superimposed to show the effects of varying the modeling parameters: b) crustal thickness $h = 9$ to 19 km in 2 km increments; c) crustal velocity $\alpha = 6.2$ to 7.2 km/s in 0.2 km/s increments (β and ρ vary accordingly); and d) differential attenuation $\delta t^* = 2.96$ to 0.68 s. The best fitting parameters for this example describe a two-layer crust overlain by 2.1 km of water, h α, and δt^* are: 14.0 km at 6.8 km/s, 3.5 km at 5.05 km/s, and 1.78 s.

Modeling *sP* Waveforms

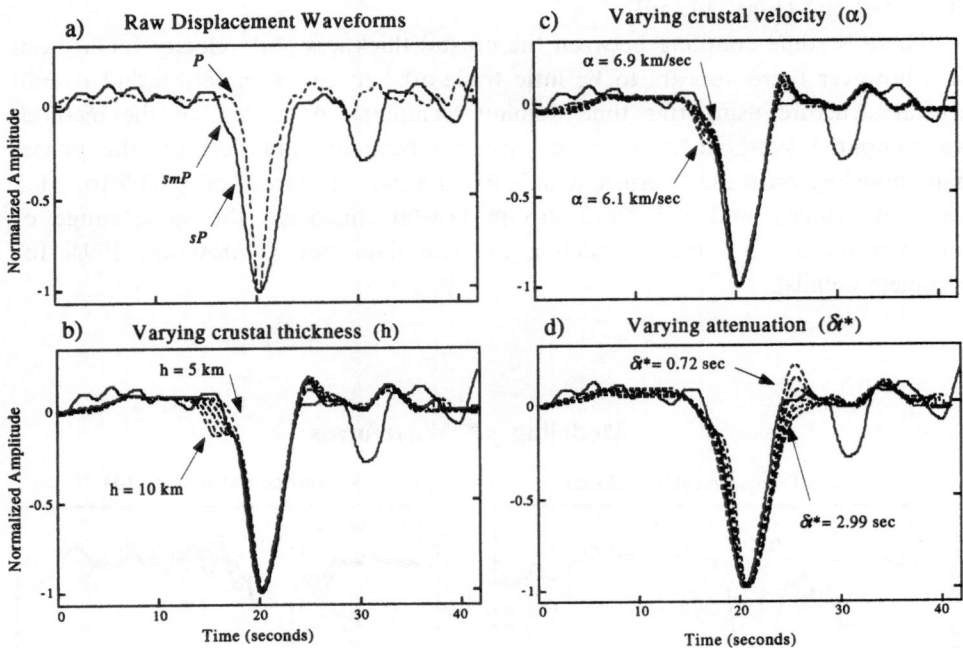

Figure 5

Example of wave-form modeling *sP*; the results of convolving the direct *P* pulse with different crustal transfer functions and attenuation operators to match the *sP* pulse. Data are stacked North American stations from Event 7 which bounce in the Lau Basin. a) Normalized displacement *P* (dashed) and *sP* (solid) wave forms showing the $s_m P$ precursory phase. Wave forms are superimposed to show the effects of varying the modeling parameters: b) crustal thickness $h = 5$ to 10 km in 1 km increments; c) crustal velocity $\alpha = 6.1$ to 6.9 km/s in 0.15 km/s increments (β and ρ vary accordingly); and d) differential attenuation $\delta t^* = 2.99$ to 0.72 s. The best fitting parameters for this example describe a two-layer crust overlain by 1.6 km of water, h, α, and δt^* are: 2.5 km at 4.5 km/s, 8.0 km at 6.9 km/s, and 1.26 s with $h_w = 1.6$ km.

Frequency Domain Measurement of Attenuation

We use the body-wave equalization method presented in FLANAGAN and WIENS (1994) to measure the differential attenuation (δt^*) between the *pP-P*, *sP-P*, and *sS-S* phase pairs. In this way the spectral ratio of *pP/P*, *sP/P*, *sS/S* provides a vertical average of the attenuation in the localized region above the earthquake source (see Fig. 1). The merit of this method lies in the canceling of unpredictable, common, frequency dependent factors (e.g., source, receiver function, and instrument response) by means of the spectral ratio and avoidance of geometrical effects on absolute amplitude by measuring the lograithmic frequency-derivative of the spectral ratio.

For a given phase-pair the far-field source spectrum, lower mantle propagation effects, and receiver response will be identical, thus the spectrum of the reflected $sP(\omega)$ and direct $P(\omega)$ wave forms is related by: $sP(\omega) = C \cdot P(\omega) \cdot R(\omega) \cdot A(\omega)$ where C, $R(\omega)$, $A(\omega)$ are defined as above. The complex attenuation operator $(A(\omega) = e^{-\omega dt^*/2})$ is obtained from a spectral division:

$$A(\omega) = \frac{sP(\omega)}{P(\omega) \cdot R(\omega) \cdot C}. \tag{1}$$

The natural logarithm of this spectral ratio is a linear function of frequency, and its slope yields the differential attenuation, δt^*:

$$\ln[|A(\omega)|] = \frac{\omega}{2}(t^*_{sP} - t^*_P) = \frac{\omega}{2}\delta t^*. \tag{2}$$

The δt^* estimates are measured in the frequency range 0.05 to 0.5 Hz, and can then be related to the quality factor (Q) by

$$\delta t^* = \int \frac{\Delta T}{Q} \tag{3}$$

where ΔT is the differential travel time between P and sP, and Q represents the frequency-independent quality factor of the medium.

An example of this spectral method is shown in Figure 6 using an sP-P phase pair from event 7. The broadband seismograms are first processed by deconvolving the instrument response over the bandwidth 0.01 to 2.0 Hz. The wave forms are then windowed and transformed into the frequency domain, and the spectra of the downgoing phase is multiplied by the crustal transfer function, obtained from modeling the crustal precursor as described above. The differential attenuation, δt^*, is then readily determined from the slope of the log-amplitude spectrum of the complex operator $A(\omega)$ (Fig. 6, bottom).

An optimum window size was chosen to give the clearest amplitude spectra, but because crustal multiples and water reverberations were present after some wave forms, the length of the time window varied. Spectral techniques are significantly less sensitive to the effect of crustal reflections (KANAMORI, 1967; REVENAUGH and JORDAN, 1991), a result corroborated by FLANAGAN and WIENS (1994) and SIPKIN and REVENAUGH (1994). As an additional check, we examine a random sample of amplitude spectra from several crustal transfer functions we computed to search for large spectral holes or other highly nonlinear features in our bandwidth of interest; indeed, the spectra are essentially flat. While frequency domain techniques appear more immune to the effects of crustal reverberations, this is only true if all the energy remains within the time window chosen for analysis. Thus our window selection aims to capture the crustal reverberations. There is also trade-off between a large window length with good stability but many spectral holes, and a short window length that eliminates spectral holes but can render the δt^* measurement

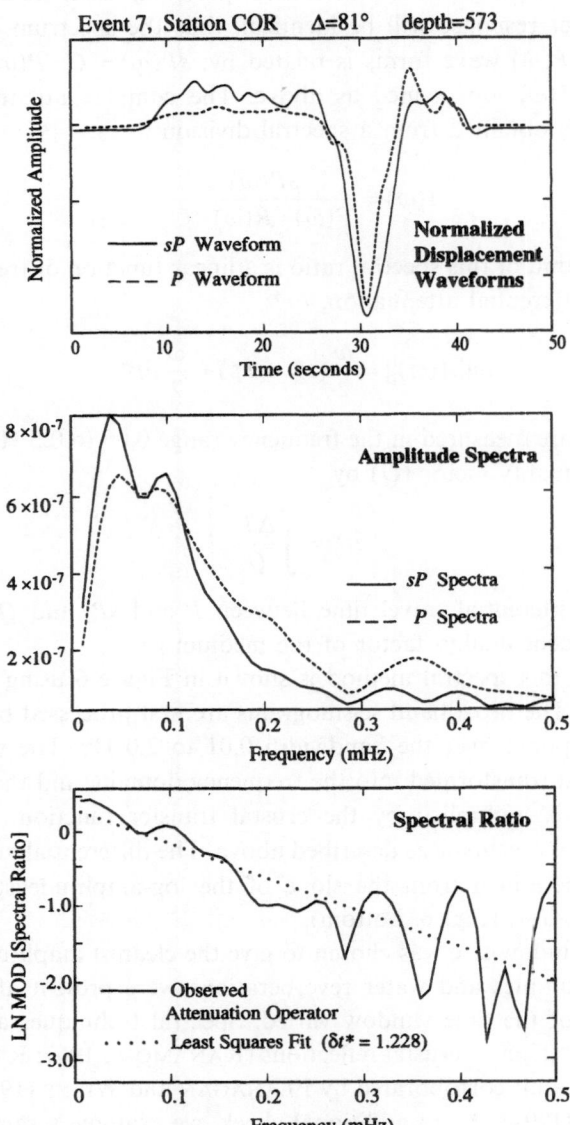

Figure 6

Measurement of δt^* for a broadband sP-P phase pair using the spectral ratio technique. (*top*) Wave forms with the instrument response removed are aligned to show the greater attenuation of the sP phase. A synthetic crustal response function has been applied to the P pulse to simulate the crustal interaction at the sP reflection point. (*center*) Amplitude spectra of the sP phase (solid trace) and P phase (dotted trace). (*bottom*) Natural log of the amplitude spectrum of the derived attenuation operator (solid) and the least-squares fit (dotted) to the slope which yields $\delta t^* = 1.228$.

highly variable. The quality of the δt^* determination is commonly checked by computing the phase spectrum of $A(\omega)$, but those for the broadband data were not as stable as those from the previous analyzed long-period data. Consequently we assume that the spectral stacking described below averages out many of these effects.

When measuring δt^* a spectral bandwidth must also be chosen; the frequency band selected is based not only on the instrument response function but on the SNR of the amplitude spectra of each phase. A reasonable choice is to select a bandwidth that includes as much of the usable spectrum as possible. This is accomplished by using a variable bandwidth that is based on the background noise level of the trace; Figure 7 illustrates this by comparing the spectra of the three different phase-pairs. Whereas the pP-P and sP-P phases contain significant energy up to 0.5 Hz, the sS-S energy falls to the noise level at approximately 0.2 Hz. Thus we cannot make all δt^* measurements over the same bandwidth and use 0.05–0.5 for pP and sP; and 0.05–0.2 Hz for sS. At this point we also look for systematic changes in the spectral decay slope which may indicate a frequency dependence of Q within the bandwidth of this study (i.e., if Q increases strongly with ω, it may be manifested as a concave upward shape of the spectra, BOWMAN, 1988). No such change is seen in either the individual or stacked spectra; they all appear to be quite linear. This lack of spectral curvature has been observed by both BOWMAN (1988) and ROTH et al. (1998) and indicates that within the narrow bandwidths of each study Q appears frequency independent, but when the results of several studies covering four decades of frequency are combined we see clear trends in the Q results (discussed below).

Rather than use individual phase-pair measurements demonstrated in Figure 6, we use the spectral stacking procedure outlined in FLANAGAN and WIENS (1994) to combine differential attenuation operators $A(\omega)$ of several phase pairs from a single event. Stacking increases the SNR, stabilizes the δt^* estimates, and helps to average out the effects of heterogeneity within the region under study. In this procedure we measure the individual complex operators $A(\omega)$ and normalize them to a common low-frequency amplitude prior to stacking. This normalization ensures that all attenuation operators are weighted equally in the stack, and aids in smoothing out any spectral holes which may be present in the individual spectra. An average δt^* for phase-pairs from a single event, having similar bounce points (e.g., Fig. 3) is then measured from the inverse slope of the log-amplitude spectrum of the stacked operators, as above.

All of the stacks show reasonably linear slopes, yet the phase spectra were not as stable as for the long-period data. As expected, the δt^* parameters measured from the stacked attenuation operators are intermediate to the individual observations, and we feel that the Q determined from the stacked traces are the more reliable.

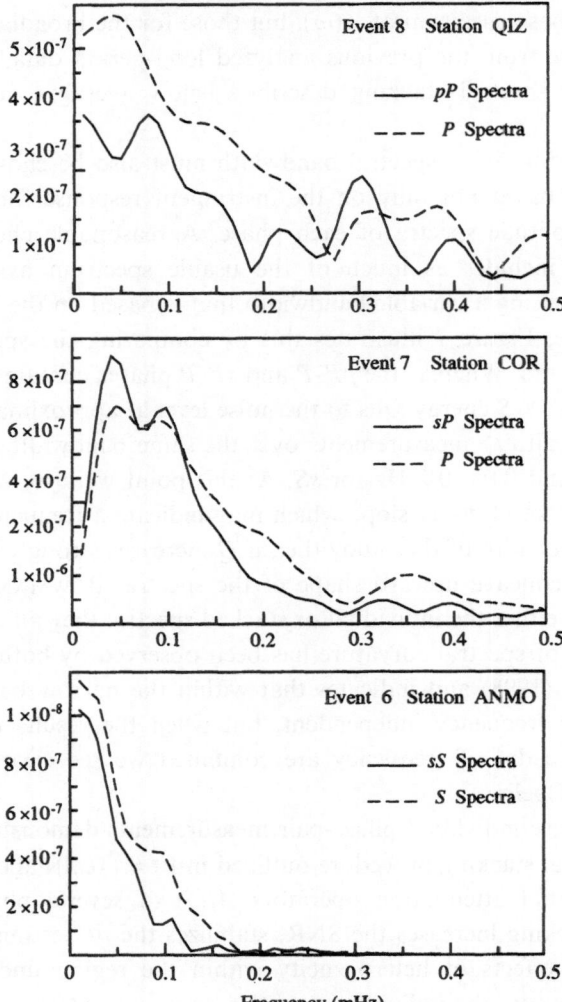

Amplitude Spectra of *P* and *S* Phase Pairs

Event 8   Station QIZ

—— *pP* Spectra
- - - *P* Spectra

Event 7   Station COR

—— *sP* Spectra
- - - *P* Spectra

Event 6   Station ANMO

—— *sS* Spectra
- - - *S* Spectra

Frequency (mHz)

Figure 7

Comparison of amplitude spectra showing the difference in frequency content of three broadband phase-pairs: *pP-P*, *sP-P*, and *sS-S*. The loss of energy at higher frequencies in the shear waves limits the δt^* measurement to 0.05–0.2 Hz while the vertical components contain energy up to 0.5 Hz. Also, note the common structure in the spectra of the three phase pairs, indicating uniform source spectra for both the upgoing (depth phases *pP*, *sP*, or *sS*) and downgoing phases (direct *P* or *S*).

Observations of P- and S-wave Attenuation

Quantitative analysis has been performed on the 22 wave-form stacks shown in Figures 8 and 9, and the δt^* measurements are summarized in Table 2. As dictated by event-receiver geometries, we do not have even sampling of the entire Lau backarc region. Most of the ray paths sample the northern Lau Basin and northern Lau Ridge and to some extent the Fiji Islands; there is also limited coverage of the central and southern part of the Lau Basin (Figs. 8 and 9). Typically, we were able to make two stacks for any given event: phases with reflection points at azimuths to North American stations, and phases with reflection points at azimuths to Asian stations. The detailed velocity structures derived from modeling the crustal precursors are presented in FLANAGAN (1994); the resulting crustal thicknesses are listed in Table 2 and discussed only briefly here.

A visual inspection of the stacked wave forms which sample the remnant arc and inner basin demonstrates significant differences with regard to the timing and

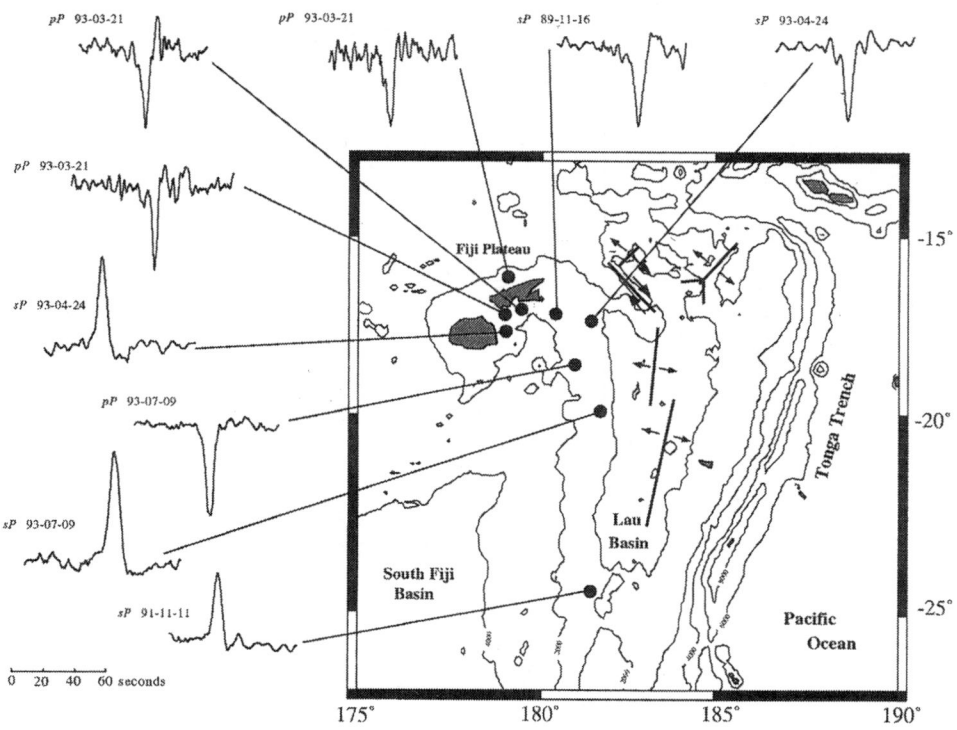

Figure 8

Mercator projection of the Tonga-Fiji region showing the surface bounce points (black filled circles) and wave-form stacks of *pP*, *sP*, and *sS* used for measuring δt^* and modeling of the crustal structure. The crustal precursor s_mP or p_mP appears quite strong under the thick Fijian crust (~ 20 km) and decreases slightly to the south along the Lau Ridge (crust $13 \sim 15$ km thick). Bathymetry is shown at 2000 m depth intervals; tectonic features (bold lines with arrows) taken from PARSON and HAWKINS (1994) and HAWKINS (1995).

Table 2

δt^* Measured from stacked spectra of broadband sS, sP, and pP wave forms

Event	Stations/ # records	Phase pair	Depth (km)	Q Time domain	Q Stacked spectra	Q Uncertainty range†	δt^* (sec)	Bandwidth (Hz)	Region sampled	Crustal thickness (km)
4	N.A./2	sS-S	194.0	80	45	32–80	1.837	0.01 – 0.2	S. Basin	9.1
6	N.A./3	sS-S	278.0	80	67	47–119	1.707	0.01 – 0.2	C. Basin	7.0
10	N.A./5	sS-S	412.0	100	90	63–155	1.840	0.01 – 0.2	C. Basin	8.5
2	Asia/2	sP-P	232.0	100	100	84–124	0.7986	0.01 – 0.5	N. Basin	7.0
6	N.A./7	sP-P	278.0	100	98	75–140	0.9671	0.01 – 0.5	C. Basin	7.0
10	Asia/12	sP-P	412.0	80	135	111–189	0.9966	0.01 – 0.5	Lau Ridge	17.5
1	Asia/3	sP-P	531.0	120	176	145–224	0.9646	0.01 – 0.5	Fiji	15.0
11	N.A./2	sP-P	556.0	120	154	125–201	1.132	0.01 – 0.5	N. Basin	7.5
7	N.A./4	sP-P	573.0	140	154	129–190	1.149	0.01 – 0.5	N. Basin	10.5
8	N.A./2	sP-P	584.0	140	178	155–211	1.022	0.01 – 0.5	N. Basin	9.5
9	N.A./2	sP-P	600.0	160	204	169–255	0.9245	0.01 – 0.5	Lau Ridge	13.0
9	Asia/5	sP-P	600.0	120	177	156–221	1.058	0.01 – 0.5	Fiji	13.0
10	Asia/12	pP-P	412.0	160	174	153–218	0.5083	0.01 – 0.5	Lau Ridge	15.5
11	N.A./11	pP-P	556.0	300	303	256–369	0.4165	0.01 – 0.5	N. Basin	8.0
8	Asia/2	pP-P	584.0	200	292	261–334	0.4119	0.01 – 0.5	Fiji	20.0
9	N.A./9	pP-P	600.0	300	316	293–347	0.3939	0.01 – 0.5	Lau Ridge	17.0

† Uncertainties are one standard deviation $(Q^{-1} \pm 1\sigma)^{-1}$ based on the linear regression of stacked spectra.

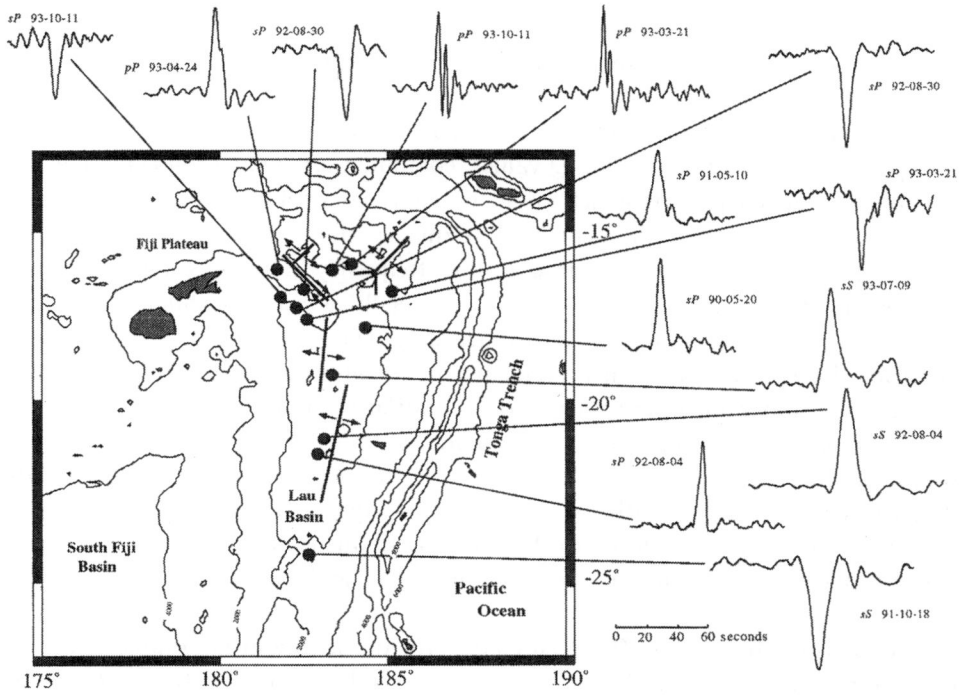

Figure 9

Mercator projection of the Tonga-Fiji region showing the surface reflection points (black-filled circles) and wave-form stacks of *pP*, *sP*, and *sS* used for measuring δt^* and modeling of the crust beneath the central and northern Lau Basins. The crustal precursors are small due to the thin oceanic crust in the basin, yet consistent velocity structures are found from modeling which show a thickness of about 7 km in the central basin, increasing to 11 km toward the north. Note the more broad appearance of these wave-form stacks relative to those in the previous figure, indicating significantly higher attenuation beneath the Lau Basin compared to Fiji. Bathymetry is shown at 2000 m depth intervals; tectonic features (black lines with arrows) taken from PARSON and HAWKINS (1994) and HAWKINS (1995).

amplitude of the Moho reflectors and the obvious differences in attenuation. Figure 8 shows the bounce points and the stacked wave forms used in the modeling of phases bouncing at the Fiji Plateau and the Lau Ridge. Events 1, 8 and 9 have *pP* and *sP* wave forms which sample Fiji most directly, and yield somewhat consistent crustal structures which range in total thickness from up to 20 km directly beneath Fiji, thinning to 13 km towards the northern Fiji Basin and to 15 km towards the east in the transition to the Lau Ridge. The best-fitting models for the northern, central, and southern Lau Basins have overall crustal thickness of 7 to 11 km (e.g., Events 6 and 7). The northern Lau Basin is perhaps the most densely sampled and the results from modeling several types of stacked wave forms yield excellent agreement of crustal structure. For pP bounces beneath a substantial water layer (i.e., greater than 2000 m), prominent later arrivals are produced (e.g., Events 8, 9, and 11) but can be modeled remarkably well using the known bathymetry and the time domain technique.

We obtain δt^* measurements for three sS-S pairs, four pP-P pairs, and nine sP-P pairs, and the corresponding Q values, computed using equation (3), are listed in Table 2. Stacked records which could be analyzed with both methods showed good agreement between the time and frequency domain Q results. The time domain Q estimates are often close to the frequency domain values although typically do not have as good resolution especially when Q is high (i.e., when attenuation is low it is more difficult to measure). The δt^* values listed in Table 2 are from the spectral ratio measurements along with the bandwidth used, the number of seismograms in the stack and the station-group (N.A. or Asia), and the region sampled by the cluster of bounce points of (see Figs. 8, 9). Standard errors computed from the linear regression (Fig. 6, bottom) are also listed. For ease of comparison to an earlier work we have chosen to quote values of Q, rather than Q^{-1}, the actual quantity determined by both the frequency and time methods such that the uncertainty bounds of Q are asymmetric and stated as: $(Q^{-1} \pm 1\sigma)^{-1}$ (Table 2).

The interpretation of the Q measurements for the sS-S and pP-P phase pairs is straightforward as we are able to measure Q_β and Q_α directly from δt^* using equation (3). If we have Q observations for both S and P waves from the same event (hence, sampling the same depth range) and with similar bounce point locations, then we can compute our own empirical Q_α/Q_β ratio. The only possible data we have for such a calculation is the pP-P and sS-S phase pairs of event 10 (Table 2), which give a ratio of $Q_\alpha/Q_\beta = 1.93$. Since the pP bounces in Lau Ridge and the sS bounces in the Lau Basin, this ratio may be slightly high. Previous models of Q_β structure in this region show that for the same depth range (412 km), seismic waves sampling the Lau Basin will have a lower vertically integrated Q than waves sampling the Lau Ridge due to the strong lateral variations in the shallow mantle Q_β structure. Therefore, the Q_β of 90 is likely lower than it it was measured from an sS-S phase pair sampling the Lau Ridge (as the pP phase does) and thus makes the $Q_\alpha/Q_\beta = 1.93$ slightly high. In fact, this ratio may be lower and closer to the value of 1.75 found by ROTH et al. (1998) using locally recorded P and S waves, however, further teleseismic data are required to corroborate this. Ideally, pP and sS waves from the same event and having the same bounce point location are needed to make a more accurate measure of the Q_α/Q_β ratio.

If dissipation is all in the shear modulus the ratio Q_α/Q_β takes a value of 9/4 (discussed in the next section), however, if the dominant dissipation mechanism is thermoelastic or is due to relaxation in the bulk modulus associated with a first-order phase change then the ratio could be substantially lower (ANDERSON, 1989). There is a suggestion that Q_α approaches Q_β at short periods (CLEMENTS,

1982), and in any case the ratio of 9/4 gives a minimum estimate of Q_β and maximum estimate of Q_α. This might suggest that shear dissipation has a lesser effect at shorter periods, but this convergence of compressional and shear attenuation at short periods could also be due to the scattering of *P*-wave energy (CORMIER, 1982).

Interpretation of Q_β and Q_α Measured from sP-P Phase Pairs

The interpretation of the differential attenuation of the *sP-P* phase pairs requires some analysis because *sP* propagates both as an *S* wave and *P* wave (see Fig. 1). In order to retrieve a physically meaningful *Q* estimate from these δt^* measurements we must make assumptions about the relationship between Q_β and Q_α.

The results of several studies have indicated that compressional losses in the earth are negligible compared to shear losses (SAILOR and DZIEWONSKI, 1978; DZIEWONSKI and ANDERSON, 1981). The lack of bulk or volume attenuation likely indicates that shear mechanisms dominate the attenuation process, particularly in the upper mantle. We adopt the hypothesis of no bulk dissipation (i.e., $Q_K = \infty$, where K is the bulk modulus) (ANDERSON and ARCHAMBEAU, 1964) which yields the relationship: $Q_\beta/Q_\alpha = 4/3(\beta^2/\alpha^2)$. If we further assume an isotropic Poisson solid ($\alpha = \sqrt{3}\beta$) the relationship becomes: $Q_\alpha = 9/4 Q_\beta$. Recall that δt^* represents the difference in attenuation between any two phases and can be expressed as the path integral:

$$\delta t^* = t_{sP}^* - t_P^* = \int_{sP} \frac{dT}{Q} - \int_P \frac{dT}{Q}. \tag{4}$$

The path by which the *sP* and *P* phases differ is the up- and downgoing legs of *sP* above the focal depth. Recall that this phase travels as an *S* wave on the upward segment and as a *P* wave on the downward segment of the path such that for a focal depth of 600 km:

$$\delta t^* = \int_{600\ km}^{0\ km} \frac{dT_S}{Q_\beta} - \int_{0\ km}^{600\ km} \frac{dT_P}{Q_\alpha}. \tag{5}$$

To retrieve the Q_β values we combine the above relationships which effectively reduces the problem back to a single unknown (Q_β):

$$\delta t^* = \int_{600\ km}^{0\ km} \frac{dT_S}{Q_\beta} - \frac{4}{9} \int_{0\ km}^{600\ km} \frac{dT_P}{Q_\beta}. \tag{6}$$

The integrated travel times for the upgoing (*S* wave) and downgoing (*P* wave) legs are calculated by ray tracing through PREM (DZIEWONSKI and ANDERSON, 1981). The Q_β values are then calculated from the nine δt^*_{sP-P} measurements using equation (6) and are listed in Table 3; the Q_α are determined simply by multiplying

Table 3

Q_β and Q_α calculated from $sP-P$ phase pairs

| Event | Depth (km) | Q_sP | δt^*_{sP} (sec) | $Q_\alpha/Q_\beta = 1.93$ | | $Q_\alpha/Q_\beta = 2.25$ | | Region sampled |
				Q_β	Q_α	Q_β	Q_α	
2	232.0	100	0.7986	79	154	77	174	N. Basin
6	278.0	98	0.9671	76	145	74	167	C. Basin
10	412.0	135	0.9966	107	207	103	232	Lau Ridge
1	531.0	176	0.9646	139	268	134	302	Fiji
11	556.0	154	1.132	118	228	115	260	N. Basin
7	573.0	154	1.149	122	235	117	264	N. Basin
8	584.0	178	1.022	138	266	135	304	N. Basin
9	600.0	204	0.9245	161	309	155	349	Lau Ridge
9	600.0	177	1.058	139	267	134	301	Fiji

Q_β by the factor 9/4 and are listed for comparison with *pP-P* observations. Also listed in Table 3 for comparison are Q_β estimates calculated using a Q_α/Q_β ratio of 1.93; the value determined in this study. As expected, the Q values determined from the *sP-P* are intermediate to those measured from *sS-S* and *pP-P* phase pairs. It is difficult to determine which of the Q_α/Q_β ratios produces values closest to those measured directly. Often both computed values of Q_β and Q_α in Table 3 are within the error bounds of the Q observations listed in Table 2, and the data are too few to make a conclusive statement at this time.

A plot of Q_β and Q_α against focal depth shows that the deeper events have a lower mean path attenuation than the shallower events (Fig. 10). We observe an increase of Q with increasing depth as the values from the deep events are larger than from the shallow events, but our data are too sparse to perform a formal inversion for three-dimensional Q structure as in (FLANAGAN and WIENS, 1998; ROTH *et al.*, 1998). Although a detailed interpretation of lateral variations in Q would be premature, it is encouraging that higher Q values are seen for bounce points under Fiji and the Lau Ridge relative to the lower Q values found in the Lau Basin even though some of the error bars indicated in Figure 10 are large for some measurements.

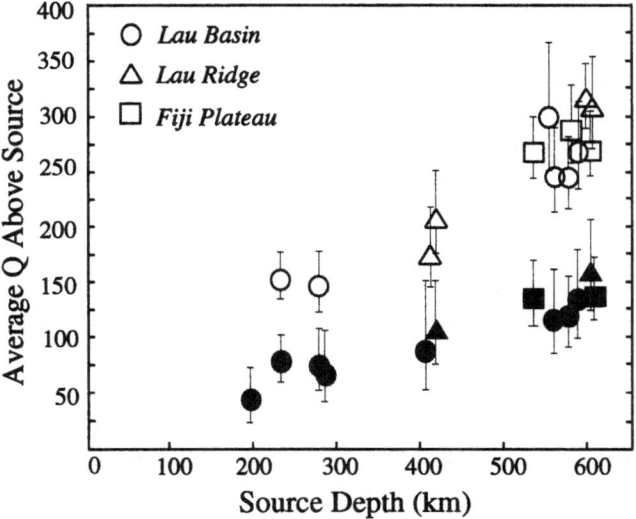

Figure 10

Our 16 individual broadband Q measurements plotted as a function of source depth. Both Q_β (filled) and Q_α (open) increase with increasing focal depth in agreement with previous Q studies, indicating attenuation decreases rapidly with depth in the upper mantle. The error bounds represent the uncertainties calculated from Table 2.

Table 4

Comparison of long-period and broadband Q_β and Q_α measurements

Depth (km)	Long-period Q_β† (0.01–0.083 Hz)	Broadband Q_β (0.01–0.20 Hz)	Broadband Q_α (0.01–0.50 Hz)	Region sampled
0–200	25–38	45–57	102–121	Lau Basin
0–400	46–53	90	–	Lau Basin
0–600	72–87	118–138	303	Lau Basin
0–200	53–72	–	–	Fiji/Lau Ridge
0–400	66–79	107	174–207	Fiji/Lau Ridge
0–600	88–104	139–161	267–316	Fiji/Lau Ridge

† FLANAGAN and WIENS (1998).

Discussion: Evidence for Frequency Dependent Q

The broadband Q_β values presented here are higher than the long-period Q_β models derived in FLANAGAN and WIENS (1998) for the Lau backarc by about a factor of two but do show similar trends both laterally and with depth. Both the direct Q measurements and the interpreted Q values are summarized in Table 4 along with the long-period results, and together they demonstrate the trends with depth. The upper 200 km beneath the central and northern Lau Basins display a Q_β of 45 to 57 and the upper 400 km display a Q_β of 90, both measured directly from sS-S phase pairs (Table 2, Fig. 10); the upper 600 km beneath the basin has a Q_β of 118–138 calculated from the sP-P phase pairs using $Q_\alpha/Q_\beta = 1.93$ (Table 3). The Lau Ridge and Fiji Islands show slightly lower attenuation (higher Q) with a Q_β of 107 in the upper 400 km and Q_β of 139–161 in the upper 600 km, all calculated from sP phases. We have only one direct pP-P measurement of the Lau Basin yielding $Q_\alpha = 303$ in the upper 556 km and three measurements for the Lau Ridge and Fiji Plateau giving a Q_α of 267–316 in the upper 600 km of the mantle and a Q_α of 174–207 in the upper 412 km.

Because our broadband measurements are few, we do not attempt to derive a Q_α or Q_β model for the region. The differences in the results of the study presented here and in FLANAGAN and WIENS (1998) are attributed to the differing frequency bandwidth of the data sets, and this comparison suggests that an increase in Q occurs at frequencies less than 1.0 Hz. In an attempt to quantify this dependence we combine our results with other Q measurements made in different bandwidths and examine them as a function of frequency.

Other investigations of the attenuation in the Tonga Subduction Zone are numerous and involve both S- and P-wave measurements in a variety of frequency bands. Whereas these studies vary significantly in their approach, observations of reduced amplitudes, loss of high frequencies, and delayed arrival times of both P and S phases traveling through the Lau backarc are remarkably robust. The absolute values of Q_α and Q_β reported by these studies are quite different, but trends in the variation of Q both laterally and with depth are clear. We summarize in Table 5 and Figure 11 the published Q results of several studies of surface waves (BUSSY et al., 1993; DUREK et al., 1993; ROMANOWICZ, 1995), teleseismic body waves (REVENAUGH and JORDAN, 1991; FLANAGAN and WIENS, 1998; BHAT-TACHARYYA et al., 1998), and locally recorded body waves (BOCK and CLEMENTS, 1982; BOWMAN, 1988; ROTH et al., 1998) in the Fiji-Tonga area. While sensitivity to anelastic structure is different for surface waves (good radial resolution) and body waves (good lateral resolution) data sets, data at overlapping frequencies can be used together to develop Q models more complete than with individual data sets alone.

For the greatest consistency, we plot Q results from these studies which are restricted to the Fiji Plateau and represent the vertically averaged Q in the entire upper 600 km of the mantle (Fig. 11, Table 5). This sometimes required calculating Q by ray tracing through the layered models, or extracting regional values from published three-dimensional maps, as in the case of the surface waves models. This

Table 5

Summary of Tonga Q results at different frequencies

Shear Wave Attenuation:

Q_β

Depth range 0–400 km	Depth range 0–600 km	Center Frequency (Hz)	Bandwidth (Hz)	Reference
120–340	170–370	3.0[§]	0.500–7.6	BOWMAN (1988)
–	–	2.0	0.500–3.5	BOCK and CLEMENTS (1982)
116–169	201–215	1.0	0.300–1.5	ROTH et al. (1998)
103	139–161	0.3	0.050–0.5	this study
66–79	88–104	0.05	0.010–0.083	FLANAGAN and WIENS (1998)
81–97	106	0.03	0.020–0.1	BHATTACHARYYA et al. (1998)
–	127	0.025	0.005–0.06	REVENAUGH and JORDAN (1991)
85	197	0.01	0.005–0.016	BUSSY et al. (1993)
171	332	0.005	0.004–0.008	ROMANOWICZ (1995)
136	194	0.004	0.003–0.006	DUREK et al. (1993)

Compressional Wave Attenuation:

Q_α

Depth range 0–400 km	Depth range 0–600 km	Center frequency (Hz)	Bandwidth (Hz)	Reference
203–481	399–1020	3.0[§]	1.00–7.6	BOWMAN (1988)
120–200	370–560	2.0	0.50–3.5	BOCK and CLEMENTS (1982)
203–297	352–376	1.5	0.75–3.5	ROTH et al. (1998)
174–207	267–316	0.3	0.05–0.5	this study

§ Center frequency not reported by authors.

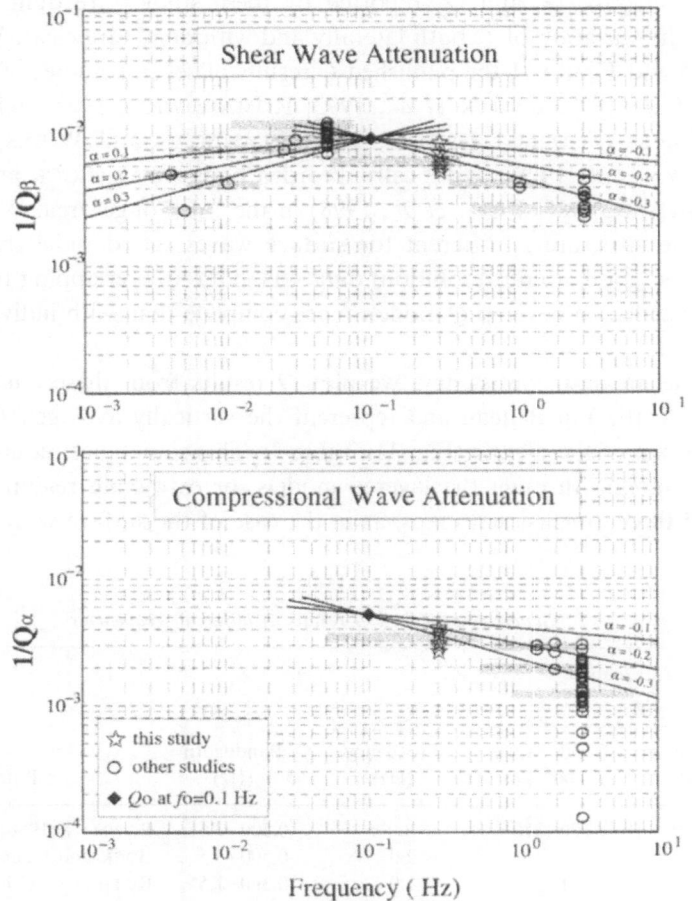

Figure 11

Summary of constraints on the frequency dependence of Q in the Tonga-Fiji region. Though they are plotted as attenuation ($1/Q$), the values represent the vertically averaged Q_β or Q_α in the upper 600 km of the mantle beneath the Fiji Plateau only (to avoid bias from the extremely low Q in the Lau backarc). Gray bars indicate bandwidth over which the Q measurement is made in each study while the $1/Q$ values themselves are plotted at the center frequency reported in Table 3. The solid lines represent predictions of a power-law expression for Q of the form $Q^{-1}(\omega) = Q_0^{-1}(\omega_0)(\omega/\omega_0)^{-\alpha}$ where $\alpha = \pm 0.1, \pm 0.2, \pm 0.3$ as labeled, and $Q_0 = 100$ for shear or $Q_0 = 190$ compression. We observe significant frequency dependence of Q_β well below 1.0 Hz; Q_α however, is inconclusive.

is done to minimize the trade-off between lateral and depth variations in Q within Tonga-Fiji (e.g., the anomalously low Q of the backarc basin) as well as variations between this and other subduction zones or oceanic and continental settings; as a result, we must exclude several high-quality studies of other geographic regions. The Q_α and Q_β are plotted at their center frequency but are actually measured over a range of frequencies (indicated by the gray bars); the complete list of these data is given in Table 5. The center frequency was provided by the authors of each study (typically given as the dominant period of their data) except in one case. Thus for the BOWMAN (1988) results we simply choose the midpoint in log-frequency as the center frequency. Although Q in a shallower portion of the uppermost mantle is given in Table 5 (0–400 km), we plot only the whole upper mantle Q (0–600 km) in Figure 11. The other data are listed to demonstrate the striking correlations between these different studies.

The theory of anelastic attenuation on mantle materials has been investigated by several workers. LIU *et al.* (1976) and MINSTER and ANDERSON (1981) proposed that distributions of relaxation times ($\tau_0 = 1/\omega_0$) in an anelastic material yield the weak frequency dependence of internal friction (Q^{-1}) observed in the earth but that this dependence may be difficult to estimate because of the trade-off with the depth variation of Q^{-1}. ANDERSON and GIVEN (1982) proposed an absorption band Q model with more complex frequency dependence including a broad peak of attenuation in the upper mantle where $Q(\omega)$ is proportional to ω^α in the band, and proportional to ω and $1/\omega$ at higher and lower frequencies, respectively. As an example to compare with the data plotted in Figure 11, we follow the power-law formulation of ANDERSON and MINSTER (1979) to describe the variation of Q with frequency: $Q^{-1}(\omega) = Q_0^{-1}(\omega_0)(\omega/\omega_0)^{-\alpha}$ where $Q_0^{-1}(\omega_0)$, ω_0, and α are constant. By the form of this equation, as ω both increases and decreases, $Q^{-1}(\omega)$ becomes small with a maximum value at the reference frequency $Q_0(\omega_0)$.

We plot attenuation ($1/Q$) against frequency since, in these coordinates, a single frequency dependence renders a straight line where the slope is the parameter α. For the shear-wave attenuation (Figure 11, top) the trend illustrates that both the high- and low-frequency observations show less attenuation (higher Q_β) than the intermediate-frequency body wave studies, and the inflection appears to be between 0.05 and 0.1 Hz. The solid lines are predictions from the power-law model where the parameter α is equal to ± 0.1, ± 0.2, ± 0.3 as labeled, $\omega_0 = 0.1$ Hz, and $Q_0(\omega_0) = 100$ for shear or $Q_0(\omega_0) = 190$ compression. The reference frequency ω_0 depends on the overall bandwidth in question as well as location in the earth (i.e., upper versus lower mantle). Previous investigations of the frequency dependence of Q assert that any increase in Q at high frequencies does not begin until the frequency exceeds about 1.5 Hz (KANAMORI, 1967; BURDICK, 1985; SIPKIN, 1988), and in such modeling the reference frequency is assumed to be 1.0 Hz. Given that we are examining the upper 600 km of the mantle between 0.001 and 10 Hz, we choose $\omega_0 = 0.1$ Hz, although changing the starting values Q_0 and ω_0 will not affect the slopes α in any case. The trend of the data in Figure 11 implies a change in behavior of attenuation with frequency well below 1.0 Hz. Note at low frequencies, $1/Q$ is proportional to ω and at high frequencies $1/Q$ is proportional to $1/\omega$ just as absorption band models predict (e.g., ANDERSON and GIVEN, 1982). Although we plot the predictions for Q_α as well (Figure 11, bottom), the lack of data at frequencies below 0.1 Hz makes any evaluation inconclusive. It is encouraging, however, to see that below 0.1 Hz the Q_α does show an overall frequency dependence with $\alpha = -0.2$ to -0.3.

We do not assert that the frequency dependence is entirely described by this simple model, as it is likely that a distribution of relaxation peaks exists and perhaps an absorption band model of Q would follow the trends just as well. The data in Figure 11 are appreciably scattered and may give us an indication of the overall spatial resolution of the various data sets. For example, the Q^{-1} at longer periods is taken from models with larger averaging lengths than those measured in this study. This fact could mean that Q^{-1} is underestimated due to some accidental sampling of the strongly attenuating Lau Basin which would lead to erroneously larger absolute α values. The strongest observation seen by viewing the data collectively is that Q changes as a function of ω at less than 1.0 Hz, a result not previously reported.

Comparison with Mineral Physics Experiments

Attempts to use seismic results to characterize the state of the earth's interior have been hampered by the lack of relevant experimental data taken under controlled laboratory conditions, especially at the low frequencies of teleseismic wave propagation (e.g., KARATO and SPETZLER, 1990). Often such experimentally determined high-frequency relationships must be extrapolated to the seismic band, however recent improvements in analytic techniques yield new empirical relationships for attenuation at seismic frequencies which can be compared with our seismic observations.

The results of GETTING et al. (1997) show that shear attenuation is quite small at high frequency (10 Hz) and low temperature (1000 K), but increases with increasing temperature and decreasing frequency. They model the attenuation as a thermally activated process with a power-law frequency dependence (BERCKHEMER et al., 1982), and their experimentally determined parameter of $\alpha = 0.25$ for $\omega = 0.003–30$ Hz. GUEGUEN et al. (1989), in experiments with forsterite single crystals, found $Q^{-1} \sim \omega^{-0.2}$ at high temperature and seismic frequencies, while SATO et al. (1989), in a study at high temperature and pressure but ultrasonic frequencies, found a weaker frequency dependence of $\alpha < -0.2$. JACKSON and PATERSON (1993) made measurements at 1000°C and 300 MPa over the frequency range 0.03 to 1 Hz, and found yet a weaker dependence with $\alpha = 0.19$ to 0.15.

Overall these estimates of the frequency dependence parameter (α) are consistent with the trends observed in the $1/Q$ values plotted in Figure 11, as the data lie within the experimentally determined range of $0.15 < \alpha < 0.3$. However, physical mechanisms of internal friction are not well understood or quantified, and this prevents an unambiguous application of laboratory data to earth materials. More work needs to be accumulated in both the seismic observations and experimental constraints before more advanced relationships between mechanisms of attenuation and well-characterized seismic waves can be interpreted. Quantifying the intrinsic attenuation and relating it to more fundamental material properties such as thermochemical state, stress state, volatile content, and crystal lattice defects (including melts) will ultimately help tell us about the tectonic history of the upper mantle in this subduction environment.

Summary

Although the data coverage in this study is limited, there is a clear indication of variation in Q_β and Q_α for different propagation paths in the Tonga-Fiji region. The similarity of trends observed among several different studies indicates that the teleseismic wave-form method is capable of imaging upper mantle anelastic structure, and now that the broadband data span more than three decades in frequency they should be helpful in evaluating the various mechanisms of attenuation proposed for seismic waves (e.g., JACKSON and ANDERSON, 1970; ANDERSON and MINSTER, 1979; MINSTER and ANDERSON, 1981). With regard to attenuation beneath active backarc basins, the use of *pP-P* and *sS-S* waves together will help to define zones of partial melt as the geometry and concentration of interstitial melt will affect shear and compressional waves differently.

The intermediate frequency content of teleseismic broadband data can place strong constraints on possible Q structures in the upper mantle. However, the data set used in this study lack sufficiently high SNRs to examine the frequency dependence of attenuation (especially for *S* waves) extensively between 0.05 and 0.5 Hz, where it appears to vary the most. Further observations would provide additional information in this crucial frequency band, and the use of *P* and *S* phases in concert will allow a more direct measurement of the Q_P/Q_S ratio (1.93 in this study) which may vary for different mechanisms of attenuation and/or with different geographic locations. Nonetheless the results presented here, while preliminary, provide compelling evidence of a variation of Q with frequency between 0.1 and 1.0 Hz, and having the power-law form: $Q \propto \omega^\alpha$ with α between -0.1 and -0.3, consistent with experimentally derived values.

Acknowledgments

We thank J. Bhattacharyya, R. Bowman, J. Durek, B. Romanowicz, E. Roth, A. Sheehan, S. Sipkin, and M. Weber for many individual discussions about all aspects of this subject. Comments from H. Bungum and an anonymous reviewer greatly improved the clarity of this manuscript. This research was partially supported by the National Science Foundation under grant EAR 99117669; M.P.F. is supported by an NSF Postdoctoral Fellowship, and the Cecil H. and Ida M. Green Foundation.

REFERENCES

ANDERSON, D. L., *Theory of the Earth* (Blackwell Scientific, Oxford 1989) 366 pp.

ANDERSON, D. L., and ARCHAMBEAU, C. B. (1964), *The Anelasticity of the Earth*, J. Geophys. Res. *69*, 2071–2084.

ANDERSON, D. L., and GIVEN, J. W. (1982), *Absorption Band Q Model for the Earth*, J. Geophys. Res. *87*, 3893–390.

ANDERSON, D. L., and MINSTER, J. B. (1979), *The Frequency Dependence of Q in the Earth and Implications for Mantle Rheology and Chandler Wobble*, Geophys. J. R. Astron. Soc. *58*, 431–440.

BACHE, T. C., BRATT, S. R., and BUNGUM, H. (1986), *High-frequency P-wave Attenuation from Five Teleseismic Paths from Central Asia*, Geophys. J. R. Astr. Soc. *85*, 505–522.

BERCKHEMER, H., KAMPFMANN, W., AULBACH, E., and SCHMELING, H. (1982), *Shear Modulus and Q of Forsterite and Dunite near Partial Melting from Forced-oscillation Experiments*, Phys. Earth Planet. Int. *29*, 30–41.

BHATTACHARYYA, J., MASTERS, G., and ROMANOWICZ, B. (1998), *Global Lateral Variations of Shear-wave Attenuation in the Mantle*, J. Geophys. Res., in preparation.

BOCK, G., and CLEMENTS, J. R. (1982), *Attenuation of Short-period P, PcP, ScP, and pP Waves in the Earth's Mantle*, J. Geophys. Res. *87*, 3905–3918.

BOWMAN, J. R. (1988), *Body-wave Attenuation in the Tonga Subduction Zone*, J. Geophys. Res. *93*, 2125–2139.

BURDICK, L. J. (1985), *Estimation of the Frequency Dependence of Q from ScP and ScS Phases*, Geophys. J. R. Astr. Soc. *80*, 35–55.

BUSSY, M., MONTAGNER, J. P., and ROMANOWICZ, B. (1993), *Tomographic Study of Upper Mantle Attenuation in the Pacific Ocean*, Geophys. Res. Lett. *20*, 663–666.

CLEMENTS, J. R. (1982), *Intrinsic Q and its Frequency Dependence*, Phys. Earth Planet. Int. *27*, 286–299.

CORMIER, V. F. (1982), *The effect of Attenuation on Seismic Body Waves*, Bull. Seismol. Soc. Am. *72*, S169–S200.

DER, Z. A., McELFRESH, T. W., and O'DONNELL, A. (1982), *An Investigation of the Regional Variations and Frequency Dependence of Anelastic Attenuation in the Mantle under the United States in the 0.5–4.5 Hz Band*, Geophys. J. R. Astron. Soc. *69*, 67–100.

DING, X.-Y., and GRAND, S. (1993), *Upper Mantle Q Structure Beneath the East Pacific Rise*, J. Geophys. Res. *98*, 1973–1985.

DUREK, J. J., RITZWOLLER, M. H., and WOODHOUSE, J. H. (1993), *Constraining Upper Mantle Anelasticity Using Surface Wave Amplitude Anomalies*, Geophys. J. Int. *114*, 249–272.

DZIEWONSKI, A. M., and ANDERSON, D. L. (1981), *Preliminary Reference Earth Model*, Phys. Earth Planet. Int. *25*, 297–356.

FLANAGAN, M. P., *Upper Mantle Attenuation of Back Arc Basins Using Differential Shear Wave Measurements*, Ph.D., 229 pp. (Washington University, St. Louis, MO, 1994).

FLANAGAN, M. P., and WIENS, D. A. (1990), *Attenuation Structure Beneath the Lau Backarc Spreading Center from Teleseismic S Phases*, Geophys. Res. Lett. *17*, 2117–2120.

FLANAGAN, M. P., and WIENS, D. A. (1994), *Radial Upper Mantle Attenuation Structure of Inactive Backarc Basins from Differential Shear Wave Attenuation Measurements*, J. Geophys. Res. *99*, 15,469–15,485.

FLANAGAN, M. P., and WIENS, D. A. (1998), *Inversion for Three-dimensional Shear Wave Attenuation Structure of the Lau Backarc Basin*, Phys. Earth Planet. Int., in preparation.

FUCHS, K., and MÜLLER, G. (1971), *Computation of Synthetic Seismograms with the Reflectivity Method and Comparison with Observations*, Geophys. J. R. Astr. Soc. *23*, 417–433.

FUTTERMAN, W. I. (1962), *Dispersive Body Waves*, J. Geophys. Res. *67*, 5279–5291.

GETTING, I. C., DUTTON, S. J., BURNLEY, P. C., KARATO, S.-i, and SPETZLER, H. A. (1997), *Shear Attenuation and Dispersion in MgO*, Phys. Earth Planet. Int. *99*, 249–257.

GUEGUEN, Y., DAROT, M., MAZOT, P., and WOIRGARD, J. (1989), Q^{-1} *of Forsterite Single Crystals*, Phys. Earth Planet. Int. *70*, 254–258.

HAWKINS, J. W., *Evolution of the Lau Basin—Insights from ODP Leg 135*. In *Active Margins and Marginal Basins of the Western Pacific* (eds. B. Taylor and J. Natland) (American Geophys. Union, Washington D.C. 1995) pp. 125–173.

JACKSON, D. D., and ANDERSON, D. L. (1970), *Physical Mechanisms of Seismic Wave Attenuation*, Rev. Geophys. Space Phys. *8*, 1–63.

JACKSON, I., and PATERSON, M. S. (1993), *A High-pressure High-temperature Apparatus for Studies of Seismic Wave Dispersion and Attenuation*, Pure appl. geophys. *141*, 445–466.

KANAMORI, H. (1967), *Spectrum of Short-period Core Phases in Relation to the Attenuation in the Mantle*, J. Geophys. Res. *72*, 2181–2186.

KARATO, S., and SPETZLER, H. A. (1990), *Defect Microdynamics in Minerals and Solid-state Mechanisms of Seismic Wave Attenuation and Velocity Dispersion in the Mantle*, Rev. Geophys. *28*, 399–421.

KENNETT, B. L. N., *Seismic Wave Propagation in Stratified Media*, (Cambridge University Press, Cambridge 1983) 339 pp.

KJARTANSSON, E. (1979), *Constant Q-wave Propagation and Attenuation*, J. Geophys. Res. *84*, 4737–4748.

KROEGER, G. C., *Syntheses and Analysis of Teleseismic Body Wave Seismograms*, Ph.D. (Stanford Universtiy, Palo Alto, 1987).

LIU, H. P., ANDERSON, D. L., KANAMORI, H. (1976), *Velocity Dispersion due to Anelasticity; Implications for Seismology and Mantle Composition*, Geophys. J. R. Astr. Soc. *47*, 41–58.

LUNDQUIST, G. M., and CORMIER, V. C. (1980), *Constraints on the Absorption Band Model of Q*, J. Geophys. Res. *85*, 5244–5256.

MENKE, W., *Geophysical Data Analysis: Discrete Inverse Theory* (Academic Press, San Diego, Calif. 1989) 289 pp.

MINSTER, J. B., and ANDERSON, D. L. (1981), *A Model of Dislocation-controlled Rheology for the Mantle*, Phil. Trans. R. Soc. Lon. *299*, 319–356.

MITCHELL, B. J. (1995), *Anelastic Structure and Evolution of the Continental Crust and Upper Mantle from Seismic Surface Waves Attenuation*, Rev. Geophys. *33*, 441–462.

PARSON, L. M., and HAWKINS, J. W., *Two-stage ridge propagation and the geological history of the Lau backarc basin*. In *Proc. ODP, Scientific Results 135* (eds J. W. Hawkins, L. Parson, and J. Allen) (Ocean Drilling Program, College Station, TX 1994) pp. 819–828.

REVENAUGH, J. S., and JORDAN, T. H. (1991), *Mantle Layering from ScS Reverberations 1. Wave-form Inversion of Zeroth-order Reverberations*, J. Geophys. Res. *96*, 19,749–19,762.

ROMANOWICZ, B. (1995), *A Global Tomographic Model of Shear Wave Attenuation in the Upper Mantle*, J. Geophys. Res. *100*, 12,375–12,394.

ROTH, E., WIENS, D. A., DORMAN, L., HILDEBRAND, J., and WEBB, S. (1998), *Attenuation Structure of the Lau Backarc from Regional Broadband Wave Forms*, J. Geophys. Res., in press.

SAILOR, R. V., and DZIEWONSKI, A. M. (1978), *Measurement and Interpretation of Normal Mode Attenuation*, Geophys. J. R. Astr. Soc. *53*, 559–581.

SATO, H., SACKS, I. S., MURASE, T., MUNCILL, G. E., and FUKOYAMA, H. (1989), Q_P-melting Temperature Relation in Peridotite at High Pressure and Temperature: Attenuation Mechanism and Implications for the Mechanical Properties of the Upper Mantle, J. Geophys. Res. *94*, 10,647–10,661.

SHEEHAN, A. F., and SOLOMON, S. C. (1992), *Differential Shear Wave Attenuation and its Lateral Variation in the North Atlantic Region*, J. Geophys. Res. *97*, 15,339–15,350.

SIPKIN, S. A. (1988), *Estimation of the Attenuation Operator for Multiple ScS Waves Using Digitally Recorded Broadband Data*, Geophys. Res. Lett. *15*, 832–835.

SIPKIN, S. A., and JORDAN, T. H. (1979), *Frequency Dependence of Q_{ScS}*, Bull. Seismol. Am. *69*, 1055–1079.

SIPKIN, S. A., and REVENAUGH, J. (1994), *Regional Variation of Attenuation and Travel Time in China from Analysis of Multiple-ScS Phases*, J. Geophys. Res. *99*, 2687–2699.

WALCK, M. C. (1988), *Spectral Estimates of Teleseismic P-wave Attenuation to 15 Hz*, Bull. Seismol. Soc. Am. *78*, 726–740.

ZHANG, Z., and LAY, T. (1993), *Investigation of Upper Mantle Discontinuities near Northwestern Pacific Subduction Zones Using Precursors to sSH*, J. Geophys. Res. *98*, 4389–4405.

(Received March 5, 1998, revised August 21, 1998, accepted August 25, 1998)

Pure appl. geophys. 153 (1998) 377–398
0033–4553/98/040377–22 $ 1.50 + 0.20/0

❙Pure and Applied Geophysics

Three-dimensional Mapping of Magma Source and Transport Regions from Seismic Data: The Mantle Wedge beneath Northeastern Japan

HIROKI SATO,[1] KENICHI MURO[1] and AKIRA HASEGAWA[2]

Abstract—We investigate the distribution of partial melt in island arc using the seismic velocity structure of the mantle wedge beneath northeastern Japan. The comparison of the seismic tomography with laboratory velocity data on a partially-molten mantle rock yields estimates of melting zones in three dimensions. We employ experimental data on the degree of partial melt in hydrous peridotite to give constraints on the melt fraction and temperature. Melting and magma-rich zones derived from the velocity structure coincide with observed low Q zones. The results of the three-dimensional mapping indicate that the source of magma in island arc is diapir-like melting patches localized within the low velocity zones of the mantle wedge. Extensive volcanic activity along the volcanic front is due to the presence of vast magma-rich zones just beneath the Moho. Those melting zones in the uppermost mantle may, in turn, cause melting of lower crustal materials and produce felsic magma. Melt appears to stay at and beneath the Moho, where crystallization fractionation may proceed. Melt exists at greater depths in the back-arc region, which may correlate with across-arc variations of chemical compositions of the volcanic rocks observed in northeastern Japan. We suggest that magma migration in the ductile lower crust may cause low-frequency microearthquakes, and magma penetration into the brittle upper crust may produce mid-crustal S-wave reflectors.

Key words: Magma distribution, three dimensions, velocity structure, seismic Q, mantle wedge, northeastern Japan.

1. Introduction

The seismic velocity and attenuation (Q^{-1}) structure of the northeastern Japan arc has been extensively studied by, for example, HASHIDA and SHIMAZAKI (1987), ZHAO et al. (1992, 1994), HASEGAWA et al. (1993, 1994), HASEGAWA and ZHAO (1994), SATO et al. (1996) and TSUMURA et al. (1996). Their results have generally shown low velocity and low Q zones in the mantle wedge dipping downwards to the

[1] Department of Earth and Space Science, Osaka University, Matikane-yama, Toyonaka, Osaka 560-0043, Japan. Fax: 81-6-850-5541. E-mail: roki@ess.sci.osaka-u.ac.jp
[2] Research Center for Prediction of Earthquakes and Volcanic Eruptions, Tohoku University, Aoba-ku, Sendai 980-8578, Japan.

back-arc side, and extremely low velocities immediately beneath the volcanic front. ZHAO *et al.* (1992) analyzed 14,045 arrival times including *P* and *S* first arrivals and later phases of converted waves. Complex seismic velocity discontinuities are taken into account in their three-dimensional velocity inversion. Their results (Fig. 1) have shown low velocity zones in the crust beneath the volcanoes and in the central part of the mantle wedge, and up to 6% high velocity in the subducting Pacific plate. HASEGAWA *et al.* (1993) found that low-frequency microearthquakes occur close to the low velocity zones around the Moho discontinuity, and that mid-crustal

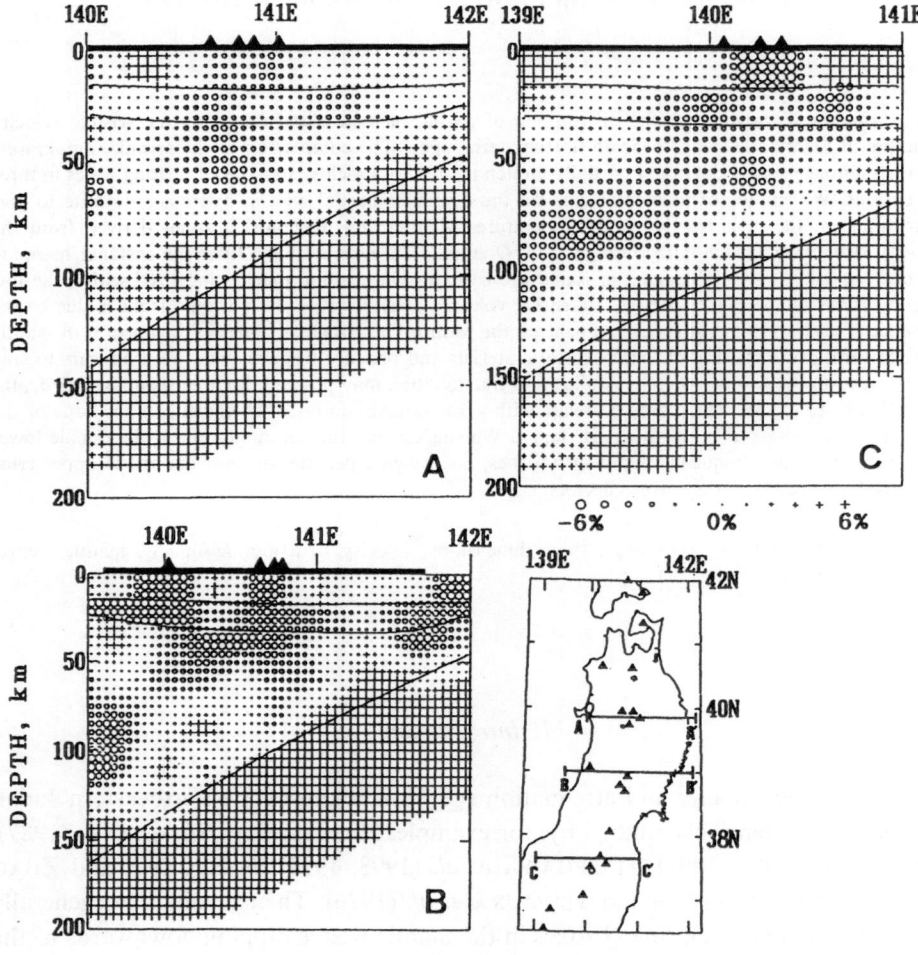

Figure 1

East–west vertical cross sections of *P*-wave velocity perturbations (in percent) beneath northeastern Japan (after ZHAO *et al.*, 1992). The location of each cross section is shown in the insert map (A, B and C along the lines AA′, BB′ and CC′, respectively). Triangles are active volcanoes, and heavy horizontal lines on the top denote land areas. The Conrad and Moho discontinuities and the upper plane of the subducted slab are shown by thin solid lines.

S-wave reflectors exist in the vicinity of the low velocity zones beneath active volcanoes. These observations indicate the close relationship between magmatism, slow velocity anomalies, low-frequency microearthquakes and S-wave reflectors.

We have used laboratory velocity and attenuation data of dry mantle peridotite determined at high pressure to estimate the melt fraction and temperature of the low velocity and low Q zones in the upper mantle (SATO and SACKS, 1989; SATO et al., 1989; SATO and RYAN, 1994). SOBOLEV et al. (1996) also investigated partial melting in a dry mantle composition in detail from combined seismic tomography and laboratory velocity data, by taking the effect of anelasticity on seismic velocity into account. In these previous studies, the effects of volatiles on seismic data were not considered. In subduction zones, petrological and geochemical studies have indicated the presence of slab-derived H_2O in the island arc mantle (e.g., BROWN et al., 1982; KUSHIRO, 1987; TATSUMI, 1989; PEACOCK, 1990; ISHIKAWA and NAKAMURA, 1994). Here we consider changes of seismic velocity with H_2O content, and estimate the partial melt fractions and temperatures in the low velocity zones beneath northeastern Japan. The partial melting zones obtained in our approach are shown in two and three dimensions, and are discussed in relation to the low Q zones, the occurrence of low-frequency microearthquakes and the presence of mid-crustal S-wave reflectors. From the results we suggest possible magma ascent pathways beneath the Japan arc.

2. Seismic Velocity in the Hydrous Mantle Wedge

In order to estimate partial melt fractions and temperatures in the mantle wedge, we use the three-dimensional P-wave velocity structure as determined by ZHAO et al. (1992). The comparison of their velocity structure with laboratory velocity and elasticity data yields constraints on the thermal structure beneath northeastern Japan. We first calculate velocities in dry solid peridotite and in basalt melt as a function of pressure and temperature, from mineral and melt elasticity data. We then obtain velocities in partially molten dry peridotite, following the method described by MAVKO (1980) and SCHMELING (1985a). We finally incorporate the effect of H_2O on velocity, referring to laboratory velocity data on hydrous peridotites determined at high pressure and temperature. Following this procedure, we estimate seismic velocities in the hydrous mantle wedge, and make comparisons with the observed seismic structure.

2.1. Calculation of Velocity in Dry Peridotite

A fertile spinel lherzolite such as pyrolite is selected as a typical mantle peridotite, which consists of olivine 56%, orthopyroxene 25%, clinopyroxene 17%, and spinel 2% by weight. For each mineral (for elastically isotropic mineral aggregate), compressional-wave velocities at high pressure and temperature are computed from laboratory elasticity data on isotropic mineral aggregate. First, we calculate densities, elastic constants and their pressure derivatives at a given temperature and one atmosphere, using the method described by DUFFY and ANDERSON (1989). We consider the effect of anharmonicity on seismic velocity. The one-atmosphere physical properties calculated at high temperature are then extrapolated adiabatically to high pressure. Third-order finite strain theory yields P- and S-wave velocities Vp and Vs, respectively, at higher pressure P as (e.g., DAVIES, 1974; DUFFY and ANDERSON, 1989)

$$\rho Vp^2 = (1 - 2\varepsilon)^{5/2} L[1 + (5 - 3KL'/L)\varepsilon]$$

$$\rho Vs^2 = (1 - 2\varepsilon)^{5/2} G[1 + (5 - 3KG'/G)\varepsilon] \qquad (1)$$

$$P = -3\varepsilon(1 - 2\varepsilon)^{5/2} K[1 + (6 - 3K'/2)\varepsilon]$$

with $\varepsilon = [1 - (\rho/\rho_0)^{2/3}]/2$ and $L = K + 4G/3$, where ρ is the density, ρ_0 is the density at one atmosphere, ε is the strain, K is the adiabatic bulk modulus, and G is the shear modulus. Primes refer to the pressure derivatives. Temperature, T, is calculated as a function of pressure through the equation

$$T(P) = T_0 + \int_0^P (\gamma T/K)\, dP \qquad (2)$$

where T_0 is a given temperature and γ is the Grüneisen parameter. Empirically, the Grüneisen parameter is found to decrease with pressure as $\gamma = \gamma_0(\rho/\rho_0)^{-q}$ ($q \cong 1$) (JEANLOZ and KNITTLE, 1986). However, the density increase is no more than 3%, when pressures increase from 0 to 3 GPa (the pressure range of the present calculation). Thus, the pressure dependence of the Grüneisen parameter does not cause errors (much less than 0.5%) in calculations of the adiabatic temperature from equation (2). Here we employ $\gamma = 1.1$ regardless of pressure (e.g., ISAAK, 1992).

Using equations (1) and (2) and elasticity data given by DUFFY and ANDERSON (1989), P-wave velocities in olivine (($Mg_{0.9}Fe_{0.1})_2SiO_4$), orthopyroxene and clinopyroxene are computed as a function of pressure and temperature. The velocity of peridotite is then calculated by taking a Voigt-Reuss-Hill (VRH) average (HILL, 1952). We have neglected spinel because of its small amounts. This causes errors of less than 1% in the velocity calculation. The velocity at high pressure was calculated to ~3 GPa (~100 km depth), where spinel lherzolite is stable (no garnet is formed). The calculated P-wave velocities in peridotite are shown as a function of depth in Figure 2. The present results agree well (less than 1% difference) with the laboratory velocities measured at 0.5 and 1 GPa for a dry peridotite by MURASE and KUSHIRO (1979). There is only a minor compositional difference between the peridotite samples used for their velocity measurements and those considered for the present numerical calculations. Because differences in the bulk moduli and

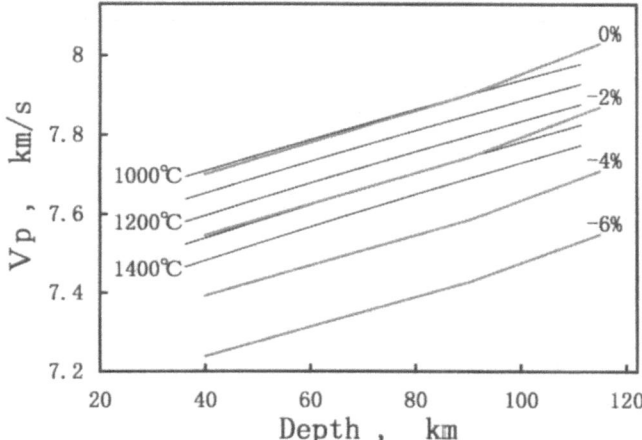

Figure 2
Calculated *P*-wave velocity in dry peridotite as a function of depth (thin lines of 1000 to 1400°C), with initial velocity model and velocity perturbations (thick lines of 0 to −6%) by ZHAO *et al.* (1992).

rigidities of olivine, orthopyroxene and clinopyroxene are small, the compositional variations in peridotite do not cause significant changes in the velocity estimates (less than 1%; cf. SOBOLEV *et al.*, 1996). For comparison, the initial (average) velocity model and velocity perturbations given by ZHAO *et al.* (1992) are also shown in Figure 2.

2.2. *Calculation of Velocity in Partially Molten Hydrous Peridotite*

Velocities in partially molten rock depend on the geometry of the intercrystalline melt phase. Experimental observations (WAFF and BULAU, 1979; DAINES and RICHTER, 1988) and thermodynamic considerations (BULAU *et al.*, 1979) have

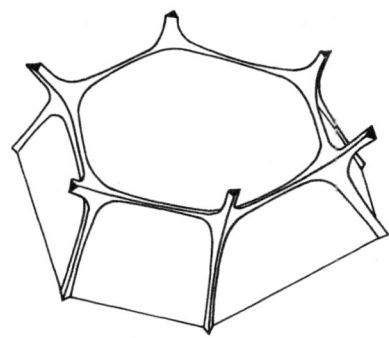

Figure 3
Schematic illustration of melt tubes along grain edges (after SMITH, 1964).

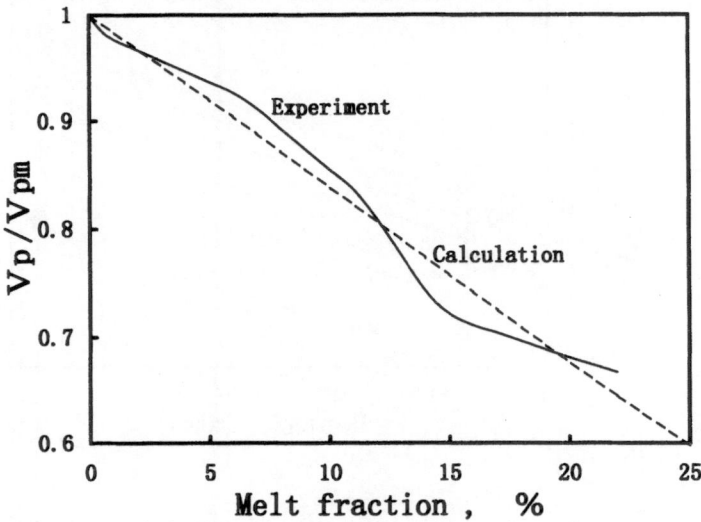

Figure 4
P-wave velocity in dry peridotite normalized by the velocity at the solidus temperature as a function of partial melt fraction. (Experiment: velocity measured at 1 GPa by MURASE and KUSHIRO (1979). Calculation: velocity calculated for the tube model at 1 GPa.)

indicated that melt tubes along grain edges (Fig. 3) are expected to be in equilibrium in most silicate partial-melt systems at high pressures. SCHMELING (1985b) pointed out that melt in the form of tubes can explain the seismic and conductivity data of the oceanic asthenosphere most easily, as compared with other geometries. Experiments on the equilibrium distribution of $H_2O–CO_2$ fluids in dunite and quartzite, for example, have shown that a fluid phase cannot exist as a thin film along grain boundaries in a rock (WATSON and BRENAN, 1987). Previous studies have generally indicated that the equilibrium partial melt texture in the upper mantle is a tube geometry. Here we employ the tube model to calculate *P*-wave velocity in a partially molten peridotite. This model is consistent with the laboratory velocity data determined at high pressure, as described later (cf. Fig. 4).

To calculate the *P*-wave velocity in a partially molten rock, we follow the method described by SCHMELING (1985a). We also calculate the density and the bulk modulus of basalt melt as a function of pressure, by using third-order finite strain theory (i.e., equation (1)). The bulk modulus, its pressure derivative and the density of melt at ambient conditions used here are 18 GPa, 4.3 and 2.7 g/cm³, respectively (e.g., MURASE and MCBIRNEY, 1973; SATO and MANGHNANI, 1985; MANGHNANI *et al.*, 1986; TANIGUCHI, 1989; AGEE and WALKER, 1993). Temperature effects on these parameters are not considered, since they are known to be small (e.g., MURASE, 1982; TANIGUCHI, 1989). Because melt fractions determined in this study range only up to 2% (see later section 3), the changes in melt elasticity data, due to changes in chemical composition and H_2O content, cause no difference in the results of this study. Taking the pressure dependence of elasticity of basaltic melt into account, and employing the tube model (SCHMELING, 1985a), we obtain the P-wave velocity in partially molten dry peridotite. The results of this calculation at 1 GPa are shown in Figure 4, by normalizing Vp with the P-wave velocity at the solidus Vpm. The tube model fits well to the laboratory velocity data determined at 1 GPa for a partially molten dry peridotite by MURASE and KUSHIRO (1979).

Finally, we consider the H_2O effect on seismic velocity. Velocities in hydrous peridotites, including serpentine and brucite (up to 11 wt.% H_2O), were measured at 1 GPa by ITO and MATSUSHIMA (1989) and ITO (1990). These results are the only available data sources for hydrous systems at high pressures and temperatures, and show a systematic velocity decrease with increasing H_2O content as given by

$$Vp/Vpo = 1 - aw \qquad (3)$$

where Vp is the P-wave velocity in hydrous peridotite, Vpo is the P-wave velocity in dry peridotite, w is the weight percent of H_2O, and $a = 9 \times 10^{-5}T - 0.02$ (T in degrees centigrade). Here we use equation (3) to incorporate the H_2O effects on seismic velocity.

From a knowledge of the phenocryst assemblages of volcanic rocks and the solubility data of H_2O, SAKUYAMA (1983) estimated the maximum H_2O content in parental basaltic magmas beneath island arcs to be about 2.5 wt.%, and the overall H_2O content in the arc mantle to be at most 0.5 wt.%. Later KUSHIRO (1987) reported 0.07–0.13 wt.% H_2O in most parts of the mantle wedge beneath Japan, and a possible local enrichment of H_2O in certain portions of the mantle wedge. In addition, analyses of H_2O in basaltic glasses have indicated a bulk earth H_2O content of about 0.055 to 0.19 wt.% (JAMBON and ZIMMERMANN, 1990). The amount of CO_2 in the mantle beneath island arcs may be considerably less than that of H_2O (e.g., PEACOCK, 1990). Finally, accretion models for the solar system yield average H_2O contents for the mantle of 0.1 wt.% or less (e.g., RINGWOOD, 1966; DREIBUS and WANKE, 1989). We here consider hydrous peridotites including 0.1 and 0.2 wt.% H_2O. From data by ITO (1990) (equation (3)), the corresponding velocity drops are 1% and 2%, respectively, over the temperature range of hydrous partial melt (1100–1400°C; see Table 1). After correcting for these values, we finally

Table 1

Temperatures of partial melt zones in the mantle wedge beneath northeastern Japan for batch and fractional melting regimes

| Depth, km | Temperature, °C | | | |
| | 0.1 wt.% H_2O | | 0.2 wt.% H_2O | |
	Batch	Fractional	Batch	Fractional
40	1200	1230	1120	1170
80	1290	1330	1150	1230

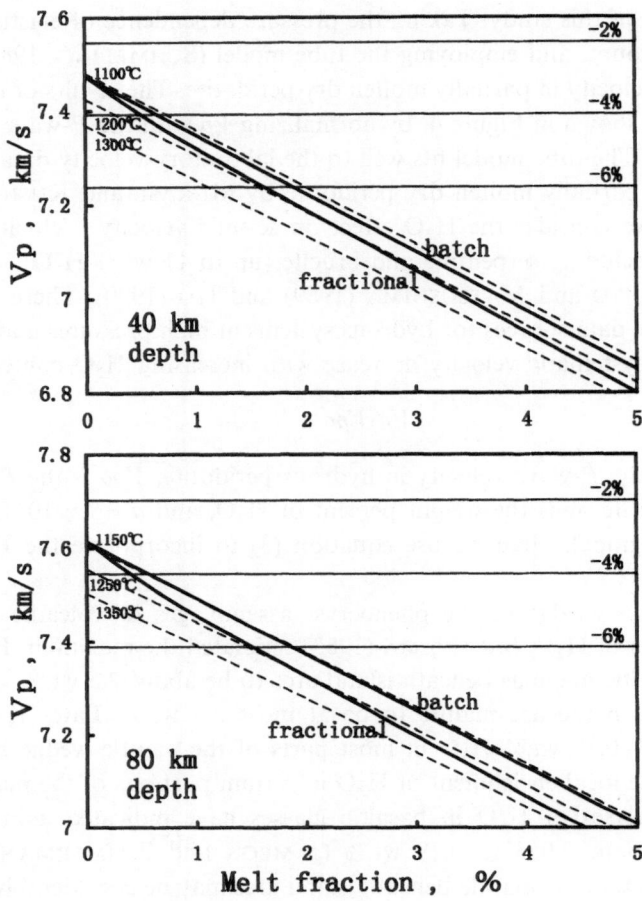

Figure 5

Diagrams for estimating temperature and partial melt fraction from seismic and laboratory velocity data at 40 km and 80 km depths for 0.2 wt.% H_2O (dashed line: calculated P-wave velocity in partially molten hydrous peridotite as a function of melt fraction and temperature, thin solid line: P-wave velocity for -2 to -6% perturbations by ZHAO *et al.* (1992), thick solid line: P-wave velocity as a function of melt fraction for batch and fractional melting (cf. Figure 6)).

obtain the estimates of P-wave velocities in a partially molten hydrous peridotite. The results of the present calculations are shown as a function of partial melt fraction and temperature in Figure 5, together with P-wave velocity perturbations ($-2\% \sim -6\%$) in the mantle wedge as given by ZHAO *et al.* (1992).

3. Estimate of Partial Melt Fraction and Temperature

Velocities calculated at a given depth as a function of melt fraction and temperature (Fig. 5) are now compared with phase equilibria data. This comparison yields estimates of partial melt fraction and temperature. Melting phase relations and melt fractions of hydrous peridotites were investigated in detail by IWAMORI *et al.* (1995). They combine available laboratory data to allow estimates of melt fractions at pressures to ~8 GPa and temperatures to the liquidus of peridotite including H$_2$O contents to 0.2 wt.% (Fig. 6). They consider two extreme cases of melting: fractional and batch melting. In the case of fractional melting, most melt is removed from the source region and ascends to shallower depth soon after melting is initiated. Therefore, thermal and chemical isolation of melt occurs. As melt is removed, the system tends to become depleted and the solidus and liquidus temperatures increase. Higher temperatures are necessary to cause partial melting in the more depleted system. This effect is investigated by IWAMORI *et al.* (1995), using a model of two-phase flow with relative movement between the melt and residue during the melt segregation process. On the other hand, no melt migration occurs and the entire melt produced stays within the source region in the case of batch melting. This is the case determined from laboratory experiments on high-pressure melting phase equilibrium studies on peridotites. We consider here these two cases to estimate temperatures and partial melt fractions. The two cases provide the upper and lower bounds, respectively, for the temperature estimate.

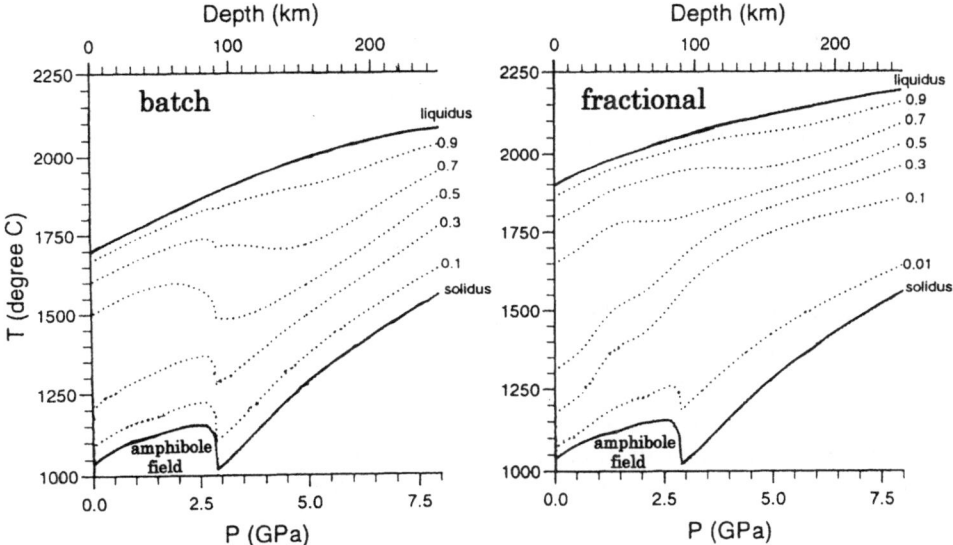

Figure 6

Partial melt fraction in hydrous peridotite including 0.2 wt.% H$_2$O as a function of pressure and temperature for batch and fractional melting. Numerals are fractions of partial melt, and amphibole stability field, solidus and liquidus are shown (modified after IWAMORI *et al.*, 1995).

Degrees of partial melt as a function of temperature at 40 km and 80 km depths (= 1.1 and 2.5 GPa) as given by IWAMORI *et al.* (1995) are redrawn, for batch and fractional melting, in the $Vp(T)$ vs. melt fraction diagram in Figure 5. Fractional melting requires higher temperature and thus lower P-wave velocity than batch melting. The intersection of the lines of batch (or fractional) melting and the velocity given by ZHAO *et al.* (1992) yields estimates of temperature and melt fraction for the case of no (or complete) chemical isolation of melt. Temperatures thus determined for the two melting cases of peridotite, including 0.1 and 0.2 wt.% H_2O in the partial melt zones (1–2% melt) beneath northeastern Japan, are given in Table 1. Higher temperatures are obtained for smaller H_2O contents. Temperatures should also be higher for fractional melting vis-à-vis batch melting. The melt fraction obtained in the mantle wedge is relatively small (up to 2% corresponding to a velocity perturbation to −6%), and the difference in the H_2O content between 0.1 and 0.2 wt.% produces no significant difference in the melt fraction. However, it does cause some difference in the temperature estimate (Table 1).

Because of the errors in velocity calculations, the presence of less than 1% melt is not resolved. As can be seen from Figure 5, 1% error in velocity yields ∼0.5% and ∼20°C uncertainty in the estimates of melt fraction and temperature, respectively. Because we use constraints from melting phase relations for hydrous peridotite (i.e., Fig. 6), the temperature uncertainties in this study are relatively small if the melt fractions were determined accurately at high pressure. Small amounts of melt (up to 2%) determined in this study do not imply that the melt fraction in the mantle wedge beneath northeastern Japan does not exceed 2%. Because of the limited spatial resolution of the seismic studies, regions of high melt fraction (i.e., regions of extremely low velocities) localized in less than $5^3 \sim 10^3$ km^3, are not determined. Localized magma reservoirs were not resolved from the seismic studies by ZHAO *et al.* (1992).

The seismic structure of ZHAO et al. (1992) determines some low velocity zones in the fore-arc mantle wedge (cf. Fig. 1), where the surface heat flow is low, the cold slab is subducted, and no volcanic activities have been observed. In general, temperatures estimated from heat flow or from numerical calculations on wedge flow are substantially lower than the H_2O-saturated solidus. Therefore, no partial melting is expected in the fore-arc mantle. In this region, fluids released from the subducted sediments may cause low velocities. A large amount of H_2O may be released through the dehydration of serpentine in the fore-arc mantle (e.g., TATSUMI, 1989; PEACOCK, 1996). The presence of more slab-derived fluids at shallower depths has been suggested for the generation of arc magmas in the southern Andes (HICKEY *et al.*, 1986). Pb, Sr and Nd isotope studies have also indicated that the fluids decrease with increasing distance from the Japan trench, and that sedimentary components are consumed at relatively shallow depths (e.g., NOHDA and WASSERBURG, 1981; NAKAMURA *et al.*, 1985). It is large amounts of volatiles released at the wedge apex near the trench that can cause low velocity in the fore-arc region—not high temperature. We describe below partial melting in the low velocity zones beneath the volcanic front and the back-arc region, where the surface heat flow is high and magmatic activity is expected.

4. Two- and Three-dimensional Mapping of Partial Melt Zones

ZHAO et al. (1992) determined the velocity structure beneath northeastern Japan in three dimensions; therefore, it is possible to map the partial melting and magma-rich zones obtained in this study in two and three dimensions. Figure 7 shows east-west vertical cross sections of the melting zones in the mantle wedge including 0.1–0.2 wt.% H_2O. As described earlier, the partial melt fraction and thus the melting zones do not change over this range of H_2O content. As can be seen from Figures 1 and 7, partial melt is obtained only in the fairly low velocity zones (velocity perturbations below −4%). Figure 8 shows the results of the three-dimensional mapping of the melting zones. The most extensive partial melting occurs in the uppermost mantle just beneath the volcanoes, which may account for the largest eruptive volumes in the volcanic front. Note that extensive regions of magma do not exist generally in the mantle wedge, but exist locally beneath the sites of volcanoes and dip downwards to the west. The occurrence of melt at greater depth towards the west of the volcanic front may correlate with the observed increases in incompatible elements in the back-arc region (e.g., KUNO, 1966; DICKINSON and HATHERTON, 1967).

Contours of −3% velocity perturbation given by ZHAO et al. (1992) are also shown in Figures 7 and 8. Within the contours, the velocity perturbations are below −3%, and thus higher temperatures than the averages are expected. High temperature regions appear to extend to greater depths in the mantle wedge, suggesting a plume-like uprising of hot mantle material. In general, high-temperature low-velocity zones are wide in the direction along the volcanic front but narrow in the east-west direction, though a few exceptions exist. Three major low-velocity zones exist below ~60 km depth, and branch off at the uppermost mantle, eventually underplating the Moho just below the volcanoes. Partial melting zones obtained in this study localize within these low-velocity zones. Especially the two large low-velocity zones are observed in the mantle wedge at around 40–41°N 139.5–140.5°E and 37.5–38.5°N 139–140°E. The former low-velocity zone exists below ~60 km depth and extends to the direction of the inverted L-shaped melting zone above ~60 km depth. The latter low-velocity zone branches off in three directions above ~50 km depth, and also underplates the Moho directly beneath the volcanic front, i.e., beneath Zao, Azuma and Adatara volcanoes (cf. Fig. 10). Several other low-velocity zones are also observed beneath the volcanic front and in the intermediate depth of the mantle wedge. Melting zones localized within hot mantle material thus correlate with volcanic activity in the northeastern Japan arc.

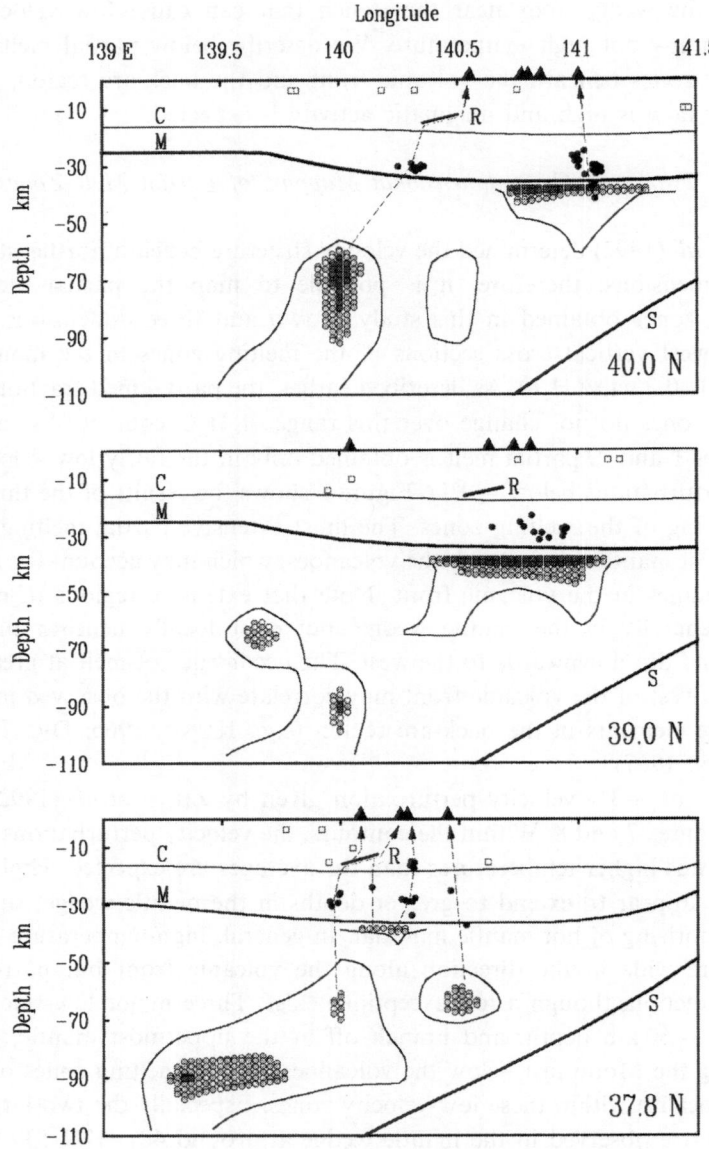

Figure 7

Mapping (across-arc vertical cross section) of partial melt zones in the upper mantle beneath northeastern Japan along 37.8°N, 39.0°N and 40.0°N from 139°E to 141.5°E (R: mid-crustal *S*-wave reflector, C: Conrad, M: Moho, S: upper plane of subducted slab, triangles: active volcanoes, solid circles: hypocenters of low-frequency microearthquakes, squares: hypocenters of large crustal earthquakes ($M \geqq 6$), hatched circles: regions of 1% melt, dark hatched squares: regions of 2% melt, contour lines: lines of -3% velocity perturbation, thin dashed lines with arrow: possible magma ascent pathways). Vertical scale is exaggerated 1.1 times. In between volcanoes and partial melting zones, mid-crustal *S*-wave reflectors exist, and low-frequency microearthquakes and large crustal earthquakes occur. It is observed that partial melt stays directly beneath the Moho at the volcanic front, which may cause melting of lower-crustal materials and produce felsic magma. Fractional crystallization may also proceed in the melting zones beneath the Moho.

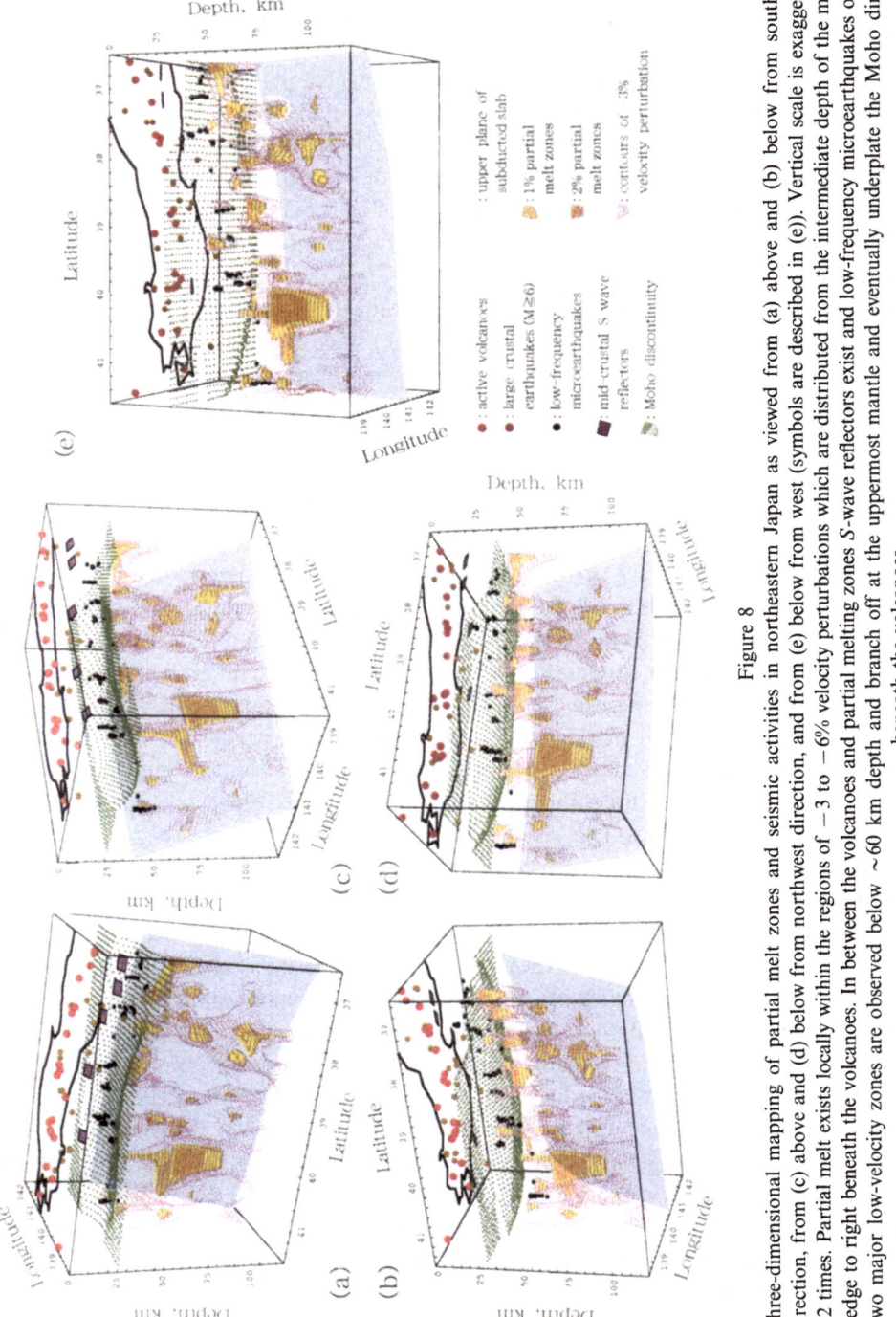

Figure 8

Three-dimensional mapping of partial melt zones and seismic activities in northeastern Japan as viewed from (a) above and (b) below from southwest direction, from (c) above and (d) below from northwest direction, and from (e) below from west (symbols are described in (e)). Vertical scale is exaggerated 3.2 times. Partial melt exists locally within the regions of -3 to -6% velocity perturbations which are distributed from the intermediate depth of the mantle wedge to right beneath the volcanoes. In between the volcanoes and partial melting zones S-wave reflectors exist and low-frequency microearthquakes occur. Two major low-velocity zones are observed below ~ 60 km depth and branch off at the uppermost mantle and eventually underplate the Moho directly beneath the volcanoes.

Beneath the volcanoes, mid-crustal S-wave reflectors as well as magma-rich zones are inferred, and low-frequency microearthquakes occur in the crust as also shown in Figures 7 and 8. The close relationship of the inferred magma-rich zones with S-wave reflectors and microearthquakes indicates that magma ascent from the uppermost mantle to the ductile lower crust may cause low-frequency microearthquakes, and subsequent magma penetration into the brittle upper crust may produce mid-crustal reflectors. Seismological studies have indicated that S-wave reflectors represent the upper surface of thin magma bodies (e.g., MIZOUE, 1980; MATSUMOTO and HASEGAWA, 1996). Magma movement in the brittle upper crust may locally increase stress (or at least change the state of stress) in the crust, and may contribute to the occurrence of crustal earthquakes (including destructive earthquakes). Large crustal earthquakes ($M \geq 6$, since 1931) occur in relatively high velocity and high Q regions located around the volcanoes and the melting zones (Figures 7 and 8).

5. Seismic Q Structure Beneath Japan

HASHIDA and SHIMAZAKI (1987) and HASHIDA (1989) determined the three-dimensional Q structure beneath Japan using seismic intensity data, and obtained low Q in the mantle wedge beneath the active volcanoes and high Q in the subducting Pacific plate. A similar attenuation structure has been found beneath the Kanto-Tokai area in central Japan by SEKIGUCHI (1991). He obtained fairly low Q values for P waves ($Qp = 125$) in the crust and in the mantle wedge at ~ 75 km depth beneath the active volcanoes. In northeastern Japan, UMINO and HASEGAWA (1984) determined the three-dimensional Q structure by using spectral amplitude ratios of P and S waves. They obtained a high Q for S waves ($Qs \cong 1500$) in the descending lithosphere, an intermediate Q ($Qs \cong 500-800$) in the mantle wedge on the fore-arc side of the volcanic front, and a low Q ($Qs \cong 50$) in the mantle wedge on the back-arc side of the volcanic front. To avoid uncertain assumptions on P- to S-wave spectral ratio, TSUMURA *et al.* (1996) used an inversion method which allows simultaneous estimates of source spectra, average attenuation factor and site effects. Applying the method to northeastern Japan, they obtained a high Q ($Qp > 500$) in the descending slab and a low Q ($Qp < 250$) in the mantle wedge (Fig. 9). Regions of fairly low Q values ($Qp = 125$) are also determined directly beneath the volcanoes and downwards to the west. Every seismic study has shown that low Q zones, in general, coincide with the low velocity zones (e.g., compare Figs. 1 and 9). High seismic wave attenuation has generally been observed beneath volcanic fields, suggesting the presence of magma. In the upper mantle, Q is typically low in tectonically active regions, and regional variations in Q may be explained by differences in temperature (MITCHELL, 1995). Partial melting zones determined in this study correspond to low Q zones as well.

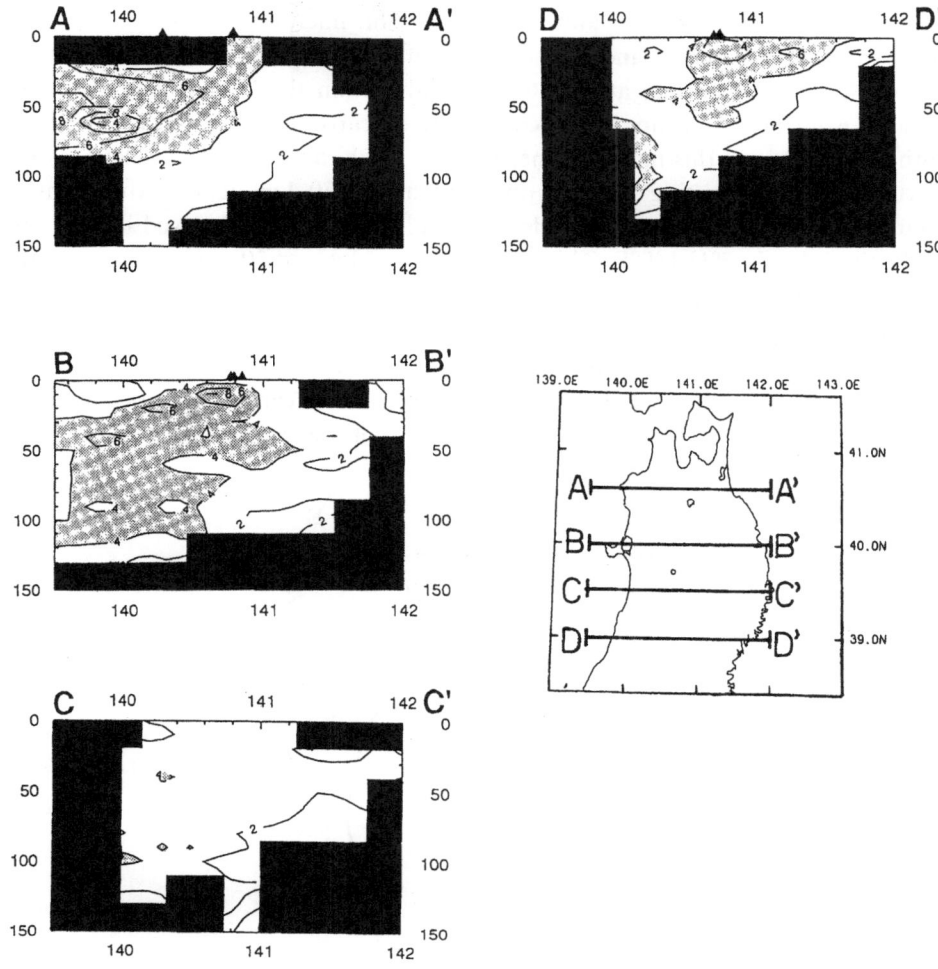

Figure 9

East–west vertical cross sections of the Q structure of P waves beneath northeastern Japan (after TSUMURA *et al.*, 1996). The location of each cross section is shown in the insert map. Numerals are Q^{-1} values multiplied by 1000. Shaded areas indicate low Q zones ($Q^{-1} > 0.004$). Triangles are active volcanoes. Low Q is determined right beneath the volcanoes and downwards to the west. Along the line CC' no volcanoes exist and the mantle wedge shows a relatively high Q.

6. Partial Melting, Volcanoes, Low Q Zones and Seismic Activities

The interrelationship between partial melting, volcanoes, low Q zones, S-wave reflectors and low-frequency microearthquakes is described in detail below. A fine Q structure determined by TSUMURA *et al.* (1996) has resolved several characteristic low Q zones beneath northeastern Japan (Fig. 9). Here we discuss their Q structure as well as seismic activities, in relation to the melting zones obtained in this study.

Volcanoes, *S*-wave reflectors and hypocenters of microearthquakes have been shown in Figures 7 and 8, and are summarized in Figure 10.

A melting zone appears at 40–46 km depth within the low-velocity zone just beneath Osoreyama volcano. Low-frequency microearthquakes occur at 36–47 km depth in and around this melting zone. Beneath Iwaki and Hakkoda volcanoes the inverted L-shaped melting zone is observed at 40–70 km depth. An extensive diapir-like melting zone which includes up to 2% melt is also observed below ∼60 km depth at 40–40.5°N 140°E. This melting zone is located 30 ∼ 40 km west of the inverted L-shaped melting zone, and both are separated. These two melting zones are wide (to ∼100 km) in the direction of longitude, but their thickness only extends to ∼20 km. An extensive low Q zone ($Qp < 170$) is also determined in the mantle wedge of this region (Fig. 9). Two large crustal earthquakes occur in the fore-arc area between Hakkoda and Osoreyama volcanoes (Fig. 10).

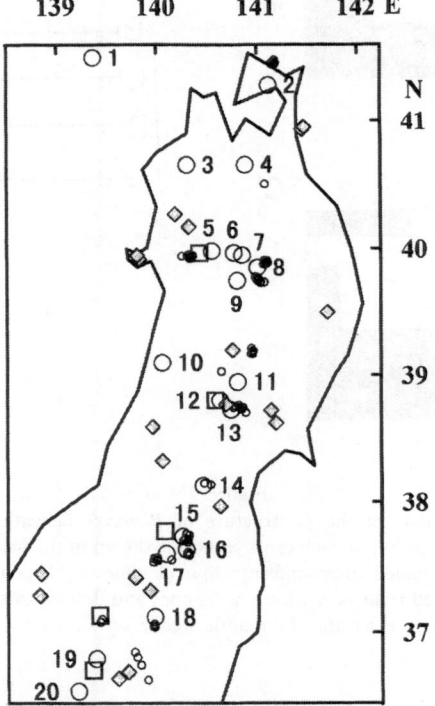

Figure 10

Volcanoes and seismic activities in northeastern Japan (circles with number: active volcanoes, 1: Oshima-Oshima, 2: Osoreyama, 3: Iwaki, 4: Hakkoda, 5: Moriyoshi, 6: Akita-Yakeyama, 7: Hachiman-tai, 8: Iwate, 9: Akita-Komagatake, 10: Chokai, 11: Kurikoma, 12: Onikobe, 13: Narugo, 14: Zao, 15: Azuma, 16: Adatara, 17: Bandai, 18: Nasu, 19: Nikko-Shirane, 20: Akagi, squares: mid-crustal *S*-wave reflectors, hatched diamonds: hypocenters of large crustal earthquakes ($M \geq 6$), small circles: hypocenters of low-frequency microearthquakes).

A melting zone is observed directly beneath Akita-Yakeyama and Hachimantai volcanoes, and some melting regions are also in the intermediate depth. In the back-arc of this region exists the extensive diapir-like melting zone described above. Q is generally low ($Qp < 250$) and localized low Q zones ($Qp < 170$) are observed in the crust and the upper mantle beneath these volcanoes (Fig. 9). A mid-crustal S-wave reflector exists close to one of the localized low Q zones beneath Moriyoshi volcano, and many hypocenters of low-frequency microearthquakes are observed at 29–32 km depth. Microearthquakes are also distributed vertically at 24–37 km depth in the lower crust beneath Iwate volcano. Four large crustal earthquakes occur in the upper crust of the back-arc side of Moriyoshi volcano since 1931.

In the back-arc area of Chokai volcano small melting zones are observed at 60–70 km and ~90 km depths, and the region, including 2% melt, appears in the uppermost mantle at 40–50 km depth. Low-frequency microearthquakes have not been observed yet around Chokai volcano. Right beneath Kurikoma and Narugo volcano area an extensive melting zone, including 2% melt in the center, appears at 40–50 km depth. In this region low Q is also determined in the crust and the upper mantle dipping downwards to the west. Beneath Onikobe a mid-crustal S-wave reflector exists, and low Q ($Qp < 170$) is obtained below the reflector. Near this low Q zone many hypocenters of low-frequency microearthquakes are distributed vertically at 23–31 km depth beneath Narugo, and at 27–30 km depth beneath the north of Kurikoma volcano. Several large crustal earthquakes occur around this region. A 2% melting zone also appears at 40–50 km depth directly beneath Zao volcano. Two low-frequency microearthquakes are observed at 25 km and 37 km depths, and large crustal earthquakes also occurred near Zao.

Around Azuma and Bandai volcanoes several melting zones appear in the intermediate depth of the mantle wedge of the back-arc region at 40–45 km, 60–70 km and ~90 km depths, and a relatively extensive melting zone at 80–100 km. Beneath Azuma volcano a mid-crustal S-wave reflector exists, and many hypocenters of low-frequency microearthquakes are distributed at 15–33 km depth. Beneath Nasu, Nikko-Shirane and Akagi volcanoes, extensive melting zones including 2% melt appear at 40–50 km depth. These melting zones are connected in the uppermost mantle. At 60–70 km depth beneath Nasu volcano a melting zone reappears. Mid-crustal S-wave reflectors exist beneath Hinoemata in the west of Nasu volcano and beneath Nikko-Shirane volcano. Many hypocenters of low-frequency microearthquakes are observed in the lower crust at 24–30 km depth beneath Hinoemata and at 27–29 km depth beneath Nikko-Asio. Large crustal earthquakes also occurred around this region.

7. Generation and Ascent of Magma

Generation and ascent of magma as inferred from Figures 7 and 8 are summarized as follows (see also Fig. 11). The lowering of the solidus temperature

Longitude

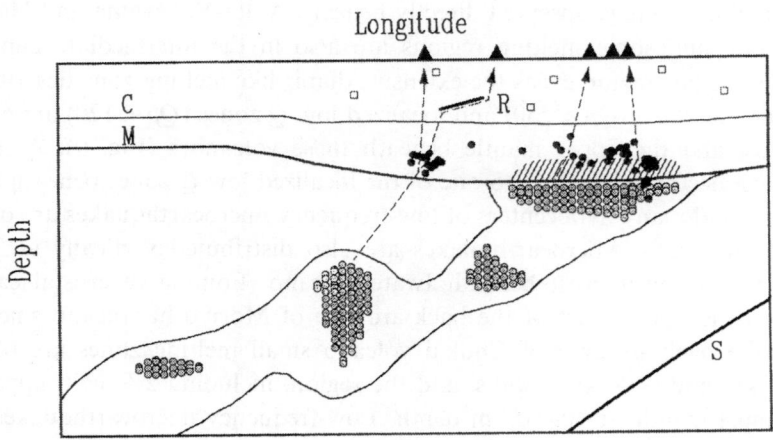

Figure 11

Generation and ascent of magma in the mantle wedge beneath northeastern Japan (R: mid-crustal *S*-wave reflector, C: Conrad, M: Moho, S: upper plane of subducted slab, solid triangles: active volcanoes, open squares: hypocenters of large crustal earthquakes, solid circles: hypocenters of low-frequency microearthquakes, shading area: melting of the lower crust, hatched circles: partial melt zones, lines: contours of -3% velocity perturbation, dashed lines with arrow: inferred magma ascent pathways).

by adding H_2O to the mantle wedge may cause melting and produce magma beneath northeastern Japan. Localized diapir-like melting zones of a few tens of kilometers are produced within the high temperature regions. A part of magma released from the melting zones at greater depths may ultimately cause volcanic activity in the back-arc of northeastern Japan. Fractional crystallization may occur within magma stagnated just below the Moho. Such stagnant magma may cause the melting of lower-crustal materials and generate felsic magma. Lower-crustal sections just above the melting zones generally show low velocities (compare Figs. 1 and 7), which is consistent with these melting scenarios. Some magma will segregate from the melting zone and ascend through the ductile lower crust, producing low-frequency microearthquakes. Magma penetration into the brittle upper crust may cause the opening of the fault planes and/or cracks, forming mid-crustal reflectors, and finally melt ascends to magma reservoirs. The continuous supply of magma from melting zones beneath the Moho may cause the surface volcanic activity of the island arc. However, most of the magma will cool at depth and form the island arc crust. The local existence of melting zones observed in this study may account for the lifetime of the island arc volcanoes. Once a melting zone cools down volcanic activity will cease. New volcanic activity may restart after another diapir in the mantle wedge rises to the Moho discontinuity.

8. Discussion

Numerical studies on the mantle wedge have indicated that high temperature regions can be derived through the upwelling of hot mantle material which is induced by wedge flow associated with subduction (e.g., HONDA, 1985; DAVIES and STEVENSON, 1992; FURUKAWA, 1993; IWAMORI et al., 1995). HONDA and UYEDA (1983) suggested that advective transport of heat plays a more important role than does shear heating in elevating the mantle wedge temperature to melting conditions. HONDA (1985) indicated that a magma migration rate of ~ 0.05 cm/year is necessary to explain the high heat flow in the Japan arc (>120 mW/m^2). FURUKAWA (1993) estimated the temperature and stress state beneath arcs induced by the subducting slab. Calculated deviatoric stress is more than a few tens of megapascals in partially molten mantle which is large enough to open cracks (magma ascent pathways) and to aid magma segregation. The stress state is compressive on the trench side, whereas it becomes tensile on the back-arc side. He suggested that the change in stress field causes the volcanic front. The overall temperature distributions in partial melting zones in this study (Table 1) are compatible with previous numerical models on the subduction zones.

Several petrological and geochemical studies have indicated that the temperature of the top of the slab at depth will be above the stability field of hydrous minerals such as serpentine, amphibole and lawsonite (e.g., TATSUMI, 1989; PEACOCK, 1996). Dehydration occurs when hydrous minerals decompose at a depth of the phase boundary. This may cause the local existence of volatiles, i.e., forming of hydrous columns in the mantle wedge (TATSUMI, 1989; DAVIES and STEVENSON, 1992). The present results indicate the local existence of hot melting regions.

9. Conclusions

We investigate partial melting in the mantle wedge beneath northeastern Japan from seismic velocity data and melting phase relations on mantle peridotite. For the H$_2$O content of 0.1 and 0.2 wt.%, respectively, temperatures in 1–2% melting zones are estimated to be about 1200°C and 1150°C at 40 km depth, and about 1300°C and 1200°C at 80 km depth. We have shown upper mantle melting zones in three dimensions. The results enable us to understand the interrelationship between magma-rich regions, active volcanoes, low Q zones, S-wave reflectors and hypocenters of low-frequency microearthquakes. This, in turn, allows us to investigate three-dimensional magma ascent pathways in the subduction zone.

Acknowledgments

We are very grateful to Dr. H. Iwamori and Dr. D. Zhao for helpful suggestions and comments, Dr. D. Zhao for kindly providing the *P*-wave velocity data, and Dr. M. P. Ryan for his thoughtful reviews and critical reading of the manuscript.

REFERENCES

AGEE, C. B., and WALKER, D. (1993), *Olivine Flotation in Mantle Melt*, Earth Planet. Sci. Lett. *114*, 315–324.

BROWN, L., KLEIN, J., MIDDLETON, R., SACKS, I. S., and TERA, F. (1982), *[10]Be in Island-arc Volcanoes and Implications for Subduction*, Nature *299*, 718–720.

BULAU, J. R., WAFF, H. S., and TYBURCZY, J. A. (1979), *Mechanical and Thermodynamic Constraints on Fluid Distribution in Partial Melts*, J. Geophys. Res. *84*, 6102–6108.

DAINES, M. J., and RICHTER, F. M. (1988), *An Experimental Method for Directly Determining the Interconnectivity of Melt in a Partially Molten System*, Geophys. Res. Lett. *15*, 1459–1462.

DAVIES, G. F. (1974), *Effective Elastic Moduli under Hydrostatic Stress—I. Quasi-harmonic Theory*, J. Phys. Chem. Solids *35*, 1513–1520.

DAVIES, J. H., and STEVENSON, D. J. (1992), *Physical Model of Source Region of Subduction Zone Volcanics*, J. Geophys. Res. *97*, 2037–2070.

DICKINSON, W. R., and HATHERTON, T. (1967), *Andesitic Volcanism and Seismicity around the Pacific*, Science *157*, 801–803.

DREIBUS, G., and WANKE, H., In *Origin and Evolution of Planetary and Satellite Atmospheres* (eds. Atreya, S. K., Pollack, J. B., and Matthews, M. S.) (Univ. of Arizona Press, Tucson 1989) pp. 268–288.

DUFFY, T. S., and ANDERSON, D. L. (1989), *Seismic Velocities in Mantle Minerals and the Mineralogy of the Upper Mantle*, J. Geophys. Res. *94*, 1895–1912.

FURUKAWA, Y. (1993), *Magmatic Processes under Arcs and Formation of the Volcanic Front*, J. Geophys. Res. *98*, 8309–8319.

HASEGAWA, A., and ZHAO, D., *Deep structure of island arc magmatic regions as inferred from seismic observations*. In *Magmatic Systems* (ed. Ryan, M. P.) (Academic Press, San Diego 1994) pp. 179–195.

HASEGAWA, A., HORIUCHI, S., and UMINO, N. (1994), *Seismic Structure of the Northeastern Japan Convergent Margin: A Synthesis*, J. Geophys. Res. *99*, 22295–22311.

HASEGAWA, A., YAMAMOTO, A., ZHAO, D., HORI, S., and HORIUCHI, S. (1993), *Deep Structure of Arc Volcanoes as Inferred from Seismic Observations*, Phil. Trans. R. Soc. Lond. *A342*, 167–178.

HASHIDA, T. (1989), *Three-dimensional Seismic Attenuation Structure beneath the Japanese Islands and its Tectonic and Thermal Implications*, Tectonophys. *159*, 163–180.

HASHIDA, T., and SHIMAZAKI, K. (1987), *Determination of Seismic Attenuation Structure and Source Strength by Inversion of Seismic Intensity Data: Tohoku District, Northeastern Japan Arc*, J. Phys. Earth *35*, 67–92.

HICKEY, R. L., FREY, F. A., GERLACH, D. C., and LOPEZ-ESCOBAR, L. (1986), *Multiple Sources for Basaltic Arc Rocks from the Southern Volcanic Zone of the Andes (34°–41°S): Trace Element and Isotopic Evidence for Contributions from Subducted Oceanic Crust, Mantle, and Continental Crust*, J. Geophys. Res. *91*, 5963–5983.

HILL, R. (1952), *The Elastic Behavior of a Crystalline Aggregate*, Proc. Phys. Soc. London *A65*, 349–354.

HONDA, S. (1985), *Thermal Structure beneath Tohoku, Northeast Japan—A Case Study for Understanding the Detailed Thermal Structure of the Subduction Zone*, Tectonophys. *112*, 69–102.

HONDA, S., and UYEDA, S., *Thermal process in subduction zones—a review and preliminary approach on the origin of arc volcanism*. In *Arc Volcanism: Physics and Tectonics* (eds. Shimozuru, D. and Yokoyama, I.) (Terra Scientific Pub. Co., Tokyo 1983) pp. 117–140.

ISAAK, D. G. (1992), *High-temperature Elasticity of Iron-bearing Olivines*, J. Geophys. Res. *97*, 1871–1885.

ISHIKAWA, T., and NAKAMURA, E. (1994), *Origin of the Slab Component in Arc Lavas from Across-arc Variation of B and Pb Isotopes*, Nature *370*, 205–208.

ITO, K. (1990), *Effects of H₂O on Elastic Wave Velocities in Ultrabasic Rocks at 900°C under 1 GPa*, Phys. Earth Planet. Inter. *61*, 260–268.

ITO, K., and MATSUSHIMA, S. (1989), *Compressional and Shear Wave Velocities in Ultrabasic Rocks with Hydrous Minerals to 900°C under 1 GPa*, J. Seismol. Soc. Japan *42*, 171–181.

IWAMORI, H., MCKENZIE, D., and TAKAHASHI, E. (1995), *Melt Generation by Isentropic Mantle Upwelling*, Earth Planet. Sci. Lett. *134*, 253–266.

JAMBON, A., and ZIMMERMANN, J. L. (1990), *Water in Oceanic Basalts: Evidence for Dehydration of Recycled Crust*, Earth Planet. Sci. Lett. *101*, 323–331.

JEANLOZ, R., and KNITTLE, E., *Reduction of mantle and core properties to a standard state by adiabatic decompression. In Chemistry and Physics of the Terrestrial Planets* (ed. Saxena, S. K.) (Springer-Verlag, New York 1986) pp. 275–309.

KUNO, H. (1966), *Lateral Variation of Basalt Magma Type Across Continental Margins and Island Arcs*, Bull. Volcanol. *29*, 195–222.

KUSHIRO, I., *A petrological model of the mantle wedge and lower crust in the Japanese island arcs. In Magmatic Processes: Physicochemical Principles* (ed. Mysen, B. O.) (Lancaster Press, Pennsylvania 1987) pp. 165–181.

MANGHNANI, M. H., SATO, H., and RAI, C. S. (1986), *Ultrasonic Velocity and Attenuation Measurements on Basalt Melts to 1500°C: Role of Composition and Structure in the Viscoelastic Properties*, J. Geophys. Res. *91*, 9333–9342.

MATSUMOTO, S., and HASEGAWA, A. (1996), *Distinct S Wave Reflector in the Midcrust beneath Nikko-Shirane Volcano in the Northeastern Japan Arc*, J. Geophys. Res. *101*, 3067–3083.

MAVKO, G. M. (1980), *Velocity and Attenuation in Partially Molten Rocks*, J. Geophys. Res. *85*, 5173–5189.

MITCHELL, B. J. (1995), *Anelastic Structure and Evolution of the Continental Crust and Upper Mantle from Seismic Surface Wave Attenuation*, Rev. Geophys. *33*, 441–462.

MIZOUE, M. (1980), *Deep Crustal Discontinuity Underlain by Molten Material as Deduced from Reflection Phases on Microearthquake Seismograms*, Bull. Earthq. Res. Inst. Univ. Tokyo *55*, 705–735.

MURASE, T. (1982), *Density and Compressional Wave Velocity of Basaltic and Andesitic Liquids*, Carnegie Inst. Wash. YB *81*, 303–305.

MURASE, T., and KUSHIRO, I. (1979), *Compressional Wave Velocity in Partially Molten Peridotite at High Pressures*, Carnegie Inst. Wash. YB *78*, 559–562.

MURASE, T., and MCBIRNEY, A. R. (1973), *Properties of Some Common Igneous Rocks and their Melts at High Temperatures*, Geol. Soc. Am. Bull. *84*, 3563–3592.

NAKAMURA, E., CAMPBELL, I. H., and SUN, S.-S. (1985), *The Influence of Subduction Processes on the Geochemistry of Japanese Alkaline Basalts*, Nature *316*, 55–58.

NOHDA, S., and WASSERBURG, G. J. (1981), *Nd and Sr Isotopic Study of Volcanic Rocks from Japan*, Earth Planet. Sci. Lett. *52*, 264–276.

PEACOCK, S. M. (1990), *Fluid Processes in Subduction Zones*, Science *248*, 329–337.

PEACOCK, S. M., *Thermal and petrologic structure of subduction zones. In Subduction: Top to Bottom* (eds. Bebout, G. E., Scholl, D. W., Kirby, S. H., and Platt, J. P.) (Am. Geophys. Union, Washington D.C. 1996) pp. 119–133.

RINGWOOD, A. E., In *Advances in Earth Sciences* (ed. Hurley, P. M.) (M.I.T. Press, Cambridge 1966) pp. 287–356.

SAKUYAMA, M. (1983), *Petrology of Arc Volcanic Rocks and their Origin by Mantle Diapirs*, J. Volcano. Geotherm. Res. *18*, 297–320.

SATO, H., and MANGHNANI, M. H. (1985), *Ultrasonic Measurements of Vp and Qp: Relaxation Spectrum of Complex Modulus on Basalt Melts*, Phys. Earth Planet. Inter. *41*, 18–33.

SATO, H., and RYAN, M. P., *Generalized upper mantle thermal structure of the western United States and its relationship to seismic attenuation, heat flow, partial melt, and magma ascent and emplacement. In Magmatic Systems* (ed. Ryan, M. P.) (Academic Press, San Diego 1994) pp. 259–290.

SATO, H., and SACKS, I. S. (1989), *Anelasticity and Thermal Structure of the Oceanic Upper Mantle: Temperature Calibration with Heat Flow Data*, J. Geophys. Res. *94*, 5705–5715.

SATO, H., SACKS, I. S., and MURASE, T. (1989), *The Use of Laboratory Velocity Data for Estimating Temperature and Partial Melt Fraction in the Low Velocity Zone: Comparison with Heat Flow and Electrical Conductivity Studies*, J. Geophys. Res. *94*, 5689–5704.

SATO, T., KOSUGA, M., and TANAKA, K. (1996), *Tomographic Inversion for P Wave Velocity Structure beneath the Northeastern Japan Arc Using Local and Teleseismic Data*, J. Geophys. Res. *101*, 17597–17615.

SCHMELING, H. (1985a), *Numerical Models on the Influence of Partial Melt on Elastic, Anelastic and Electric Properties of Rocks. Part I: Elasticity and Anelasticity*, Phys. Earth Planet. Inter. *41*, 34–57.

SCHMELING, H. (1985b), *Partial Melt below Iceland: A Combined Interpretation of Seismic and Conductivity Data*, J. Geophys. Res. *90*, 10105–10116.

SEKIGUCHI, S. (1991), *Three-dimensional Q Structure beneath the Kanto-Tokai District, Japan*, Tectonophys. *195*, 83–104.

SMITH, C. S. (1964), *Some Elementary Principles of Polycrystalline Microstructure*, Met. Rev. *9*, 1–48.

SOBOLEV, S. V., ZEYEN, H., STOLL, G., WERLING, F., ALTHERR, R., and FUCHS, K. (1996), *Upper Mantle Temperatures from Teleseismic Tomography of French Massif Central Including Effects of Composition, Mineral Reactions, Anharmonicity, Anelasticity and Partial Melt*, Earth Planet. Sci. Lett. *139*, 147–163.

TANIGUCHI, H. (1989), *Densities of Melts in the System $CaMgSi_2O_6$-$CaAl_2Si_2O_8$ at Low and High Pressures, and their Structural Significance*, Contrib. Mineral Petrol. *103*, 325–334.

TATSUMI, Y. (1989), *Migration of Fluid Phases and Genesis of Basalt Magmas in Subduction Zones*, J. Geophys. Res. *94*, 4697–4707.

TSUMURA, N., HASEGAWA, A., and HORIUCHI, S. (1996), *Simultaneous Estimation of Attenuation Structure, Source Parameters and Site Response Spectra—Application to the Northeastern Part of Honshu, Japan*, Phys. Earth Planet. Inter. *93*, 105–121.

UMINO, N., and HASEGAWA, A. (1984), *Three-dimensional Qs Structure in the Northeastern Japan Arc*, J. Seismol. Soc. Japan *37*, 217–228.

WAFF, H. S., and BULAU, J. R. (1979), *Equilibrium Fluid Distribution in an Ultramafic Partial Melt under Hydrostatic Stress Conditions*, J. Geophys. Res. *84*, 6109–6114.

WATSON, E. B., and BRENAN, J. M. (1987), *Fluids in the Lithosphere, 1. Experimentally-determined Wetting Characteristics of CO_2—H_2O Fluids and their Implications for Fluid Transport, Host-rock Physical Properties, and Fluid Inclusion Formation*, Earth Planet. Sci. Lett. *85*, 497–515.

ZHAO, D., HASEGAWA, A., and HORIUCHI, S. (1992), *Tomographic Imaging of P and S Wave Velocity Structure beneath Northeastern Japan*, J. Geophys. Res. *97*, 19909–19928.

ZHAO, D., HASEGAWA, A., and KANAMORI, H. (1994), *Deep Structure of Japan Subduction Zone as Derived from Local, Regional, and Teleseismic Events*, J. Geophys. Res. *99*, 22313–22329.

(Received December 12, 1997, revised April 2, 1998, accepted April 23, 1998)

Pure appl. geophys. 153 (1998) 399–417
0033–4553/98/040399–19 $ 1.50 + 0.20/0

▌**Pure and Applied Geophysics**

Comparison between Time-domain and Frequency-domain Measurement Techniques for Mantle Shear-wave Attenuation

Joydeep Bhattacharyya[1]

Abstract—We study the frequency- and time-domain techniques which have been used to measure shear attenuation in the mantle using long-period body waveforms. In the time-domain technique, waveform modeling is carried out and the attenuation model that best fits the data is chosen. In the frequency-domain technique, we solve for the attenuation model that best fits the spectra of the seismic waveforms. Though theoretically both these techniques are equivalent, modeling assumptions and measurement biases associated with each technique can give rise to different results. In this study, we compare these two techniques in terms of their accuracy in obtaining mantle shear attenuation. Specifically, we estimate the biases in constraining attenuation from differential $SS - S$ and absolute S waveforms. We carry out these tests using realistic synthetic seismograms and we follow this with an analysis of recorded data to verify the results from the synthetic tests. For the $SS - S$ waveforms, the primary biasing factors are interference with seismic phases due to mantle discontinuities and due to crustal reverberation under the SS bounce point. These factors can affect the t^* measurements by up to 0.5 s in the frequency domain and more than 1.5 s in the time domain. For the S waveforms, the frequency-domain measurements are accurate to 0.3 s while the time-domain measurements can vary by more than 2.0 s from the predicted values. These errors are also manifested in the t^* measurements made using teleseismically recorded waveforms and lead to comparatively larger noise levels in the time-domain measurements. Based on these results, we propose that in long-period body-wave attenuation studies, frequency-domain techniques should be the method of choice.

Key words: Body waves, seismic Q, S waves, $SS - S$ waveform.

1. Introduction

Several body-wave attenuation studies involve the analysis of long-period waveforms to constrain Q_β variations in the mantle (Sipkin and Jordan, 1980; Revenaugh and Jordan, 1991; Bhattacharyya *et al.*, 1996, 1998; Lay and Wallace, 1983, 1988; Chan and Der, 1988; Sheehan and Solomon, 1992; Ding and Grand, 1993; Reid and Woodhouse, 1997; Sipkin and Revenaugh, 1994; Flanagan and Wiens, 1990, 1994). The attenuation of seismic body waves is typically obtained from their pulse shapes. It can be measured using both

---

[1] Department of Geology and Geophysics, Yale University, New Haven, CT 06511, U.S.A. E-mail: joydeep@hess.geology.yale.edu; Fax: (203) 432-3134.

frequency- and time-domain techniques and can be quantified by the parameter t^* which is defined as the integrated measure of attenuation along the raypath. In the time-domain technique, the t^* value corresponding to the best fitting attenuation operator which maps the observed waveform to the predicted waveform is used. For the frequency-domain technique, the spectral ratio between the observed and the predicted waveforms can be used to obtain the attenuation operator. Though we expect to obtain similar results using the two techniques, various biasing factors can give rise to different t^* values. This difference is primarily due to the fact that attenuation measurements can be sensitive to the seismic source and lateral variations in the velocity structure. For instance, multiple waveforms arriving at nearly the same time will give rise to complicated seismograms which can lead to erroneous attenuation measurements. The effect of these biasing factors can be different and distinct for the time- and frequency-domain techniques. Therefore, to properly assess our measurements, we must estimate the errors which can be associated with each of these techniques.

In this study, we attempt to compare these two techniques as applied to the analysis of shear attenuation in the mantle from teleseismic, long-period transverse component seismograms. We investigate the measurement bias on both differential $(SS - S)$ and absolute waveforms (S), testing for the effects due to interference from nearby waveforms and for various modeling approximations, e.g., absorption band model (LIU *et al.*, 1976; LUNDQUIST and CORMIER, 1980). These tests are carried out using synthetic seismograms computed for realistic, one-dimensional mantle models. To verify the applicability of these tests to seismic data, we follow this with an analysis of a large data set of globally recorded seismograms. In the following sections, we will first describe the attenuation measurements. Next, we will present estimates of measurement bias in each of these techniques. Finally, we will compare these estimates and choose the technique most appropriate for measuring seismic Q from long-period body waves.

2. Method

To estimate the effects of each of the biasing factors, we first construct simple radially symmetric earth models. These models are derived from published one-dimensional earth models. We then compute synthetic seismograms at different distance ranges and source depths for each of these models, using a toroidal-mode summation method (GILBERT and DZIEWONSKI, 1975). These seismograms are then convolved with a typical instrument response to have similar spectral characteristics of long-period data. We use the response of the *SRO* seismometer in this study. Next, we attempt to recover the best-fitting attenuation operator, using both frequency- and time-domain techniques, which maps the reference waveform to the attenuated waveform. For example, the reference phase for the $SS - S$ measurements is the Hilbert transformed S and the t^* value measures the attenuation difference between the waveforms. We use the Hilbert transform to compensate for the propagation effects of SS. CHOY and RICHARDS (1975) have demonstrated the applicability of the Hilbert transform in modeling the pulse-shape distortion of long-period SS waveforms caused by the ray touching a caustic surface. These best-fit t^* values are then compared with theoretically predicted (computed using ray-tracing through the model) values for that seismogram. The bias in the measurement is defined as the difference between these two values.

The frequency-domain technique adopted in this study is presented in detail in BHATTACHARYYA et al. (1996). Briefly, we use the multiple-taper technique of THOMSON (1982) to compute robust spectral estimates of the reference and the attenuated waveforms. We then compute t^* from the shape of the ratio of the two spectra.

The time-domain technique used in our study involves the forward modeling of seismic waveforms. Figure 1 describes the time-domain attenuation measurement technique that we adopt for the S waveforms. A synthetic waveform (in this case the instrument response), is first convolved with a series of probable t^* operators. The convolved waveforms are then cross-correlated with the S waveform an the t^* operator that provides the best fit between the two waveforms (parameterized by the correlation coefficient) is chosen for the seismogram. For the $SS - S$ measurements, the S waveform is used as the synthetic waveform. We first align the S and the Hilbert transformed SS waveform. We then convolve a suite of different attenuation operators to the S waveform and thereafter cross-correlate each of them with SS. The t^* value associated with the waveform with the largest correlation coefficient is chosen for each seismogram.

3. Attenuation Measurements for $SS - S$ Waveforms

BHATTACHARYYA et al. (1996) presented measurements of the $SS - S$ differential t^* values which are primarily sensitive to the variations of Q_β in the upper mantle. This study demonstrates that the amount of scatter in the t^* values is significant and the error estimates for individual measurements can be large. In the following subsections, we estimate the accuracy of these values by considering the primary factors which can bias our measurement technique.

3.1 Testing the t^* Model

For this experiment, we construct a simple earth model with a quadratic velocity and density profile, constant attenuation (Q_β) and dispersion (at a reference period of 1.0 s). The model has no core, mantle discontinuities or a crust. Thus there are no waveforms interfering with the S or the SS waveforms. The synthetics are computed for a surface source, thereby avoiding the depth phases. Variations of this model are constructed with attenuation values of $Q_\beta = 50$ and 100. Figures 2a and 2b show the shear-wave velocity profile and the travel times of the S and the SS phases, respectively, for the more attenuating model ($Q_\beta = 50$). Figure 2c shows the record section of the synthetic seismograms computed using this model. The S, SS, SSS phases and the surface waves are clearly visible in these seismograms.

In Figure 3, we present the results of t^* measurements utilizing the synthetics shown in Figure 2. For the time-domain measurements, the best-fit t^* values are

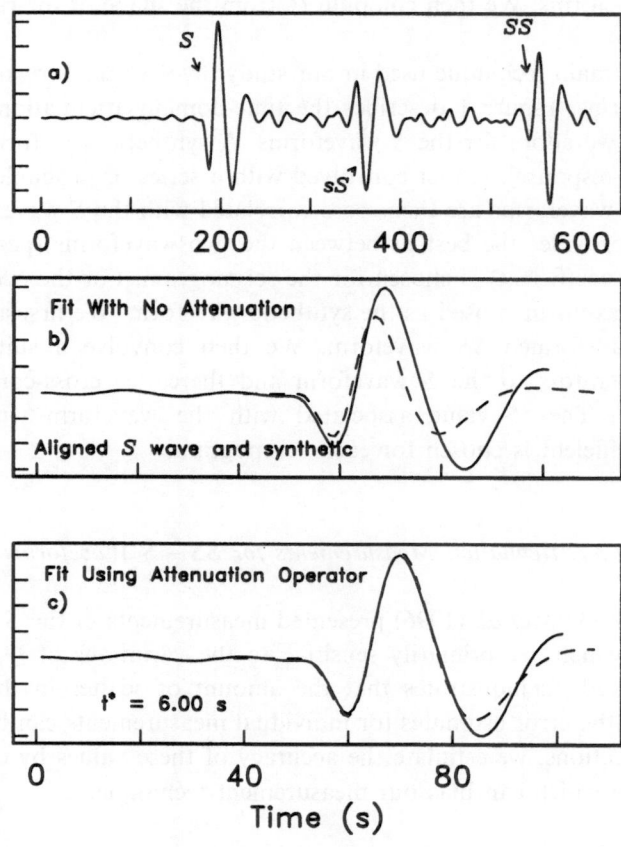

Figure 1

An illustration of the time-domain measurement technique using an S waveform. a) A long-period seismogram with a source depth of 120 km showing the S, sS, and the SS phases. b) The data, i.e., the S waveform, is aligned with the synthetic, i.e., the instrument response. c) The S waveform and the synthetic are aligned after the latter is convolved with a t^* operator of 6.0 s. This t^* operator provides the best mapping between the data and the synthetic and is selected as the measurement for this seismogram.

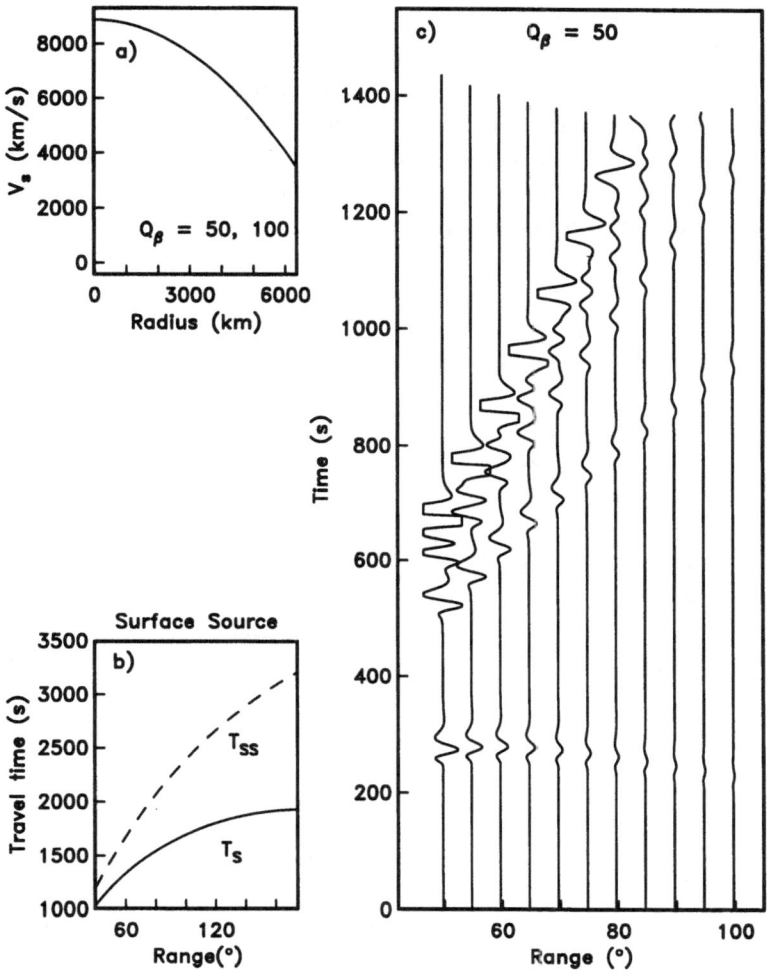

Figure 2

Simple radially symmetric earth model constructed to test the validity of the frequency-independent absorption band model. a) The quadratic velocity profile and the constant shear-wave attenuation of this model. b) Travel-time curves for the S and SS waves for the model. c) Record section of synthetic seismograms computed for the model. The surface waves have been clipped to show details in the body waveforms.

close to the theoretical predictions for the model with $Q_\beta = 50$ (Fig. 3a). For $Q_\beta = 100$, in the range of $55°-70°$, there is discrepancy between the measured and the predicted values for the time-domain measurements which can reach 0.75 s. This might be due to inadequate data windowing in the fitting process. From Figure 3b we can conclude that the measured values which use the frequency-domain technique are within 0.25 s of the model predictions.

We can conclude that for simple earth models, nearly accurate estimates of t^* values can be made which employ both time- and frequency-domain techiques. We note that the frequency-domain technique supplies better estimates, i.e., closer to the predicted values. The strength of the time-domain technique lies in the fact that we can estimate t^* using only a small-part of the seismic waveform and therefore can remove interfering waveforms from the measurement window. On the other hand, the frequency-domain technique requires a window length of at least 35–40 s to

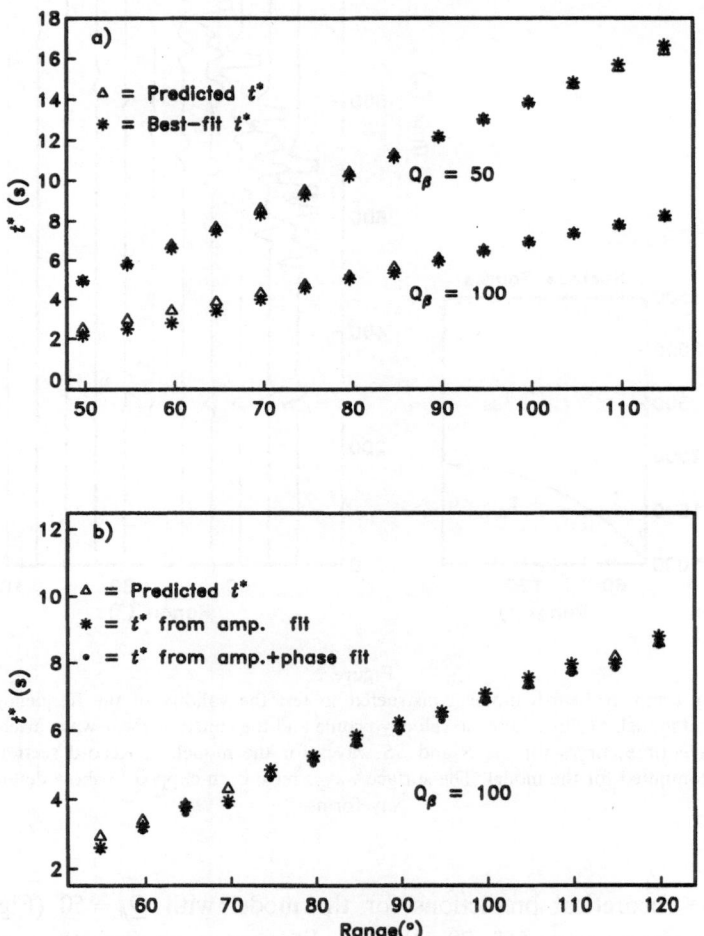

Figure 3
a) Time-domain t^* measurements using synthetic seismograms as described in section 3.1. The predicted t^* values are adequately recovered for both the $Q_\beta = 50$ and $Q_\beta = 100$ models. b) Frequency-domain measurements using both the amplitude fitting and the transfer function fitting techniques as described in BHATTACHARYYA et al. (1996). The measured values agree well with the predicted values.

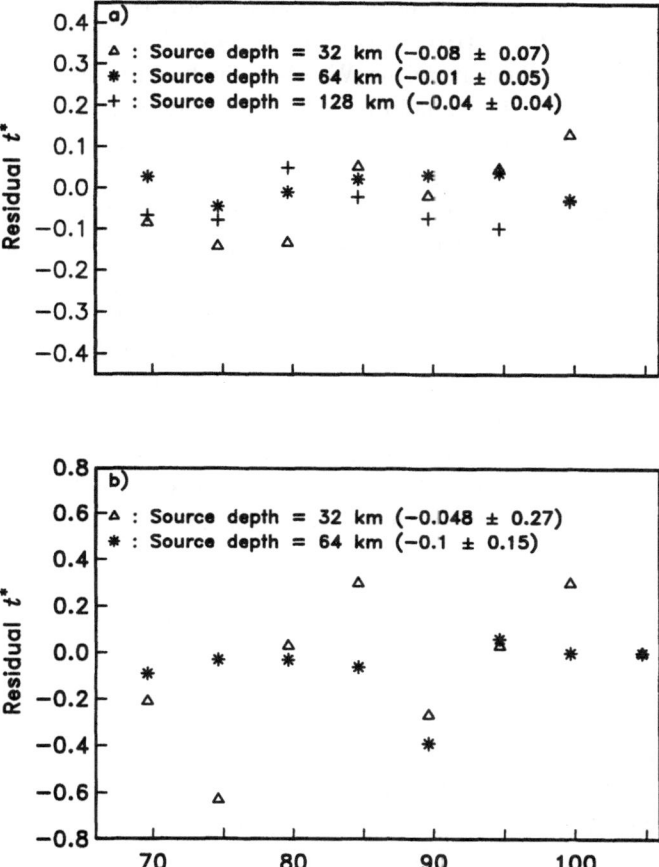

Figure 4

Effect of the depth phases on measured t^*. a) Frequency-domain measurements with the source depths at 32 km, 64 km, and 128 km. The bias can be up to ± 0.15 s. b) Time-domain measurements with source depths at 32 km and 64 km. The bias can be up to 0.6 s. In both cases, the bias decreases as source depth increases.

construct robust estimates of the narrow band (15–25 s period) seismic data, sampled at 1 Hz.

3.2 The Depth Phases, sS and sSS

We study the bias from the depth phase using non-attenuating, non-dispersive, quadratic (i.e., without a core, crust, or discontinuities) earth models. We use four earthquakes at depths of 32, 64, 128, and 193 km. We find that the bias on t^*

measurements decreases with increasing source depth. This is because, for shallower sources, the depth phase arrives closer to the main phase and thereby increases the interference between the waveforms. We show the bias for the frequency-domain t^* measurements for the three shallower depths in Figure 4a. The bias is usually less than ± 0.1 s and can reach 0.15 s for shallower sources. The time-domain measurements (measurements from shallower sources are shown in Fig. 4b), though, are more biased compared to the frequency-domain measurements. For seismograms with source depths at 64, 128, and 193 km, the best-fit t^* values approach the expected values (i.e., residual $t^* = 0$–0.4 s). This discrepancy is more pronounced for seismograms with source depths of 32 km, where it can differ by up to ± 0.6 s from predicted values. Therefore, we can conclude that the bias from depth phases increases for shallower source depths. This is because: (a) the depth phase is closer in time and therefore can superpose on the direct phase; (b) the depth phases are usually more attenuated for deeper sources. We note that globally most earthquakes are shallow (i.e., ≤ 50 km source depth). Therefore, the effect of this bias can be significant for time-domain $SS - S$ t^* measurements.

3.3 Crustal Reverberations under the SS Bounce-point

We investigate the effect of the crust applying two models differing in their crustal thicknesses. Single-layer crustal models are considered. These models are modifications of the non-attenuating, non-dispersive, quadratic earth model described in section 3.1. The first model has a 35-km-thick crust which is the total crustal thickness of model *PEMC* (DZIEWONSKI *et al.*, 1974). The crustal thickness of the second model is equal to 20 km. The impedance contrast at the Moho is similar to that in *PEMC*, i.e., approximately equal to 0.17. We show the results from the frequency-domain measurements in Figure 5a. The bias is ≤ 0.2 s for the model with a 35-km crust and that from a 20-km crust is larger and can extend to -0.4 s. The negative bias in the latter case is mainly due to the interference from the underside reflection, i.e., precursor, from the Moho, which arrives just before the SS phase. In the time-domain measurements (Fig. 5b) for the 35-km model there is a slightly negative bias (< 1.0 s) between 60°–70° range. Above 70° the bias is ≈ 0.4 s. In the 20 km case we could only get acceptable fits if we use the first half of S and SS waveforms. In both time- and frequency-domain, the bias slightly increases with distance. At ranges below 90° there is a slight positive bias (< 0.5 s) and at higher ranges, the negative bias is ≈ -1.5 s.

We observe that the effect of a thin crust (less than 20 km thick) can be a significant source of bias for $SS - S$ t^* measurements. In the synthetic tests, the effect of the crust is pronounced when the fitting window includes the back-end of the S and SS waveforms. This is because the postcursor from the topside Moho reflection is harder to separate when constructing the fitting window than the underside Moho reflection, i.e., the precursor, since the latter can appear as a distinct arrival fronting the *SS* phase. Thus, the biasing effect of the crust can be minimized if the crustal precursor can be effectively removed from the fitting window. This is quite difficult for real data as this feature can be hard to detect in the presence of pre-event noise. Therefore, in realistic situations, the effect of superposed arrivals due to the crust under the *SS* bounce-point is expected to be more pronounced than that which we have documented here. Conversely, explicitly modeling for the crust under the *SS* bounce-point, using results from a global

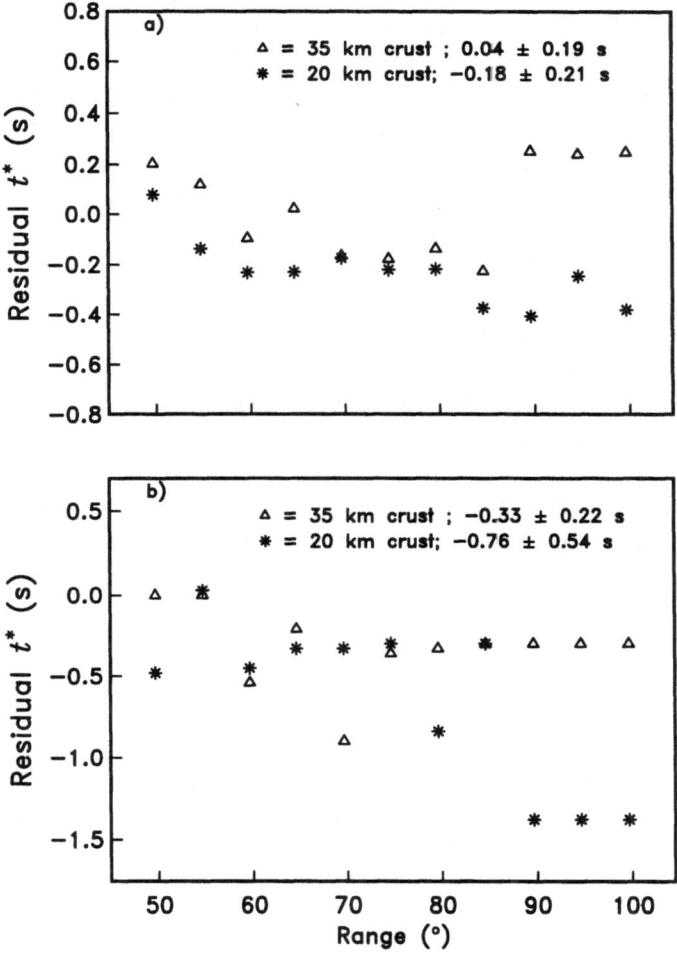

Figure 5

Residual t^* measured from synthetics for earth models with two different crustal thicknesses. a) Frequency-domain measurements. The measurements can be biased by up to 0.4 s due to a shallow crustal layer under the *SS* bounce-point. b) Time domain measurements. The measured t^* value may differ by up to 1.5 s from the predicted value.

compilation of crustal thicknesses (e.g., MOONEY *et al.*, 1998), can potentially decrease this bias. Given an accurate crustal velocity model, this bias can be effectively removed by correctly matching the *S* and the *SS* waveforms.

3.4 Upper Mantle Discontinuities

The topside and underside reflections from the upper mantle discontinuities can generate several phases which are precursors to *SS* and postcursors to *S*. Depending on the location of the discontinuities, these phases can have arrival times close to the *SS* waveform. Discontinuity phases related to *ScS*, though of smaller amplitude, can also interfere with the *SS* waveform. Though a standard earth model, e.g., *PEMC*, exhibits a number of upper mantle discontinuities, the converted phases are strongest for the 670-km and the 420-km discontinuities. We use two earth models; one with just 670 and 420 km discontinuities (model 4.1) and the other with upper mantle discontinuities at 120 km, 220 km, 420 km, and 660 km depths, as in *PEMC* (model 4.2). Neither of the models has a crust. We tested the bias due to these arrivals using synthetics for both attenuating and non-attenuating dispersive models and the results are similar.

The impedance contrasts at the 670 and 420 discontinuities in model 4.1 are as in *PEMC*, i.e., 0.09 and 0.05, respectively. We make the measurements for seismograms in the distance range 65° to 100° where the waveforms are clearly observed. The measurements show a strong variation from the theoretical predictions in the distance range 70°–85° as they differ by as much as 0.6 s for the frequency domain and up to 1.5 s for the time domain from the predicted values for both attenuating models (Fig. 6). At ranges $\geq 90°$, where both *S* and *SS* are sufficiently separated from the discontinuity phases, we recover the best-fit t^* values to within experimental accuracy (0.25 s) of the predicted values.

The measurements from models 4.1 and 4.2 can be compared to explore the effects of the 120-km and 220-km discontinuities. Though the impedance contrasts at these discontinuities are smaller than those for 660-km and 420-km discontinuities (as in *PEMC*), their arrival times are closer to the *SS* phase, which can lead to a significant bias in $SS - S$ t^* measurement. The best-fit t^* values are found to be quite similar for both models (Fig. 6). Therefore, we can conclude that the biasing effects of the shallow upper mantle discontinuities, at 120-km ad 220-km depths, are probably insignificant.

3.5 Effect of ScS

To investigate this factor, we use a realistic earth model which is smooth through the mantle. This allows us to study the range-dependent effects due to *ScS* and also remove unwarranted complications from mantle discontinuities. For this purpose, we choose the 1066*A* (GILBERT and DZIEWONSKI, 1975) model without

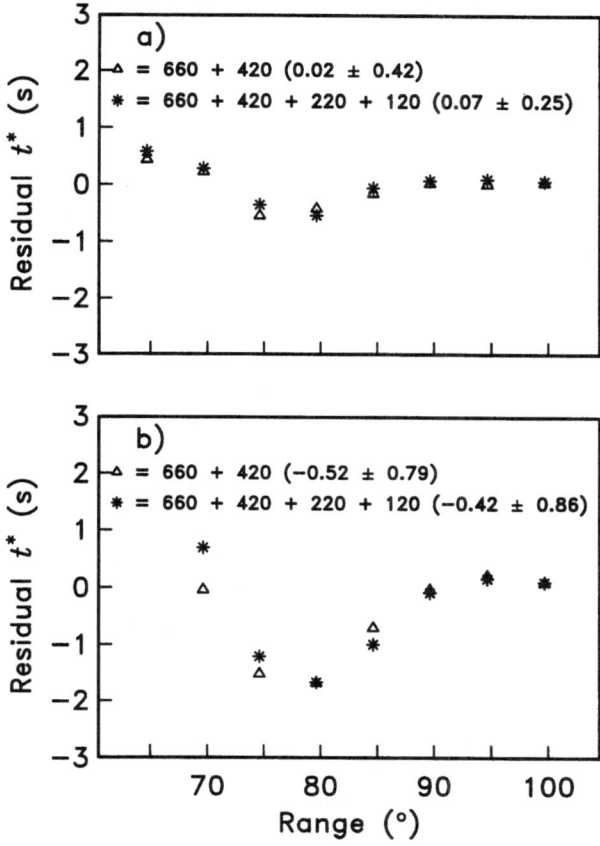

Figure 6
Residual t^* values showing the bias from the discontinuity phases. a) Frequency-domain measurements can be biased up to ± 0.5 s. b) Time-domain measurements can be biased by up to 1.9 s.

the crustal layer. In this model, the S velocity gradient varies considerably through the mantle with a distinct break in the mid-mantle. Also, this model has a low-velocity zone in the uppermost mantle. We slightly change this model and choose a quadratic mantle model with gradient $= 0$ at the core-mantle boundary. This removes the mid-mantle discontinuity and insures that ScS is the only arrival between S and SS in the synthetics.

In Figure 7 we display the $SS - S$ residual t^* values (measure t^*—predicted t^*) due to ScS. The effect of ScS on these waveforms is more at higher ranges ($\geq 90°$). This is because at these ranges long-period S and ScS have arrival time differences considerably smaller than the dominant period of the waveforms, and they effectively overlay each other. Using the frequency-domain method, we can recover the predicted values with an accuracy of ± 0.15 s. The bias is considerably higher in the time-domain measurements, and it can reach to 2.0 s for some distance ranges.

3.6 Improving $SS - S\ t^*$ Measurements

We infer from the synthetic tests that interfering waveforms can bias $SS - S\ t^*$ measurements made using both frequency- and time-domain techniques. Careful choice of fitting windows can allow us to isolate and discard biased waveforms. In the frequency-domain technique the interference appears as characteristic holes in the spectra. This instability in the spectra is clearly observable, thereby allowing us to: (a) select the part of the waveforms which gives us stable spectra, or, (b) remove the corrupted t^* measurements from subsequent analysis. Observing and discriminating against this interfering waveforms is much more difficult in time-domain. In the latter case, we can minimize the effect of interference only if we observe a sharp break in the waveform. For long-period waveforms, with a dominant period ≈ 20 s, this waveform distortion is smooth and therefore difficult to identify. This leads to higher levels of bias in time-domain measurements. Therefore, we suggest that frequency-domain measurement techniques should be preferred in similar attenuation studies. Our experiments suggest that we might expect a scatter of up to 0.5 s from the frequency-domain measurements.

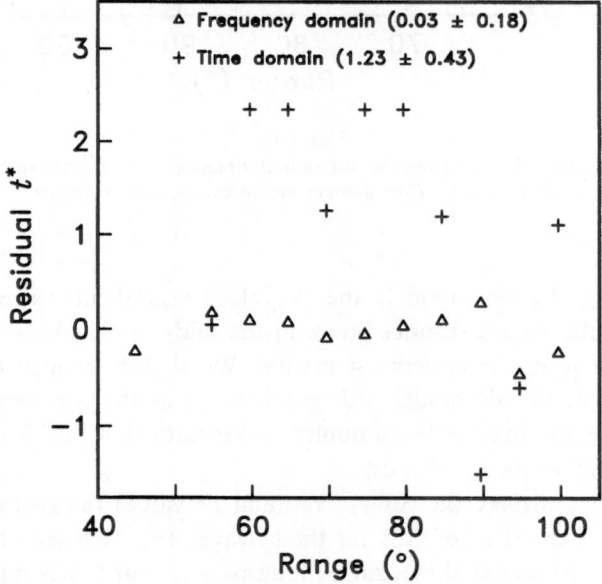

Figure 7

Bias on $SS - S\ t^*$ measurements due to ScS. Though the bias is small for frequency-domain measurements (± 0.25 s), it can exceed 2.0 s for time-domain measurements.

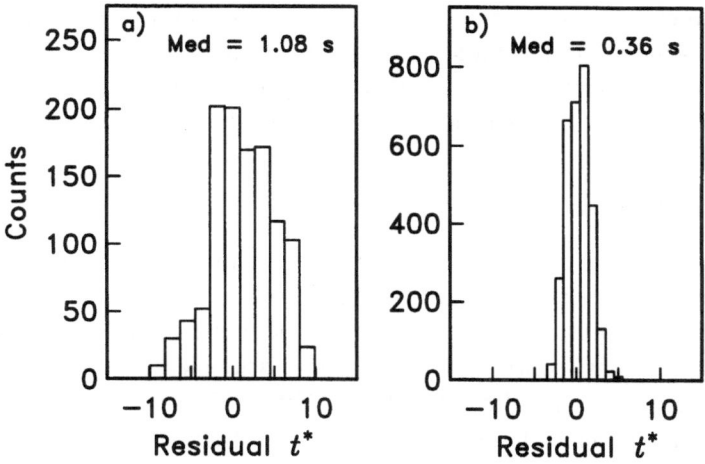

Figure 8

Distribution of $SS - S$ t^* from an analysis of globally recorded long-period shear waveforms using different techniques. a) Time-domain measurements; b) frequency-domain measurements. The time-domain measurements have a wider distribution, probably due to scatter from noisy measurements.

3.7 Time-domain versus Frequency-domain t^* Measurements for Globally Recorded Teleseismic $SS - S$ Waveforms

The differential pulse widths between the SS and the S waveforms are used to model for attenuation in the upper mantle under the SS bounce-point. In this experiment, we use both the techniques to measure t^* from more than 2800 high quality seismograms in $45°-100°$ distance range. (An analysis of the frequency-domain measurements has been presented in BHATTACHARYYA et al., 1996). For a comparison of the methods, we present the distribution of the t^* values measured using both techniques (Fig. 8). We observe that the time-domain measurements have considerably larger variability, with a higher average value. This difference in variability suggests to us that much of the scatter in the time-domain measurements is probably due to noise, i.e., interference from nearby phases, rather than due to three-dimensional attenuation structure. This is similar to the conclusion derived from analyzing synthetic seismograms which indicates that the time-domain $SS - S$ t^* values are less reliable than the corresponding frequency-domain values.

4. Attenuation of S Waveforms

In this section, we investigate the measurement biases on the attenuation measurements for S waveforms utilizing both time- and frequency-domain techniques. The measurements are made using a t^* model for seismograms at $20°-100°$

distance range. The methodology is similar to the earlier synthetic experiments: compute seismograms for radially symmetric simple earth models; model for a frequency-independent t^* operator with the reference pulse being the instrument response; and compare this measured t^* with predicted values for that model.

4.1 Measurement Bias Estimated from Synthetic Tests

The measured S t^* value is sensitive to attenuation variations along the S raypath through the entire mantle. We have identified the discontinuity phases (the topside and the bottomside reflections from the upper mantle discontinuties) and ScS to be the primary source of bias in these measurements. We note that the depth phase, sS, can strongly affect the S waveform. To remove this bias, we only consider seismograms with a source depth greater than 80 km. Moreover, to avoid source complexity, we use earthquakes with magnitude (M_b) between 5.5 and 6.3. Since the interfering waveforms have different moveout velocities with respect to S, the bias on measured t^* values is expected to be range dependent. Consequently, we expect that the discontinuity phases will have a stronger effect at shorter distance ranges and ScS will interfere at higher ranges with the S waveform.

To constrain range-dependent effects, we need to use seismograms computed for a realistic earth model. We use the $PEMC$ model, without crust, for this purpose. Figure 9 shows the synthetic seismograms computed for this model at 28° to 106° distance range. For the synthetic computation, the source is fixed at a nominal depth of 140 km. As we can observe from this figure, the back-end of the S waveform can be very complicated for distances below 35°. This is because the arrival times of the discontinuity phase, the Love wave, and the direct S waveform are close. This waveform complication severely biases the t^* measurements at these distance ranges. Also, for the time-domain measurements, we fit only the first half of the waveform. This reduces the bias from crustal reverberations and finite source size. Figure 10 delineates the predicted and the measured frequency- and time-domain t^* values for synthetics at 62°–102° ranges. One obvious conclusion we can draw from this figure is that the frequency-domain measurements are exceedingly more precise than those of time-domain. We note that the bias due to ScS at ranges 84°–92° for frequency-domain measurements can be up to 0.4 s and for time-domain measurements can be up to 2.25 s at 82°–100°. Destructive interference due to ScS arriving at the back-end of the S pulse is the cause of this bias. At ranges greater than 92°, ScS overlies the S waveform and both phases have similar raypaths, leading to negligible bias in the frequency-domain measurements. For ranges $\geq 100°$, we observe a positive bias in the measured t^* values. This is probably due to the bias from the diffracted phase around the core mantle boundary, S_{diff}. We can conclude from this test that for absolute S t^* measurements the frequency-domain technique, at ranges 35°–100°, is accurate to within ≈ 0.3 s.

On the other hand, similar measurements using a time-domain technique can be biased by more than 2.0 s.

4.2 Measurement Bias Estimated for Teleseismic S Waveforms

We analyze more than 2700 absolute S waveforms, using long-period Global Seismic Network (GSN) data collected during the years 1976–1987. These are high

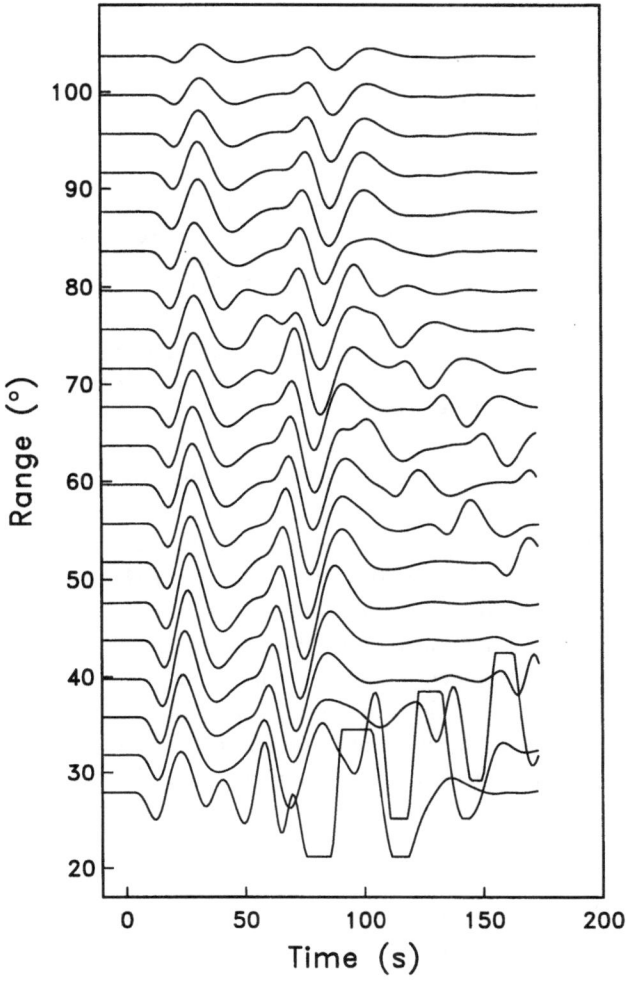

Figure 9

Synthetic seismograms for the PEMC model without the crustal layer. We note that the S waveform is quite complicated at shorter distance ranges, i.e., less than 35°. The arrival of ScS at the back-end of the S pulse, at ranges near 85°–92°, also causes distortion to the S pulse. We have clipped the surface waves at distances ≤35° to accentuate the body waveforms.

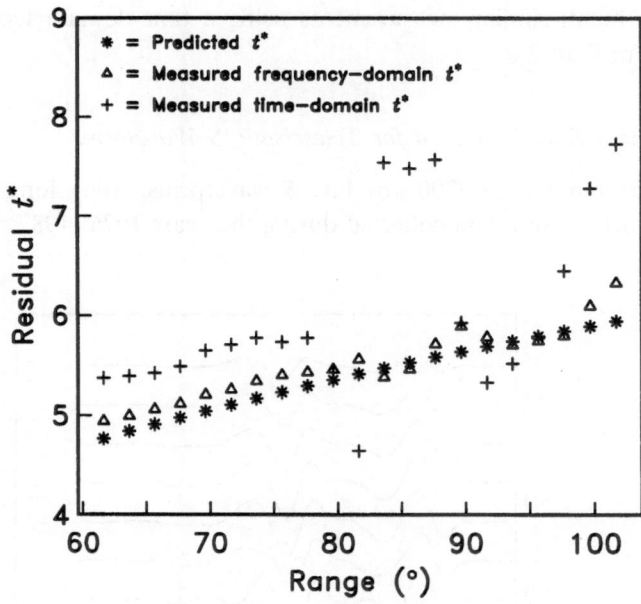

Figure 10

Frequency- and time-domain absolute S measured t^* values along with the predicted t^* values which the synthetics show in Figure 9. Time-domain technique can furnish severely biased estimates of t^* value while the frequency-domain measurements can retrieve the predicted values nearly accurately.

signal-to-noise ratio transverse component recordings of earthquakes with source depths greater than 80 km. Figure 11 shows the distribution of the measured t^* values using both time- and frequency-domain techniques. In this figure, we present the measured residual t^* values with respect to *PREM* (DZIEWONSKI and ANDERSON, 1981). The time-domain values, which are a subset of the frequency-domain measurements, have a much wider distribution and also a higher average value. Since both sets of measurements constrain the same physical quantity, i.e., variations of shear-wave attenuation in the mantle, this larger variability is most probably due to increased measurement biases in the time-domain t^* values. This is similar to the conclusion we reached from the analysis of synthetic seismograms in Section 4.1.

5. Frequency-domain Technique versus Time-domain Technique: Which One to Choose?

As we have mentioned in the preceding sections, both synthetic and real data suggest that the time-domain technique can give us more biased estimates of attenuation compared to frequency-domain measurements. Seismic waveforms of smaller or comparable amplitude which arrive close to the phase analyzed, can give rise to subtle waveform effects which are hard to detect and thereby, difficult to isolate, in the time-domain measurements. In the frequency-domain technique, these interfering arrivals usually give rise to instability in the spectra and thus can be identified and their effect reduced. Moreover, high-frequency components of a seismic waveform can be selectively de-emphasized in the time-domain fitting process, both due to source spectral rolloff and seismic attenuation. Therefore the t^* values, which are mostly determined from the low-frequency components, can be biased. On the other hand, the spectral ratio technique approximately weighs each of the spectral components equally, thereby reducing this bias. The synthetic tests suggest that the bias from the time-domian technique can exceed 2.0 s while the bias from the frequency-domain technique hardly exceeds 0.5 s for both S and $SS - S$ t^* measurements. Hence, we prefer the frequency-domain measurements to constrain the attenuation of shear waves in the mantle. Typically, the range of measured frequency-domain t^* values (with respect to $PREM$) is -2.5 to 2.5 s (see Figs. 8 and 11) and therefore these measurements have a signal-to-noise ratio of 3–4. We emphasize that interference from superposed arrivals will bias the t^* measurements and therefore these measurements should be removed. We find that the frequency-domain technique is comparatively more efficient than the time-domain technique in identifying these inaccurate t^* values.

In this study, we have chosen a frequency-independent attenuation model which is equivalent to those used in similar analyses of long-period seismic data (BHAT-TACHARYYA et al., 1996, 1998; DING and GRAND, 1993; FLANAGAN and WIENS,

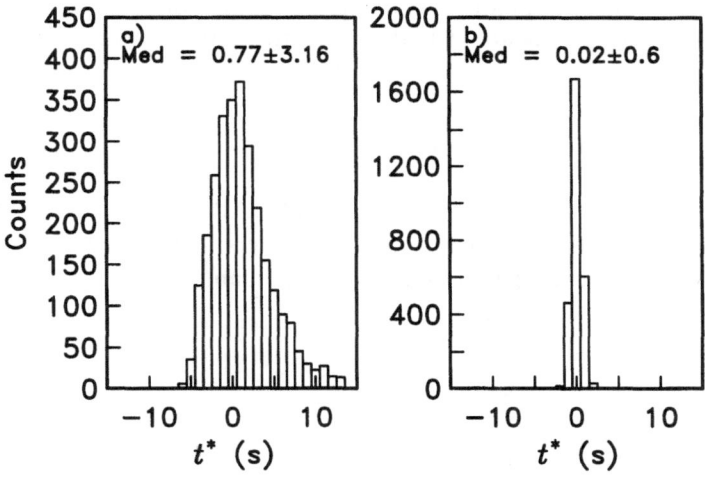

Figure 11

A comparison between time-domain (a) and frequency-domain (b) measurements for teleseismic S waveforms. We observe considerably more scatter in the time-domain measurements, probably due to biased picks.

1990, 1994). As described in BHATTACHARYYA et al. (1993), t^* values can be biased if we have either a low-frequency or a high-frequency absorption band. This bias is expected to be small for the narrow-band data (15–25 s) used in this study.

6. Conclusions

In this study, we have explored the various biasing factors which can effect the commonly used time- and frequency-domain attenuation techniques. We have used an absorption band model to measure seismic attenuation (LIU et al., 1976; LUNDQUIST and CORMIER, 1980). We can draw the following conclusions from this analysis:

1. The frequency independent t^* model is adequate to estimate mantle attenuation from long-period seismograms.
2. Interference from nearby phases is the primary source of bias in long-period body-wave attenuation measurements.
3. Time-domain technique can give severely biased estimates of attenuation. Subtle effects from interfering waveforms, in some cases, can affect the measured t^* value by more than 2.0 s.
4. Frequency-domain measurements are comparatively more capable of retrieving the predicted t^* values. Using this technique, we can expect a bias extending to 0.5–0.75 s for the $SS - S$ measurements and reaching 0.2–0.4 s for the S measurements.
5. Frequency-domain techniques, compared to time-domain techniques, produce more robust measurements of body-wave attenuation in the mantle.

Acknowledgments

This research has been supported by grant EAR94–05948 from the National Science Foundation. Suggestions from Guy Masters, Peter Shearer, and an anonymous reviewer have considerably improved this manuscript.

REFERENCES

BHATTACHARYYA, J., SHEARER, P. M., and MASTERS, G. (1993), Inner Core Attenuation from Short-period PKP (BC) versus PKP (DF) Waveforms, Geophys. J. Int. 114, 1–11.
BHATTACHARYYA, J., MASTERS, G., and SHEARER, P. M. (1996), Global Lateral Variations of Shear-wave Attenuation in the Upper Mantle, J. Geophys. Res. 101, 22,273–22,289.
BHATTACHARYYA, J., MASTERS, G., and ROMANOWICZ, B. (1998), Global Lateral Variations of Shear-wave Attenuation in the Mantle, Geophys. J. Int., in review.
CHAN, W. W., and DER, Z. A. (1988), Attenuation of Multiple ScS in Various Parts of the World, Geophys. J. Int. 92, 303–314.

CHOY, G. L., and RICHARDS, P. G. (1975), *Pulse Distortion and Hilbert Transformation in Multiply Reflected and Refracted Body Waves*, Bull. Seismol. Soc. Am. *65*, 55–70.

DING, X.-Y., and GRAND, S. P. (1993), *Upper Mantle Q Structure under the East Pacific Rise*, J. Geophys. Res. *98*, 1973–1985.

DZIEWONSKI, A. M., HALE, A. L., and LAPWOOD, E. R. (1974), *Parametrically Simple Earth Models Consistent with Geophysical Data*, Phys. Earth Planet. Inter. *10*, 12–48.

DZIEWONSKI, A. M., and ANDERSON, D. L. (1981), *Preliminary Reference Earth Model*, Phys. Earth Planet. Inter. *25*, 297–356.

FLANAGAN, M. P., and WIENS, D. A. (1990), *Attenuation Structure beneath the Lau Backarc Spreading Center from Teleseismic S Phases*, Geophys. Res. Lett. *17*, 2117–2120.

FLANAGAN, M. P., and WIENS, D. A. (1994), *Radial Upper Mantle Structure of Inactive Backarc Basins from Differential Shear Wave Measurements*, J. Geophys. Res. *99*, 15,469–15,485.

GILBERT, F., and DZIEWONSKI, A. M. (1975), *An Application of Normal Mode Theory to the Retrieval of Structural Parameters and Source Mechanisms from Seismic Spectra*, Phil. Trans. Roy. Soc. London, Ser. A *278*, 187–269.

LIU, H., ANDERSON, D. L., and KANAMORI, H. (1976), *Velocity Dispersion due to Anelasticity: Implications for Seismology and Mantle Composition*, Geophys. J. R. Astron. Soc. *47*, 41–58.

LAY, T., and WALLACE, T. C. (1983), *Multiple ScS Travel Times and Attenuation beneath Mexico and Central America*, Geophys. Res. Lett. *10*, 301–304.

LAY, T., and WALLACE, T. C. (1988), *Multiple ScS Attenuation and Travel Times beneath Western North America*, Bull. Seismol. Soc. Am. *78*, 2041–2061.

LUNDQUIST, G. M., and CORMIER, V. C. (1980), *Constraints on the Absorption Band Model of Q*, J. Geophys. Res. *85*, 5244–5256.

MOONEY, W. D., LASKE, G., and MASTERS, T. G. (1998), *CRUST 5.1: A Global Crustal Model at $5° \times 5°$*, J. Geophys. Res. *103*, 727–747.

REID, F., and WOODHOUSE, J. H. (1994), *Measurement of Differential Travel Times and t* and the Structure of the Upper Mantle*, EOS, Trans. AGU *78*, F485.

REVENAUGH, J. S., and JORDAN, T. H. (1991), *Mantle Layering from ScS Reverberations, 1. Waveform Inversion of Zeroth-order Reverberations*, J. Geophys. Res. *96* 19,749–19,762.

SHEEHAN, A. F., and SOLOMON, S. C. (1992), *Differential Shear-wave Attenuation and its Lateral Variation in the North Atlantic Region*, J. Geophys. Res. *97*, 15,339–15,350.

SIPKIN, S. A., and JORDAN, T. H. (1980), *Regional Variations of Q_{ScS}*, Bull. Seismol. Soc. Am. *70*, 1071–1102.

SIPKIN, S. A., and REVENAUGH, J. S. (1994), *Regional Variation of Attenuation and Travel Time in China from Analysis of Multiple-ScS Phases*, J. Geophys. Res. *99*, 2687–2699.

THOMSON, D. J. (1982), *Spectrum Estimation and Harmonic Analysis*, IEEE Proc. *70*, 1055–1096.

(Received March 26, 1998, revised May 6, 1998, accepted May 20, 1998)

Pure appl. geophys. 153 (1998) 419–438
0033–4553/98/040419–20 $ 1.50 + 0.20/0

❘Pure and Applied Geophysics

A Model to Study the Bias on Q Estimates Obtained by Applying the Rise Time Method to Earthquake Data

SALVATORE DE LORENZO[1]

Abstract—Among Q-estimation methods a simple linear technique consists of the evaluation of the increasing rise time of body waves with the increase of their travel time. This method, known as the rise time method, was theoretically justified for an impulsive source time function (Dirac delta function). WU and LEES (1996), throughout finite difference calculations, showed that, when considering finite source time functions, characterized by a cut-off frequency around 20 Hz, the rise time method can be satisfactorily applied to invert earthquake data.

In order to establish the applicability of the rise time method to an arbitrary earthquake source we analytically solved the problem of the propagation, throughout an anelastic medium, of a signal generated by a finite dimensions seismic source: the shear dislocation fault of BRUNE (1970). Analyzing theoretical rise time vs. travel-time curves, we were able to distinguish two different corner frequency ranges in which the trend is different. When corner frequency is below 10 Hz the discrepancies with the rise time method increase with a decrease of the corner frequency. When corner frequency is above 10 Hz no meaningful differences are observed.

The application of the model to a synthetic data set, based on the sources-receivers configuration of the 15 November 1995 Border Town, Nevada, earthquake sequence, shows that a significant bias affects Q estimates obtained with the rise time method, for seismic events characterized by a Brune corner frequency less than 5 Hz.

Key words: Brune source function, rise time, quality factor, ω^{-2} model.

1. Introduction

When a seismic wave propagates through the earth's interior it is subjected to a fractional loss of energy, mainly due to degradation in the elastic properties of the rocks (shear and compressional moduli). To quantify the energy lost by a wave traversing an imperfectly elastic material, seismologists introduced the quality factor Q, defined as:

$$1/Q = 2\pi(\Delta E/E) \tag{1}$$

[1] Università di Bari, Dipartimento di Geologia e Geofisica, Campus Universitario, Via E. Orabona, 4, 70125-Bari, Italy. E-mail: giandelo@inopera.it

where ΔE is the energy lost by a periodic wave in a cycle and E is the maximum of the energy in that cycle (AKI and RICHARDS, 1980). The amount of anelastic attenuation suffered by a wave traveling a distance x is given by $t^* = T/Q$ where T is the travel time of the wave.

There are several reasons why detailed knowledge of the absorbing properties of materials is required. First, measurements of the quality factor Q constitute a fundamental tool in the evaluation of the physical state of the rocks. In the uppermost crust low-Q anomalies are usually associated to fractured systems fluid-filled (ZUCCA et al., 1994) or to magmatic bodies (SANDERS and NIXON, 1995), particularly in volcanic areas.

Based on the strong decrease of Q with increasing temperature observed in laboratory experiments (KAMPFMANN and BERCKEMER, 1985), MITCHELL (1995) explained the decrease of Q with increasing depth, in the mantle, in terms of differences in the temperature.

Second, detailed knowledge of the path effects due to anelastic attenuation is required when correcting displacement spectra to estimate source parameters of tectonic earthquakes (ABERCROMBIE, 1995, 1997).

Finally, in engineering seismology an adequate comprehension of the absorbing properties of the most surficial layers is required. ANDERSON and HOUGH (1984) suggested that the spectral decay at high frequencies observed on Californian accelerograms is related to attenuation within the earth; they attributed it to attenuation close to the accelerograph site.

An important problem in studying attenuation in the earth consists of the Q dependence on frequency. Several authors (CAPUTO, 1967; KJARTANSSON, 1979) demonstrated that a linear dissipative model (i.e., Q almost independent on frequency) can satisfactorily explain intrinsic absorption of body waves in materials. An exactly constant Q was used both in earthquake sources studies (ABERCROMBIE, 1995, 1997) and in several tomographic inversions of the attenuation structures in volcanic areas (SANDERS and NIXON, 1995; WU and LEES, 1996).

It was pointed out (FEHLER et al., 1992) that losses due to scattering from heterogeneities could explain the observed frequency dependence of Q in various regions of the earth. AKI (1986) observed that Q is usually high and poorly dependent on frequency for old stable continents; on the contrary, for young active regions, Q is low and shows a strong frequency dependence.

Several authors (e.g., O'NEILL and HEALY, 1973; ANDERSON, 1986) studied the problem of the bias affecting source parameters estimates when anelastic attenuation is not well determined. Clearly, the same problem occurs when the purpose consists of determining anelastic attenuation from earthquakes data when source parameters are not well determined.

This paper deals with the problem of the bias affecting Q estimates derived by the rise time method when applied to earthquakes data. In order to quantify the bias we deduced an analytical expression of the wave field generated by a finite source time function throughout an anelastic media and compared our results with those relative to the impulsive source time function.

2. The Model

A simple and powerful Q estimation technique is the rise time method. This method, proposed by RICKER (1953), was successively applied by GLADWIN and STACEY (1974) to ultrasonic pulses propagating in the walls of hard-rock tunnels. The rise time τ of a pulse (Fig. 1) is defined as the ratio between its maximum amplitude and its maximum slope. The theoretical background of this method was given by KJARTANSSON (1979). He demonstrated that when considering the propagation of a pulse (Dirac's delta) throughout an anelastic medium characterized by an exactly constant Q, the relation between τ and t^* is linear:

$$\tau = \tau_0 + C(Q)t^* \tag{2}$$

where τ_0 is the rise time at the source. The slope $C(Q)$ of the straight line (2) is a function of Q. When $Q > 20$ the slope is independent on Q; its value is different when considering displacement seismograms ($C = 0.485$) or velocity seismograms ($C = 0.298$), or acceleration seismograms ($C = 0.145$).

There are two fundamental reasons why the rise time method is often preferred to other techniques. First, the rise time does not depend on the arrival time of the pulse, therefore it avoids uncertainty in the uniqueness of the wavelet arrival. Secondly, since the rise time measurement requires only the first quarter of the

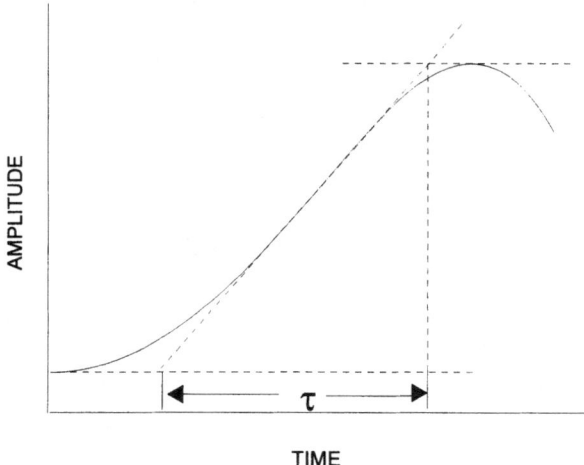

Figure 1
Geometrical picture of the rise time τ of the signal, defined by the maximum amplitude divided by the maximum slope.

wavelet, the errors on the rise time estimates are poorly influenced by secondary arrivals due to waves diffracted by heterogeneities in the medium.

The main problem which develops when using the rise time method to estimate anelastic attenuation from signals generated by earthquakes, consists in the actual difficulty of establishing the validity of equation (2) for signals containing a finite source time function. When observing that the amplitude of the Fourier transform of the Dirac delta function is the unit constant, it can be easily derived that, theoretically, the rise time method cannot be applied to cases in which a spectral falloff at finite frequencies is observed on displacement spectra, i.e., in all cases in which we account for finite dimensions of seismic source.

This is confirmed by results of BLAIR and SPATIS (1982). They stated that, when improperly applied, the rise time method leads to unacceptable results. Analyzing ultrasonic wave trains generated by two types of explosives and recorded by an array of accelerometers, they established that the different initial rise time of the pulses lead to different values of ultrasonic attenuation along the same path. They concluded that the source spectrum had to be considered when inferring ultrasonic attenuation from changes in rise time.

A point source at which stress is instantaneously relieved radiates P- and S-waves displacement pulses that are Dirac delta functions. On the contrary, when stress is relieved over a finite length of faulting, the radiated pulses are broadened and the rise time of the pulse is proportional to the finite rupture time. The effect of the finite dimensions of faulting reveals itself in a spectral falloff of the amplitude displacement spectra at high frequencies, above an assigned value of the frequency. Various earthquake source theories predict a far-field displacement spectrum that is constant at low frequencies and inversely proportional to the square of the frequency at high frequencies (ω^{-2} model) (AKI, 1967; BRUNE, 1970; MADARIAGA, 1977). A spectrum's corner frequency is usually defined as that frequency where the high and low-frequency asymptote intersect (MOLNAR et al., 1983).

When quality factor is very low along the ray path, the shape of the spectrum of an earthquake can be altered so that source corner frequency can be masked by an apparent lower corner frequency due to Q. This phenomenon is referred to as the f_{max} effect (HANKS, 1982). PAPAGEORGIU and AKI (1983) explained f_{max} in terms of a source effect thus implying a departure from self-similarity. By analyzing the source spectrum of various earthquakes recorded from boreholes, ABERCROMBIE (1995) pointed out that there is also no breakdown in self-similarity when considering fault length less than 10 m and that f_{max} is an artifact of severe attenuation along the ray path. Moreover she showed that the ω^{-2} source model can be satisfactorily used to fit earthquake spectrum. Our model (whose theoretical derivation is reported in the Appendix) is concerned with an approach to the problem that follows the point of view of ANDERSON and HOUGH (1984) and of ABERCROMBIE (1995).

Our interest is mainly confined to establish the cases in which the rise time method, formulated to study the propagation of pulses (Dirac's delta) in an anelastic medium, can also be applied to the signals generated by a finite source time function. Using a finite difference calculation, Wu and Lees (1996) showed that, when considering finite source time functions (Ricker, Ricker derivative, Gaussian, Erf) characterized by a cut-off frequency at $f_0 = 20$ Hz, no meaningful differences with the rise time method are obtained, both for the homogeneous and the heterogeneous anelastic media.

Clearly, results by Wu and Lees cannot be generalized at any finite dimension seismic source. The same authors, in fact, stated that the linearity of rise time of body waves vs. t^* for arbitrary sources has not been theoretically modeled and its application to real earthquake sources is unknown.

In order to obtain a general comprehension of the problem, it is necessary to solve the problem using a completely arbitrary cut-off frequency at the source, and this is possible only when we are able to establish the exact expression of the time domain far-field displacement generated in an anelastic medium by a finite source time function. Based on this, we solved the problem of the propagation of a signal containing a finite source time function throughout an anelastic homogeneous medium. We used the Brune (1970) source function to model tectonic earthquake sources. In the Brune model, seismic source is modeled as a shear dislocation that is equivalent to a circular fault along which stress is instantaneously relieved. The finite length of faulting reveals itself on a seismic spectrum leading to a corner frequency in the spectrum given by

$$f_0 = 2.34 \frac{v}{\pi r} \tag{3}$$

where v is the phase velocity of the radiated waves and r is the radius of the circular fault. Scaling law for a Brune source can be derived by applying the relation

$$\Delta\sigma = \frac{7M_0}{16r^3} \tag{4}$$

where $\Delta\sigma$ is the stress drop and M_0 the seismic moment.

In determining the analytical form of the wave field generated by a Brune circular source in an anelastic homogeneous medium, we adopted the Strick (1970) attenuation operator. The integration was carried out in the complex plane of a z variable, whose real part coincides with frequency. Details of the integration are described in the Appendix.

The obtained wave forms (eq. A.14) were plotted and compared with those obtained by Strick (1970), relative to the propagation throughout the anelastic medium of a pulse (Dirac delta function) (Fig. 2). The comparison allows us to state that, when considering a Brune source there is an additional delay of the arrival time of the energy peak with respect to the impulsive case (Fig. 2a) and an additive low-pass filtering in the ascending part of the signal which adds to the low-pass filtering caused by attenuation. The decay of the peak amplitude with distance and the broadening of the wave packet for the far-field displacement given by (A.14) is shown in Figure 2b. As can be seen (Fig. 2c), the changes in the shape of the signal and the delay of peak arrival time are very small when the cut-off

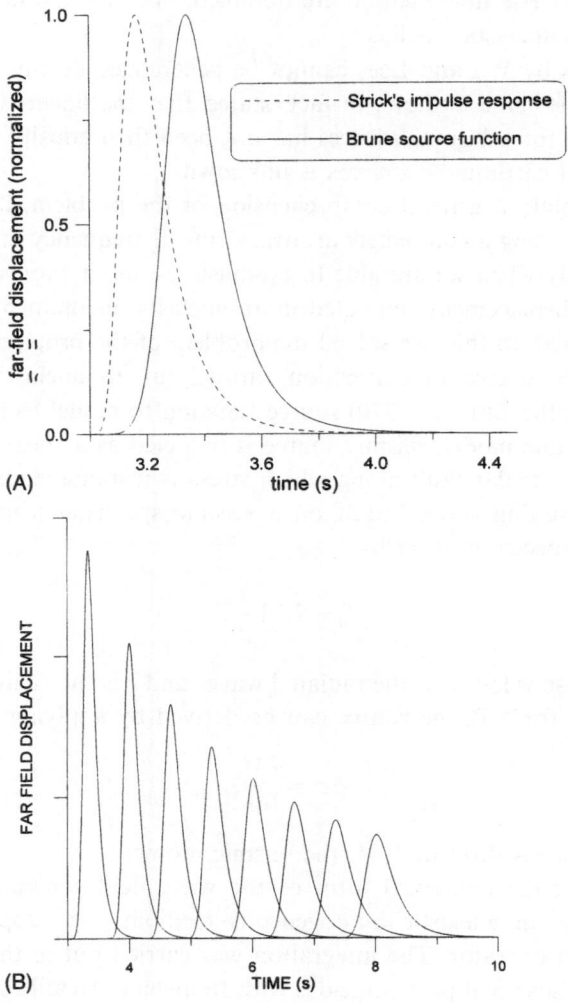

Figure 2
(A) Comparison between the wave shape of the far-field impulse response proposed by STRICK (1970) and the far-field displacement resulting from the Brune source function for the case $c(\omega_0 = 1 \text{ Hz}) = 3$ km/s and $Q = 20$. (B) Far-field displacement caused by the source time function (A.14) for different distance source-receivers, in the case $c(\omega_0 = 1 \text{ Hz}) = 3$ km/s and $Q = 20$. (C) The effect of different values of the cut-off frequency ω_0 on the shape of the far-field given by equation (A.14). For high values of the corner frequency ω_0 the filtering effect due to the source spectrum is strongly reduced.

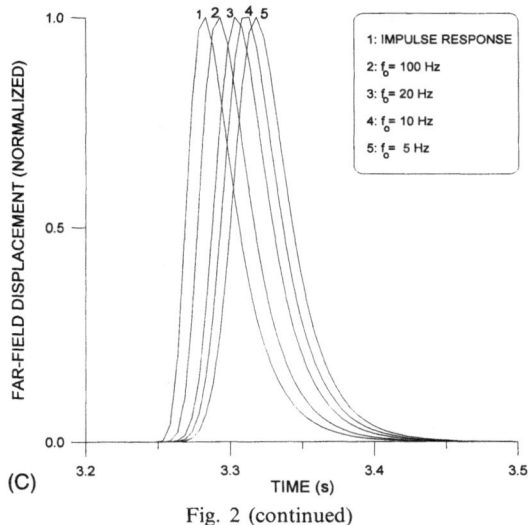

Fig. 2 (continued)

frequency ω_0 is increased. This is not surprising, because, in the limit $\omega_0 \to \infty$, the expression (A.14) coincides with the far-field impulse response (STRICK, 1970).

3. Discussion

In this section we are concerned with establishing the bias affecting Q estimates obtained with the rise time method when analyzing data from earthquake sources.

The first step in our revision of the rise time method to account for finite source dimensions consists of the calculation of rise time vs. travel-time curves.

In Figure 3 we compared our results with those obtained by using impulsive source solutions (STRICK, 1970). We assumed $c = 3$ km/s at reference frequency $\omega_0 = 1$ Hz.

The rise time dependence on travel time exhibits a deviation (Fig. 3) from the linear trend. The results are summarized in Table 1. A plot in logarithmic scale according to the following equation

$$\log \tau = A + B \log T \tag{5}$$

seems to provide a more accurate description of the rise time vs. travel-time curves. This deviation from a linear trend is very small (in the range 10^{-3}–10^{-2} s) and decreases with increasing Q; Table 1 shows an asymptotic tendency of the curves to the linear trend when $Q \to \infty$. Because of the uncertainty that affects seismograms, it could develop that the rise time data are equally well fitted utilizing the assumption of a linear law as given by (2) or of a power law as given by (5) so that the discussion regarding the effective trend of theoretical rise time curves becomes useless.

The most important question regards, on the contrary, the different estimates of the attenuation that can be obtained using appropriate source time functions instead of an impulsive source function. In fact, as shown in Figures 3 and 4, a systematic gap between rise time vs. travel-time curves for the impulse response case and those relative to the source time function (A.7) exists. This indicates that the source effects on the attenuation cannot be neglected. As can be seen in Figure 4, the theoretical rise time curve, relative to the Brune source and corresponding to an assigned quality factor Q_1, results very close to that relative to the impulse response case but corresponding to another value of the quality factor Q_2, with $Q_2 < Q_1$. The difference between Q_2 and Q_1 increases with the increase of Q_1. This is a significant result because the sufficiently marked differences between Q_2 and Q_1 (up to $Q_1 - Q_2 = 35$ in Fig. 4) imply that a significant underestimate of Q could occur using impulse response solutions instead of the appropriate source function. In

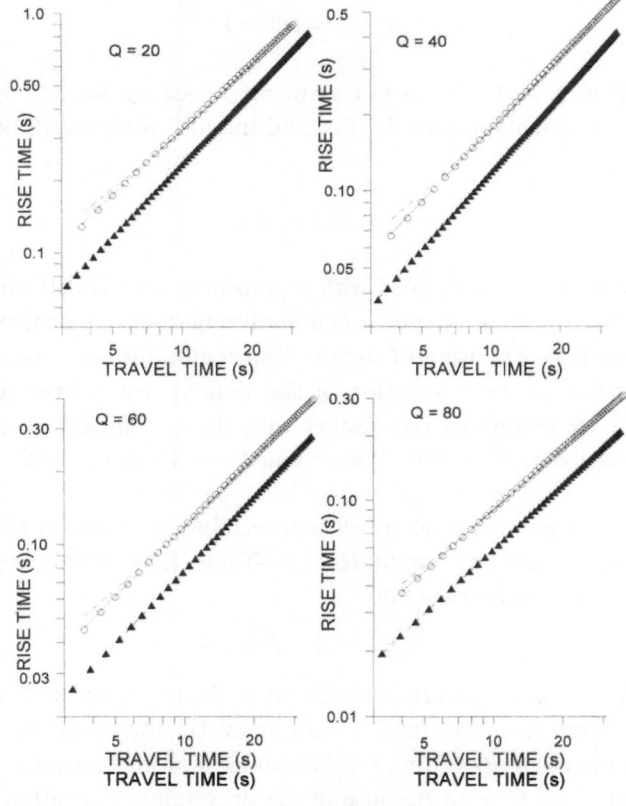

Figure 3

Comparison between the rise time vs. travel-time curves for the impulse response case (triangles) and those relative to the Brune source function (open dots). The dashed line represents the best-fitting straight line for both the cases. The solid line represents the regression curve (for the ω-square case) described by equation (5). The comparisons refer to the case $c(\omega_0 = 1\ \text{Hz}) = 3\ \text{km/s}$.

Table 1

Slope B and intercept A values of the regression curves of the rise time vs. travel time
$$\log \tau = A + B \log T$$
$$c(\omega_0 = 1) = 3 \text{ km/s}$$

Q	A	B
20	−3.209	0.906
40	−3.897	0.930
60	−4.307	0.944
80	−4.599	0.952

order to evaluate the effect of the corner frequency ω_0 on the rise time, we calculated theoretical rise time vs. Q^{-1} curves for different values of ω_0. As can be seen (Fig. 5), the deviation of the curves with respect to the impulse response case increases with decreasing ω_0, reaching high values up to 0.3 s for the case $\omega_0 = 0.1$ Hz. With the increase of ω_0 the misfit reduces, vanishing for $\omega_0 \geq 10$ Hz.

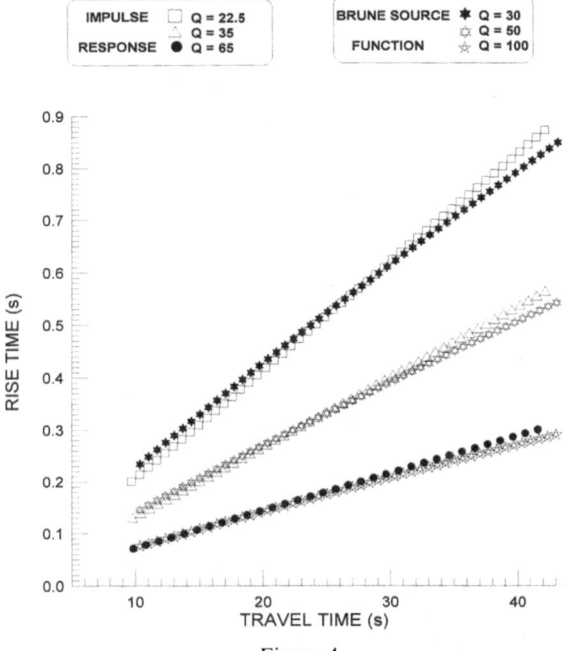

Figure 4

Rise time vs. travel-time curves both for the impulse response case and for the ω-square case, for different values of Q, shown in the squares on the top of the figure. The rise time curve relative to the impulse response case, corresponding to an assigned Q value, almost coincides with that obtained using equation (A.14) but corresponding to a greatest Q value. This result indicates that an underestimate of the quality factor could occur if we neglect source effects.

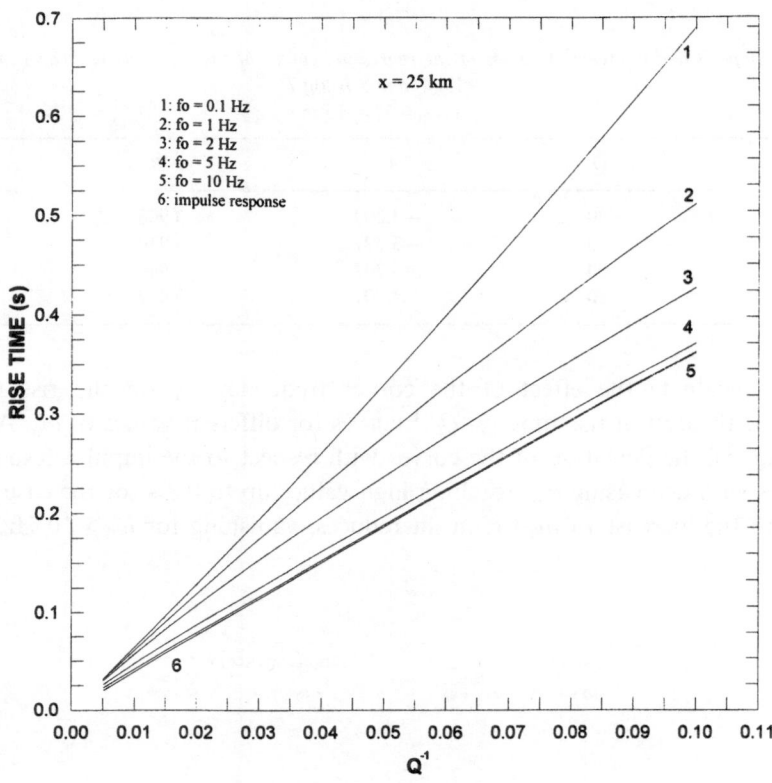

Figure 5

Rise time vs. Q^{-1} curves for the ω-square case, at distance $x = 25$ km from the source. The different curves refer to different values of the cut-off frequency ω_0, shown in the square. The curve corresponding to $\omega_0 = 10$ Hz almost overlaps with that relative to the impulse response case. The curves refer to the case $c(\omega_0) = 3$ km/s.

To establish for which corner frequencies the rise time method can be applied and in which cases it should be modified to account for source properties we neglect the small deviations from linearity of τ vs. t^* data.

We carried out several numerical simulations of the wave field, given by (A.14) varying the corner frequency of the Brune source in the range 1–10 Hz and Q in the range 100–500, and computed theoretical τ vs. t^* curves described by equation (2) measuring τ on velocigrams. The result is that $C(Q)$ depends on the source spectrum. This is clearly shown in Figure 6 where the slope $C(Q)$ of the straight lines as a function of the Brune corner frequency is plotted, for different Q values.

In the range 1–5 Hz $C(Q)$ is both a function of Q and of the corner frequency. With an increase of the corner frequency, $C(Q)$ tends to 0.292 (approximately the same value obtained by KJARTANSSON (1979) for velocity seismograms), independently on Q.

The plot of Figure 6 compels us to affirm that above 10 Hz $C(Q)$ is practically independent both on Q and on frequency. In this way we can confirm previous results by Wu and Lees (1996) that used a constant value of $C(Q)$ to invert rise time data with seismic sources characterized by cut-off frequencies around 20 Hz. On the contrary, when seismic spectra are characterized by corner frequencies less than 10 Hz, a bias affects Q estimates obtained with the rise time method.

4. Example of Application

The techniques actually used to jointly estimate source parameters and attenuation are concerned with the fitting of the seismic spectrum of radiated waves. Anderson and Hough (1984) inverted acceleration spectra of some Californian earthquakes using a theoretical form of the acceleration spectrum given by

$$A(f) = A_0 \exp(-\pi k f) \tag{6}$$

where A_0 represents the long-period amplitude of acceleration spectrum and k accounts for both site and attenuation effects. Theoretically, expression (6) is obtained by assuming that seismic spectra are represented by the convolution of a ω-square source model with a constant attenuation operator.

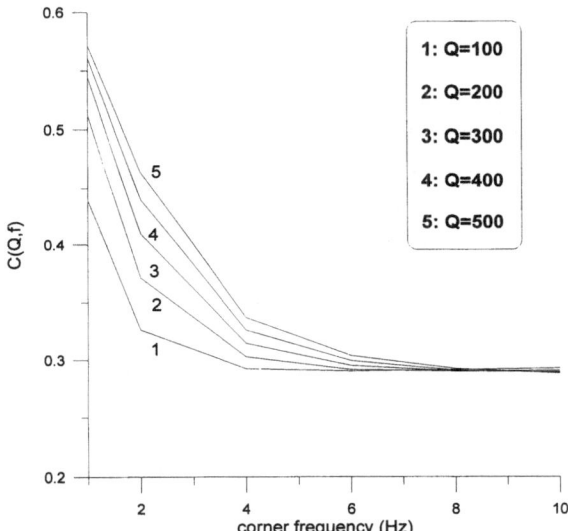

Figure 6

Plot of the slope $C(Q)$ of the straight line used to interpolate rise time vs. travel time for velocity seismograms corresponding to the wave field generated by a Brune source function in an anelastic medium, as a function of the corner frequency of the spectrum. The different curves refer to different Q values, shown in the square. The curves refer to the case $c(\omega_0) = 3$ km/s.

A refining of Anderson and Hough's model was proposed by ABERCROMBIE (1995). The displacement spectra were fitted with a theoretical function of the form

$$\Omega(f) = \frac{\Omega_0 \, e^{-(\pi f t/Q)}}{\left[1 + \left(\dfrac{f}{f_c}\right)^n\right]^{1/\gamma}} \tag{7}$$

where Ω_0 is the long-period amplitude, f is the frequency, f_c is the corner frequency, t is the travel time of the body wave, Q is the frequency independent quality factor, n the high frequency fall-off rate and γ is a constant. The choice of the values of n and γ depends on the source model used to fit data. When using nonlinear optimization techniques, amplitude spectra can be inverted to obtain source parameters and attenuation. When $\gamma = 2$ and $n = 2$ in equation (7) the formulation of Abercrombie is equivalent to that of Anderson and Hough.

Based on the approach proposed by ABERCROMBIE (1995), ICHINOSE et al. (1997) inverted source parameters and attenuation for the 15 November, 1995 Border Town, Nevada, earthquake sequence. They determined both Brune corner frequencies, $k = t/Q$ and M_w ($1.8 < M_w < 4.5$) of 15 earthquakes using $\gamma = 2$ and $n = 2$ in equation (7).

Our model is theoretically equivalent of the Anderson and Hough model (or of Abercrombie model for $\gamma = 2$ and $n = 2$). The only difference consists of the data chosen to estimate model parameters. In the Anderson and Hough model the data are the seismic spectra; in our model they consist of rise time measurements of body waves.

Here we are interested in quantifying the bias on Q estimates obtained by the application of the classical rise time method on signals generated by earthquake sources. In order to quantify this effect we computed synthetic velocity seismograms for the source receivers configurations of the 15 November 1995 Border Town, Nevada, earthquake sequence using as Q values those obtainable by k of ICHINOSE et al. (1997) throughout the equation:

$$Q = T/k \tag{8}$$

where T is the travel time of the waves computed in the velocity model of ICHINOSE et al. (1997). Also phase velocity β of S is obtained from this velocity model averaging on the β velocity of single layers.

Because the Brune source function does not account for directivity effects of seismic radiation, our model does not account for this effect, so that, our results, conceptually, are true only in a neighbouring of the receivers.

For each source-receiver couple studied by ICHINOSE et al. (1997) (reported in Table 2) we computed synthetic velocigrams by using equation (A.14) along an assigned profile containing 10 stations located in an angular neighbouring of 10° around the receiver. On each synthetic seismogram rise time was measured and rise time vs. travel time was successively fitted by applying equation (2), in order to

Table 2

Source parameters and attenuation k value for 15 events of the 15 November Border Town, Nevada, earthquake sequence (Δ = epicentral distance; r = radius of the circular fault in the Brune model; f_c = corner frequency; k = attenuation factor $Q = T/k$; Q (rise time = estimates of Q obtained with the rise time method; ΔQ = bias affecting Q estimates). The first six columns were deduced by ICHINOSE *et al.* (1997)

Event Number	Station	Δ (km)	f_c (Hz)	k (msec)	r (m)	Q	Q (rise time)	ΔQ
02	WCN	45.6	1.5	62	859	229	160	69
12	BRD1	10.6	9.3	27	141	122	119	3
13	BRD1	10.6	11.1	27	118	122	120	2
13	BRD2	12.6	8.8	28	148	140	135	5
14	BRD1	7.7	9.9	28	131	86	85	1
14	BRD2	9.8	18.9	26	69	117	117	0
15	BRD1	10.6	5.0	28	260	118	105	13
15	BRD2	12.6	5.1	27	258	145	123	22
16	BRD1	10.0	15.6	26	84	120	119	1
16	BRD2	11.9	13.6	26	96	143	142	1
17	BRD1	10.3	13.0	29	101	111	110	1
17	BRD2	12.2	16.0	28	81	136	135	1
18	BRD1	10.8	21.7	26	60	129	129	0
18	BRD2	12.9	18.4	26	71	155	155	0
19	BRD1	10.5	23.7	26	55	126	126	0
19	BRD2	12.2	20.2	26	64	146	146	0
20	BRD1	10.8	14.0	27	93	125	125	0
20	BRD2	12.7	14.7	32	89	124	124	0
21	BRD1	11.2	7.6	28	171	125	115	10
21	BRD2	13.0	8.6	28	152	145	130	15
22	BRD1	11.2	9.7	28	135	124	121	3
22	BRD2	12.7	8.0	30	163	132	125	7
23	BRD1	11.7	3.8	29	346	126	100	26
23	BRD2	13.1	4.5	28	291	146	125	21
24	BRD1	10.2	19.8	26	66	122	122	0
24	BRD2	12.0	16.2	26	80	144	142	2
27	BRD1	8.1	15.1	26	86	97	96	1
27	BRD2	9.7	15.9	26	82	116	115	1

compute the estimate of Q obtained with the rise time method. Results are reported in Table 2. The differences between exact Q values and those obtained with the rise time method were plotted as a function of the Brune corner frequency in Figure 7. The differences are sufficiently high in the range 1–5 Hz (corresponding to $2.6 < M_w < 4.5$) and decays to zero when corner frequency increases beyond 10 Hz. These results are clearly a consequence of the above explained dependence of $C(Q)$ on source spectrum (Fig. 6).

5. Conclusions

By using a simple seismic source model we have shown that when corner frequencies are sufficiently high ($f_0 > 10$ Hz) rise time is still a powerful technique to estimate attenuation, also from signals generated by earthquakes.

In all other cases, the dependence of $C(Q)$ on source spectrum implies that the classical rise time method can lead to biased estimates of attenuation when applied to earthquake data (e.g., for a Brune source characterized by a radius $r = 859$ m and a corner frequency $f_0 = 1.5$ Hz we determined a difference $\Delta Q = 69$ when $Q = 229$).

The approach we presented is the natural extension of the ANDERSON and HOUGH (1984) frequency method at the time domain. In fact, in both models seismograms are interpreted in terms of the convolution of an attenuation operator with an ω-square source function. Some differences in estimating attenuation could be obtained by the two methods since, in practice, spectra of body waves are often computed by including coda of body waves so that Q estimates could, in principle, differ.

Results of this work, obtained by considering only the general properties of the seismic spectra, must be extended to take into account other properties of faulting processes (e.g., the role played by a finite rupture velocity along the fault) in order to also model the directivity effect of the seismic radiation.

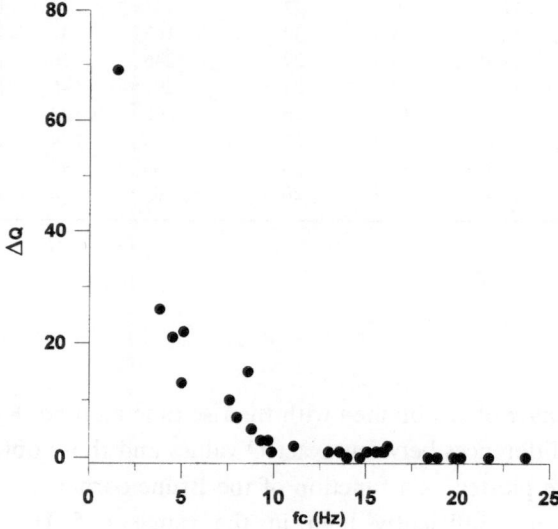

Figure 7
The bias ΔQ affecting Q estimates derived by the rise time method determined on a synthetic data set using as source-receivers configuration the 15 November, Border Town, Nevada, earthquake sequence (ICHINOSE et al., 1997), as a function of the Brune corner frequency.

Appendix

The description of seismic attenuation is usually carried out throughout the quality factor Q, that can be defined (KJARTANSSON, 1979) as

$$\frac{1}{Q} = \tan \delta \qquad (A.1)$$

where δ is the phase angle between stress and strain. As can be theoretically demonstrated (AKI and RICHARDS, 1980) we need to admit a frequency dependence of Q to satisfy the causality requirements. Since, in some cases, a constant Q was obtained by various seismic analyses, some authors (e.g., STRICK, 1970) proposed a power-law for attenuation, that gives a Q essentially constant over a wide frequency range.

In the frequency domain, the far-field displacement of seismic waves propagating in an attenuating medium is given by

$$A(x, \omega) = A_0(x, \omega)B(x, \omega)/x \qquad (A.2)$$

where $A_0(x, \omega)$ represents the source spectrum and $B(x, \omega)$ the anelastic response of the medium to a seismic pulse (Dirac's delta). Here x represents the source-receiver distance.

In this paper we utilize the attenuation law (STRICK, 1970)

$$B(x, \omega) = \exp(-\gamma(p)x) \qquad (A.3)$$

where $p = i\omega$, and $\gamma(p)$ is the complex propagation function, given by

$$\gamma(p) = \tau p + K(s)p^s \qquad (A.4)$$

where τx represents the travel time for the purely elastic case, and s is a real number. As discussed in detail by STRICK (1970) the above choice of the propagation function (A.4) allows us to have an essentially constant Q. Moreover, if

$$0 < s < 1$$

the casuality requirements are satisfied. As can be deduced by KJARTANSSON's paper (1979)

$$K(s) = \frac{1}{c_s}$$

and

$$c_s = \sqrt{\frac{M_0}{\rho}} \, \omega_0^{-\gamma}$$

where M_0 is the modulus at reference frequency ω_0 and ρ is the density. Moreover, the introduction of the complex propagation function (A.4) takes to a phase velocity dispersion law given by

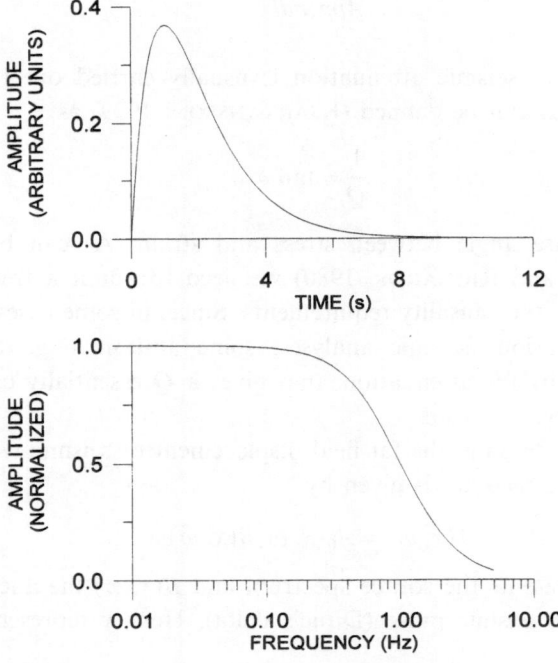

Figure A.1

Far-field displacement generated by a Brune source function (on the top) and its amplitude spectra (on the botton) for the case $\omega_0 = 1$ Hz

$$c(\omega) = c_0 \left| \frac{\omega}{\omega_0} \right|^\gamma \tag{A.5}$$

where c_0 is the phase velocity at reference frequency ω_0 and

$$\gamma = 1 - s = \frac{1}{\pi} \tan^{-1} \left(\frac{1}{Q} \right) \approx \frac{1}{\pi Q}. \tag{A.6}$$

We consider now a BRUNE (1970) source time function. It can be expressed as

$$a_0(t) = u \omega_0^2 t \, \exp(-\omega_0 t). \tag{A.7}$$

In equation (A.7) $\omega_0 = 2.21 \; \beta/r$, where r is the radius of the equivalent circular dislocation surface and β is the phase velocity of the S waves. Moreover

$$u = \frac{1}{\omega_0^2} f \frac{r}{R} \left(\frac{\sigma}{\mu} \right) \beta \tag{A.8}$$

where R is the distance of the observation point, σ the effective stress, μ is the rigidity and f is a term used to take into account geometrical spreading.

Using the theory of analytical functions (residue theory) (SVESNICHOV and TICHONOV, 1984) it can be easily shown that the source time function (Fig. A.1) given by (A.7) is the Fourier transform of the following function in the frequency domain

$$A_0(\omega) = \frac{u}{(1 + i\omega/\omega_0)^2} \tag{A.9}$$

whose modulus is given by

$$|A_0(\omega)| = \frac{u\omega_0^2}{\omega^2 + \omega_0^2}.$$

When we substitute (A.9) and (A.3) in (A.2) and transform the result in the time domain, we obtain

$$A(x, t) = \int_{-i\infty}^{i\infty} F(p) \exp(xf(p))\, dp \tag{A.10}$$

where

$$F(p) = \frac{u}{2\pi i x (1 + (p/\omega_0))^2} \tag{A.11}$$

and

$$xf(p) = -Kxp^s + (t - \tau x)p. \tag{A.12}$$

To evaluate $A(x, t)$ let us consider the following integral:

$$\oint_C F(z) \exp(xf(z))\, dz \tag{A.13}$$

where the imaginary part of the complex variable z is p and C (Fig. A.2) is a close contour in the z plane given by

$$C = \gamma_1 \cup \gamma_2$$

where γ_1 is the oriented segment described by

$$\gamma_1 = \{z : [\text{Re}(z) = 0; -R \leq \text{Im}\, z \leq R]\}$$

and γ_2 is the oriented half-circle described by equation

$$z = Re^{i\varphi} \quad -\frac{\pi}{2} < \varphi < \frac{\pi}{2}.$$

The pole of $F(z)$: $z = -\omega_0$ lies externally to the integration contour (Fig. A.2) so that $F(z)$ is an analytical function inside C.

We use the steepest descent method (SVESNICHOV and TICHONOV, 1984; EWING et al., 1957) to evaluate (A.13). Observing that the only saddle point of $xf(z)$ is given by

$$z_1 = \left[\frac{skx}{t - \tau x} \right]^{1/(1-s)}$$

in which $xf(z)$ and its second derivative assume that values

$$xf(z_1) = -\left(\frac{1-s}{s} \right)(t - \tau x)\left(\frac{sKx}{t - \tau x} \right)^{1/(1-s)}$$

$$x \frac{d^2 f}{dz^2} \bigg|_{z_1} = (1-s)(t - \tau x)\left(\frac{t - \tau x}{sKx} \right)^{1/(1-s)}$$

and that

$$F(z_1) = \frac{u}{2\pi i x \left(1 + \dfrac{1}{\omega_0}\left(\dfrac{sKx}{t - \tau x} \right)^{1/(1-s)} \right)^2}$$

we obtain

$$\oint_C F(z) \exp(xf(z))\, dz = \frac{u \, \exp\left\{ -\left(\dfrac{1-s}{s} \right)t'\left[\dfrac{sKx}{t'} \right]^{1/(1-s)} \right\}}{\left((1-s)(t')\left(\dfrac{t'}{sKx} \right)^{1/(1-s)} \right)^{1/2}\left(1 + \dfrac{1}{\omega_0}\left(\dfrac{sKx}{t'} \right)^{1/(1-s)} \right)^2}$$

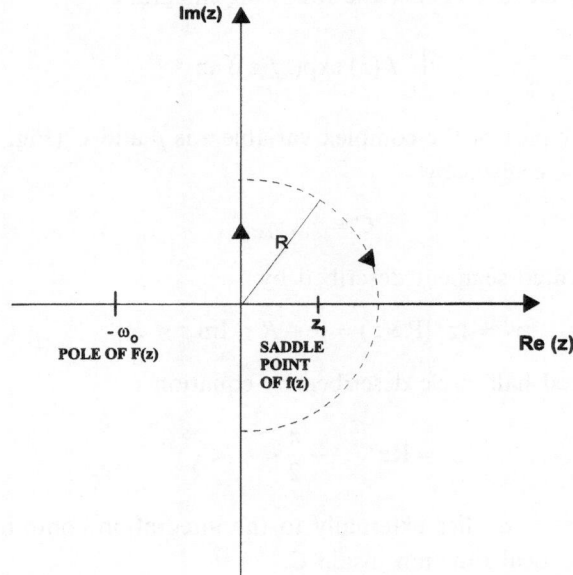

Figure A.2
The close contour along which is carried out the integration of the integral (A.13). Also are shown the positions of the saddle point of $f(z)$ and of the pole of $F(z)$.

where

$$t' = t - \tau x.$$

Being $F(z)$ uniformly tending to zero for $R \to \infty$, in the limit $R \to \infty$, utilizing Jordan's Lemma (Svesnichov and Tichonov, 1984) it can be easily shown that

$$\lim_{R \to \infty} \int_{\gamma_2} F(z) \exp(xf(z))\, dz = 0$$

so that the far-field displacement is given by

$$A(x, t) = \frac{u \exp\left\{-\left(\dfrac{1-s}{s}\right)t'\left[\dfrac{sKx}{t'}\right]^{1/(1-s)}\right\}}{\left((1-s)(t')\left(\dfrac{t'}{sKx}\right)^{1/(1-s)}\right)^{1/2}\left(1+\dfrac{1}{\omega_0}\left(\dfrac{sKx}{t'}\right)^{1/(1-s)}\right)^2} + O(x^{-3/2}). \tag{A.14}$$

Acknowledgments

I wish to thank Prof. F. Mongelli for his encouragement, Prof. A. Zollo for very helpful suggestions concerning scaling law and f_{max}, Prof. M. Dragoni for his critical reading of the manuscript, and Dr. P. Harabaglia for discussions regarding seismic source. Finally, I am sincerely grateful to the Editor, R. Madariaga, and to an anonymous reviewer for general comments and suggestions that greatly enhanced the paper.

References

Abercrombie, R. E. (1995), *Earthquake Source Scaling Relationship from −1 to 5 M_L Using Seismograms Recorded at 2.5-km Depth*, J. Geophys. Res. *100*, 24,015–24,036.

Abercrombie, R. E. (1997), *Near-surface Attenuation and Site Effects from Comparison of Surface and Deep Boreholes Recordings*, Bull. Seismol. Soc. Am. *87*, 731–744.

Aki, K. (1967), *Scaling Law of Seismic Spectrum*, J. Geophys. Res. *72*, 1217–1232.

Aki, K. (1986), *Physical theory of earthquakes*. In *Seismic Hazard in Mediterranean Regions* (J. Bonnin, M. M. Cara, A. Cisternas, and R. Fantechi, eds.) (Kluwer Academic Publishers, Netherlands 1986) pp. 3–35.

Aki, K., and Richards, P. G., *Quantitative Seismology: Theory and Methods*, Vols. I and II (W. H. Freeman, San Francisco 1980).

Anderson, J. G. (1986), *Implication of attenuation for studies of the earthquake source*. In *Earthquake Source Mechanics* (M. Ewing, ed.) (Geophysical Monograph 37, American Geophysical Union 1986) pp. 311–317.

Anderson, J. G., and Hough, S. E. (1984), *A Model for the Shape of the Fourier Amplitude Spectrum at High Frequencies*, Bull. Seismol. Soc. Am. *74*, 1969–1993.

Blair, D. P., and Spathis, A. T. (1982), *Attenuation of Explosion Generated Pulse in Rock Masses*, J. Geophys. Res. *87*, 3885–3892.

Brune, J. N. (1970), *Tectonic Stress and the Spectra of Seismic Shear Waves from Earthquakes*, J. Geophys. Res. *75*, 4997–5009.

CAPUTO, M. (1967), *Linear Models of Dissipation whose Q is Almost Frequency Independent—II*, Geophys. J. Royal Astron. Soc. *13*, 529–539.

EWING, W. M., JARDETSKJ, W. S., and PRESS, F., *Ealstic Waves in Layered Media* (McGraw-Hill, New York 1957) 380 pp.

FEHLER, M., HOSHIBA, M., SATO, H., and OBARA, K. (1992), *Separation of Scattering and Intrinsic Attenuation for the Kanto-Tokai Region, Japan, Using Measurements of S-wave Energy versus Hypocentral Distance*, Geophys. J. Royal Astron. Soc. *108*, 787–800.

GLADWIN, M. T., and STACEY, F. D. (1974), *Anelastic Degradation of Acoustic Pulses in Rock*, Phys. Earth Planet. Inter. *8*, 332–336.

HANKS, T. C. (1982), f_{max}, Bull. Seismol. Soc. Am. *62*, 561–589.

ICHINOSE, G. A., SMITH, K. D., and ANDERSON, J. G. (1997), *Source Parameters of the 15 November 1995 Border Town, Nevada, Earthquake Sequence*, Bull. Seismol. Soc. Am. *87*, 652–667.

KAMPFMANN, W., and BERCKEMER, H. (1985), *High Temperature Experiments on the Elastic and Anelastic Behaviour of Magmatic Rocks*, Phys. Earth Planet. Inter *40*, 223–247.

KJARTANSSON, E. (1979), *Constant Q-wave Propagation and Attenuation*, J. Geophys. Res. *84*, 4737–4758.

MADARIAGA, R. (1977), *High-frequency Radiation from Crack (Stress Drop) Models of Earthquake Faulting*, Geophys. J. Royal Astron. Soc. *51*, 625–651.

MITCHELL, B. J. (1995), *Anelastic Structure and Evolution of the Continental Crust and Upper Mantle from Seismic Surface Wave Attenuation*, Rev. Geophys. *33*, 441–462.

MOLNAR, P., TUCKER, B. E., and BRUNE, J. N. (1973), *Corner Frequencies of P and S Waves and Models of Earthquake Sources*, Bull. Seismol. Soc. Am. *63*, 2091–2104.

O'NEILL, M. E., and HEALY, J. H. (1973), *Determination of Source Parameters of Small Earthquakes from P-wave Rise Time*, Bull. Seismol. Soc. Am. *63*, 599–614.

PAPAGEORGIU, A., and AKI, K. (1983), *A Specific Barrier Model for the Quantitative Description of Inhomogeneous Faulting and the Prediction of Strong Ground Motion, I, Description of the Model*, Bull. Seismol. Soc. Am. *73*, 953–978.

RICKER, N. H. (1953), *The Form and Laws of Propagation of Seismic Wavelets*, Geophys. *18*, 10–40.

SANDERS, C. O., and NIXON, L. D. (1995), *S-wave Attenuation Structure in Long Valley Caldera, California, from Three-component S-to-P Amplitude Ratio Data*, J. Geophys. Res. *100*, 12,395–12,404.

STRICK, E. (1970), *A Predicted Pedestal Effect for Pulse Propagation in Constant Q Solids*, Geophys. *35*, 387–403.

SVESNICHOV, A. G., and TICHONOV, A. N., *Teoria delle funzioni di una variabile complessa* (Editori Riuniti, Roma 1984) 320 pp.

TONN, R. (1989), *Comparison of Seven Methods for the Computation of Q*, Phys. Earth Planet. Inter. *55*, 259–268.

WU, H., and LEES, M. (1996), *Attenuation Structure of Coso Geothermal Area, California, from Wave Pulse Widths*, Bull. Seismol. Soc. Am. *86*, 1574–1590.

ZUCCA, J. J., HUTCHINGS, L. J., and KASAMEYER, P. W. (1994), *Seismic Velocity and Attenuation Structure of the Geysers Geothermal Field, California*, Geothermics *23*, 111–126.

(Received July 2, 1997, revised May 20, 1998, accepted August 7, 1998)

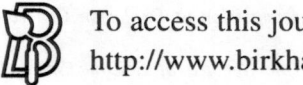

To access this journal online:
http://www.birkhauser.ch

Part II
Crustal Studies

Pure appl. geophys. 153 (1998) 441–473
0033–4553/98/040441–33 $ 1.50 + 0.20/0

Pure and Applied Geophysics

Seismic-frequency Laboratory Measurements of Shear Mode Viscoelasticity in Crustal Rocks II: Thermally Stressed Quartzite and Granite

Cao Lu[1]* and Ian Jackson[1]

Abstract—Forced torsional oscillation techniques have been used to explore the seismic-frequency shear mode viscoelasticity of specimens of two crustal rocks (Cape Sorell quartzite and Delegate aplite), cycled between room temperature and 700°C under conditions of moderate confining pressure. The anisotropy and intergranular inhomogeneity of thermal expansivity in these materials give rise to large deviatoric stresses, resulting in thermal cracking at temperatures above a pressure-dependent threshold temperature, associated with the onset of very pronounced temperature sensitivity of the shear modulus, in general accord with the predictions of fracture mechanics models. For Delegate aplite in particular, the shear modulus behaves reproducibly during multiple thermal cycles at different confining pressures, consistent with the notion that the thermal cracks are of low aspect ratio (minimum/maximum dimension), and are therefore readily closed by the prevailing confining pressure once the thermal stresses are removed. Marked frequency-dependent dissipation of shear strain energy is observed on heating each rock to temperatures $\geq 500°C$, although the attenuation varies significantly with prior thermal history, probably as a result of progressive dehydration and relaxation of deviatoric stresses. Temperature and pressure dependent crack densities for Delegate aplite have been estimated by comparison of the observed shear moduli with those expected for a crack-free aggregate. In parallel with the forced oscillation tests, measurements have been made of the rate at which (argon) pore pressure equilibrium is re-established following a perturbation. Combination of these results, which provide a proxy for permeability, with the inferred crack densities indicates that the variation of permeability with crack density is well described by a percolation model with a threshold crack density of ~ 0.2.

Key words: Seismic velocity, attenuation, quartz α/β transition, thermal microcracking, permeability.

Introduction

The elastic wave speeds of crustal rocks have been experimentally studied under conditions of high pressure and temperature by many authors (e.g., Fielitz, 1971; Kern, 1978; Christensen, 1966, 1979). The effect of confining pressure on velocities has been extensively investigated, while relatively fewer data are available

---

[1] Research School of Earth Sciences, The Australian National University, Canberra, ACT 0200, Australia, Fax: +61 6 279 8253. E-mail: Ian.Jackson@anu.edu.au
* Deceased 18 December 1996.

on the influence of temperature. It is commonly observed that increasing pressure increases the wave velocities, whereas increasing temperature has the opposite effect. Elastic wave propagation through dry rocks is known to be very sensitive to the state of microcracking of the polycrystalline aggregates. The variations of velocity with pressure and temperature are generally believed to be associated mainly with the presence of pre-existing and thermally-induced microcracks within the rocks, especially at low confining pressure (<100 MPa). Observations suggest that, on average, about 100 MPa pressure for every 100°C of temperature change is needed to suppress thermal cracking (e.g., JACKSON, 1991). CHRISTENSEN (1974) measured ultrasonic wave velocities in crystalline rocks at pressures up to 3000 MPa, and demonstrated that changes in velocity with pressure do not approach those of crack-free polycrystals until the confining pressure is above 1000 MPa. He pointed out that even at high pressures (1000–3000 MPa), solid contact between the mineral components is presumably only approximate because some porosity has originated from anisotropic thermal contraction of the minerals. The evolution of (thermally-induced) microcracks is thus one of the important issues involved in laboratory studies of rock properties at high pressure and temperature.

Crack porosity, surface area, and connectivity control the mass and energy transfer through the rock as well as its thermoelastic response at low pressure and temperature (e.g., HARLOW and PRACHT, 1972; PEARSON et al., 1983; WANG et al., 1989). Understanding of the relationship between elastic properties and microcracks is of importance for seismic applications in energy and mineral resource recovery as well as in the evaluation and monitoring of possible sites for underground nuclear waste repositories.

Rock properties related to thermally-induced microcracking have long been a field of active research. IDE (1937) reported irreversible reduction of compressional wave velocity as a result of heating specimens of Sudbury norite and Quincy granite in air to peak temperatures below 400°C. He suggested that differential thermal expansion between adjacent mineral grains creates new microcracks responsible for the decrease in velocity. Measurements of the velocity as a function of temperature at elevated pressures suggested that confining pressure inhibits the formation of thermal microcracks to some degree. Thermal expansion of rocks has been studied at 1 atm (e.g., RICHTER and SIMMONS, 1974; SIMMONS and COOPER, 1978; BAUER and JOHNSON, 1979) and at elevated pressures (e.g., WONG and BRACE, 1979; HEARD, 1980; VAN DER MOLEN, 1981; PAGE and HEARD, 1981; HEARD and PAGE, 1982). The influence of pressure and temperature on permeability of rocks has also been studied (BRACE et al., 1968; ZOBACK and BYERLEE, 1975; SUMMER et al., 1978; KRANZ et al., 1979; TRIMMER et al., 1980; DOYEN, 1987). They concluded that thermally-induced microcracks strongly influence the physical properties of rocks. The experimental techniques used to detect thermally-induced microcracking include acoustic emission (JOHNSON et al., 1978; BAUER and JOHNSON, 1979; ATKINSON et al., 1984; MAJER and MCEVILLY, 1985; GLOVER et al., 1995), differential strain analysis (KOWALLIS and WANG, 1983; CARLSON and WANG, 1986), scanning electron microscopy (SPRUNT and BRACE, 1974; HOMAND-ETI-ENNE and HOUPERT, 1989; WANG *et al.*, 1989), and measurements, so far exclusively at ultrasonic frequencies, of elastic wave velocity or dynamic modulus (JOHNSON *et al.*, 1978; WANG and HEARD, 1985; WANG *et al.*, 1989). However, there has been no systematic laboratory study on the effects of thermal cracking on elastic wave velocities or moduli undertaken at seismic frequencies.

For rocks exposed to sufficiently high pressure and temperature, there is inevitably a competition between brittle and plastic processes. Thermal cracking therefore occurs only under favourable circumstances. For example, previous experimental studies have demonstrated the dominance of plastic deformation over thermal cracking in limestones and marbles (FREDRICH and WONG, 1986; LU and JACKSON, 1996).

In this study, selected "crystalline" crustal rock specimens have been tested dry under conditions of high confining pressure and temperature in order to investigate the relationship between thermally generated crack porosity and their elastic/anelastic properties at seismic frequencies. Firstly, the conditions required for the production of stable populations of thermal microcracks in crustal rocks need to be explored if the influence of the cracks on their elastic properties is to be understood. In a further paper (LU and JACKSON, in preparation), an experimental study of the seismic properties of fluid-saturated rocks and their *in situ* fluid transport properties will be presented. The ultimate goal of this experimental project is to explore the relationship between fluid transport properties (permeability) and seismic properties under conditions of varying confining pressure, and the relationship between total crack porosity and crack connectivity, through complementary information from the seismic properties and the fluid-flow measurements.

Rock Specimens and Experimental Procedure

Rock Specimens

Two rocks have been chosen for this study: a quartzite (Cape Sorell, Tasmania) and a fine-grained granite (Delegate aplite, N.S.W.). Delegate aplite was chosen for its availability, relatively fine grain size, and homogeneity. Moreover, its elastic properties have previously been studied under conditions of pressure and temperature with an ultrasonic method (VAN DER MOLEN, 1979), and under conditions of moderate confining pressure and room temperature with the seismic-frequency torsional forced oscillation method (JACKSON *et al.*, 1984). Delegate aplite has been described by VAN DER MOLEN and PATERSON (1979) as a fine-grained granitic rock consisting of roughly equal fractions of quartz (31%), plagioclase (31%) and K-feldspar (35%) with 3% biotite and chlorite. It is isotropic and equigranular in appearance with 0.5 mm average and 1.0 mm maximum grain size, and has a bulk density of 2.57 g/cm^3 and a total porosity of 2.3%, of which about one third is accessible to water saturation. Weight loss of 0.25% following heating in air to 800°C was attributed to strongly adsorbed moisture, possibly augmented by the dehydration of K-feldspar alteration products. One of the cylindrical specimens of Delegate aplite (#1) for this study had been previously used in a series of experiments at room temperature and pressures up to 170 MPa (JACKSON et al., 1984). The remaining specimens (#3–#6) were newly cored from the same block.

Cape Sorell quartzite is a nearly pure quartz rock which represents a close approximation to a genuinely monomlnerallic aggregate. Samples were collected from a quarry at Cape Sorell (near Strahan on the west coast of Tasmania, Australia). It is a metamorphosed sandstone with an average grain size of about 0.5 mm, and is light-grey to translucent in appearance. The quartzite comprises 99% by volume of quartz with less than 1% muscovite on the grain boundaries. The quartz grains are slightly deformed with a weak preferred orientation. Microscopical study of thin sections (Fig. 1) showed that there is a second population of quartz grains, considerably smaller in size (≤ 50 μm), and located at grain boundaries and sealed

Cape Sorell quartzite

Figure 1
Micrograph of Cape Sorell quartzite: transmitted light, crossed polarisers, width of field 2.9 mm.

fractures, which is interpreted as the result of limited recrystallisation. The grain boundaries are sutured and there are traces of healed fractures and old grain boundaries. The density and (connected) porosity of the quartzite are respectively 2.637 g/cm^3 and 0.3%, determined by saturating the specimen (15 mm in diameter and 145 mm in length) with distilled water in vacuum. Comparison of the measured density with the X-ray density for quartz 2.648 g/cm^3 (SMYTH and MCCORMICK, 1995) gives an estimate of 0.4% for the total porosity.

Experimental Procedure

The seismic properties (shear modulus G and associated strain energy dissipation Q^{-1} or internal friction $\delta = \tan^{-1}(Q^{-1})$ have been measured with a locally developed forced-torsional-oscillation apparatus. A detailed description of the instrument has been given by JACKSON *et al.* (1984) and its further development has been described by JACKSON and PATERSON (1993). A brief account will therefore suffice here. A steel elastic standard, of known elastic modulus and negligible internal friction, which is maintained at room temperature, and a compound specimen assembly described below are connected mechanically in series and are therefore exposed to the same alternating torque, which is generated by the electromagnetic drivers located within the pressure vessel. The time-varying angular distortions induced by the applied torque in the specimen assembly alone, and in the combination of specimen assembly and elastic standard, are measured by pairs of sensitive capacitance displacement transducers. The relative amplitude and phase of these sinusoidal displacement-versus-time signals constrain the complex torsional compliance of the assembly containing the rock specimen.

This specimen assembly consists of a cylindrical rock specimen mounted within the hot zone of the internal furnace between two hollow alumina torsion rods (15 mm O.D. and 2 mm I.D.) which occupy the regions of strong temperature gradient. Both the specimen and the alumina torsion rods are isolated from the argon pressure medium by enclosure within a thin-walled (0.25 mm) iron jacket sealed with an O-ring against a steel torsion rod at each end.

In order to derive the shear modulus and associated strain energy dissipation of the rock specimen from the torsional compliance of the compound specimen assembly, a further measurement is performed under identical conditions on another assembly in which the rock specimen is replaced by a reference specimen (here Degussa Duramic 99.7% alumina) of identical dimensions and known modulus. The shear modulus and strain energy dissipation for the rock specimen are then obtained by parallel processing of the forced oscillation data for these two assemblies. This procedure requires knowledge of the temperature-dependent (and potentially complex) shear moduli appropriate for the polycrystalline alumina and for the iron jacket. The behaviour of the Duramic alumina departs negligibly from the elastic ideal for temperatures $\leq 700°C$; accordingly, an ultrasonically-determined value for its room-temperature shear modulus (151.5) GPa) is used along with the value of $\partial G/\partial T$ derived from published single-crystal elasticity data. Significantly viscoelastic behaviour of the iron jacket above 400°C is modelled with the aid of unpublished data from forced oscillation experiments on an alloy of similar composition (JACKSON et al., in preparation).

Robust a priori estimation of uncertainties in the quantities G and Q^{-1} derived from the raw experimental data is problematical, mainly because the noise which contaminates the output of the displacement transducers tends to be strongly correlated between the two channels. For these reasons, the precision of the data is best judged from the reproducibility of duplicate determinations and the consistency of the trends presented below. The largest single source of uncertainty arises in calibration of the displacement transducers, leading under the least favourable conditions to error of about $\pm 2\%$ in G.

All rock specimens were precision ground into cylinders of 15.0 mm diameter and 145 mm length, and lapped to optical flatness within $\lambda/2$ at each end. They were oven dried at 80°C (110°C for Delegate aplite samples #1 and #2) for more than two days and then incorporated within the specimen assembly. Thermal cycling of the rock specimens was conducted under various constant confining pressures P_c ranging from 50 to 200 MPa. Each specimen was heated at about 4°C per minute to each of a series of temperatures, typically spaced at intervals of 100°C, reaching a maximum of 700°C. At each such temperature, forced oscillation measurements were undertaken to determine the shear modulus and associated dissipation. Measurements were also made during staged cooling from 700°C down to room temperature. Smaller temperature intervals, either 25°C or 50°C, were occasionally used within a temperature range of particular interest. Changes of confining pressure were made at room temperature between successive thermal cycles.

Access to each end of the jacketted rock specimen, which is normally vented to atmosphere, has allowed the development of a pore-fluid system which provides independent monitoring and control of the pore fluid pressures in reservoirs respectively, "upstream" and "downstream" from the rock specimen. This system, to be described in detail by LU and JACKSON (in preparation), allows us to control the pore fluid pressure P_f, and thus the effective pressure $P_e = P_c - P_f$ acting on the rock specimen under investigation. It also provides us with the capability to measure the crack porosity and permeability of the rock. With the pore fluid system in equilibrium (i.e., uniform fluid pressure throughout), a small increment in pore pressure is applied either to the upstream or to the downstream reservoir. The return to equilibrium is observed by monitoring the pressure variation for the upstream reservoir with elapsed time. The time taken for a new equilibrium state to be established provides a measure of the permeability of the rock specimen. In this study, measurements were made by applying a pore pressure increment δP_f of 5 to

Table 1

History of thermal cycling of the specimen of Cape Sorell quartzite

Cycle #	1	2	3	4	5	6	7	8	9
P_c (MPa)	200	100	50	200	50	100	200	100	50
T_{max} (°C)	630	500	400	600	575	600	700	700	700

10 MPa to the upstream reservoir at zero initial pore pressure in the system. The argon pore fluid gradually flows through the rock specimen into the downstream reservoir where the pore fluid pressure changes only imperceptibly because of its markedly larger storage capacity (about 35:1). Examples of the measured time dependence of the upstream pore pressure P_u are given in the following section.

Experimental Results

Thermally Cycled Quartzite (Cape Sorell, Tasmania).

Temperature and pressure dependence of the shear modulus. The sequence of thermal cycles conducted on the one specimen of Cape Sorell quartzite is summarised in Table 1, which gives the (constant) confining pressure and peak temperature T_{max} for each thermal cycle. Figure 2 shows the temperature dependence of shear modulus at various confining pressures ranging from 50 to 200

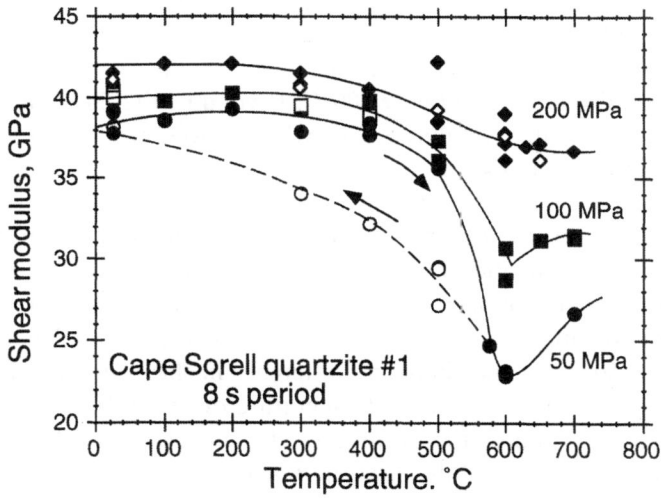

Figure 2

Variation of the shear modulus G of Cape Sorell quartzite with temperature at various confining pressures (circle—50 MPa; square—100 MPa; diamond—200 MPa; solid symbols—heating; open symbols—cooling). Uncertainties in individual determinations of G are comparable with the size of the plotting symbols. Significant hysteresis is observed only at 50 MPa.

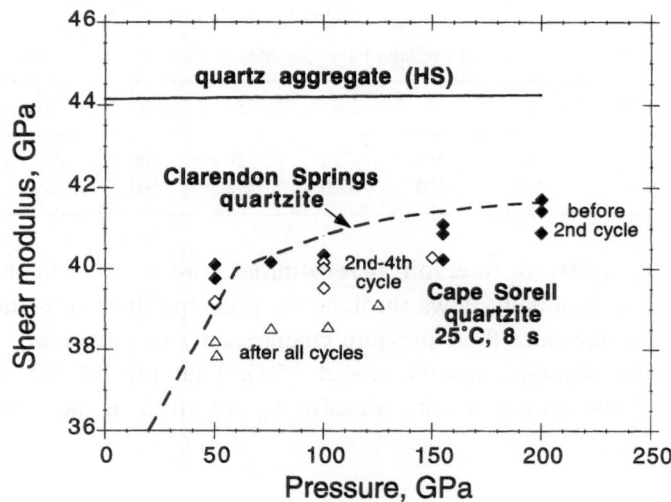

Figure 3
Variation of the shear modulus of Cape Sorell quartzite with confining pressure at room temperature.
Data obtained prior to the second thermal cycle (see Table 1)—solid diamonds; data obtained during
second, third and fourth cycles—open diamonds; data obtained following all thermal cycles—triangles.
For comparison, ultrasonic data for Clarendon Springs Quartzite (CHRISTENSEN, 1966) are also
displayed (broken line) along with the average of the Hashin-Shtrikman bounds for polycrystalline
quartz calculated from single-crystal elasticity data (collated by GEBRANDE, 1982). For the spacing of
the Hashin-Shtrikman bounds, see Figure 10.

MPa. In general, the shear modulus is relatively temperature-insensitive for temperatures below 400°C. At higher temperatures, the modulus decreases markedly with increasing temperature, especially at the lower confining pressures, for which a minimum modulus is observed near 600°C. Beyond 600°C the temperature sensitivity of the shear modulus is reduced (200 MPa) or even reversed in sign—this latter effect being most pronounced at 50 MPa. The shear modulus behaved reversibly during thermal cycling at confining pressures of 100 and 200 MPa. However, substantial hysteresis was observed in the variation of shear modulus with temperature during the fifth and the ninth thermal cycles, which were performed to relatively high peak temperatures at 50 MPa confining pressure (Fig. 2). The maximum discrepancy in shear modulus between heating and cooling sequences occurs at 400–500°C, while almost no difference in shear modulus between heating and cooling was observed at room temperature or at 600°C.

At room temperature, the shear modulus increases steadily with increasing confining pressure. The modulus measured at room temperature and 50–150 MPa confining pressure is observed to decrease consistently throughout the course of the thermal cycling experiments, whereas the pressure sensitivity of the modulus appears to increase marginally (Fig. 3). Following the last (ninth) thermal cycle at 50 MPa confining pressure, the cumulative reduction in the shear modulus measured at 50 MPa and room temperature amounted to about 5%.

Modulus dispersion. The values of the shear modulus measured at temperatures up to and including 500°C are essentially independent of oscillation periods between 1 and 100 seconds (Fig. 4). At higher temperatures (≥ 500°C), the shear modulus of Cape Sorell quartzite decreases slightly and systematically with increasing oscillation period. At 700°C and 100 MPa (where it is most precisely determined), the magnitude of the dispersion between 3 and 100 s reaches 1.1%.

Strain energy dissipation. The measured values of the dissipation (Q^{-1}) are more scattered and apparently more dependent upon prior thermal history than are the values of the shear modulus. The dissipation data set is most extensive at 100 MPa (Fig. 5), where it is evident that Q^{-1} is low, generally < 0.003, and essentially frequency-independent for $T \leq 500$°C. Significant frequency dependence is observed at the highest temperature of this study (700°C) and 100 MPa where Q^{-1} varies with oscillation period T_0 as T_0^n with $n = 0.42$. The results obtained at 50 and 200 MPa were fewer and more scattered than at 100 MPa. Anomalously high values of dissipation (~ 0.02–0.03) for 28 s period and temperatures of 500–600°C were observed fairly consistently during heating half-cycles at 200 MPa, but were absent from the corresponding cooling half-cycles. This formed part of a more general pattern whereby the dissipation or internal friction measured at a given temperature and pressure tended to be low following prior heat treatment to comparable or greater temperature.

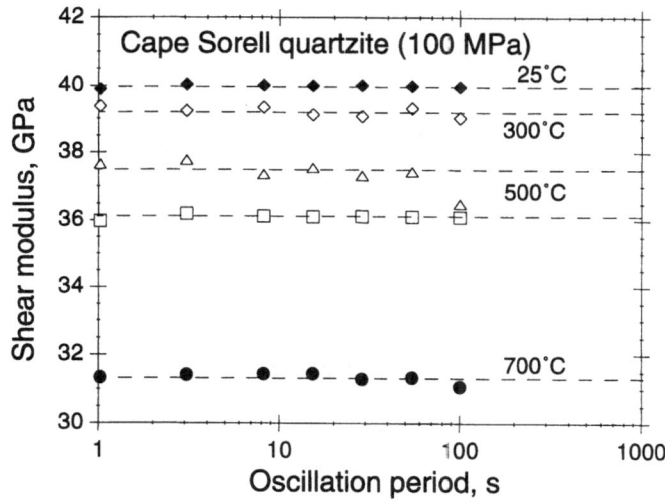

Figure 4
The variation of shear modulus with oscillation period for Cape Sorell quartzite at a confining pressure of 100 MPa and representative temperatures.

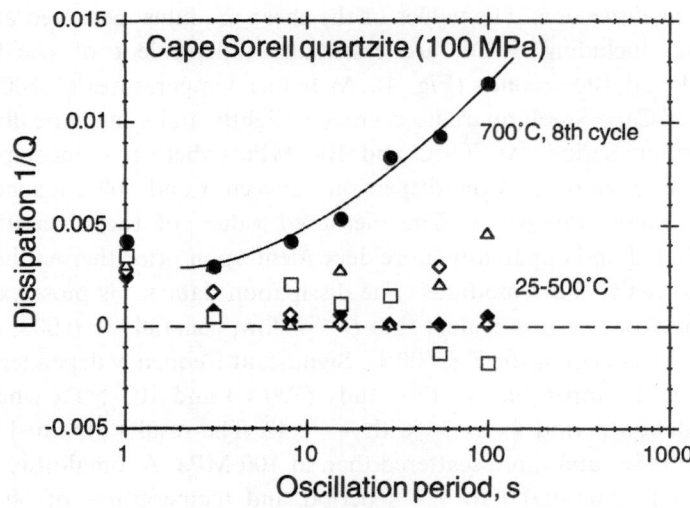

Figure 5

The variation of strain energy dissipation Q^{-1} with oscillation period and temperature at a confining pressure of 100 MPa. The difference between dissipation Q^{-1} and internal friction $\delta = \tan^{-1}(Q^{-1})$ is negligible for the values of Q^{-1} (everywhere < 0.03) measured in this study.

Thermally Cycled Granite (Delegate Aplite)

Shear modulus. Figure 6 shows a representative example (for specimen #5) of the variation of shear modulus with temperature at confining pressures of 50, 100

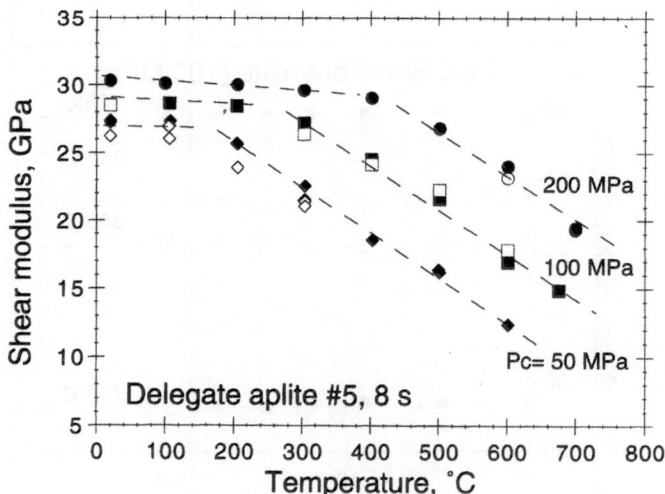

Figure 6

The variation of shear modulus with temperature at different confining pressures for Delegate aplite (circle—200 MPa; square—100 MPa; diamond—50 MPa; solid symbols—heating; open symbols—cooling).

and 200 MPa for Delegate aplite. The shear modulus decreases markedly with increasing temperature during the thermal cycles at 50 MPa confining pressure, with a 55% reduction between room temperature and 600°C. At confining pressures of 100 and 200 MPa, there is a threshold temperature (~ 250°C at 100 MPa, ~ 400°C at 200 MPa) below which the shear modulus decreases only slightly with increasing temperature, and above which the temperature dependence of shear modulus is more pronounced. No significant hysteresis was observed in the variation of shear modulus during thermal cycling confining pressures ranging from 50 to 200 MPa. The measurements of shear modulus were reproducible at a given temperature and confining pressure for all of the thermal cycles up to 700°C.

Dissipation. As for the Cape Sorell quartzite, the strain energy dissipation (Q^{-1}) of Delegate aplite is low and relatively insensitive to both temperature and oscillation period below a threshold temperature near 400°C (Fig. 7a). At 50 MPa, Q^{-1} increases systematically as the rock specimen is heated to progressively higher temperatures above 400°C, with broadly consistent results for specimens #5 and #6 (Fig. 7b). However, sustained exposure of specimen #6 to 600°C at 50 MPa for periods of many hours, results in considerably lower values of Q^{-1} (results labelled "600ti" with $i = 1$, 2 and 3 in Fig. 7b). Data subsequently obtained at progressively higher confining pressures indicate that values of Q^{-1} significantly greater than 0.005 were obtained only when the rock was heated to a higher temperature than it had previously experienced (676°C at 100 MPa (Fig. 7c) and 700°C at 200 MPa (Fig. 7d)). The only exception to this rule is the behaviour of specimen #6 ("rej") tested at 600°C and 100 MPa following exposure to laboratory atmosphere during replacement of its iron jacket. At 700°C and 200 MPa, temporal evolution towards lower values of Q^{-1} is again observed (Fig. 7d).

For the highest temperatures reached at each pressure, the internal friction becomes significantly frequency-dependent, varying with oscillation period T_0 as T_0^n with values of the exponent n near 0.3 (0.31 for #6 at 500°C and 50 MPa, 0.33 for #5 at 700°C and 200 MPa). Under these conditions only is there significant dispersion of the shear modulus which amounts to -2.1 and -3.5%, respectively, for the period range 3–100 s.

Measurement of Fluid Transport Properties

Representative experimental data documenting the decay of an increment δP_f in pore fluid pressure superimposed on the steady pore fluid pressure P_f of the upstream reservoir are shown in Figure 8. These results for a specimen of Delegate aplite exposed to 600°C at various confining pressures demonstrate the expected trend of relatively faster equilibration (higher permeability) through fluid flow at low confining pressures. Such data are adequately described by an exponential decay function:

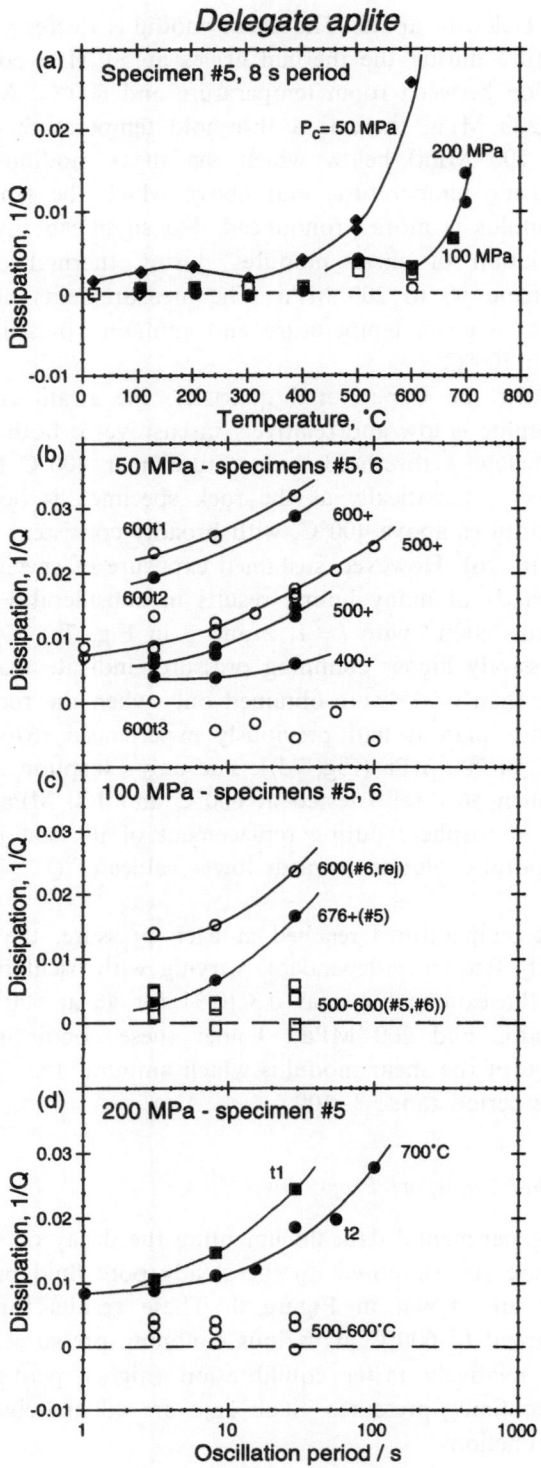

Delegate aplite

(a) Specimen #5, 8 s period

(b) 50 MPa - specimens #5, 6

(c) 100 MPa - specimens #5, 6

(d) 200 MPa - specimen #5

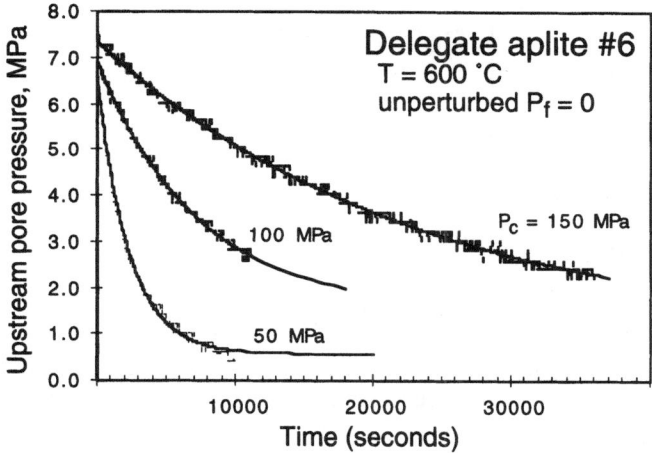

Figure 8

Pressure decay in the upstream reservoir resulting from flow of argon pore fluid through the rock specimen. Fits to equation (1) (solid lines) are superimposed on raw data from the transient flow experiments on Delegate aplite at 600°C and various indicated confining pressures.

Table 2

Rate constant A and time constant τ for pore pressure equilibration derived from least-squares fits to the P_u-time data for Delegate aplite

P_c (MPa)	P_f (MPa)	T (°C)	A (10^{-5} s^{-1})	error* (10^{-5} s^{-1})	τ (s)	R***
50	0–7	25	2.83	2.98	35,300	0.9975
50	0–6	204	2.20	15.0	45,500	0.9862
50	0–5	303	3.23	8.35	30,960	0.9958
50	0–6	500	11.2	7.92	8,900	0.9983
50	0–7	600	44.2	66.0	2,300	0.9970
103	0–6	402	2.64	3.07	37,900	0.9969
102	0–7	600	14.3	39.5	7,000	0.9974
127	0–6	25	1.76	7.04	56,800	0.9877
150	0–8	600	4.36	6.29	22,900	0.9977
150	0–6	600	5.35	7.32	18,700	0.9805

* standard errors calculated from the variance-covariance matrix.
** correlation coefficient.

Figure 7

The variation of strain energy dissipation Q^{-1} for Delegate aplite with temperature (a; solid symbols— heating; open symbols—cooling; diamonds—50 MPa; squares—100 MPa; circles—200 MPa), pressure and oscillation period (b, c and d). In panels (b)–(d), curves are labelled with the temperature in °C. The plus symbol attached to the temperature label denotes data obtained at the indicated temperature during a heating half-cycle. The influence of increasing duration of exposure to the *P-T* conditions is illustrated by multiple data sets labelled "t1", "t2", etc. in (b) and (d), whereas the label "rej" relates to measurements made following the rejacketting of the specimen.

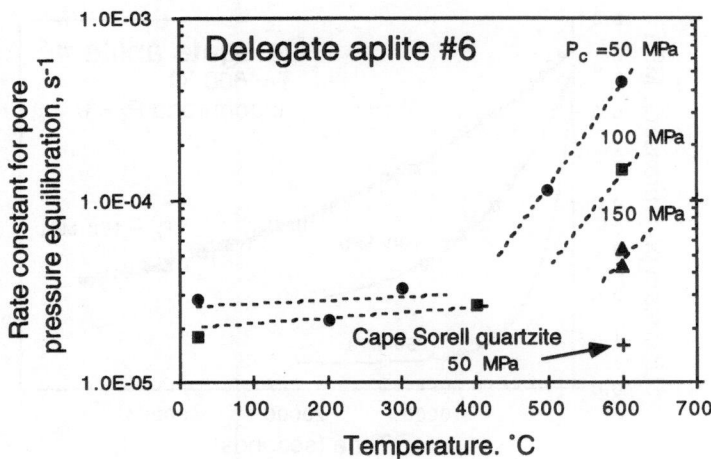

Figure 9
Temperature dependence of the rate constant A for pore pressure equilibration in Delegate aplite (from fits to equation (1)), at confining pressures of 50, 100 and 150 MPa. The result for the considerably less permeable Cape Sorell quartzite at 50 MPa and 600°C is included for comparison.

$$P_u = \delta P_f \exp(-At) + P_f. \tag{1}$$

The values of the rate constant A and the time constant $\tau = 1/A$ for pore pressure equilibration thus derived for all such experiments are presented in Table 2. Least-squares fits of the experimental data to equation (1) are indicated by the continuous lines in Figure 8. The adequacy of the fits to these and other data is indicated by values near unity for the correlation coefficient (Table 2).

The rate constant A depends on the permeability of the rock specimen, the viscosity of the argon pore fluid, and the storage capacities of the specimen and upstream reservoir, which are expected to vary markedly across the range of low pore pressures (0–10 MPa) encountered in these experiments, because of the strong pressure sensitivity of the compressibility of argon. The simple transient flow theory (BRACE et al., 1968; HSIEH et al., 1981) is therefore not strictly applicable in this case (LU and JACKSON, in preparation). Nevertheless, the value of A is a useful quantitative indicator of the rate of pore pressure equilibration within the rock specimen under various confining pressures. Measured time constants for Delegate aplite vary from 2300 s (∼40 min) at 50 MPa and 600°C to 56,800 s (∼16 hr) at 127 MPa and 25°C.

A further series of transient flow experiments conducted a higher pore pressures, where the compressibility of argon is much less pressure dependent, provides for the robust determination of permeability (LU and JACKSON, in preparation). At a representative effective pressure $P_{eff} = P_c - P_f$ of 50 MPa, the permeability of Delegate aplite remains essentially constant at 1×10^{-19} m^2 as the temperature is increased from 20 to 300°C, thereafter increasing markedly with increasing T to about 1×10^{-17} m^2 at 650°C. The variation of the rate constant A with temperature for the conditions $P_c = 50$ MPa and $P_f \sim 0$ reflects this marked change (~ 2 orders of magnitude, in permeability.

Figure 9 is a semi-logarithmic plot of the variation of the rate constant A with temperature for Delegate aplite (sample # 6) at confining pressures of 50, 100 and 150 MPa. A is low and nearly constant below 400°C at all confining pressures. It begins to increase above a threshold temperature which is apparently pressure dependent. The results in Figure 9 suggest that this temperature increases from 350 ± 50°C for 50 MPa, to 450 ± 50°C for 100 MPa and 550 ± 50°C for 150 MPa. For comparison, the rate constant A (equivalent to a time constant τ of ~ 16 hr) for Cape Sorell quartzite at confining pressure of 50 MPa and 600°C has been included in Figure 9. The low value of A for Cape Sorell quartzite suggests that its permeability is very low (comparable to that of Delegate aplite at room temperature and confining pressure of 100 MPa) even under the most favourable conditions for fluid flow, i.e., confining pressure of 50 MPa and temperature of 600°C. Transient flow tests on the Cape Sorell quartzite specimen thus indicate that the quartzite has very low permeability and can be considered impermeable on normal experimental time scales.

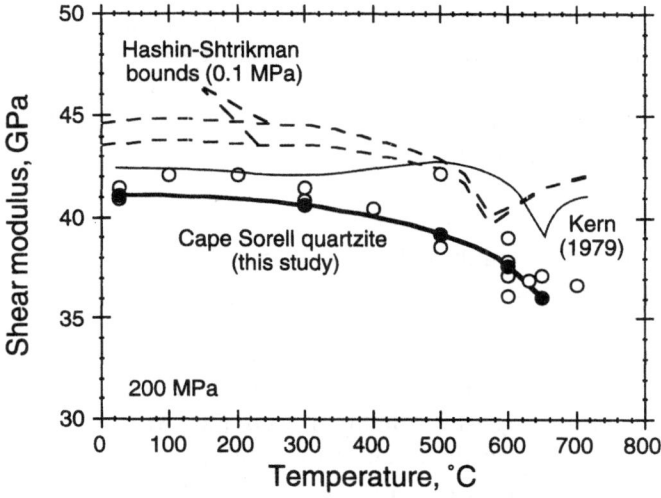

Figure 10
Temperature dependence of the shear modulus for Cape Sorell quartzite (bold curve and filled circles—cooling; open circles—heating) compared with data of KERN (1979, continuous curve) for a Finnish quartzite, both measured at 200 MPa confining pressure. Also shown are bounds on the modulus for an ideal (zero-porosity) quartz polycrystal (broken curves) calculated from single-crystal elasticity data.

Discussion

Elasticity and Microcracks

The interpretation of experimental data concerning the temperature and pressure dependence of the elastic moduli for rock specimens requires a robust framework comprising the following related elements. Firstly, the elastic moduli to be expected in the absence of porosity, which provide an upper bound on measured properties of real rocks, are well approximated as the average of the generally closely spaced upper and lower Hashin-Shtrikman bounds, which may be calculated from the relevant single-crystal elastic constants (e.g., WATT et al., 1976). Secondly, the effect of any porosity upon the elastic properties will depend in principle upon both the volume fraction of pores ϕ and the distribution of their aspect ratios α and orientations. However, theoretical descriptions of the elasticity of cracked solids (reviewed by JACKSON, 1991, for example) typically yield expressions for the effective elastic moduli which are functions of a single suitably defined crack density parameter ε. For randomly oriented ellipsoidal cracks of a given aspect ratio α (minimum/maximum dimension), the crack density parameter of O'CONNELL and BUDIANSKY (1974) becomes

$$\varepsilon = (3/4\pi)\,(\phi/\alpha). \tag{2}$$

The effective shear modulus G of the cracked medium is then given by

$$G/G_0 = 1 - n(v^*)\varepsilon, \tag{3}$$

(O'CONNELL and BUDIANSKY, 1974). Here $n(v^*)$ is a function of Poisson's ratio v^* for the cracked medium, and v^* in turn is related to v and ε by an auxiliary relationship which must be satisfied for self-consistency. The third relevant consideration is the magnitude of the confining pressure P_c required to close a crack with aspect ratio α, which is given by

$$P_c \approx E \cdot \alpha, \tag{4}$$

where E is the Young's modulus (WALSH, 1965).

The Hashin-Shtrikman bounds on the shear modulus for an ideal quartz polycrystal (with random orientation of crystallites and zero porosity) calculated from quartz single-crystal elasticity data (collated by GEBRANDE, 1982), are compared in Figure 10 with the shear modulus for Cape Sorell quartzite measured in the present study at the highest confining pressure (200 MPa). The measured shear modulus is consistently lower than that of the ideal quartz aggregate (by about 8% at room temperature), but varies with temperature along a trend sub-parallel to that for the ideal quartz aggregate for temperatures up to 500°C. If the modulus deficit of 8% were attributed through equation (3) to the influence of porosity (0.4%), an average aspect ratio for the pores of about 0.02 would be inferred. The closure pressure (~ 2 GPa) for such cracks calculated from equation (4) with $E \sim 100$ GPa far exceeds the confining pressures employed in this study, consistent with the inference that these cracks remain open throughout the course of the measurements.

The shear modulus calculated for the quartz aggregate attains a minimum value at 580°C, the approximate temperature of the quartz α/β phase transition at ambient pressure. The fact that the Hashin-Shtrikman bounds become substantially more closely spaced above the α/β transition temperature is a consequence of the lesser elastic anisotropy of the β phase. The present results are also closely consistent with the ultrasonic data of KERN (1979) for a Finnish quartzite measured at the same confining pressure (200 MPa). The magnitude of the reduction in modulus for each of the quartzites at the minimum (near 650°C at 200 MPa) is comparable with that caused by the α/β transition in the ideal (zero-porosity) quartz polycrystal. Thus for these quartzites at 200 MPa confining pressure, the microcracking invoked by KERN (1979), in explaining the observed temperature dependence of the shear modulus, probably does not occur. However, the much larger reductions in modulus observed in the Cape Sorell quartzite when tested at lower confining pressures, and the results for Delegate aplite, do require consideration of the conditions under which thermal stresses might result in microcracking.

For both Cape Sorell quartzite and Delegate aplite, the increase of room-temperature shear modulus with increasing confining pressure below 200 MPa (Figs. 3 and 6) is considerably more pronounced than for a fully dense aggregate (Fig. 3) and is interpreted to reflect the progressive closure of pre-existing microcracks with aspect ratios $< 3 \times 10^{-3}$. The very marked temperature sensitivity of G for Cape Sorell quartzite above 500°C at 50–100 MPa (Fig. 2), and for Delegate aplite above the observed pressure-dependent threshold temperature (Fig. 6) is interpreted below to reflect microcracking in response to thermal stresses. The strong pressure sensitivity of G within each of these regimes requires crack closure pressures in the range 50–200 MPa for much of the thermally generated crack porosity and thus aspect ratios $< 3 \times 10^{-3}$ for $E = 72$ and 96 GPa for Delegate aplite and quartzite respectively (VAN DER MOLEN, 1979; KERN, 1979).

Thermal Microcracks

Microcracking occurs when the thermally induced stress exceeds the local strength of the rock (e.g., SIMMONS and RICHTER, 1976). The thermal cracking inferred to occur in Cape Sorell quartzite at the lower confining pressures of the present study must be attributed to stresses associated with the substantial anisotropy in the thermal expansion of quartz. In addition to thermal expansion anisotropy, thermal expansion mismatches among the different constituent minerals will also produce thermal stress heterogeneity on a grain-to-grain scale. Both factors presumably contribute to the thermal cracking in Delegate aplite, observed beyond appropriate threshold temperatures at confining pressures of 50–200 MPa.

Two-dimensional fracture mechanics models have been developed in which the spatially varying normal stress caused by the intragranular anisotropy or intergranular mismatch of thermal expansivity is calculated and used in an integration along the length of the crack to determine the stress intensity factor K_I for mode I (tensional) cracking (e.g., FREDRICH and WONG, 1986). For the influence of anisotropy $\Delta\alpha$ in thermal expansivity acting over a temperature interval ΔT upon pre-existing intergranular cracks of length a in an array of hexagonal grains of side L at ambient pressure, EVANS (1978) thus obtained

$$K_I = E\Delta\alpha \ \Delta T \ L^{1/2} \ f(a/L)/(1 + v) \qquad (5)$$

where E is the Young's modulus and v is Poisson's ratio of the (single-phase) material and f is a function of a/L, very well approximated for the interval $0.02 < a/L < 0.2$ by

$$f(a/L) = 1.15 \ (a/L)^{0.27}. \qquad (6)$$

For given L, the calculated stress intensity factor thus increases mildly with increasing size of the pre-existing crack. The influence of confining pressure P is readily incorporated through superposition on the thermal stress field of a normal compressive stress (FREDRICH and WONG, 1986), yielding a contribution

$$\Delta K_I = -P(\pi a/2)^{1/2} \qquad (7)$$

which reduces the stress intensity factor calculated from the thermal stress field. It follows that the influence of confining pressure is to increase the temperature increment which is needed if the critical stress intensity factor K_{IC} for crack growth is to be exceeded.

The variation of shear modulus with temperature for Cape Sorell quartzite (Fig. 2) suggests that appreciable thermally induced microcracking in the rock did not occur until temperature reached about 450°C and then only for the lower (50–100 MPa) confining pressures. It is therefore of interest to calculate the stress intensity factor, as a function of the size a of pre-existing cracks, under the conditions $\Delta T = 430$°C and $P = 100$ MPa, and to compare the resulting values with published data concerning the critical stress intensity factor for quartzites. For the temperature interval 20–450°C, average values of E, $\Delta\alpha$ and v for quartz are approximately 90 GPa, $4 \times 10^{-6} \ K^{-1}$ and 0.05, respectively (SKINNER, 1966; KERN, 1979). The average grain of 500 μm diameter in the quartzite is approximated by a hexagonal model grain of edge length $L = 250$ μm. With these values for the various parameters, equations (5), (6) and (7) combine to yield

$$K_I \ (\text{MPa m}^{1/2}) = 2.68(a/L)^{0.27} - 1.98(a/L)^{1/2}. \qquad (8)$$

The resulting value of K_I increases only very mildly with increasing a/L from 0.65 MPa m$^{1/2}$ for $a = 5$ μm through 0.73 MPa m$^{1/2}$ for $a = 10$ μm towards a broad maximum ($K_I = 0.85$ MPa m$^{1/2}$) centred on $a = 70$ μm. These values of K_I correspond to the lowermost end of the range (0.8–2.6 MPa m$^{1/2}$) of fracture toughness K_{IC} identified for quartzites by ATKINSON (1984, Fig. 18). It is therefore concluded that the onset of thermal microcracking in Cape Sorell quartzite near 450°C at 100

MPa confining pressure can be reconciled with the lowest published values for the fracture toughness of quartzites and a wide range of initial crack sizes greater than about 10 μm.

FREDRICH and WONG (1986) have modelled both intragranular and grain-boundary cracking in response to stresses arising from either anisotropy or intergranular variability in thermal expansivity, by considering a square inclusion of side $2L$ embedded in a uniform isotropic matrix. Their result for the case of thermal expansion anisotropy and non-zero confining pressure is

$$K_I = E \; \Delta\alpha \; \Delta T \; (L/2\pi)^{1/2} \; g(a/L)/(1 - v^2) - P(\pi a/2)^{1/2}. \tag{9}$$

The function $g(a/L)$, unlike $f(a/L)$ of EVANS (1978), possesses a maximum value, of ~ 0.4 at $a/L = 0.2$, and ~ 0.5 at $a/L = 0.35$, for the cases of grain-boundary and intragranular cracking, respectively. Application to Cape Sorell quartzite with $L = 250$ μm (approximating a mean grain diameter of 500 μm), $\Delta T = 430°C$ and $P = 100$ MPa as above, results in mainly negative values of K_I, with a maximum value only marginally greater than zero at 0.04 MPa m$^{1/2}$ for $a \sim 8$ μm. A larger maximum value of K_I near 0.2 MPa m$^{1/2}$ applies for a similar initial crack size at the lower confining pressure of 50 MPa. The expressions for the zero-pressure part of K_I in the hexagonal-grain-array and square-inclusion models (equation (5) and the first term equation (9), respectively) differ by a factor of $(2\pi)^{1/2}$ in the denominator which is mainly responsible for the systematically lower values of K_I calculated from the square-inclusion model. The findings from this latter approach suggest that thermal cracking might influence the temperature dependence of shear modulus for Cape Sorell quartzite at 50 MPa (but not at 100 MPa), provided that the critical stress intensity factor were significantly lower than those reported for other quartzites. Comparison of polished surfaces of Cape Sorell quartzite specimens before and after thermal cycling (Figs. 11 (a) and (b)) provides clear evidence of a substantial increase in crack density although their connectivity, as revealed at this magnification, remains limited.

In the case of Delegate aplite, the shear modulus decreases significantly beyond a threshold temperature, which is a function of confining pressure. This critical temperature increases with rising pressure, from $\sim 150°C$ at 50 MPa, to $\sim 250°C$ at 100 MPa and $\sim 400°C$ at 200 MPa. The square-inclusion model of FREDRICH and WONG (1986) has been used to estimate the critical temperature increments expected for grain boundary thermal cracking in Delegate aplite. The contributions to the stress intensity factor from both intragranular anisotropy and intergranular mismatch in thermal expansivity have been considered, and the influence of pressure has been incorporated, as described above. The parameter values of FREDRICH and WONG (1986) for Westerly granite have been supplemented with high-pressure data ($E = 78$ GPa and $v = 0.25$ at 200 MPa and 20°C) for the same rock from SCHOCK et al. (1974). Calculations for $L = 250$ μm (consistent with an average grain size of 500 μm) and an initial crack length of 10–25 μm yield a critical temperature increment for thermal cracking at zero pressure in the range

Cape Sorell quartzite

(a)

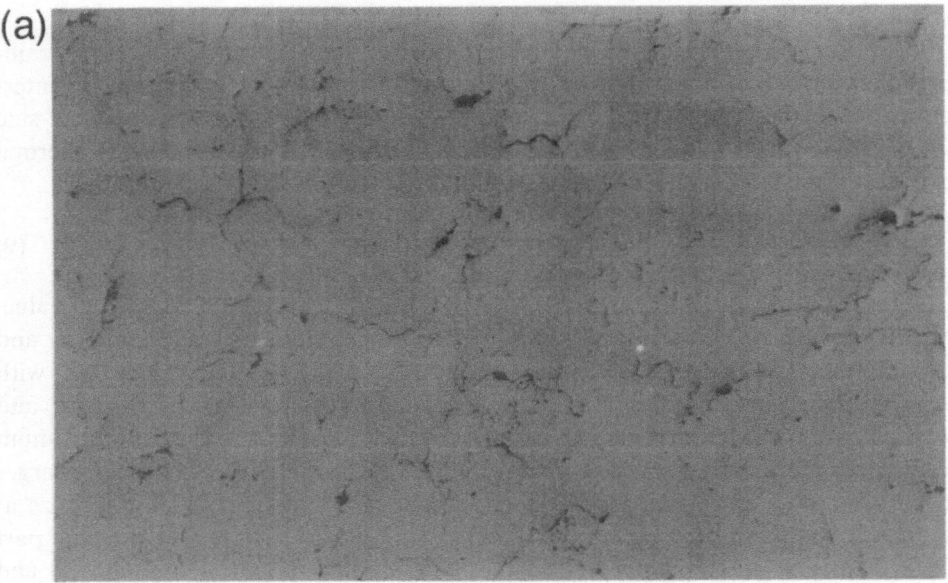

(b)

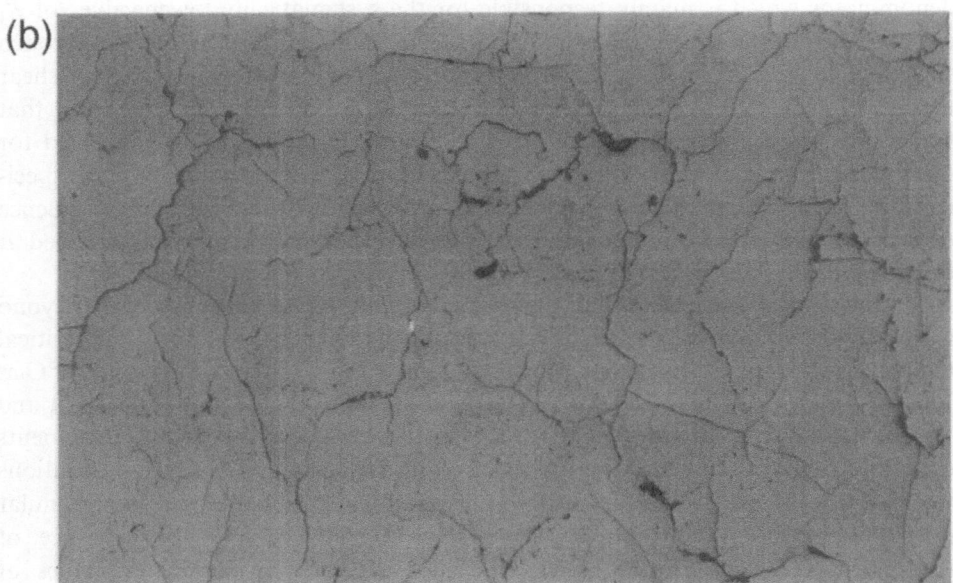

Figure 11 (a,b)

Delegate aplite

(c)

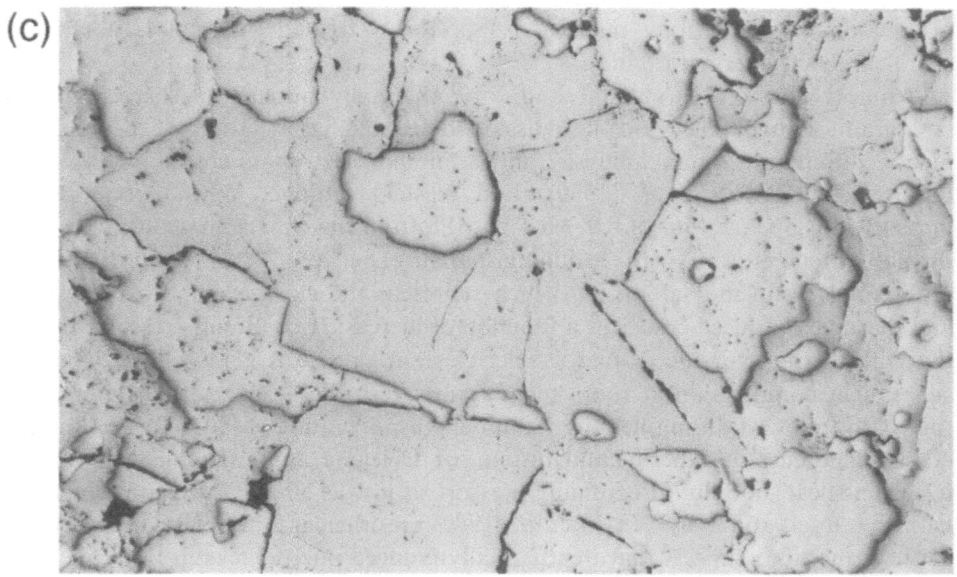

(d)

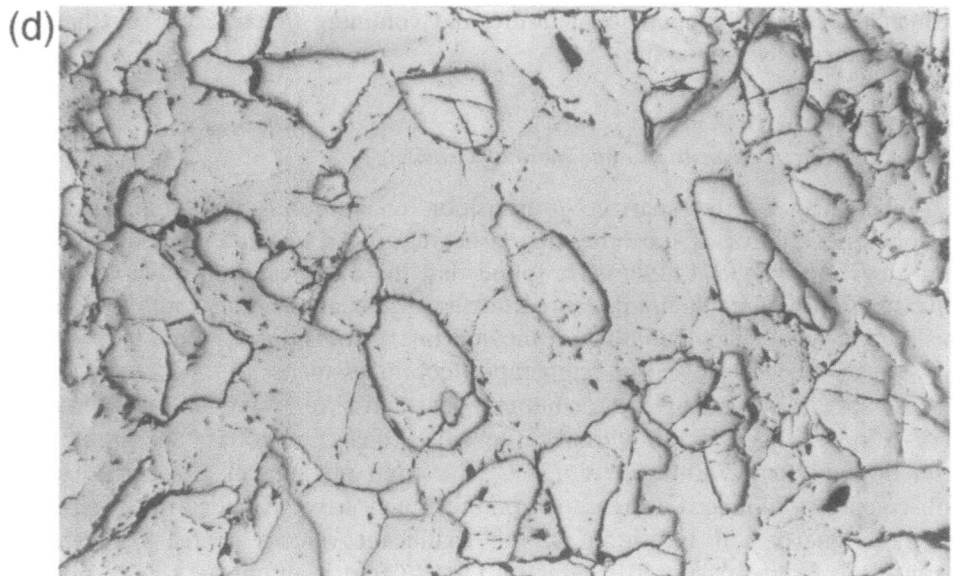

Figure 11
Reflected-light micrographs of Cape Sorell quarzite ((a) and (b)) and Delegate aplite specimens (c) and (d)), respectively before and after thermal cycling to the maximum temperatures of this study. Width of field: 2.9 mm (Cape Sorell quartzite), 1.8 mm (Delegate aplite).

90–140°C, with a pressure sensitivity $\partial T_{cr}/\partial P$ of 1.9–2.5°C/MPa, which is in fair agreement with the experimental observations ($\partial T_{cr}/\partial P \sim 1.7$°C/MPa). Examination of polished sections under reflected light (Figs. 11 (c) and (d)) suggests that the population of pre-existing microcracks in the untreated material is augmented significantly by thermal cycling.

Comprehensive experimental studies on thermally-induced microcracks in several granites involving acoustic emission detection, elastic wave speed measurements, differential strain analysis, and scanning electron microscopy have been reported by WANG et al. (1989). For Westerly granite, WANG et al. (1989) measured a slope $\partial T_{cr}/\partial P$ of 0.9 ± 0.4°C/MPa from the change in acoustic emission threshold temperature with confining pressure between 7 and 55 MPa. Their differential strain measurements, which represent the closure of both pre-existing and thermally-induced cracks as a function of increasing confining pressure, showed that the thermally-induced microcracks close at pressures below 40 MPa during subsequent compression at room temperature, regardless of the pressure-temperature path followed during heating—implying aspect ratios < 0.001 for the thermal cracks. The current experimental results for Delegate aplite demonstrate no hysteresis in shear modulus at confining pressure as low as 50 MPa, which is consistent with the observations by WANG et al. (1989) on other granites. The measurements of shear modulus suggest that the thermally-induced microcracks in Delegate aplite behave reversibly with temperature changes up to 700°C at confining pressures between 50 and 200 MPa. It is therefore inferred that a stable crack population which depends only upon temperature and confining pressure can be thermally generated in Delegate aplite.

The Quartz α/β Transition and Thermal Cracking

The effect of the quartz α-β transition on compressional and shear wave velocities of quartzites and other quartz-bearing rocks was first studied experimentally by FIELITZ (1971, 1976). He found that the compressional wave velocity V_P measured at high confining pressure decreases markedly from its room temperature value through a sharp minimum at the quartz α/β transition temperature, followed by a pronounced recovery as temperature continues to rise.

KERN (1979) measured the compressional and shear wave velocities in single-crystal quartz and several quartz-bearing rocks (quartzite, granite and granulite) as functions of temperature up to 720°C at a fixed pressure of 200 MPa. He also observed a pronounced decrease in compressional wave velocities when approaching the quartz α/β transition with a significant velocity increase beyond the transition. By comparison, the effect of the α/β transition on shear wave velocities was shown to be relatively small. He suggested that the velocity reduction is caused by the elastic softening of the structure of the constituent quartz minerals near the α/β transition and the opening of the grain-boundary cracks due to the very high thermal expansivity of quartz relative to the other component minerals. In contrast, the velocity increase in the β field may be attributed to a stiffening of the quartz structure.

Kern's experiments also confirmed that the transition temperatures inferred for the polycrystalline rocks from the location of the cusp on the velocity-temperature trajectory are generally higher than for quartz single crystals, and that the difference increases with increasing confining pressure. He concluded that the α/β transition in the quartzite is strongly influenced by the anisotropy of thermal expansion and compressibility of individual grains, and, for the granite and granulite, by the differences in thermal expansion and compressibility between component minerals. The α/β transition temperature is therefore higher than expected, and the discrepancy is more pronounced in granite and granulite than in quartzite, because the quartz grains in the former rocks are subjected to greater compressive stress than the other constituents, on account of their larger thermal expansivity. The enhancement of thermal cracking due to the α/β transition in quartz has also been monitored by GLOVER et al. (1995) using the acoustic emission method. Their acoustic emission measurements exhibit a strong peak of micro-cracking at the α/β transition temperature (573°C) for some quartz-bearing rocks (La Bresse mylonite, La Peyratte granite and amphibolite) heated at ambient pressure.

Figure 12 shows the quartz α/β transition temperatures ($T_{\alpha-\beta}$) for granites and a quartzite inferred by KERN (1979) and VAN DER MOLEN (1979) from ultrasonic V_P measurements. The present experiments on Cape Sorell quartzite reveal a marked minimum in shear modulus near 600°C under confining pressures of 50–100 MPa (Fig. 2). The shear modulus minimum is more significant at 50 MPa than at 100 MPa, and is attributed to the quartz α-β transition. At a confining pressure of 200 MPa, the cusp associated with the transition is more difficult to locate. However, a transition temperature near 650°C is suggested by the measurements made during cooling (Fig. 10). These transition temperatures for Cape Sorell quartzite are in good agreement with the results of Kern (Fig. 12). In the case of Delegate aplite (Fig. 6), no shear modulus minimum related to the quartz α/β transition was observed during the thermal cycling up to 700°C, in accord with the higher α/β transition temperatures for granites, especially Delegate aplite, inferred by KERN (1979) and VAN DER MOLEN (1979).

Frequency-dependent Strain Energy Dissipation and Associated Modulus Dispersion

Strong, frequency-dependent dissipation was often observed for both Delegate aplite and Cape Sorell quartzite at temperatures $\geq 500°C$, and high values of Q^{-1} were invariably associated with significant dispersion of the shear modulus G. As previously mentioned, however, Q^{-1} was found to be considerably more sensitive to prior thermal history than is the modulus G, making the interpretation of the Q^{-1} data necessarily more tentative. The dependence of Q^{-1} upon oscillation

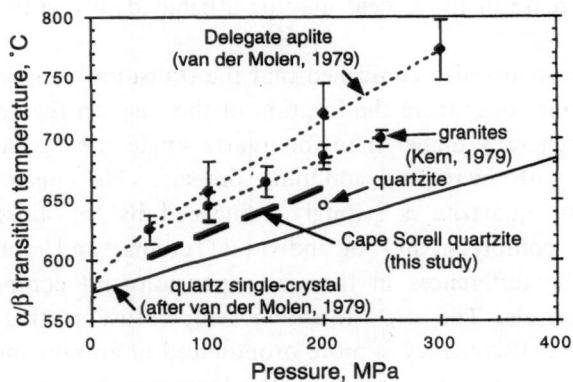

Figure 12

The quartz α/β transition temperatures inferred for Cape Sorell quartzite from the locations of the minima on the $G(T)$ curves in Figures 2 and 10, and for other quartz-bearing rocks from ultrasonic measurements of compressional wave speeds (VAN DER MOLEN, 1979; KERN, 1979).

period T_0 for both rocks at high temperatures and periods of 3–100 s is adequately represented by a power law of the form $Q^{-1} \sim T_0^n$. As mentioned above, the maximum strain amplitude 5×10^{-6} used in these high-temperature experiments lies near the upper limit of the linear regime where both shear modulus and dissipation or internal friction are strain-amplitude independent and linear theories of attenuation of seismic waves apply. The frequency-dependent dissipation is associated with noticeable dispersion of the shear modulus over the same frequency range at these high temperatures (e.g., Fig. 4). The measured dispersion of shear modulus has been compared with that calculated from the measured dissipation through the Kramers-Kronig relation of the linear theory. For this purpose, we used an approximate form of the Kramers-Kronig relation derived by BRENNAN and SMYLIE (1981) for a linear-viscoelastic medium for which the attenuation coefficient varies with the angular frequency as ω^m $(0 \leqslant m < 1)$, or equivalently, to first order, $Q^{-1} \sim (T_0)^n$ with $m = 1 - n$

$$G(T_0) = G(T_{0,\text{ref}})\{1 - \tan[\pi(1 - n)/2][Q^{-1}(T_0) - Q^{-1}(T_{0,\text{ref}})]\}. \tag{10}$$

This approximate relationship holds for the case $Q^{-1} \ll 1$ and for oscillation periods T_0 in the neighbourhood of the reference oscillation period $T_{0,\text{ref}}$. The measured dispersion of shear modulus over the range of oscillation period from 3 to 100 seconds is always fairly consistent with expectations from linear theory. For example, for Cape Sorell quartzite tested at 700°C and 100 MPa (Figs. 4 and 5), the dispersions directly observed and calculated from equation (10) are respectively -1.1 and -1.2%. The higher values of Q^{-1} measured on Delegate aplite at 500°C and 50 MPa (Fig. 7 (b)) and at 700°C and 200 MPa (Fig. 7 (d)) yield calculated dispersions of -3.0%, compared with observed dispersions of -2.1 and -3.5%, respectively.

It is tempting to link much of the variation of Q^{-1} with temperature and prior thermal history to the occurrence of cracking and to the stress concentrations responsible for it. A relatively short duration of exposure to temperatures higher than those encountered during previous heat treatment at the same pressure is evidently most conducive to intense strain energy dissipation. These conditions apply, for example, to the Cape Sorell quartzite specimen tested at 700°C and 100 MPa during the eighth thermal cycle (Table 1, Fig. 5). For the Delegate aplite specimens cycled in temperature, first at 50 MPa and subsequently at 100 and 200 MPa, Q^{-1} increases systematically with increasing temperature for the interval 400–600°C at 50 MPa (Figs. 7 (a, b)). However, exposure of specimen # 6 to these conditions for a period of several hours resulted in progressively much lower values of Q^{-1} (Fig. 7 (b)). A similar, though less dramatic example of temporal evolution of Q^{-1} is that of Delegate aplite specimen # 5 tested at 700°C and 200 MPa (Fig. 7 (d)).

The strong, frequency-dependent dissipation observed at high temperatures is thus, at least in part, a transient phenomenon, plausibly linked to stress concentrations which are caused by intragranular anisotropy and intergranular variability in thermal expansivity. Each stage of heating of the virgin rock between successive temperatures increases the stress intensity factors for each pre-existing and previously thermally grown crack. For a subset of the population of cracks, the additional temperature increment will be sufficient to increase the stress intensity factor to the critical value. Depending upon the form of the functional dependence of the stress intensity factor upon crack size, each such crack will either increase in size to a new stable value (less than the grain size) for which the stress intensity factor is sub-critical or propagate unstably until it encounters conditions less favourable for growth in adjacent grains (e.g., FREDRICH and WONG, 1986). High confining pressure (~ 100 MPa) reduces the effective stress intensity factor more for relatively large cracks and therefore should have a stabilising influence favouring the former scenario.

Over periods of hours under constant temperature-pressure conditions, this equilibrium may be threatened either by sub-critical crack growth (e.g., ATKINSON, 1984), or by plastic deformation ahead of the crack tip. Any stress relaxation associated with the latter would reduce the stress intensity factor at the crack tip and conceivably cause the crack to shrink in size. Dislocations which are generated and multiply under favourable conditions ahead of the crack tip may well be sufficiently mobile at temperatures of 500–700°C to account for some of the frequency-dependent attenuation observed at high temperatures. This picture is further complicated by any asperities across which the opposing faces of pre-existing cracks are in contact, and a possible role for volatiles as discussed below. A serious attempt at a mechanistic interpretation clearly requires detailed information concerning crack morphologies and statistics from future studies.

Possible Effect of Moisture Content on Strain Energy Dissipation

Previous experiments on thermally cycled rocks (KISSELL, 1972; TODD and SIMMONS, 1973; WARREN, 1973; JOHNSON et al., 1978; TITTMANN et al., 1979; JOHNSTON and TOKSÖZ, 1980; CLARK et al., 1981), tested at or below atmospheric pressure, have revealed an initial decrease in Q^{-1} with increasing temperature in spite of the increase in crack population, which they attributed to the removal of adsorbed water vapour from the rocks. JOHNSTON and TOKSÖZ (1980) suggested that the dependence of attenuation on temperature in the thermally cycled rocks is the result of two competing effects: outgassing, which decreases attenuation, and thermal cracking, which might increase attenuation due to increase in crack porosity and the density of intergranular frictional contacts.

Sources of moisture in the rocks tested in this study include adsorbed water not removed by oven-drying at 80°C, water released by decrepitation of fluid inclusions, and any free water produced by dehydration of the minor hydrous layer silicate phases (1% muscovite in Cape Sorell quartzite, minor chlorite associated with feldspar alteration in Delegate aplite). The venting to atmosphere of the interior of the specimen assembly at each end of the jacketted rock specimen allows the gradual escape of volatiles from the pore space of the specimen. Such progressive dehydration of the specimen, especially under conditions in which the aqueous pore fluid pressure generated within the rock approaches or exceeds the confining pressure, may contribute to the temporal and history-dependent behaviour of the internal friction described above.

The systematically higher values of internal friction measured during the first thermal cycle for Delegate aplite specimen #5 at 50 MPa (Fig. 7 (a)) would then be associated with the presence of water, whereas the later lower values of Q^{-1} would reflect drier conditions. The relatively high values of Q^{-1} obtained on Delegate aplite specimen #6 following exposure to laboratory atmosphere and rejacketting might also reflect the involvement of volatiles (Fig. 7 (c)). The notion that moisture might have a strong influence upon the mechanical properties of aggregates of contacting grains of silica (and possibly other siliceous minerals) derives support from observations of time-dependent adhesional and frictional behaviour of amorphous silica surfaces which is attributed to the presence of a thin gel-like layer (VIGIL et al., 1994).

The Relationship Between Crack Density and Crack Connectivity

The values for the rate constant A for pore pressure equilibration between upstream and downstream reservoirs, derived from the pore pressure decay curves (e.g. Fig. 8) and displayed in Figure 9, demonstrate that the "onset" of fast argon flow occurs at about 350°C for confining pressure of 50 MPa, 450°C for 100 MPa, and 550°C for 150 MPa. Although the permeability may not be rigorously calculated from A with the theory of the transient flow method because of the high and strongly pressure-dependent compressibility of argon at low P_f, we may expect that the permeability increases with increasing A.

Indeed, as mentioned above, a parallel series of transient flow measurements at higher pore pressures and an effective pressure $P_{eff} = P_c - P_f$ of 50 MPa, demonstrates that the permeability increases by two orders of magnitude from 10^{-19} to 10^{-17} m^2 as the temperature is increased from the threshold for appreciable connectivity near 350°C to 650°C. The low-temperature value is comparable with that (0.7×10^{-19} m^2) reported for the fine-grained Westerly granite under similar conditions (BRACE et al., 1968). At the relatively low pore pressures (<8 MPa) employed in the present measurements, the presence of the argon pore fluid is not expected to have a significant influence on the microstructure of the rock. The transport properties measured under these conditions therefore closely approximate those of the dry rock under the same confining pressure and temperature.

Examination of the temperature dependence of the shear modulus in dry Delegate aplite (Fig. 6) reveals similar values of 0.65–0.71 for the ratio ($G_T/G_{25°C}$) at the threshold temperatures for fast pore fluid flow at each of the three values (50, 100 and 150 MPa) of the confining pressures ($G_{25°C}$ has been taken as the value of shear modulus measured at room temperature and the highest confining pressure (200 MPa) in the experiment). For each such value of $G_T/G_{25°C}$, the increase in crack density may be estimated from equation (3). In Figure 13, the measured rate constant for pore pressure equilibration (a proxy for permeability) is plotted against the crack density inferred from the shear modulus measurements under the same conditions.

As discussed previously, the onset of thermal cracking in Delegate aplite is indicated by a marked increase in the temperature sensitivity of the shear modulus, at temperatures of approximately 150°C, 250°C and 400°C for confining pressures of 50, 100 and 200 MPa, respectively (Fig. 6), however substantially higher temperatures are required for relatively rapid pore pressure equilibration (Fig. 9). This is emphasised in Figure 13 where it is seen that A is a single-valued function of the inferred density of thermal cracks, and that A increases significantly only beyond a threshold crack density of 0.18 ± 0.02. This is because the permeability depends on the interconnected crack porosity within the rock, while the shear modulus is affected by the total crack porosity. Increased isolated crack porosity does not contribute to the increase of permeability. This kind of phenomenon in transport properties of fractured or porous rocks has been well described by percolation models (e.g., GUEGUEN and DIENES, 1989; BALBERG et al., 1991; GÓMEZ et al., 1995).

Percolation models provide universal laws which determine the geometrical and physical properties of a system:

$$Y \sim (x - x_c)^\beta \qquad (11)$$

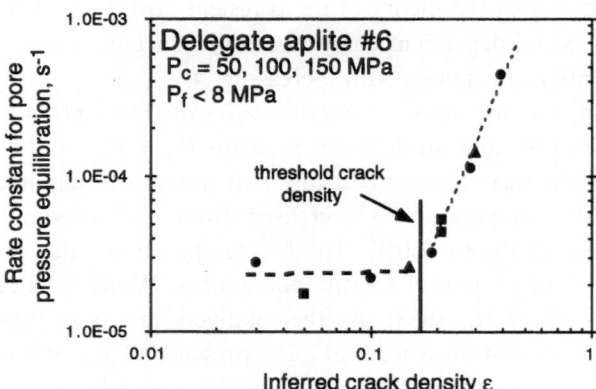

Figure 13
The rate constant A for pore pressure equilibration (a proxy for permeability) versus thermally-induced crack density for various confining pressures (circles—50 MPa; triangles—100 MPa; squares—150 MPa). The crack density is inferred from the observed shear modulus deficit through the theory of O'CONNELL and BUDIANSKY (1974).

where Y is a geometrical or physical quantity, x is the fractional volume of the conducting phase (e.g., porosity or crack density), x_c is the critical value for the onset of percolation, i.e., of system-wide connectivity, and β is an exponent specific to the quantity Y. The assumption of these percolation models is that the spatial distribution of cracks inside the rock is random, a reasonable approximation for thermal cracking in a rock subject to uniform temperature under hydrostatic stress. The degree of crack connectivity depends only upon the probability (x) of the presence of cracks. When x reaches the critical value of x_c, the internal network of connected cracks acquires a macroscopic size, and as a consequence, conduction properties (e.g., permeability) will appear (GÓMEZ et al., 1995). Such a model with $Y = \log A$ and $x = \log \varepsilon$ provides an excellent representation of the variation of the rate constant A with crack density ε; the critical value of crack density for the onset of percolation is evidently 0.18 ± 0.02 and the exponent β adopts the value 3.4 (Fig. 13).

Conclusions

In this study, experiments have been conducted on two crystalline crustal rocks, Cape Sorell quartzite and Delegate aplite in order to investigate the effects of thermally generated crack porosity on the shear-mode seismic properties measured at seismic frequencies, under conditions of varying confining pressure (50–200 MPa) and temperature (25–700°C) with the following findings:

(1) The measured temperature dependence of the shear modulus G shows that for a given confining pressure, there is a threshold temperature T_{cr} beyond which G decreases markedly. Below T_{cr}, the $G(T)$ trend for the quartzite is convincingly parallel to that calculated for a zero-porosity quartz polycrystal, suggesting that the thermal stresses arising from the anisotropy and heterogeneity of thermal expansion are insufficient to cause cracking. For Delegate aplite, the threshold temperature increases with increasing confining pressure with $\partial T_{cr}/\partial P_c \sim 1.7°C/MPa$. The experimental results are broadly consistent with expectations based on fracture mechanics models.

(2) The consistent temperature dependence of the shear modulus for Delegate aplite, and for Cape Sorell quartzite at higher confining pressures (>100 MPa), demonstrates that thermal cracking in these rocks is fully reversible, in the sense that the thermally generated crack porosity is eliminated by modest confining pressures when the supporting thermal stresses are removed. Under these circumstances, the inferred crack density is a function only of temperature and confining pressure, and is not dependent upon thermal history, in marked contrast to our experience with Carrara marble (LU and JACKSON, 1996). For this reason, and because of the more restricted temperature range for thermal cracking in Cape Sorell quartzite, further experiments on the influence of (argon) pore-fluid saturation upon the properties of thermally cracked rocks have been restricted to Delegate aplite (LU and JACKSON, in preparation).

(3) The quartz α/β transition temperature, inferred from the location of the cusp on the $G(T)$ trajectory for the thermally cycled quartzite, increases with increasing confining pressure in accord with the findings of Fielitz and Kern. The modest reduction in G observed at 200 MPa confining pressure is attributed entirely to the anomalous behaviour of the elastic moduli in the neighbourhood of the α/β transition. The fact that the cusps associated with the transition are far more pronounced at lower confining pressures (50–100 MPa), indicates a substantial role for thermal cracking caused by the very large and strongly anisotropic thermal expansivity of the α phase, especially in the vicinity of the transformation. Significant hysteresis in $G(T)$ for Cape Sorell quartzite was observed during the thermal cycling to peak temperatures $\geq 575°C$ at 50 MPa. For Delegate aplite, however, no significant $G(T)$ cusp suggestive of the quartz α/β transition was observed during the thermal cycling up to 600°C at 50 MPa and up to 700°C at $P_c \geq 100$ MPa. This observation is consistent with the findings of VAN DER MOLEN (1979) on the influence of pressure on the α/β transition temperature in Delegate aplite.

(4) The time scales for the flow of argon at low pore pressure (<8 MPa) through the rock specimens were measured during thermal cycling. For Cape Sorell quartzite, the results indicate that the rock specimen is essen-

tially impermeable even under the experimental conditions most favourable for fluid flow (50 MPa confining pressure and 600°C), although the measured shear modulus suggests that significant thermal cracking occurs. This implies that the thermally induced microcracks in the rock specimen must remain poorly connected. The relationship between the argon-flow timescale (representing permeability or crack connectivity) and the crack density inferred from the modulus measurements suggests that there is a threshold crack density $(0.18 \pm 0.02$ for Delegate aplite) beyond which connectivity and thus permeability increase markedly. Evidently the critical crack density for establishment of crack connectivity is not reached in the experiments on the Cape Sorell quartzite.

(5) The strain energy dissipation Q^{-1} for both rocks is generally low at temperatures below 500°C, but increases noticeably at higher temperatures. At temperatures > 500°C, Q^{-1} becomes frequency-dependent, increasing with decreasing frequency (increasing oscillation period T_0). As for Carrara marble (LU and JACKSON, 1996), the frequency dependence of Q^{-1} for Cape Sorell quartzite and Delegate aplite is well described by the power law $Q^{-1} \sim T_0^n$, with values of n in the range 0.3–0.4. Q^{-1} decreases very substantially with increasing exposure to high temperature conditions, possibly as a consequence of the progressive relaxation of thermal stresses and/or dehydration.

Acknowledgements

Stephen Cox is thanked for assistance with field sampling of the Cape Sorell quartzite and for advice concerning the operation of pore-fluid systems. Constructive reviews by Paul Glover and Teng-fong Wong are gratefully acknowledged. The technical assistance of Harri Kokkonen and Graeme Horwood is greatly appreciated.

REFERENCES

ATKINSON, B. K. (1984), *Sub-critical Crack Growth in Geological Materials*, J. Geophys. Res. *89*(B6), 4077–4114.

ATKINSON, B. K., MACDONALD, D., and MEREDITH, P. G., *Acoustic response and fracture mechanics of granite subjected to thermal cycling experiments*. In *Third Conference on Acoustic Emission/Microseismic Activity in Geologic Structures and Materials* (Trans. Tech. Publications, Clausthal, West Germany 1984), pp. 5–18.

BALBERG, S. J., BERKOWITZ, B., and DRACHSLER, G. E. (1991), *Application of a Percolation Model to Flow in Fractured Hard Rocks*, J. Geophys. Res. *96*(B6), 10015–10021.

BAUER, S. J., and JOHNSON, B. (1979), *Effects of Slow Uniform Heating on the Physical Properties of Westerly and Charcoal Granites*, Proc. 20th U.S. Symp. Rock Mech., 7–18.

BRACE, W. F., WALSH, J. B., and FRANGOS, W. T. (1968), *Permeability of Granite under High Pressure*, J. Geophys. Res. *73*, 2225–2236.

BRENNAN, B. J., and SMYLIE, D. E. (1981), *Linear Viscoelasticity and Dispersion in Seismic Wave Propagation*, Rev. Geophys and Space Phys. *19*(2), 233–246.

CARLSON, S. R., and WANG, H. F. (1986), *Microcrack Porosity and in situ Stress in Illinois Borehole UPH 3*, J. Geophys. Res. *91*(10), 10421–10428.

CHRISTENSEN, N. I. (1966), *Shear Wave Velocities in Metamorphic Rocks at Pressures to 10 Kilobars*, J. Geophys. Res. *71*, 3549–3556.

CHRISTENSEN, N. I. (1974), *Compressional Wave Velocities in Possible Mantle Rocks to Pressures of 30 Kilobars*, J. Geophys. Res. *79*(2), 407–412.

CHRISTENSEN, N. I. (1979), *Compressional Wave Velocities in Rocks at High Temperatures and Pressures, Critical Thermal Gradient, and Crustal Low-velocity Zones*, J. Geophys. Res. *84*, 6849–6857.

CLARK, V. A., SPENCER, T. W., and TITTMANN, B. R. (1981), *The Effect of Thermal Cycling in the Seismic Quality Factor Q of Some Sedimentary Rocks*, J. Geophys. Res. *86*(B8), 7087–7094.

DOYEN, P. M. (1987), *Crack Geometry in Igneous Rocks: A Maximum Entropy Inversion of Elastic and Transport Properties*, J. Geophys. Res. *92*(B8), 8169–8181.

EVANS, A. G. (1978), *Microfracture from Thermal Expansion Anisotropy—I. Single Phase Systems*, Acta Metallurgica *26*, 1845–1853.

FIELTZ, K. (1971), *Elastic Wave Velocities in Different Rocks at High Pressure and Temperature up to 750°C*, Z. Geophys. *37*, 943–956.

FIELITZ, K., *Compressional and shear wave velocities as a function of temperature in rocks at high pressure*. In *Exploration Seismology in Central Europe* (eds. Giese P., Prodehl C., and Stein A.) Springer Verlag, Berlin-New York 1976) pp. 40–44.

FREDRICH, J. T., and WONG, T. F. (1986), *Micromechanics of Thermally-induced Cracking in Three Crustal Rocks*, J. Geophys. Res. *91*(B12), 12743–12764.

GEBRANDE, H., *Elastic wave velocities and constants of elasticity of rocks and rock forming minerals*. In *Landolt-Börnstein Tables* (ed. G. Angenheister) new series, Vol. 1b (Springer Verlag, Berlin 1982) pp. 1–98.

GLOVER, P. W., BAUD, P., DAROT, M., MEREDITH, P. G., BOON, S. A., LeRAVALAC, M., ZOUSSI, S., and REUSCHLÉ (1995), *α/β Phase Transition in Quartz Monitored Using Acoustic Emissions*, Geophys. J. Int. *120*, 775–782.

GÓMEZ, J. B., PACHECO, A. F., and SEGUÍ-SANTONJA, A. J. (1995), *A Model for Crack Connectivity in Rocks, A Discussion*, Math. Geol. *27*(1), 23–39.

GUEGUEN, Y., and DIENES, J. (1989), *Transport Properties of Rocks from Statistics and Percolation*, Math. Geol. *21*(1), 1–13.

HARLOW, F. W., and PRACHT, W. E. (1972), *A Theoretical Study of Geothermal Energy Extraction*, J. Geophys. Res. *77*, 7038–7048.

HEARD, H. C. (1980), *Thermal Expansion and Inferred Permeability of Climax Quartz Monzonite to 300°C and 27.6 MPa*, Int. J. Rock Mech. Min. Sci. and Geomech. Abstr. *17*, 289–296.

HEARD, H. C., and PAGE, L. (1982), *Elastic Moduli, Thermal Expansion, and Inferred Permeability of Two Granites to 350°C and 55 Megapascals*, J. Geophys. Res. *87*(B11), 9340–9348.

HOMAND-ETIENNE, F., and HOUPERT, R. (1989), *Thermally-induced Microcracking in Granites: Characterization and Analysis*, Int. J. Rock Mech. Min. Sci. Geomech. Abstr. *26*(2), 125–134.

HSIEH, P. A., TRACY, J. V., NEUZIL, C. E., BRESEHOEFT, J. D., and SILLIMAN, S. E. (1981), *A Transient Laboratory Method for Determining the Hydraulic Properties of "Tight" Rocks—I. Theory*, Int. J. Rock Mech. Min. Sci. and Geomech. Abstr. *18*, 245–252.

IDE, J. M. (1937), *The Velocity of Sound in Rocks and Glasses as a Function of Temperature*, J. Geol. *45*, 689–716.

JACKSON, I., PATERSON, M. S., NIESLER, H., and WATERFORD, R. M. (1984), *Rock Anelasticity Measurements at High Pressure, Low Strain Amplitude and Seismic Frequency*, Geophys. Res. Lett. *12*, 1235–1238.

JACKSON, I. (1991), *The Petrophysical Basis for the Interpretation of Seismological Models for the Continental Lithosphere*, Geol. Soc. Aust. Spec. Publ. *17*, 81–114.

JACKSON, I., and PATERSON, M. S. (1993), *A High-pressure, High-temperature Apparatus for Studies of Seismic Wave Dispersion and Attenuation*, Pure appl. geophys. *141* (2/3/4), 445–466.

JOHNSTON, D. H., and TOKSÖZ, M. N. (1980), *Thermal Cracking and Amplitude-dependent Attenuation*, J. Geophys. Res. *85*(B2), 937–942.

JOHNSON, B., GANGI, A. F., and HANDIN, J. (1978), *Thermal Cracking of Rock Subjected to Slow, Uniform Temperature Changes*, Proc. 21st U.S. Symp. Rock Mech., Reno, Nev., 259–267.

KERN, H. (1978), *The Effect of High Temperature and High Confining Pressure on Compressional Wave Velocities in Quartz-bearing and Quartz-free Igneous and Metamorphic Rocks*, Tectonophysics *44*, 185–203.

KERN, H. (1979), *Effect of High-low Quartz Transition on Compressional and Shear Wave Velocities in Rocks under High Pressure*, Phys. Chem. Minerals *4*, 161–167.

KISSEL, F. N. (1972), *The Effect of Temperature Variation on Internal Friction in Rocks*, J. Geophys. Res. *77*, 1420–1423.

KOWALLIS, B. J., and WANG, H. F. (1983), *Microcrack Study of Granitic Cores from Illinois Deep Borehole UPH 3*, J. Geophys. Res. *88*, 7373–7380.

KRANZ, R. L., FRANKEL, A. D., ENGELDER, T., and SCHOLZ, C. H. (1979), *The Permeability of Whole and Jointed Barre Granite*, Int. J. Rock Mech. Min. Sci. Geomech. Abstr. *16*, 225–234.

LU, C., and JACKSON, I. (1996), *Seismic-frequency Laboratory Measurements of Shear Mode Viscoelasticity in Crustal Rocks I: Competition between Cracking and Plastic Flow in Thermally Cycled Carrara Marble*, Phys. Earth Planet. Int. *94*(1–2), 105–119.

MAJER, E. L., and MCEVILLY, T. V. (1985), *Acoustic Emission and Wave Propagation Monitoring at the Spent Fuel Test: Climax, Nevada*, Int. J. Rock Mech. Min. Sci. Geomech. Abstr. *22*, 215–226.

O'CONNELL, R. J., and BUDIANSKY, B. (1974), *Seismic Velocities in Dry and Saturated Cracked Solids*, J. Geophys. Res. *79*(B5), 5412–5426.

PAGE, L., and HEARD, H. C. (1981), *Elastic Moduli, Thermal Expansion, and Inferred Permeability of Climax Quartz Monzonite and Sudbury Gabbro to 500°C and 55 MPa*, Proc. U.S. 22nd Symp. Rock Mech., 97–104.

PEARSON, C. F., FEHLER, M. C., and ALBRIGHT, J. N. (1983), *Changes in Compressional and Shear Wave Velocities and Dynamic Moduli during Operation of a Hot Dry Rock Geothermal System*, J. Geophys. Res. *88*, 3468–3475.

RICHTER, D., and SIMMONS, G. (1974), *Thermal Expansion Behaviour of Igneous Rocks*, Int. J. Rock Mech. Min. Sci. and Geomech. Abstr. *11*, 403–411.

SCHOCK, R. M., BONNER, B. P., and LOUIS, H. (1974), *Collection of Ultrasonic Velocity Data as a Function of Pressure for Polycrystalline Solids*, Rep. UCRL-51508, Lawrence Livermore Natl. Lab., Livermore, Calif.

SIMMONS, G., and COOPER, H. W. (1978), *Thermal Cycling Cracks in Three Igneous Rocks*, Int. J. Mech. Min. Sci. and Geomech. Abstr. *15*, 145–148.

SIMMONS, G., and RICHTER, D., *Microcracks in rocks*. In *The Physics and Chemistry of Minerals and Rocks* (ed. Sterns, R. J. G.) (Wiley-Intersciences, New York 1976) pp. 105–137.

SKINNER, B.J., *Thermal expansion*. In *Handbook of Physical Constants* (ed. Clark, S. P.) (The Geological Society of America Memoir, 97, New York 1966) pp. 75–96.

SMYTH, J. R., and MCCORMICK, T. C., *Crystallographic data for minerals*. In *Mineral Physics and Crystallography, A Handbook of Physical Constants* (ed. T. J. Ahrens), (AGU, Washington 1995) pp. 1–17.

SPRUNT, E. S., and BRACE, W. F. (1974), *Direct Observation of Microcavities in Crystalline Rocks*, Int. J. Rock Mech. Min. and Geomech. Abstr. *11*, 139–150.

SUMMER, R., WINKLER, K., and BYERLEE, J. (1978), *Permeability Changes during the Flow of Water through Westerly Granite at Temperatures of 100°–400°C*, J. Geophys. Res. *83*(B1), 339–344.

TITTMAN, B. R., NADLER, H., CLARK, V. A., and COOMB, L. T. (1979), *Seismic Q and Velocity at Depth*, Proc. Lunar Sci. Conf. 10th, 2131–2145.

TODD, T., and SIMMONS, G. (1973), *Effect of Pore Pressure on the Velocity of Compressional Waves in Low Porosity Rocks*, J. Geophys. Res. *77*, 3731–3743.

TRIMMER, D. B., BONNER, B., HEARD, H. C., and DUBA, A. (1980), *Effect of Pressure and Stress on Water Transport in Intact and Fractured Gabbro and Granite*, J. Geophys. Res. *85*, 7059–7071.

VAN DER MOLEN, I. (1979), *Some Physical Properties of Granite at High Pressure and Temperature*, Ph. D. Thesis, Australian National University, Canberra, Australia.

VAN DER MOLEN, I. (1981), *The Shift of the α-β Transition Temperature of Quartz Associated with the Thermal Expansion of Granite at High Pressure*, Tectonophysics *73*, 323–342.

VAN DER MOLEN, I., and PATERSON, M. S. (1979), *Experimental Deformation of Partially-melted Granite*, Contrib. Mineral. Petrol. *70*, 299–318.

VIGIL, G., XU, Z., STEINBERG, S., and ISRAELACHVILI, J. (1994), *Interactions of Silica Surface*, J. Colloid Interface Sci. *165*, 367–385.

WALSH, J. B. (1965), *The Effect of Cracks on the Compressibility of Rock*, J. Geophys. Res. *70*, 381–389.

WANG, H. F., BONNER, B. P., CARLSON, S. R., KOWALLIS, B. J., and HEARD, H. C. (1989), *Thermal Stress Cracking in Granite*, J. Geophys. Res. *94*(B2), 1745–1758.

WANG, H. F., and HEARD, H. C. (1985), *Prediction of Elastic Moduli via Crack Density in Pressurized and Thermally Stressed Rock*, J. Geophys. Res. *90*(B12), 10342–10350.

WARREN, N. (1973), *Brief Note on Effects of Thermal Pulse on Elastic Moduli and Q of Rock*. Earth Planet. Sci. Lett. *20*, 280–285.

WATT, J. P., DAVIES, G. F., and O'CONNELL, R. J. (1976), *The Elastic Properties of Composite Materials.*, Rev. Geophys. Space Phys. *14*, 541–563.

WONG, T. F., and BRACE, W. F. (1979), *Thermal Expansion of Rocks: Some Measurements at High Pressure*, Tectonophysics *57*, 97–117.

ZOBACK, M. D., and BYERLEE, J. D. (1975), *The Effect of Cyclic Differential Stress on Dilatancy in Westerly Granite under Uniaxial and Triaxial Conditions*, J. Geophys. Res. *80*(11), 1526–1536.

(Received March 6, 1998, revised June 30, 1998, accepted July 25, 1998)

To access this journal online:
http://www.birkhauser.ch

Pure appl. geophys. 153 (1998) 475–487
0033–4553/98/040475–13 $ 1.50 + 0.20/0

▌Pure and Applied Geophysics

A Summary of Attenuation Measurements from Borehole Recordings of Earthquakes: The 10 Hz Transition Problem

RACHEL E. ABERCROMBIE[1,2]

Abstract—Earthquake seismograms recorded by instruments in deep boreholes have low levels of background noise and wide signal bandwidth. They have been used to extend our knowledge of crustal attenuation both in the near-surface and at seismogenic depths. Site effects are of major importance to seismic hazard estimation, and the comparison of surface, shallow and deep recordings allows direct determination of the attenuation in the near-surface. All studies to date have found that Q is very low in the near-surface (~ 10 in the upper 100 m), and increases rapidly with depth. Unlike site amplification, attenuation at shallow depths exhibits little dependence on rock-type. These observations are consistent with the opening of fractures under decreasing lithostatic pressure being the principal cause of the severe near-surface attenuation. Seismograms recorded in deep boreholes are relatively unaffected by near-surface effects, and thus can be used to measure crustal attenuation to higher frequencies (≥ 100 Hz) than surface recordings. Studies using both direct and coda waves recorded at over 2 km depth find Q to be high (~ 1000) at seismogenic depths in California, increasing only weakly with frequency between 10 and 100 Hz. Intrinsic attenuation appears to be the dominant mechanism. These observations contrast with those of the rapidly increasing Q with frequency determined from surface studies in the frequency range 1 to 10 Hz. Further work is necessary to constrain the factors responsible for this apparent change in the frequency dependence of Q, but it is clearly unwise to extrapolate Q estimates made below about 10 Hz to higher frequencies.

Key words: Crustal attenuation, site-effects, high frequency, earthquakes, borehole seismology.

Introduction

Seismic waves which propagate through the earth are attenuated both by scattering and anelastic processes. Seismometers therefore record only a depleted amount of the high frequency energy radiated from earthquakes and other seismic sources. Seismograms recorded at the earth's surface are also contaminated by seismic noise, both environmental and man-made. The combined effects of seismic

[1] Institute of Geological and Nuclear Sciences, Ltd., New Zealand.

[2] Now at: Department of Earth and Planetary Sciences, Harvard University, 20 Oxford Street, Cambridge, MA 02138, U.S.A. Tel.: 617 495 9604, Fax: 617 495 0635, E-mail: abercrombie@seismology.harvard.edu

noise and attenuation limit the frequency range of the observed signal to below a few tens of Hz at most surface sites. Consequently, the frequency range over which attenuation can be measured is also limited. This makes it difficult to link observations of attenuation in the earth with the much smaller scale (and hence higher frequency) laboratory studies of attenuation mechanisms. Additionally, investigation of the behaviour of Q with frequency requires measurements over a wide frequency range. Measurements of Q above a few tens of Hz would extend the bandwidth of data and hence provide further insight of the principal mechanisms responsible for attenuation within the crust. In addition, wider bandwidth seismograms are essential to resolve fundamental questions regarding the nature of earthquake sources. Over the last decade there has been considerable controversy concerning the scaling of earthquake source parameters, resulting from the inherent ambiguity in extracting source and attenuation parameters from surface seismograms. For example, ARCHULETA et al. (1982) proposed that earthquake self-similarity breaks down below about $M_L 3$, at a scale length of a few hundred metres. ANDERSON (1986), however, argued that this could be entirely an artifact of crustal attenuation, and it was difficult to see how such fundamental questions concerning the nature of earthquake sources could be answered with surface data.

The large distances at which seismic energy above 1 Hz is recorded are evidence that attenuation within the seismogenic and deeper crust is too low to be responsible for the effects observed by ARCHULETA et al. (1982) and many subsequent earthquake source studies. Deployments of closely spaced seismometers, however, have shown extensive variation in recorded amplitude, frequency content and coda duration over short distances, suggesting that the near-surface, near-receiver, rock has a strong effect on earthquake recordings. The site effect, as it has become known, includes amplification in the relatively low velocity, low density near-surface rocks, scattering and resonance within the shallow layers, and attenuation of higher frequency energy. Site effects are important, not just in investigating the nature of earthquakes sources, but also because they play a major role in the distribution of damage in large earthquakes (BORCHERDT et al., 1970; CELEBI et al., 1987). Strong site effects have been shown to amplify and prolong the ground shaking significantly, and consequently now form a major component of seismic hazard analyses.

Installing seismometers in even shallow boreholes decreases the level of background noise considerably, hence improving the quality of the recorded signal. The indications that significant attenuation and distortion of seismic waves are occurring at relatively shallow depths have led to the deployment of seismometers in deeper holes. The principal goals of such deployments have been to learn more about the propagation of seismic waves through the near-surface, and to record seismograms, uncontaminated by site-effects, for high resolution study of earthquake sources and the properties of the seismogenic crust. Throughout the last decade, many instruments have been deployed at depths of a few metres to a few tens of metres. Instruments at depths of a few hundred metres are also becoming more common, for example, the High Resolution Seismic Network (HRSN) at Parkfield, California (MALIN et al., 1989). The deepest instruments to date have been at depths of a few kilometres, and include the Varian well, Parkfield (operational to 1 km, e.g., JONGMANS and MALIN, 1995), Cajon Pass (to 2.9 km, ABERCROMBIE and LEARY, 1993), Long Valley (2 km, P. V. KASSEMEYER, personal communication, 1995) and the Kanto District, Japan (to 3.5 km, KINOSHITA, 1994). A high frequency sensor at 4 km recorded drilling induced events in the KTB hole, Germany (ZOBACK and HARJES, 1997), and more long-term installations are planned there. Deploying and operating borehole seismometers becomes more difficult with increasing depth, partly because the instrument is simply far-removed, and drilling and installation costs increase. More importantly, temperature increases with depth, restricting the sensor types which can be used, and the duration of recording; high temperatures are probably the principal limiting factor on the depth of installation. At 2.9 km at Cajon Pass the temperature is 120°C, similar to that at 4 km depth at KTB where the heat flow is lower. MANOV et al. (1996) provide a good overview of the logistics and constraints of deep borehole recording.

In this paper I review representative studies which use borehole recordings of earthquakes, and show how they have contributed to our knowledge of attenuation. First I consider attenuation in the near-surface, as determined by comparison between seismograms recorded at borehole and surface sensors. Secondly I describe recent measurements of crustal Q and its frequency dependence in the approximate frequency range 10–100 Hz.

Attenuation in the Near-surface

The effect on seismic waves of propagating through the near-surface rocks can be obtained by simply comparing earthquake recordings made at a surface seismometer with those from a borehole seismometer directly beneath it. In such studies the thickness of the "near-surface" is defined as the depth of the deepest seismometer. It has become relatively common practice to calculate spectral ratios between recordings at different depths and to use these ratios to estimate Q and amplification effects in the near-surface. This approach uses the same techniques and assumptions as surface studies of site effects, which compare various sites of interest to a hard rock reference site. Borehole recording is advantageous in that the reference site is a better approximation of the waveform entering the near-surface rocks at the site of interest than is a neighboring hard rock site. When using borehole recordings in spectral ratios, however, you have to take into account the fact that they include reflections from the surface and shallow layers. These surface reflections are usually very small in amplitude, especially at deep sensors (e.g.,

ABERCROMBIE, 1997), indicating high attenuation near the surface. These reflections can either be avoided by the careful choice of window lengths (e.g., ABERCROMBIE, 1997) or corrected for using simple reflectivity codes (e.g., STEIDL, 1996).

Typical values of both P wave and S wave total Q (Q_P and Q_S, respectively) in the upper 100 m are very low, around 10 in the frequency range of a few Hz to a few tens of Hz. They also appear to be almost independent of rock type. ABERCROMBIE (1997) compares the results of a study at Cajon Pass, southern California, with a number of other Californian studies (ABERCROMBIE, 1997, Table 4) including those of HAUKSSON et al. (1987) in the Los Angeles Basin, MALIN et al. (1988) at Oroville, ASTER and SHEARER (1991) at Anza, BLAKESLEE and MALIN (1991) and JONGMANS and MALIN (1995) at Parkfield, ARCHULETA et al. (1992) at Garner Valley, and GIBBS et al. (1994) in the Santa Clara Valley. These experiments encompass a wide range of rock types, but find relatively small differences in Q values in the upper few hundred metres. In the upper 500 m typical Q values are $Q_P \le 50$ and $Q_S \le 30$.

At depths of less than 500 m differences in rock type produce considerably more variation in site amplification than in Q. For example, ABERCROMBIE (1997) compares local earthquake recordings made at 2.9 km depth in the Cajon Pass borehole, at the wellhead, and at a surface granite site only 1 km away. The principal difference between the two surface sites is that the upper 500 m at the wellhead site are Miocene sandstone and at the granite site, granitic rocks. ABERCROMBIE finds $Q_P \sim 50$ and $Q_S \sim 23$ in the upper 2.9 km at the granite site which are essentially within the errors of the values obtained at the wellhead (see Fig. 1). The site amplification, however, is approximately three times higher at the wellhead than at the granite site.

Studies involving recordings made at a range of depths show that Q increases rapidly with depth, although some low Q zones are observed, for example, in the Oroville fault zone (MALIN et al., 1988). A comparison of the attenuation profile at the Cajon Pass drillhole (ABERCROMBIE, 1997) and the Varian Well (JONGMANS and MALIN, 1996; ABERCROMBIE, unpublished data) is shown in Figure 1. Both studies show low Q near the surface, increasing with depth. Despite the difference in rock types between the two sites, the Q profiles are essentially the same, within errors, to about 1 km. At greater depths, the thicker sediments and lower velocity Franciscan rocks to the NE of the San Andreas fault at Parkfield appear to have lower Q than the Mesozoic granite at similar depths at Cajon Pass.

All the studies considered to date in this review assume that near-surface Q is frequency independent. If this is the case, then the spectral ratios between different recordings depths will have linear slopes on a log-linear plot. Figure 2 displays some average spectral ratios between the wellhead and 2.5 km depth at Cajon Pass (ABERCROMBIE, 1997). They are well modelled by a straight line, indicating that near-surface Q is either frequency independent, or at least only very weakly dependent on frequency, in the frequency range of ~2 to 100 Hz. This result is in good agreement with that of many other studies which use both earthquakes (e.g., BLAKESLEE and MALIN, 1991) and artificial sources (e.g., GIBBS *et al.*, 1994). The assumption of frequency independent Q in the near-surface thus appears to be a sound one.

The reliability of a Q estimate made using the spectral ratio technique depends on the frequency range available for measuring the gradient of the spectral ratio

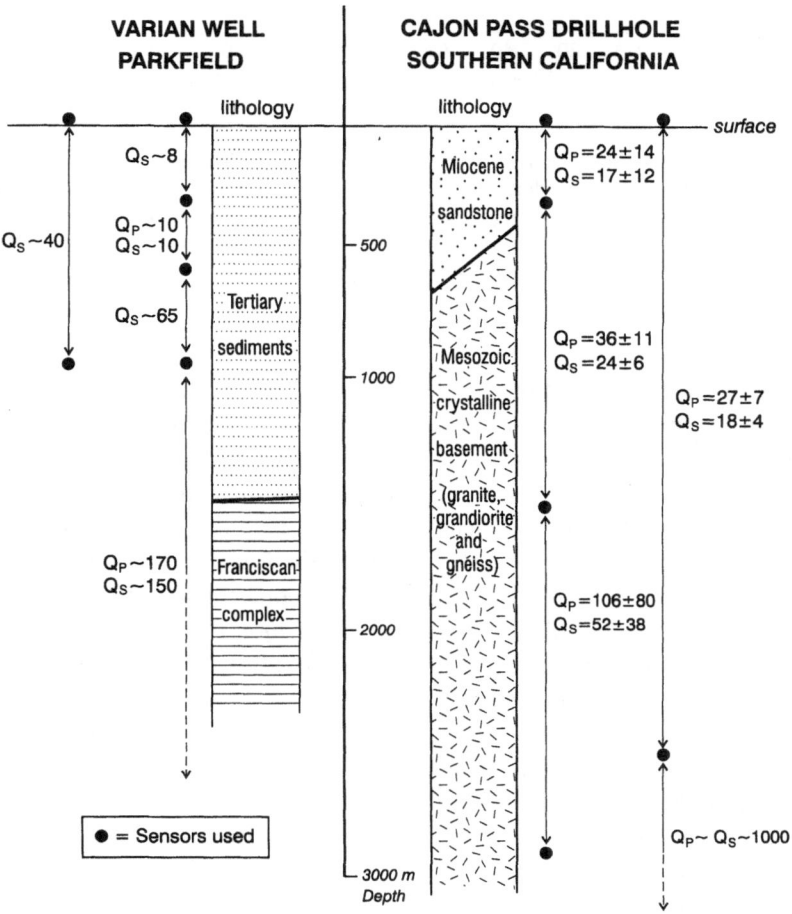

Figure 1

A comparison of the attenuation profiles with depth at the Cajon Pass scientific drillhole, southern California, and the Varian well, Parkfield, California. The Q values at Cajon Pass are estimated by ABERCROMBIE (1995, 1997) in the approximate frequency ranges 2–100 Hz (*P* waves) and 2–20 Hz (*S* waves), and errors of one standard deviation are shown (see Fig. 2). The Q values at Varian are from JONGMANS and MALIN (1996) and ABERCROMBIE (unpublished data). Both sites show very low Q near the surface, increasing with depth.

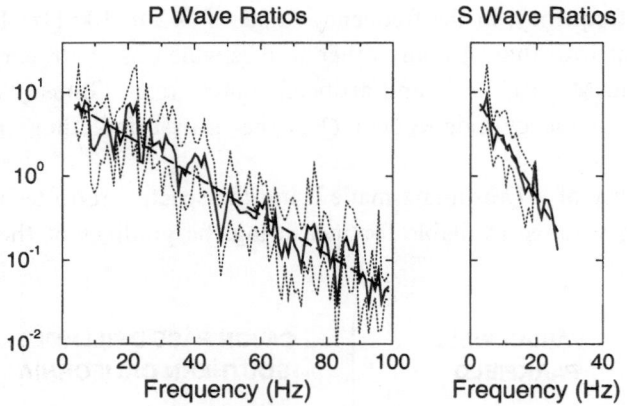

Figure 2

The mean spectral ratios (0/2.5 km) of 6 local earthquakes recorded at Cajon Pass (ABERCROMBIE, 1997). The dotted lines are 1 standard deviation and the dashed lines are the best fits, assuming Q is frequency independent ($Q_P = 27 \pm 7$, $Q_S = 18 \pm 4$). On the log-linear plots the slopes are well modelled by a straight line, implying that the assumption of frequency independent Q is reliable. Both plots are at the same scale.

(Fig. 2), and on the attenuation experienced by the waves travelling between the two sensors. Increasing the frequency range decreases the susceptibility of the gradient measurements to noise and resonance effects, and hence improves its reliability. The attenuation is directly dependent both on Q and also on the distance travelled between the two sensors. Even when the Q is very low, it is hard to constrain values over small depth ranges (say <100 m) because the total attenuation, and hence the slope of the spectral ratio, is very small. For example, compare the errors estimated over the depth ranges 0–300 m and 0–2.5 km at Cajon Pass (Fig. 1). At frequencies less than about 10 Hz, amplification and resonance effects can be large, which makes it hard to measure Q by the spectral ratio technique. At higher frequencies and over extensive depth ranges, however, the severe near-surface attenuation becomes the dominant factor and the spectral ratio method for measuring Q works well.

The borehole studies confirm that attenuation is severe in the near-surface, distorting earthquake recordings and limiting the signal bandwidth. Comparison of surface and downhole (2.5 km) recordings at Cajon Pass by ABERCROMBIE and LEARY (1993) shows clearly the effect of near-surface attenuation on measurements of earthquake source parameters. ABERCROMBIE (1995) extends this work to over 100 earthquakes and confirms that near-surface attenuation is sufficient to account for the possible breakdown in earthquake scaling below about $M_L 3$. The undistorted, high frequency seismograms recorded at 2.5 km demonstrate that earthquake self-similarly continues to at least $M_L 0$ and source sizes of less than 10 m radius. ABERCROMBIE (1996) shows that the magnitude-frequency relationship also continues unchanged to M_L0, using the small earthquakes undetected by the regional network but recorded downhole because of the low attenuation and low background noise level.

A pertinent question to earthquake source studies is "How deep must you go to remove the site effect?" As discussed in the next section, it is possible to estimate the attenuation beneath the borehole sensors, and the relevant values for the Cajon Pass and Varian Wells are included in Figure 1. ABERCROMBIE (1997) calculates that for an earthquake recorded at the Cajon Pass wellhead, with a hypocentral distance reaching 15 km, over 90% of the attenuation would occur in the upper 3 km, 80% in the upper 1.5 km, and 50% in the upper 300 m (at frequencies of a few Hz and above). The deepest instruments, over 2 km beneath the top of the granitic bedrock, are essentially beneath the site effect. However, attenuation along the relatively high Q paths to the deep instruments can still significantly affect the frequency content recorded from more distant earthquakes. Also, it cannot be ignored in very high resolution analyses which use the highest frequencies recorded, for example estimating the radiated seismic energy. At Parkfield, at least on the north-east side of the San Andreas fault, the situation is very different. Using the Q values in Figure 1, we can estimate in a similar manner, that for earthquakes at 15 km hypocentral distance from the Varian Wellhead, less than 50% of the attenuation occurs in the upper 1 km. Thus, a significant amount of attenuation occurs beneath 1 km, suggesting that at this site, the "site effect" extends to greater depth. It is interesting to compare average Q values beneath the deepest Varian sensors with those estimated beneath the stations of the HRSN, at around 200 m depth. In terms of attenuation at least, it appears that the propagation path to a sensor at 1 km depth on the northeast side of the San Andreas fault is comparable, if not lower Q, than that to a sensor at only 200 m on the SW side of the fault where the bedrock is Salinian granite (crustal attenuation and site effects at Parkfield, California, R. E. ABERCROMBIE, in preparation, 1998). It cannot, therefore, be assumed that just because a seismometer is down a borehole, all site and attenuation effects can be ignored.

To date, borehole recordings of tectonic earthquakes have been limited to frequencies below a few hundred Hz, but observations of man-made events, such as those induced by mining and drilling, have been made at higher frequencies. For example, FEUSTEL et al. (1996) recorded microseismicity produced during excavations at Strathcoma Mine (Canada) at 400 to 1600 Hz. The hypocentral distances were in the range 40–120 m, and both sources and sensors were at about 600 m depth in granitic bedrock. They use the multiple lapse time method (HOSHIBA, 1993) to estimate the intrinsic and scattering coefficients of S wave attenuation and their frequency dependence. They find $Q_S \sim 110$ (800 Hz) at this depth, and that intrinsic attenuation is approximately four times higher than scattering attenuation. The Q_S values are very similar to those obtained in the previously mentioned Californian studies at this depth. Both intrinsic and scattering Q exhibit a stronger dependence on frequency, increasing over the range 400–1600 Hz, than was observed in the earthquake studies. FEUSTEL *et al.* (1996) attribute this to local structural effects at a scale of a few metres. This very high frequency Q variation is therefore likely to be strongly site dependent.

The lack of a strong correlation between near-surface attenuation and rock type suggests that some other factor is principally responsible for the rapidly increasing attenuation towards the earth's surface. Laboratory studies of attenuation find that both Q_P and Q_S increase rapidly with increasing pressure up to about 1000 bars (~ 4 km) and then level off (JOHNSTON *et al.*, 1979). This pressure dependence of attenuation has been interpreted as resulting from the closure of cracks in the rocks with increasing confining pressure. Measurements of fracture densities in outcrops and cores, and from shear-wave splitting in the upper 500 m are about an order of magnitude larger than those made at greater depth (e.g., RENSHAW, 1997). JOHN-STON *et al.* (1979) find that at the pressures typical of the upper few kilometres, friction at cracks is the most probable dominant mechanism of intrinsic attenuation. As fractures are also major scatterers of seismic energy, scattering attenuation should also decrease with increasing depth as the fractures are closed. MORI and FRANKEL (1991) observed a significant decrease in scattering attenuation below about 5 km depth in southern California, consistent with this hypothesis. Additional support for the hypothesis that fractures are responsible for the high near-surface attenuation comes from a study of mining induced events recorded at 2 to 3 km depth in South Africa. SPOTTISWOODE (1993) found that along ray paths through solid rock Q is about 1000, similar to the values determined at similar depths at Cajon Pass. If the ray path was through a highly fractured region near the stope faces, however, then Q decreased to about 20, clearly demonstrating the importance of fractures in the attenuation of seismic waves at shallow depths. At greater depths, where fewer fractures remain open, then other attenuation mechanisms (perhaps with different frequency relationships) may be dominant.

Attenuation above 10 Hz at Seismogenic Depths

Earthquake seismograms recorded at a few hundred metres to a few kilometres beneath the earth's surface typically have a wider signal bandwidth, extending to higher frequencies than is possible with surface recorded data. Downhole seismograms thus can be used to measure crustal Q at frequencies of up to a few hundred Hz. Similar methods to those used at the surface, are used to estimate Q from downhole sensors, with the advantage that no assumptions need be made to remove strong site effects (including amplification) with poorly constrained frequency dependence. One such method is to assume that the source spectra have an ω^{-2} fall-off at high frequencies (e.g., BRUNE, 1970) and that attenuation is exponential, with Q independent of frequency (e.g., ANDERSON, 1986). ABERCROM-

BIE (1995) uses this method to analyse direct waves from over 100 earthquakes recorded at 2.5 km depth in the Cajon Pass drillhole, California, and obtains $Q_S \sim Q_P \sim 1000$ in the frequency range approximately 2 to 200 Hz. The earthquakes spanned hypocentral depths in the range 5 to 20 km, and distances of 5 to 120 km, suggesting that these Q values are averages within the seismogenic crust, at depths greater than 2.5 km. The close earthquakes (distance ≤ 20 km) have experienced negligible attenuation, however the spectra from the more distant events exhibit the exponential fall-off predicted by the frequency independent Q model used, over two decades of frequency bandwidth (ABERCROMBIE, 1995).

Studies of earthquakes also recorded at Cajon Pass, using direct and coda waves, have been performed to investigate the frequency dependence of Q_S at high frequencies in more detail. LEARY and ABERCROMBIE (1994) use seven earthquakes recorded at 2.5 km depth, and a single scattering approximation to estimate Q_S between 10 and 100 Hz. They find a weak increase in Q_S with frequency (from ~ 500 at 10 Hz to ~ 1200 at 100 Hz), and that intrinsic attenuation is the dominant mechanism; $Q_{Scatter} > 10 Q_{intrinsic}$ at all observed frequencies. ADAMS and ABERCROMBIE (1998) perform a more comprehensive analysis using over 100 earthquakes recorded at a range of depths in the Cajon Pass drillhole. They employ the multiple lapse time method (HOSHIBA et al., 1991) to determine the intrinsic and scattering contributions to Q_S and their frequency dependence. They confirm the weak frequency dependence of Q_S at high frequency ($Q_S \sim 800$ at 10 Hz increasing to ~ 1500 at 100 Hz), indicating that the assumption of frequency-independent Q_S made by ABERCROMBIE (1995) was reasonable. The absolute values of Q_S between these three studies at Cajon Pass also show good agreement (Fig. 3). ADAMS and ABERCROMBIE (1980) also find intrinsic attenuation to be the dominant mechanism. They estimate it to be approximately twice the level of scattering attenuation at seismogenic depths.

The weak frequency dependence of Q_S observed above about 10 Hz is not a simple extrapolation of the trend observed between 1 and 10 Hz (see Fig. 3). A number of studies using coda and Lg waves (e.g., JIN et al., 1994; BENZ et al., 1997) find Q_S to increase rapidly from about 30 at 1 Hz to a few hundred at 10 Hz in active tectonic areas such as the western U.S.A. HOUGH et al. (1988) use seismograms recorded on the granite batholith at Anza (California) and estimate Q from direct P and S waves. They also find that Q is essentially frequency independent from 30–70 Hz (~ 700), but lower at 10 Hz (~ 400). KINOSHITA (1994) makes use of deep borehole earthquake recordings to measure Q_S in the southern Kanto area of Japan. His results, in the frequency range 0.5 to 16 Hz, are similar to the measurements of Q_S made in California, showing an increase to about 5 Hz, and then a levelling off at higher frequencies (Fig. 3).

It is interesting to speculate on the possible causes of this change in the frequency dependence of Q at 5–10 Hz. It could result from changes in the nature of the crust at scale lengths of a few hundred metres, or alternatively, could simply be an artifact of incorrect assumptions in the models used to estimate Q. The models assume either body or surface waves, and isotropic attenuation which does not vary laterally or with depth. These assumptions are perhaps more justified for the high frequency borehole data than for the lower frequency surface recordings. Several recent studies find the coda of surface seismograms to be dominated by energy scattered near the receiver (DODGE and BEROZA, 1997) and by slow surface waves (MEREMONTE et al., 1996). It is therefore possible that the frequency behaviour of Q shown in Figure 3 is an artifact of a frequency dependence to the appropriateness of the model assumptions. Alternatively the observed variation in the frequency dependence of Q_S could be providing us with information relative to the nature of the earth's crust at different scales. AKI (1980) noted that the absolute value and frequency dependence of coda Q (0.02–25 Hz) was dependent on the level of current tectonic activity. The frequency dependence shown in Figure 3 is

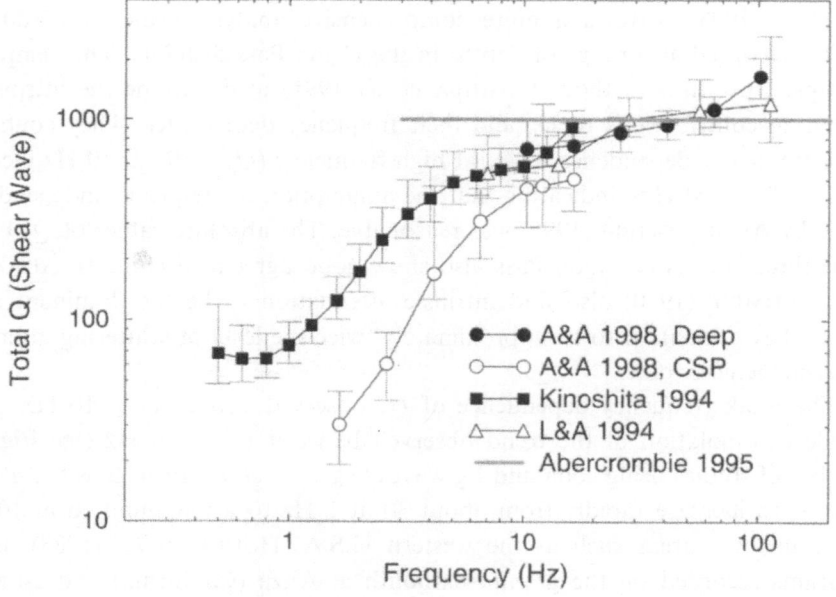

Figure 3
The frequency dependence of Q_S between 1 and 100 Hz. Various studies find Q to increase from approximately 1 to about 10 Hz. The results of KINOSHITA (1994) from the Kanto district, Japan, and ADAMS and ABERCROMBIE (1998) from a surface site in California (A & A, 1998, CSP) are plotted as examples. Above 10 Hz, however, recent borehole studies at Cajon Pass by ADAMS and ABERCROMBIE (1998) and LEARY and ABERCROMBIE (1994), labelled A & A, 1998 deep and L & A, 1994, respectively, have found only a small increase in Q with frequency. The estimate of Q, assuming frequency independence obtained from spectral fitting of direct waves travelling to the 2.5 km sensor at Cajon Pass by ABERCROMBIE (1995) is included for comparison.

also only observed in active tectonic regimes. In more stable, cratonic areas, FRANKEL and CLAYTON (1986) and BENZ et al. (1997) find Q_S to be higher and to show only a slight increase with frequency between 1 and 10 Hz. I am, unfortunately, not aware of any borehole studies of Q_S at seismogenic depths in the frequency range 10 to 100 Hz in stable cratonic areas. SPOTTISWOODE (1993), however, reports $Q \sim 1000$ in solid rock at a few kilometres depth around South African gold mines. The cause of the change in frequency behaviour of Q_S at 5–10 Hz must therefore be restricted to active crust. A thinner, warmer crust, and the presence of large crustal faults, characterized by low velocity zones (e.g., EBERHART-PHILLIPS et al., 1995; LI et al., 1994), are possible candidates.

Summary

This review demonstrates how deep borehole recordings of earthquakes have extended our knowledge of crustal attenuation both in the near-surface, and at seismogenic depths.

Q is very low in the near-surface (~ 10 in the upper 100 m), and increases rapidly with depth to a few kilometres. Unlike site amplification, near-surface attenuation appears to be almost independent of rock type. These two observations combine to suggest that the primary cause of severe near-surface attenuation is the opening of fractures with decreasing lithostatic pressure. The observed levels of near-surface attenuation are easily adequate to provide for the proposed breakdown in earthquake scaling at small magnitudes. The low correlation of this attenuation with rock type implies that it cannot be considered negligible in source studies even at hard rock sites.

Measurements of Q_S at seismogenic depths using direct and coda waves are high (~ 1000) and show only a weak increase with frequency between 10 and 100 Hz. Intrinsic attenuation appears to be the dominant mechanism; estimates of scattering Q are at least twice the value of intrinsic Q. Combining these results with those from surface studies at lower frequencies reveals a significant change in the frequency dependence of Q_S at about 10 Hz in active tectonic regions. Further work is required to determine the factors responsible for this apparent change in Q_S behaviour, but it is clearly unwise to extrapolate Q estimates made below 10 Hz to higher frequency data.

Acknowledgments

I am grateful to P. Malin, P. Shearer, M. Reyners, R. Robinson, D. Boore and J. Mori for their reviews, which significantly improved this manuscript. I also benefitted from discussion of these issues with D. Adams and R. Benites. I thank

C. Hume for draughting assistance. Funding for writing this summary was provided by IGNS, New Zealand; IGNS Contribution 1431.

REFERENCES

ABERCROMBIE, R. E. (1997), *Near-surface Attenuation and Site Effects from Comparison of Surface and Deep Borehole Recordings*, Bull. Seismol. Soc. Am. *87*, 731–744.

ABERCROMBIE, R. E. (1996), *The Magnitude-frequency Distribution of Earthquakes Recorded with Deep Seismometers at Cajon Pass, Southern California*, Tectonophys. *261*, 1–7.

ABERCROMBIE, R. E. (1995), *Earthquake Source Scaling Relationships from −1 to 5 M_L Using Seismograms Recorded at 2.5 km depth*, J. Geophys. Res. *100*, 24,015–24,036.

ABERCROMBIE, R. E., and LEARY, P. (1993), *Source Parameters of Small Earthquakes Recorded at 2.5 km Depth, Cajon Pass, Southern California: Implications for Earthquake Scaling*, Geophys. Res. Lett. *20*, 1511–1514.

ADAMS, D. A., and ABERCROMBIE, R. E. (1998), *Seismic Attenuation at High Frequencies in Southern California from Coda Waves Recorded at a Range of Depths*, J. Geophys. Res. *103*, 24,257–24,270.

AKI, K. (1980), *Scattering and Attenuation of Shear Waves in the Lithosphere*, J. Geophys. Res. *85*, 6496–6504.

ANDERSON, J. G., *Implication of attenuation for studies of the earthquake source*. In *Earthquake Source Mechanics*, Geophys. Monogr. *37* (ed. Das, S *et al.*) (AGU, Washington, DC 1986) pp. 311–318.

ARCHULETA, R. J., CRANSWICK, E., MUELLER, C., and SPUDICH, P. (1982), *Source Parameters of the 1980 Mammoth Lakes, California, Earthquake Sequence*, J. Geophys. Res. *87*, 4595–4607.

ARCHULETA, R. J., SEALE, S. H., SANGAS, P. V., BAKER, L. M., and SWAIN, S. T. (1992), *Garner Valley Downhole Array of Seismometers: Instrumentation and Preliminary Data Analysis*, Bull. Seismol. Soc. Am. *82*, 1592–1621.

ASTER, R. C., and SHEARER, P. M. (1991), *High-frequency Borehole Seismograms Recorded in the San Jacinto Fault Zone, Southern California, Part 2. Attenuation and Site Effects*, Bull. Seismol. Soc. Am. *81*, 1081–1100.

BENZ, H. M., FRANKEL, A., and BOORE, D. M. (1997), *Regional Attenuation for the Continental United States*, Bull. Seismol. Soc. Am. *87*, 606–619.

BLAKESLEE, S., and MALIN, P. (1991), *High-frequency Site Effects at Two Parkfield Downhole and Surface Stations*, Bull. Seismol. Soc. Am. *81*, 332–345.

BORCHERDT, R. D. (1970), *The Effects of Local Geology on Ground Motion near San Francisco Bay*, Bull. Seismol. Soc. Am. *60*, 29–61.

BRUNE, J. N. (1970), *Tectonic Stress and the Spectra of Seismic Shear Waves from Earthquakes*, J. Geophys. Res. *75*, 4997–5009.

CELEBI, M., PRINCE, J., DIETAL, C., ONATE, M., and CHAVEZ, G. (1987), *The Culprit in Mexico City—Amplification of Motions*, Earthq. Spectra. *3*, 315–328.

DODGE, D. A., and BEROZA, G. C. (1997), *Source Array Analysis of Coda Waves near the 1989 Loma Prieta, California Mainshock: Implications for the Mechanism of Coseismic Velocity Changes*, J. Geophys. Res. *102*, 24,437–24,458.

EBERHART-PHILLIPS, D., STANLEY, W. D., RODRIGUEZ, B. D., and LUTTER, W. J. (1995), *Surface Seismic and Electrical Methods to Detect Fluids Related to Faulting*, J. Geophys. Res. *100*, 12,919–12,936.

FEUSTEL, A. J., TRIFU, C. I., and URBANCIC, T. I. (1996), *Rock-mass Characterization Using Intrinsic and Scattering Attenuation Estimates at Frequencies from 400 to 1600 Hz*, Pure appl. geophys. *147* (2), 289–304.

FRANKEL, A., and CLAYTON, R. W. (1986), *Finite-difference Simulations of Seismic Scattering: Implications for the Propagation of Short-period Seismic Waves in the Crust and Models of Crustal Heterogeneity*, J. Geophys. Res. *91*, 6465–6489.

GIBBS, J. F., BOORE, D. M., JOYNER, W. B., and FUMAL, T. E. (1994), *The Attenuation of Seismic Shear Waves in Quaternary Alluvium in Santa Clara Valley, California*, Bull. Seismol. Soc. Am. *84*, 76–90.

HAUKSSON, E., TENG, T. L., and HENYEY, T. L. (1987), *Results from a 1500 m Deep, Three-level Downhole Seismometer Array: Site Response, Low Q Values and f_{max}*, Bull. Seismol. Soc. Am. 77, 1883–1904.

HOSHIBA, M. (1993), *Separation of Scattering Attenuation and Intrinsic Absorption in Japan with the Multiple Lapse Time Window Analysis from Full Seismogram Envelope*, J. Geophys. Res. 98, 15,809–15,824.

HOSHIBA, M., SATO, H., and FEHLER, M. (1991), *Numerical Basis of the Separation of Scattering and Intrinsic Absorption from Full Seismogram Envelope: A Monte-Carlo Simulation of Multiple Isotropic Scattering*, Papers in Meteor. and Geophys. 42, 65–91. (Meteorological Research Institute of Japan.)

HOUGH, S. E., ANDERSON, J. G., BRUNE, J., VERNON, F., BERGER, J., FLETCHER, J., HAAR, L., HANKS, T., and BAKER, L. (1988), *Attenuation near Anza, California*, Bull. Seismol. Soc. Am. 78, 672–691.

JIN, A., MAYEDA, K., ADAMS, D., and AKI, K. (1994), *Separation of Intrinsic and Scattering Attenuation in Southern California Using TERRAscope Data*, J. Geophys. Res. 99, 17,835–17,848.

JOHNSTON, D. H., TÖKSOZ, M. N., and TIMUR, A. (1979), *Attenuation of Seismic Waves in Dry and Saturated Rocks: II. Mechanisms*, Geophys. 44, 691–711.

JONGMANS, D., and MALIN, P. E. (1995), *Vertical Profiling of Microearthquake S Waves in the Varian Well at Parkfield, California*, Bull. Seismol. Soc. Am. 85, 1805–1820.

KINOSHITA, S. (1994), *Frequency-dependent Attenuation of Shear-waves in the Crust of the Southern Kanto Area, Japan*, Bull. Seismol. Soc. Am. 84, 1387–1396.

LEARY, P. C., and ABERCROMBIE, R. E. (1994), *Frequency-dependent Crustal Scattering and Absorption at 5–160 Hz from Coda Decay Observed at 2.5 km Depth*, Geophys. Res. Lett. 21, 971–974.

LI, Y., AKI, K., ADAMS, D., HASEMI, A., and LEE, W. H. K. (1994), *Seismic Guided Waves Trapped in the Fault Zone of the Landers, California, Earthquake of 1992*, J. Geophys. Res. 99, 17,705–17,722.

MALIN, P. E. BLAKESLEE, S. N., ALVAREZ, M. G., and MARTIN, A. J. (1989), *Microearthquake Imaging of the Parkfield Asperity*, Science 244, 557–559.

MALIN, P. E., WALLER, J. A., BORCHEDT, R. D., CRANSWICK, E., JENSEN, E. G., and VAN SCHAACK, J. (1988), *Vertical Seismic Profiling of Oroville Microearthquakes: Velocity Spectra and Particle Motion as a Function of Depth*, Bull. Seismol. Soc. Am. 78, 401–420.

MANOV, D. V., ABERCROMBIE, R. E., and LEARY, P. C. (1996), *Reliable and Economical High Temperature Deep Borehole Seismic Recording*, Bull. Seismol. Soc. Am. 86, 204–211.

MEREMONTE, M., FRANKEL, A., CRANSWICK, E., CARVER, D., and WORLEY, D. (1996), *Urban Seismology—Northridge Aftershocks Recorded by Multi-scale Arrays of Portable Digital Seismograms*, Bull. Seismol. Soc. Am. 86, 1350–1363.

MORI, J., and FRANKEL, A. (1991), *Depth-dependent Scattering Shown by Coherence Estimates and Regional Coda Amplitudes*, EOS Trans. AGU 72, 344.

RENSHAW, C. E. (1997), *Mechanical Controls on the Spatial Density of Opening-mode Fracture Networks*, Geology 25, 923–926.

SPOTTISWOODE, S. M., *Seismic attenuation in deep-level mines*. In *Rockbursts and Seismicity in Mines* (ed. Young, R.) (Balkema, Rotterdam 1993) pp. 409–414.

STEIDL, J. H., TUMARKIN, A. G., and ARCHULETA, R. J. (1996), *What is a Reference Site*, Bull. Seismol. Soc. Am. 86, 1733–1748.

ZOBACK, M. D., and HARJES, H. P. (1997), *Injection-induced Earthquakes and Crustal Stress at 9 km Depth at the KTB Deep Drilling Site, Germany*, J. Geophys. Res. 102, 18,477–18,491.

(Received January 5, 1998, revised June 7, 1998, accepted June 23, 1998)

To access this journal online:
http://www.birkhauser.ch

Pure appl. geophys. 153 (1998) 489–502
0033–4553/98/040489–14 $ 1.50 + 0.20/0

Pure and Applied Geophysics

Frequency-dependent Attenuation of High-frequency P and S Waves in the Upper Crust in Western Nagano, Japan

KAZUO YOSHIMOTO,[1] HARUO SATO,[1] YOSHIHISA IIO,[2] HISAO ITO,[2]
TAKAO OHMINATO[2] and MASAKAZU OHTAKE[3]

Abstract—Borehole seismograms from local earthquakes in the aftershock region of the 1984 western Nagano Prefecture, Japan earthquake were analyzed to measure the frequency-dependent characteristics of P- and S-wave attenuation in the upper crust. The records from a three-component velocity seismometer at the depth of 145 m exhibit high S/N-ratio in a wide frequency range up to 100 Hz. Extended coda normalization methods were applied to bandpass-filtered seismograms of frequencies from 25 to 102 Hz. For the attenuation of high-frequency P and S waves, our measurements show $Q_P^{-1} \simeq 0.052 f^{-0.66}$ and $Q_S^{-1} \simeq 0.0034 f^{-0.12}$, respectively. The frequency dependence of the quality factor of S waves is very weak as compared with that of P waves. The ratio of Q_P^{-1}/Q_S^{-1} is larger than unity in the entire analyzed frequency range.

Key words: Attenuation, crust, frequency dependence, high-frequency seismic waves, quality factor.

Introduction

Attenuation parameter Q^{-1} is an important factor for understanding the physical mechanism of seismic wave attenuation in relation to the composition and physical condition of the earth's interior (e.g., LIU *et al.*, 1976; SATO, 1992). For this purpose, it is necessary to investigate the frequency-dependent characteristics of both Q_P^{-1} and Q_S^{-1} in various regions and depths. Historically, the study on the crustal seismic attenuation of S waves has been more advanced than that on P waves (CASTRO *et al.*, 1990; TAKEMURA *et al.*, 1991; KATO *et al.*, 1992; KINOSHITA, 1994). It is mostly due to the requirement of the engineering seismology. Nevertheless, the information on Q_P^{-1} including Q_P^{-1}/Q_S^{-1} ratio is also

---

[1] Graduate School of Science, Tohoku University, Aoba-ku, Sendai 980-8578, Japan. Fax: +81-22-217-6783; E-mail: yoshi@zisin.geophys.tohoku.ac.jp
[2] National Research Institute for Earth Science and Disaster Prevention, Tennoudai 3-1, Tsukuba 305-0006, Japan.
[3] Geological Survey of Japan, Higashi 1-1-3, Tsukuba 305-0046, Japan.

indispensable for evaluating the strong vertical ground motion precisely in view of the earthquake-resistant design of recent huge constructions.

Recent studies on the attenuation of regional seismic waves in the crust reported roughly the same apparent attenuation ratio of P to S waves in distance; $Q_P^{-1}/Q_S^{-1} \simeq V_P/V_S$ (e.g., RAUTIAN *et al.*, 1978). This equality has been observed in the frequencies between about 1 and 30 Hz in many tectonically different regions (will be shown in Fig. 8). For frequencies below 1 Hz, previous studies on the surface wave attenuation suggested that Q_P^{-1}/Q_S^{-1} gradually decreases with decreasing frequency. For frequencies above 30 Hz, in spite of the recent studies of the high-frequency body waves in the crust (e.g., ABERCROMBIE and LEARY, 1993; FEUSTEL and YOUNG, 1994; HOLLIGER and BÜHNEMANN, 1996), high-frequency asymptote of this ratio is not yet clearly understood. In this study we measure the frequency dependence of Q_P^{-1}, Q_S^{-1} and Q_P^{-1}/Q_S^{-1} ratio for the frequency range from ten to one hundred Hertz to better understand the characteristics of high-frequency seismic wave attenuation in the crust.

Method

We use the extended coda normalization method (YOSHIMOTO *et al.*, 1993) to measure Q_P^{-1} and Q_S^{-1} in the crust. These quality factors are calculated from the apparent attenuation rate of the coda-normalized amplitude of direct waves. The basic idea of this method stands on the following proportionality among the coda spectral amplitude A_C, the source spectral amplitude of S waves S_S and the source spectral amplitude of P waves S_P:

$$A_C(f, t_C) \propto S_S(f) \propto S_P(f), \qquad (1)$$

where f is the frequency and t_C is the reference lapse time measured from the source origin time. Here we take t_C to be greater than roughly twice the travel time of the direct S wave. The second proportionality is derived from the assumption of the constant source spectral ratio of P to S waves. This assumption can be justified for regionally localized earthquakes of the same magnitude. The above equation means that $A_C(f, t_C)$ is independent of the hypocentral distance (AKI, 1980).

By using seismograms from earthquakes of different hypocentral distances, we calculate Q_P^{-1} and Q_S^{-1} by using the following equations:

$$\left\langle \ln\left[\frac{A_P(f, r)r}{A_C(f, t_C)}\right]\right\rangle_{r \pm \Delta r} = -\frac{\pi f}{Q_P(f)V_P}r + c_P(f), \qquad (2)$$

$$\left\langle \ln\left[\frac{A_S(f, r)r}{A_C(f, t_C)}\right]\right\rangle_{r \pm \Delta r} = -\frac{\pi f}{Q_S(f)V_S}r + c_S(f), \qquad (3)$$

where A_P and A_S are the amplitude spectra of the direct *P* and *S* waves, respectively. The symbol *r* means the hypocentral distance, V_P is the *P*-wave velocity, and V_S is the *S*-wave velocity. The parameters c_P and c_S include the scattering characteristics of the earth medium and the site amplification effect. The symbol $\langle\ \rangle_{r\pm\Delta r}$ denotes the average within a small hypocentral distance range of $r\pm\Delta r$. We assume here that the effects of source radiation pattern on the direct wave amplitudes are canceled by using many earthquakes of different azimuths.

Data

The data used in this study are velocity-motion records from a borehole instrument located in the aftershock region of the 1984 western Nagano Prefecture, Japan earthquake (M_S 6.9). Figure 1 shows the location of the borehole station together with the fault of the main shock which was inferred from the distance change of trilateration points (YAMASHINA and TADA, 1985). The depth of the

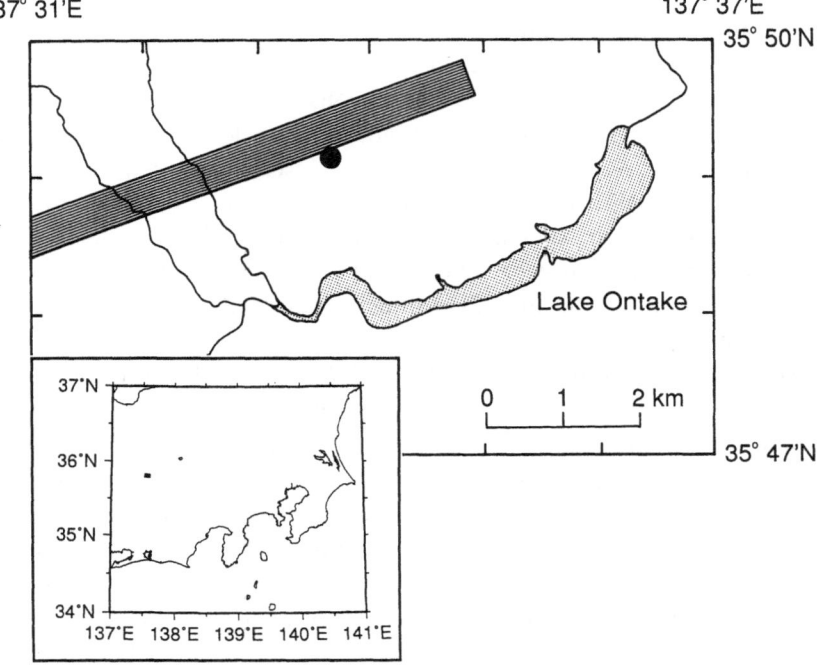

Figure 1
Map showing the location of a borehole station (solid circle). Inferred fault plane of the 1984 western Nagano Prefecture, Japan earthquake is indicated by the striped rectangle (YAMASHINA and TADA, 1985). The area of the main map is shown by the solid rectangle in the inset.

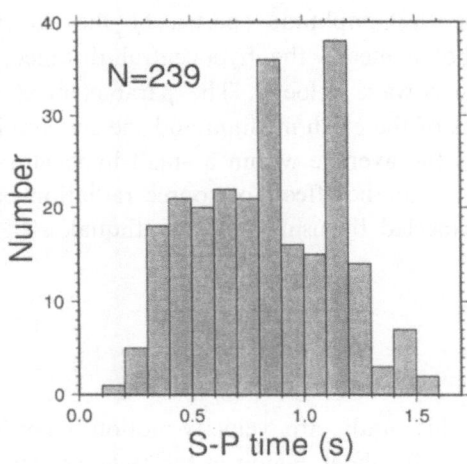

Figure 2
Histogram of *S–P* time distribution of earthquakes used for the analysis.

upper margin of estimated fault plane is 0.5 km and any surface break was not found in the focal region. The seismicity of this region is still high as of 1998. The surface geology around the borehole station consists of Paleogene rhyolite, and the *P*-wave velocity is larger than 5 km/s immediately below the free surface (HIRA-HARA *et al.*, 1992). A three-component velocity seismometer with a natural frequency of 2.2 Hz (Mark Products L22-E) is installed at a depth of 145 m in the rhyolite bedrock. Output signals were digitized with a 16-bit resolution at 1000 samples/s. Record durations were generally about 7 s. The total amplitude response including the effect of the anti-alias filter is flat from 2 to 500 Hz, at which frequency the response drops off about 3 dB.

We selected 239 seismograms with high *S/N* ratio from local earthquakes recorded in the period from October 1991 to August 1992. The magnitude is less than three for the most part. Figure 2 shows the frequency distribution of *S–P* times. Although the hypocentral locations were not determined due to the single station observation, focal depths of these events are estimated at shallower than 12 km by assuming a homogeneous velocity structure with *P*- and *S*-wave velocities of 5.5 and 3.2 km/s, respectively. Direct body waves are thus expected to penetrate only the shallow part of the crust. An example of bandpass-filtered seismograms is displayed in Figure 3. The figure demonstrates a very small background noise and the presence of high-frequency seismic signals as high as 100 Hz, which is hardly expected for surface observation due to the strong attenuation in the near-surface weathering layer. Borehole observation is of particular importance for the attenuation measurement of high-frequency seismic waves.

Analysis

We calculated Q_P^{-1} and Q_S^{-1} by measuring the apparent attenuation of the normalized root-mean-squares (rms) amplitude of *P* and *S* waves. UD and NS components were used for the analysis of *P* and *S* waves, respectively. The length of time-window applied was 0.1 s. To investigate the frequency dependence of Q_P^{-1} and Q_S^{-1}, we analyzed bandpass-filtered seismograms of 2/3 octave bandwidth (see Fig. 3). The central frequency of the four-pole Butterworth bandpass filter ranges from 16 to 102 Hz.

We took $t_C = 5.5$ s for rms coda amplitude A_C, which was calculated for NS component by using a 0.5 s time-window. The coda amplitude decay is smooth around at this t_C value. When the rms coda amplitude was smaller than twice the noise level, it was estimated from coda waves at an earlier lapse time by using a master curve of coda decay. The time-window was set at a lapse time after at least 1.5 times of the *S*-wave travel time. The master curves were obtained by taking the average of the decay rates of coda amplitudes for larger events.

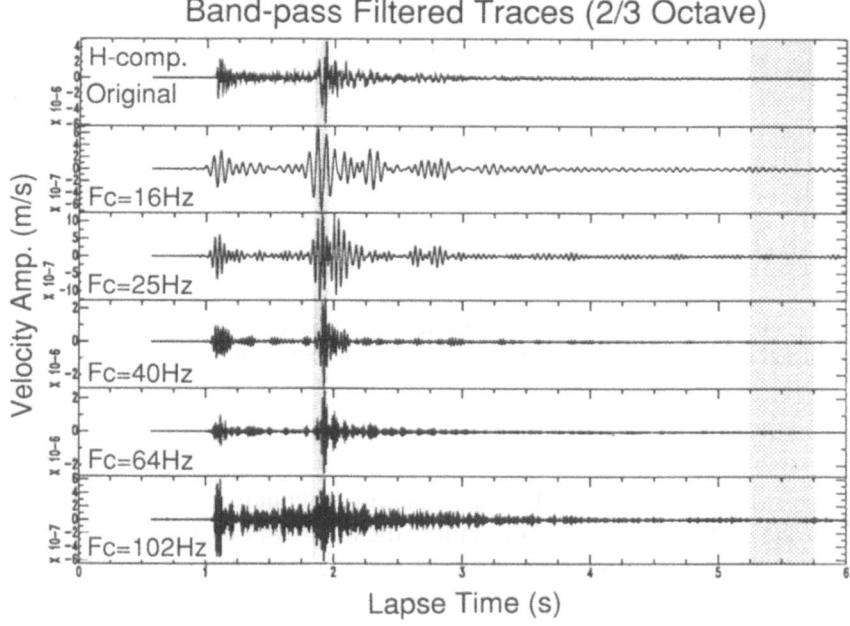

Figure 3

An example of original and bandpass-filtered velocity seismograms. Central frequency of pass-band is shown on the left of each trace. Time-windows for the measurement of the rms direct *S*-wave amplitude and the rms coda amplitude are denoted by the striped shade and the dotted shade, respectively.

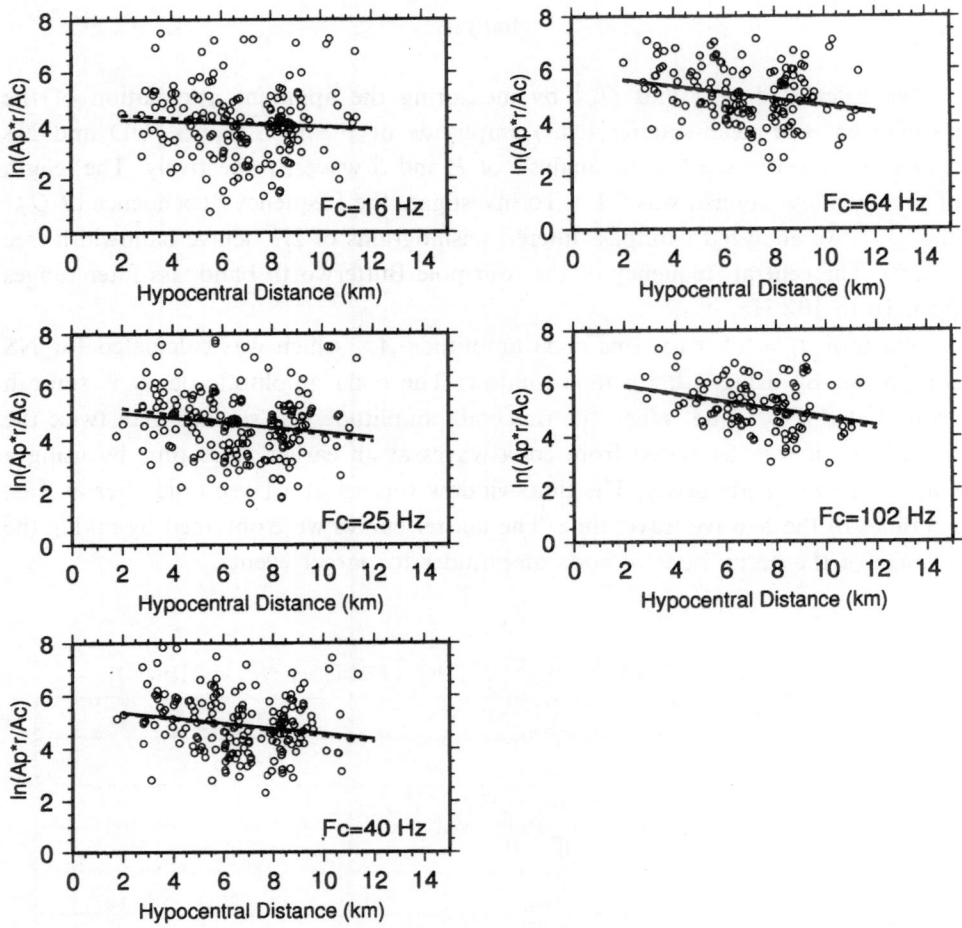

Figure 4

Apparent attenuation of the coda-normalized rms amplitude of *P* waves. The best-fit line from the least-squares estimate is indicated by the solid line. The regression line of the robust estimate is shown by the broken line.

Figures 4 and 5 illustrate the apparent attenuation of the normalized rms amplitude of *P* and *S* waves. We applied the least-squares fitting (solid line) and the robust fitting (broken line) which minimizes the sum of the absolute deviations of the data. The scatter of data around the regression lines is relatively large at low frequencies, however it decreases with increasing frequency both for *P* and *S* waves. The regression lines indicate a systematic increase of apparent attenuation with increasing frequency. At about 100 Hz, *S* wave attenuates roughly 20 dB for the hypocentral distance of 10 km. Apparent attenuation of *P* wave is similar although slightly smaller than that of *S* wave.

Results

Figure 6 shows Q_P^{-1} obtained from the apparent attenuation of the rms amplitude of the *P* waves. Our results obtained from the least-squares estimation are plotted by solid squares. The error bar indicates the standard deviation. Any significant differences larger than the error bars are not found between the least-squares estimates and the robust estimates. The value of Q_P^{-1} at 16 Hz is discarded because of the large scatter of the data. The results obtained from the least-squares estimation are also listed in Table 1. We find a clear frequency

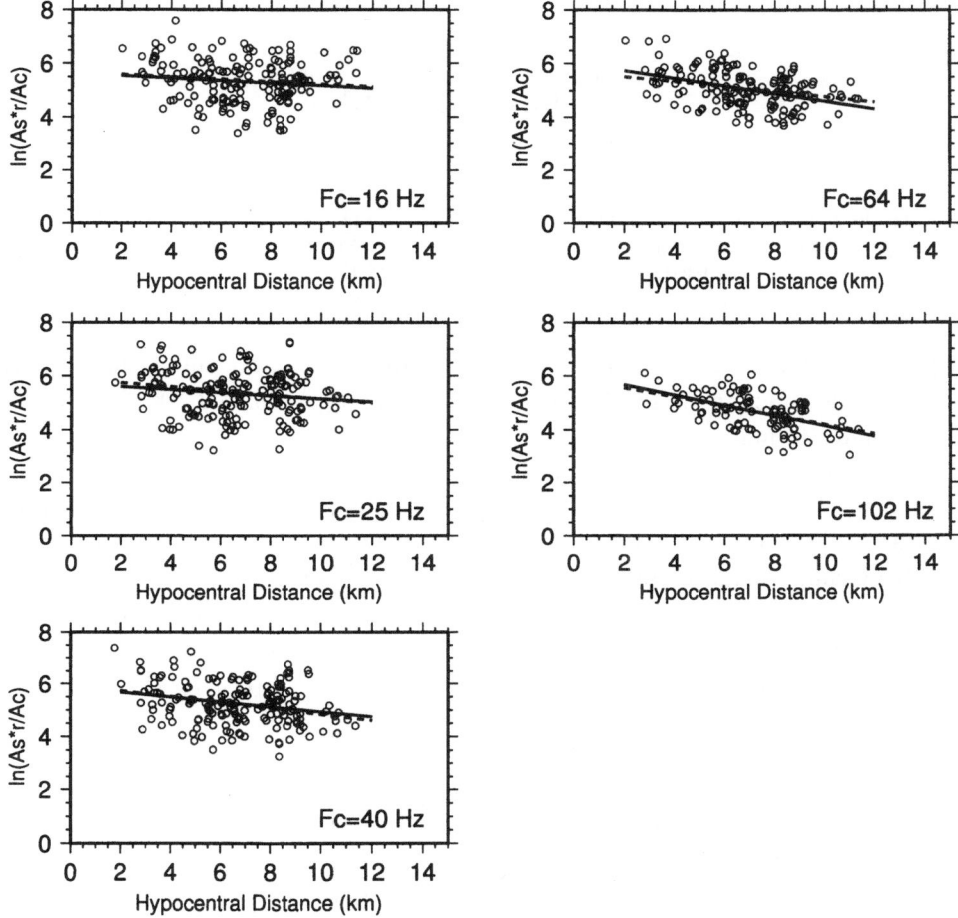

Figure 5
Apparent attenuation of the coda-normalized rms amplitude of *S* waves. The best-fit line from the least-squares estimate is indicated by the solid line. The regression line of the robust estimate is shown by the broken line.

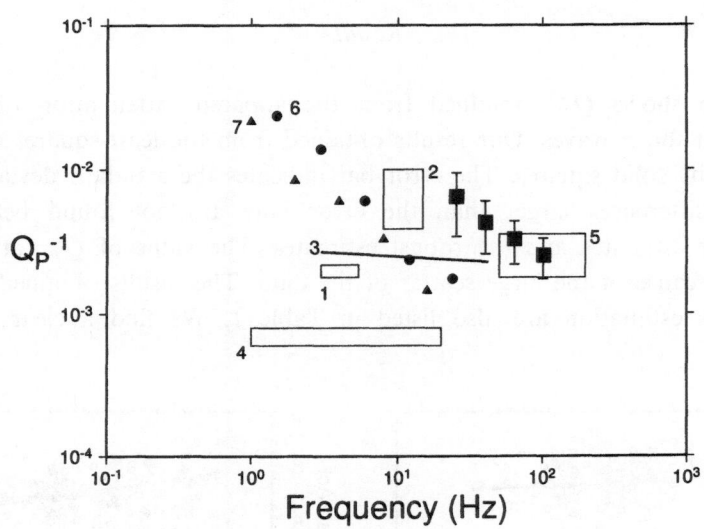

Figure 6

Q_P^{-1} in the crust, where the results of the present study are plotted by solid squares. The error bar indicates the standard deviation. Results of past studies are also plotted for comparison: 1. KOHKETSU and SIMA (1985); 2. OHTAKE (1987); 3. MASUDA (1988); 4. SEKIGUCHI (1991); 5. IIO (1992); 6. YOSHIMOTO *et al.* (1993); 7. HOSODA and KOSUGA (1995).

dependence of Q_P^{-1}, even if we allow for estimation errors. Assuming that Q_P^{-1} is a power-law function of the frequency, we obtain $Q_P^{-1} \simeq 0.052 f^{-0.66}$ by using a least-squares fitting.

In Figure 7, Q_S^{-1} obtained for the frequencies between 25 and 102 Hz are plotted by solid squares. They take the values of about 2×10^{-3}, which slightly decrease as frequency increases. We get $Q_S^{-1} \simeq 0.0034 f^{-0.12}$ by fitting a power law for this result. Frequency dependence of Q_S^{-1} is very weak in contrast to Q_P^{-1}.

The ratio of Q_P^{-1} to Q_S^{-1} is plotted by solid squares in Figure 8. Estimation errors are not sufficiently small to discuss the frequency dependence of the ratio. However, our result establishes that Q_P^{-1} is greater than Q_S^{-1} ($Q_P^{-1}/Q_S^{-1} > 1$) for the

Table 1

Q_P^{-1} and Q_S^{-1} values with standard deviations

Frequency (Hz)	Q_P^{-1}	Q_S^{-1}
25	0.0064 ± 0.0030	0.0022 ± 0.0011
40	0.0043 ± 0.0017	0.0023 ± 0.0007
64	0.0033 ± 0.0011	0.0022 ± 0.0004
102	0.0025 ± 0.0008	0.0019 ± 0.0003

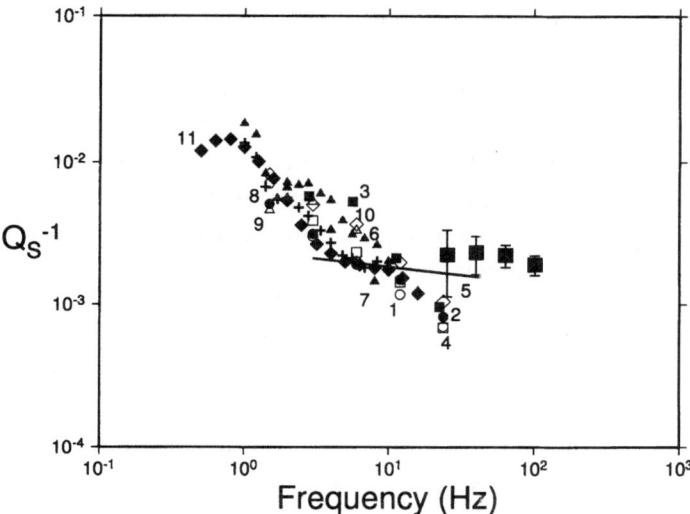

Figure 7

Q_S^{-1} in the crust, where the results of the present study are plotted by solid squares. The error bar indicates the standard deviation. Results of past studies are also plotted for comparison: 1,2. AKI (1980); 3. SATO and MATSUMURA (1980); 4. BENNET and BAKUN (1982); 5. MASUDA (1988); 6. KATO et al. (1992); 7. SCHERBAUM and SATO (1991); 8. TAKEMURA et al. (1991): 9. FEHLER et al. (1992); 10. YOSHIMOTO et al. (1993); 11. KINOSHITA (1994).

entire frequency range of 25 to 102 Hz. The ratio Q_P^{-1}/Q_S^{-1} takes the values between 1.3 and 2.9.

Discussion

In Figures 6 and 7 our results are compared with Q^{-1}, reported in the previous studies for the crustal attenuation of seismic body waves. We limit the plotted data to those obtained for crustal bedrocks in Japan, to minimize possible tectonic regionality. Data for very shallow deposit rocks are not plotted. Past studies are mostly limited to low frequencies of $f \le 30$ Hz.

As shown in Figure 6, our results of Q_P^{-1} at the frequencies of 64 and 102 Hz are consistent with that of IIO (1992) for the same region (no. 5 in the figure). At the frequency of about 30 Hz, however, Q_P^{-1} values in the western Nagano area are three to four times larger than those reported by past studies in Japan. This may be due to the depth variation of Q_P^{-1} (e.g., CARPENTER and SANFORD, 1985; HAVSKOV and MEDHUS, 1991). CARPENTER and SANFORD (1985) reported the decrease of Q_P^{-1} with increasing depth in the central Rio Grande rift, New Mexico. We note that Q_P^{-1} obtained in the present study represents the *P*-wave attenuation in the upper crust shallower than 12 km in contrast to other studies that include the whole crust. To cite another basis of the large Q_P^{-1} values in the area studied, we suggest the possible increase in attenuation due to the occurrence of the 1984 western Nagano Prefecture, Japan earthquake (OHTAKE, 1987). SUZUKI (1971) reported anomalously large Q_P^{-1} values in the area of the Matsushiro earthquake swarm, central Japan, 1965 to 1967. Those observations strongly suggest that the structural heterogeneity created by high seismic activity may bring about a remarkable increase of seismic attenuation.

In contrast to Q_P^{-1}, the frequency dependence of Q_S^{-1} obtained here is very weak and nearly frequency independent at the frequencies from 25 to 102 Hz. As shown in Figure 7, this nature seems inconsistent with the strong frequency dependence reported by other studies for lower frequencies. A plausible explanation for this discrepancy holds that Q_S^{-1} in the crust, in general, tends to constant for high-frequency seismic waves. This is suggested by laboratory experiments for ultrasonic frequencies (e.g., JOHNSTON *et al.*, 1979). For various kinds of rocks in different experimental conditions, the absolute value of Q_S^{-1} in kHz band ranges between 10^{-3} and 10^{-1}, which do not largely differ from the results of field measurements. As for the regionality of the frequency dependence of Q_S^{-1}, we need

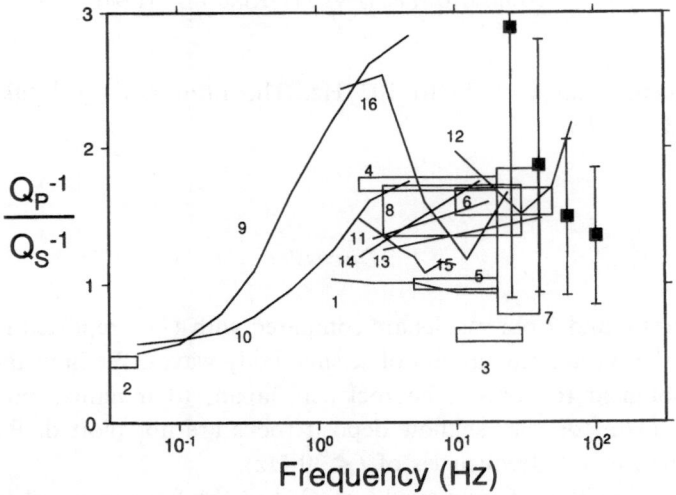

Figure 8

Q_P^{-1}/Q_S^{-1} in the crust and the upper most mantle, where the results for sedimentary deposits are not plotted. Our results are plotted by solid squares. The error bar indicates the standard deviation. Results of past studies are also plotted for comparison: 1. FEDOTOV and BOLDYREV (1969); 2. TSAI and AKI (1969); 3. BAKUN *et al.* (1976); 4. RAUTIAN *et al.* (1978); 5. FRANKEL (1982); 6. MODIANO and HATZFELD (1982); 7. HOANG-TRONG (1983); 8. CARPENTER and SANFORD (1985); 9,10. TAYLOR *et al.* (1986); 11. BUTLER *et al.* (1987); 12. HOUGH *et al.* (1988); 13. MASUDA (1988); 14. KVAMME and HAVSKOV (1989); 15. CAMPILLO and PLANTET (1991); 16. YOSHIMOTO *et al.* (1993).

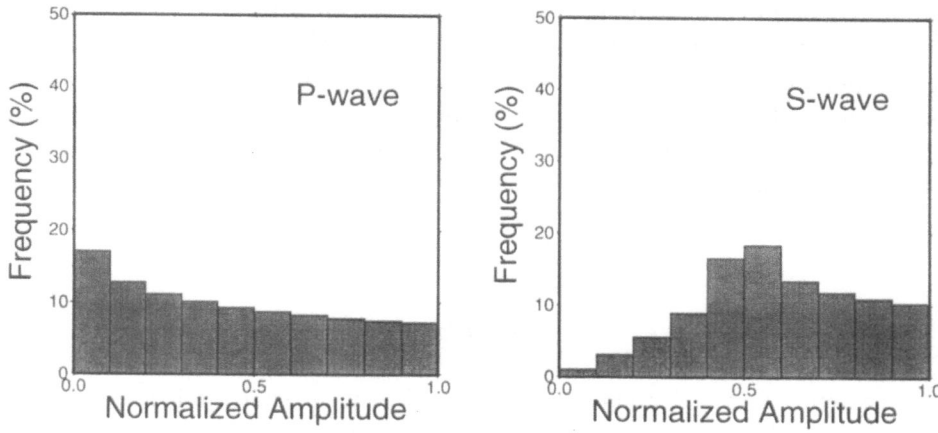

Figure 9

Histograms showing the frequency distribution of P- and S-wave amplitudes radiated from the double-couple point source. Random sampling of one million times is made on the focal sphere. The amplitude is normalized by its maximum value.

future studies of Q_S^{-1} measurement for the shallow part of the crust in different areas.

In Figure 8 we compared Q_P^{-1}/Q_S^{-1} ratio for the crust reported from various tectonically different regions in the world. The large scatter of Q_P^{-1}/Q_S^{-1} values may reflect the local regionality (e.g., LEES and LINDLEY, 1994). However, the Q_P^{-1}/Q_S^{-1} values are generally larger than unity for the frequencies from about 1 to 30 Hz. Our result indicates that the nature of $Q_P^{-1}/Q_S^{-1} \geq 1$ holds good up to the high frequency of 100 Hz. The apparent attenuation with distance is similar for P and S waves. It is quite different from that for the sedimentary deposits in the very shallow crust (e.g., CLOUSER and LANGSTON 1991).

Figures 4 and 5 show large scatter of data around the regression lines, for the P wave in particular. This may be partly attributed to the magnitude difference of earthquakes we used. In a strict sense, eq. (1) holds for the earthquakes of the same magnitude. However, the effect of magnitude difference is practically insensitive to the estimation of Q_P^{-1} as was discussed by YOSHIMOTO et al. (1993). Another plausible explanation for the large scatter of P-wave data is the difference between the source radiation patterns of P and S waves. In order to see the effect of source radiation pattern, we computed the amplitude of P and S waves at 1,000,000 points randomly distributed on the focal sphere of a point double-couple source. As is seen in Figure 9, frequency distribution of the amplitude (normalized by the maximum amplitude), exhibits a different pattern for the P wave and S wave; P-wave amplitude distributes broadly as compared with the S wave, in which about 50% of data concentrate in the 0.4 to 0.7 interval of relative amplitude. This means that scatter of P-wave amplitude is larger than that of S wave when the effect of source radiation pattern is not corrected. The nonuniform source radiation may bring about a considerable scatter of data, although the effect on Q^{-1} estimation is minimized by utilizing as many as 239 earthquakes for the analysis. We think that the station response as a function of back-azimuth may be small because of the borehole recording condition. The consistency of our result with that of IIO (1992) at 60 to 100 Hz indicates the reliability of our estimates.

Conclusions

Purposeful to investigating the frequency-dependent characteristics of seismic wave attenuation in the upper crust, we analyzed borehole seismograms of local earthquakes that took place in the aftershock region of the 1984 western Nagano Prefecture, Japan earthquake. The high quality records of three components warrant a reliable estimate of both Q_P^{-1} and Q_S^{-1} for high frequencies reaching 100 Hz. We applied the extended coda normalization method for measuring the apparent attenuation of P and S waves in the frequencies from 25 to 102 Hz. It is found that Q_P^{-1} manifests a clear frequency dependence characterized by a power-law function of $Q_P^{-1} \simeq 0.052 f^{-0.66}$. For the S wave, in contrast to P wave, the frequency dependence is very weak; $Q_S^{-1} \simeq 0.0034 f^{-0.12}$. The absolute values we obtained show relatively large attenuation of seismic waves in the upper crust of the western Nagano region. The ratio Q_P^{-1}/Q_S^{-1} takes the values between 1.3 and 2.9, and is comparable with the ratio of V_P to V_S. This indicates that the apparent attenuation with distance is roughly the same for P and S waves in this area.

Acknowledgments

The authors express appreciation to Dr. B. J. Mitchell for providing us the opportunity to write this paper. We also thank anonymous reviewers for their helpful comments and fruitful suggestions.

References

ABERCROMBIE, R., and LEARY, P. (1993), *Source Parameters of Small Earthquakes Recorded at 2.5 km Depth, Cajon Pass, Southern California: Implications for Earthquake Scaling*, Geophys. Res. Lett. *20*, 1511–1514.

AKI, K. (1980), *Attenuation of Shear Waves in the Lithosphere for Frequencies from 0.05 to 25 Hz*, Phys. Earth and Planet. Int. *21*, 50–60.

BAKUN, W. H., BUFE, C. G., and STEWART, R. M. (1976), *Body-wave Spectra of Central California Earthquakes*, Bull. Seismol. Soc. Am. *66*, 363–384.

BENNETT, H. F., and BAKUN, W. H. (1982), *Comment on Attenuation of Shear-waves in the Lithosphere for Frequencies from 0.05 to 25 Hz*, Phys. Earth and Planet. Int. *29*, 195–196.

BUTLER, R., MCCREERY, C. S., FRAZER, L. N., and WALKER, D. A. (1987), *High-frequency Seismic Attenuation of Oceanic P and S Waves in the Western Pacific*, J. Geophys. Res. *92*, 1383–1396.

CAMPILLO, M., and PLANTET, J. L. (1991), *Frequency Dependence and Spatial Distribution of Seismic Attenuation in France: Experimental Results and Possible Interpretations*, Phys. Earth and Planet. Int. *67*, 48–64.

CARPENTER, P. J., and SANFORD, A. R. (1985), *Apparent Q for Upper Crustal Rocks of the Central Rio Grande Rift*, J. Geophys. Res. *90*, 8661–8674.

CASTRO, R. R., ANDERSON, J. G., and SINGH, S. K. (1990), *Site Response, Attenuation and Source Spectra of S Waves along the Guerrero, Mexico, Subduction Zone*, Bull. Seismol. Soc. Am. *80*, 1481–1503.

CLOUSER, R. H., and LANGSTON, C. A. (1991), Q_P-Q_S *Relations in a Sedimentary Basin Using Converted Phases*, Bull. Seismol. Soc. Am. *81*, 733–750.

FEDOTOV, S. A., and BOLDYREV, S. A. (1969), *Frequency Dependence of the Body-wave Absorption in the Crust and the Upper Mantle of the Kuril Island Chain*, Izv. Akad. Nauk USSR, Earth Phys. *9*, 17–33.

FEHLER, M., HOSHIBA, M., SATO, H., and OBARA, K. (1992), *Separation of Scattering and Intrinsic Attenuation for the Kanto-Tokai Region, Japan, Using Measurements of S-wave Energy versus Hypocentral Distance*, Geophys. J. Int. *108*, 787–800.

FEUSTEL, A. J., and YOUNG, R. P. (1994), Q_β *Estimates from Spectral Ratios and Multiple Lapse Time Window Analysis: Results from an Underground Research Laboratory in Granite*, Geophys. Res. Lett. *21*, 1503–1506.

FRANKEL, A. (1982), *The Effects of Attenuation and Site Response on the Spectra of Microearthquakes in the Northeastern Caribbean*, Bull. Seismol. Soc. Am. *72*, 1379–1402.

HAVSKOV, J., and MEDHUS, J. (1991), *Q along the FENNOLORA Profile*, Geophys. J. Int. *106*, 531–536.

HIRAHARA, K., and THE MEMBERS OF THE 1986 JOINT SEISMOLOGICAL RESEARCH IN WESTERN NAGANO PREFECTURE (1992), *Three-dimensional P- and S-wave Velocity Structure in the Focal Region of the 1984 Western Nagano Prefecture Earthquake*, J. Phys. Earth *40*, 343–360.

HOANG-TRONG, P. (1983), *Some Medium Properties of the Hohenzollerngraben (Swabian Jura, W. Germany) Inferred from* Q_P/Q_s *Analysis*, Phys. Earth and Planet. Int. *31*, 119–131.

HOLLIGER, K., and BÜHNEMANN, J. (1996), *Attenuation of Broad-band (50–1500 Hz) Seismic Waves in Granitic Rocks near the Earth's Surface*, Geophys. Res. Lett. *23*, 1981–1984.

HOSODA, T., and KOSUGA, M. (1995), Q_P^{-1} *and* Q_S^{-1} *in Northern Tohoku Region Determined by Coda Normalization Method: Part 2*, Abstract of Japan Earth and Planetary Science Joint Meeting, Tokyo, 632.

HOUGH, S. E., ANDERSON, J. G., BRUNE, J., VERNON, III, F., BERGER, J., FLETCHER, J., HAAR, L., HANKS, T., and BAKER, L. (1988), *Attenuation near Anza, California*, Bull. Seismol. Soc. Am. *78*, 672–691.

IIO, Y. (1992), *Slow Initial Phase of the P-wave Velocity Pulse Generated by Microearthquakes*, Geophys. Res. Lett. *19*, 477–480.

JOHNSTON, D. H., TOKSÖZ, M. N., and TIMUR, A. (1979), *Attenuation of Seismic Waves in Dry and Saturated Rocks: II. Mechanisms*, Geophysics *44*, 691–711.

KATO, K., TAKEMURA, M., IKEURA, T., URAO, K., and UETAKE, T. (1992), *Preliminary Analysis for Evaluation of Local Site Effects from Strong Motion Spectra by an Inversion Method*, J. Phys. Earth *40*, 175–191.

KINOSHITA, S. (1994), *Frequency-dependent Attenuation of Shear Waves in the Crust of the Southern Kanto Area, Japan*, Bull. Seismol. Soc. Am. *84*, 1387–1396.

KOHKETSU, K., and SHIMA, E. (1985), Q_P *Structure of Sediments in the Kanto Plain*, Bull. Earthq. Res. Inst. *60*, 495–505.

KVAMME, L. B., and HAVSKOV, J. (1989), *Q in Southern Norway*, Bull. Seismol. Soc. Am. *79*, 1575–1588.

LEES, J. M., and LINDLEY, G. T. (1994), *Three-dimensional Attenuation Tomography at Loma Prieta: Inversion of t* for Q*, J. Geophys. Res. *99*, 6843–6863.

LIU, H. P., ANDERSON, D. L., and KANAMORI, H. (1976), *Velocity Dispersion due to Anelasticity; Implications for Seismology and Mantle Composition*, Geophys. J. R. Astr. Soc. *47*, 41–58.

MASUDA, T. (1988), *Corner Frequencies and Q Values of P and S Waves by Simultaneous Inversion Technique*, Tohoku Geophys. J., Science Rep. Tohoku Univ. *31*, 101–125.

MODIANO, T., and HATZFELD, D. (1982), *Experimental Study of the Spectral Content for Shallow Earthquakes*, Bull. Seismol. Soc. Am. *72*, 1739–1758.

OHTAKE, M. (1987), *Temporal Change of Q_P^{-1} in Focal Area of 1984 Western Nagano, Japan, Earthquake as Derived from Pulse Width Analysis*, J. Geophys. Res. *92*, 4846–4852.

RAUTIAN, T. G., KHALTURIN, V. I., MARTYNOV, V. G., and MOLNAR, P. (1978), *Preliminary Analysis of the Spectral Content of P and S waves from Local Earthquakes in the Garm, Tadjikistan Region*, Bull. Sesmol. Soc. Am. *68*, 949–971.

SATO, H. (1992), *Thermal Structure of the Mantle Wedge beneath Northeastern Japan: Magmatism in an Island Arc from the Combined Data of Seismic Anelasticity and Velocity and Heat Flow*, J. Volcanol. Geotherm. Res. *51*, 237–252.

SATO, H., and MATSUMURA, S. (1980), *Q^{-1} Value for S Waves (2–32 Hz) under the Kanto District in Japan*, Zisin *33*, 541–543 (in Japanese).

SCHERBAUM, F., and SATO, H. (1991), *Inversion of Full Seismogram Envelopes Based on the Parabolic Approximation: Estimation of Randomness and Attenuation in Southeast Honshu, Japan*, J. Geophys. Res. *96*, 2223–2232.

SEKIGUCHI, S. (1991), *Three-dimensional Q Structure beneath the Kanto Tokai District, Japan*, Tectonophysics *195*, 83–104.

SUZUKI, S. (1971), *Anomalous Attenuation of P Waves in the Matsushiro Earthquake Swarm Area*, J. Phys. Earth *20*, 1–21.

TAKEMURA, M., KATO, K., IKEURA, T., and SHIMA, E. (1991), *Site Amplification of S Waves from Strong Motion Records in Special Relation to Surface Geology*, J. Phys. Earth *39*, 537–552.

TAYLOR, S. R., BONNER, B. P., and ZANDT, G. (1986), *Attenuation and Scattering of Broadband P and S Waves across North America*, J. Geophys. Res. *91*, 7309–7325.

TSAI, Y. B., and AKI, K. (1969), *Simultaneous Determination of the Seismic Moment and Attenuation of Seismic Surface Waves*, Bull. Seismol. Soc. Am. *59*, 275–287.

YAMASHINA, K., and TADA, T. (1985), *A Fault Model of the 1984 Western Nagano Prefecture Earthquake Based on the Distance Change of Trilateration Points*, Bull. Earthq. Res. Inst. *60*, 221–230.

YOSHIMOTO, K., SATO, H., and OHTAKE, M. (1993), *Frequency-dependent Attenuation of P and S Waves in the Kanto Area, Japan, Based on the Coda-normalization Method*, Geophys. J. Int. *114*, 165–174.

(Received January 13, 1998, revised June 5, 1998, accepted June 23, 1998)

To access this journal online:
http://www.birkhauser.ch

Pure appl. geophys. 153 (1998) 503–538
0033–4553/98/040503–36 $ 1.50 + 0.20/0

▮ Pure and Applied Geophysics

Seismic Velocity and Q Structure of the Middle Eastern Crust and Upper Mantle from Surface-wave Dispersion and Attenuation

LIANLI CONG[1,2] and B. J. MITCHELL[1]

Abstract—Observed velocities and attenuation of fundamental-mode Rayleigh waves in the period range 7–82 sec were inverted for shear-wave velocity and shear-wave Q structure in the Middle East using a two-station method. Additional information on Q structure variation within each region was obtained by studying amplitude spectra of fundamental-mode and higher-mode Rayleigh waves. We obtained models for the Turkish and Iranian Plateaus (Region 1), areas surrounding and including the Black and Caspian Seas (Region 2), and the Arabian Peninsula (Region 3). The effect of continent-ocean boundaries and mixed paths in Region 2 may lead to unrealistic features in the models obtained there. At lower crustal and upper-mantle depths, shear velocities are similar in all three regions. Shear velocities vary significantly in the uppermost 10 km of the crust, being 3.21, 2.85, and 3.39 km/s for Regions 1, 2, and 3, respectively. Q models obtained from an inversion of interstation attenuation data show that crustal shear-wave Q is highest in Region 3 and lowest in Region 1. Q's for the upper 10 km of the crust are 63, 71, and 201 for Regions 1, 2, and 3, respectively. Crustal Q's at 30 km depth for the three regions are about 51, 71, and 134. The lower crustal Q values contrast sharply with results from stable continental regions where shear-wave Q may reach one thousand or more. These low values may indicate that fluids reside in faults, cracks, and permeable rock at lower crustal, as well as upper crustal depths due to convergence and intense deformation at all depths in the Middle Eastern crust.

Key words: Attenuation, Q, surface waves, Middle East.

Introduction

Seismic surface waves have been widely used to study the shear-wave velocity distribution in the earth. In recent years, high-quality data and increasing numbers of favorably located, modern, digital stations have enabled seismologists to apply tomographic techniques to surface-wave dispersion data. Numerous such studies have provided information on regional shear-velocity variations at both global and regional scales in the earth.

Studies of surface-wave attenuation have been less numerous, mainly because of well-known difficulties in measurement; surface waves may be contaminated by

---

[1] Department of Earth and Atmospheric Sciences, St. Louis University, 3507 Laclede Avenue, St. Louis, Missouri 63103. Telephone: (314) 977-3130, Fax: (314) 977-3117, E-mail: mitchell@eas.slu.edu
[2] Now at Department of Earth Sciences, Yunnan University, Kunming, Yunnan, People's Republic of China.

lateral refraction and multipathing which may focus or defocus energy and produce spurious results. It is, however, important to study attenuation because it can provide information on temperature, fluid content, phase change, and density of solid-state defects in the crust and mantle, phenomena that are not easily studied using only seismic wave velocities.

Few models of shear-wave Q (Q_μ) in the crust and upper mantle have been obtained for continental regions. Nevertheless these have given us information on how Q_μ is related to the tectonic history of various regions (MITCHELL, 1995). That work suggested that the value of Q_μ obtained in the upper crust of any region is directly proportional to the time that has elapsed there since the most recent episode of major tectonic activity.

The present study measures Rayleigh-wave dispersion and attenuation in the Middle East, a complex region of interaction between the Eurasian, Arabian, and African plates. We will invert two-station data for models of shear velocity (β) and Q_μ for three broad regions, and study smaller-scale variations of Q_μ within each region using a single-station method. The Q_μ models will then be compared to the geology and tectonic history of different portions of the Middle East.

Overview of Middle East Tectonics

The Middle East is a tectonically complex region where the Arabian, African, and Eurasian plates meet. Figure 1 shows the relative movement of the plates, the major tectonic features, and the boundaries of tectonic provinces in that region.

The Arabian plate, consisting of a shield in the west, a central shelf, and platform and basins in the east, is the most stable portion in this region. It is bounded on the northwest by the Dead Sea and East Anatolian faults, and on the northeast by the Zagros thrust belt. The North Anatolian fault extends in an east-west direction across most of Turkey. The Arabian shield consists predominantly of Precambrian gneiss and metamorphosed sedimentary and volcanic rocks that have been intruded by granites (POWERS et al., 1966; BROWN, 1972). The thin basaltic lava flows that have been emplaced along the western border of the shield since Miocene time reflect the intensity of volcanism in the region (BROWN, 1972; COLEMAN, 1977; MOKHTAR and AL-SAEED, 1994). East of the shield, the shelf is tectonically stable. A cross section through the peninsula (SEBER et al., 1997), trending roughly east-west, shows that the thickness of the sediments increases gradually across the shelf area, more rapidly to the east, reaching about 10 km in the basin region.

The Arabian and Eurasian plates are currently colliding along the Zagros suture zone in western Iran (STOCKLIN, 1974; SENGÖR and KIDD, 1979; STONELEY, 1981). The Turkish and Iranian Plateaus, resulting from the collision between the Arabian and Eurasian plates, average about 1500 km in elevation (SENGÖR and KIDD, 1979).

After a continental collision during the late Miocene, convergence between Arabia and Eurasia continues as shown by the Pliocene to recent folding and thrusting in the Zagros, the border folds of south-eastern Turkey (KETIN, 1966), and the present diffuse seismicity of the entire Turkish and Iranian Plateaus (CANITEZ and UCER, 1967; MCKENZIE, 1972; NOWROOZI, 1972; BERBERIAN, 1976).

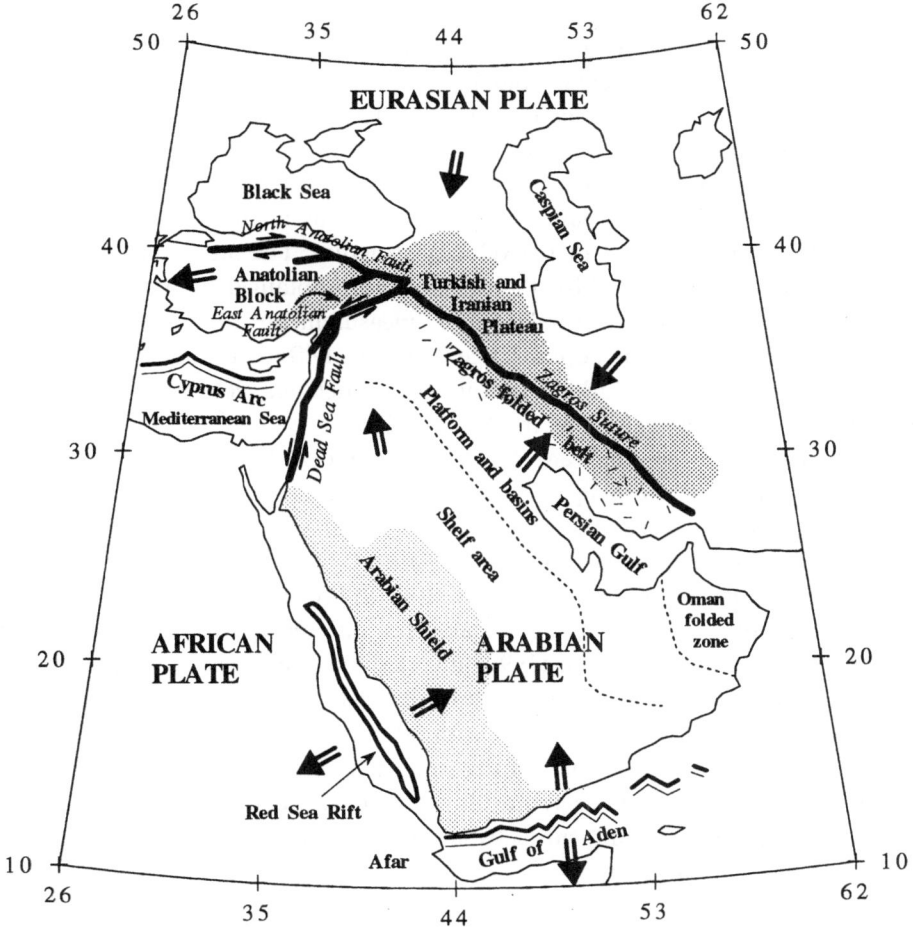

Figure 1

Tectonic map (modified from SEBER et al., 1997) showing the relative movement of the plates in the Middle East, the major tectonic features, and the tectonic provinces in the Arabian plate. Shadings delineate the Arabian Shield (lighter), and the Turkish and Iranian Plateaus (darker). For the latter region only elevations higher than 1500 m are shaded.

Previous Seismological Work in the Middle East

Although the tectonics of the Middle East is well understood, only a few reliable models of shear velocity and attenuation have been obtained there. In an early study NIAZI (1968) inverted short-period surface-wave group and phase velocities, and obtained shear velocities of 3.4 km/s for the upper crust, 3.6 km/s for the lower crust and 4.6–4.9 km/s for the upper mantle in the Arabian Peninsula. KNOPOFF and FOUDA (1975) inverted the phase velocities of Rayleigh waves for the period range 22–167 sec to obtain shear velocity models using a two-station method. The paths connecting station pairs (SHI-HLW and SHI-JER) are both located in the northern part of the peninsula, trending east-west. Their study showed a variety of solutions which can explain the data. MOKHTAR and AL-SAEED (1994) inverted group velocities of fundamental- and higher-mode Rayleigh and Love waves for velocity structure in the Arabian Peninsula. They studied paths between many epicenters and station RYD, located in the center of the Arabian Peninsula. They divided their data into three major groups and obtained different models for each group. GHALIB (1992) used data recorded at stations in the northern part of the Arabian Peninsula to invert for the velocity structure. The only models of velocity structure known for the Turkish Plateau are those obtained by MINDEVALLI and MITCHELL (1989) using the single-station measurements of Rayleigh and Love wave group velocities in the period range 8–50 sec. Different models were obtained for eastern and western Turkey. In this study, detailed velocity models will be provided for different regions of the Middle East.

Q studies in the Middle East have indicated that both regional phases and surface waves attenuate rapidly with distance. KADINSKY-CADE et al. (1981) found that regional phases were extinguished along some paths, but not along others. Numerical values for $Lg\ Q$ in Iran were obtained by NUTTLI (1980) and by JIH and LYNNES (1992). Both studies indicate that $Lg\ Q$ is low in that region, and the latter study showed that it is also quite variable for different paths across the Iranian Plateau. SEBER and MITCHELL (1992) inverted surface-wave attenuation data to obtain the Q_μ models for the upper crust of the Arabian Peninsula. They found that shear-wave Q values in the crust vary from about 60 along the margin of the Red Sea to 100–150 in the central part of the peninsula, and 65–80 in the eastern folded region. GHALIB (1992) applied the extended coda-Q technique to short-period Lg coda waves, obtaining a uniform distribution of Lg coda Q with an average value of 214 for the Arabian plate.

Regionalization and Data Acquisition

In this study we used both two- and single-station methods for both velocity and attenuation studies. In order to model the regional variation of shear velocities

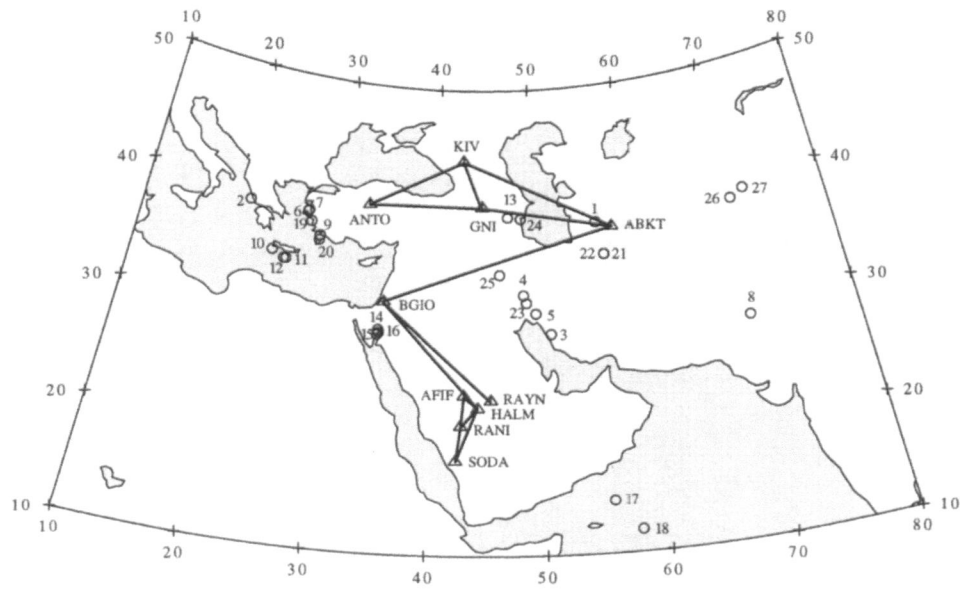

Figure 2

Paths used for two-station determinations of interstation group and phase velocities and attenuation. Triangles denote stations and circles indicate earthquakes used. Numbers refer to the events described in Table 2.

and attenuation in the Middle East, we have divided it into three regions, based upon the availability of data for two-station studies. Region 1 consists of the Turkish and Iranian Plateaus and includes the station pairs GNI-KIV and GNI-ANTO (Fig. 2). Those paths are confined strictly to continents. Paths crossing the Black and Caspian Seas are excluded from this region and are assigned to Region 2. The station pairs for that region are ANTO-KIV, GNI-ABKT, and KIV-ABKT that cover the two seas plus a large land area around them. Region 3 is the Arabian Peninsula. For that region we used data from stations within Saudi Arabia as well as one station in Israel. The station pairs for Region 3 are BGIO-RAYN, BGIO-HALM, AFIF-SODA, HALM-RANI, and HALM-SODA.

To improve regional coverage we used a multi-mode method that requires only a single station (CHENG and MITCHELL, 1981). Figure 3 shows the additional paths and improved regional coverage made possible by the single-station method.

The stations used in this study include four Incorporated Research Institution for Seismology (IRIS) stations (ANTO, ABKT, GNI, KIV), one Geophone station (BGIO), and five stations in Saudi Arabia (RANI, RAYN, HALM, AFIF, SODA), deployed temporarily by the University of California at San Diego and Boise State University. The locations of the stations are given in Table 1. All of the stations provided broadband digital data.

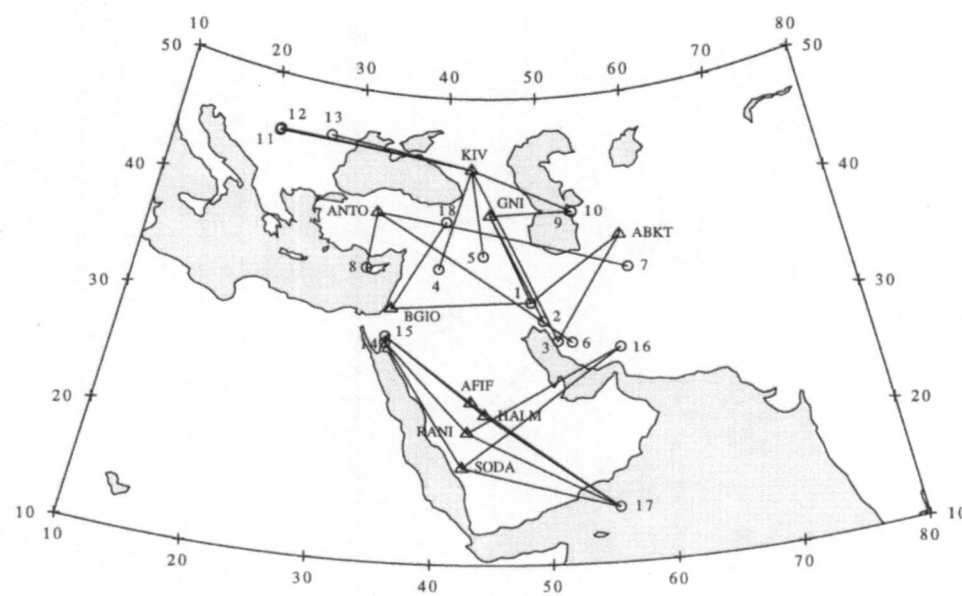

Figure 3

Paths used for single-station studies of multi-mode Rayleigh-wave attenuation. The meanings of the symbols are explained in the caption for Figure 2. Numbers refer to the events described in Table 3.

In the two-station method, 27 earthquakes were used as indicated in Table 2. The dates, origin times and locations of the earthquakes and recording stations are provided by the IRIS DMC. All but two of the earthquakes are less than 50 km in depth. The two exceptions (events 20 and 24 in Table 2) have depths of 75 and 62 km. In the study, all events lay within 8° in azimuth from great-circle paths between the two recording stations. The first station of each station pair in Table 2 is the

Table 1

Station information

Station	Latitude (°N)	Longitude (°E)
ABKT	37.9304	58.1189
ANTO	39.8689	32.7936
BGIO	31.7220	35.0878
GNI	40.0530	44.7240
KIV	43.9562	42.6888
AFIF	23.9310	43.0400
HALM	22.8454	44.3173
RANI	21.3116	42.7761
SODA	18.2921	42.3769
RAYN	23.5220	45.5008

Table 2

Earthquakes studied using the two-station method

Event	Date	Origin (h:m:s)	Lat. (°N)	Lon. (°E)	Depth (km)	Recording stations	Az-dev 1 (°)	Az-dev 2 (°)	Region
1	921006	08:57:17.7	38.42	56.52	10	GNI-ANTO	3.6	1.9	1
2	941129	14:30:28.4	38.71	20.48	21	ANTO-GNI	2.0	1.0	1
3	940329	07:56:53.9	29.10	51.26	33	GNI-KIV	7.6	5.7	1
4	940920	05:51:46.0	32.50	48.77	33	GNI-KIV	4.1	2.7	1
5	950422	00:21:48.6	30.89	49.91	25	GNI-KIV	5.8	4.1	1
6	940524	02:05:36.2	38.66	26.54	17	GNI-ABKT	7.1	4.1	2
7	940524	02:18:34.9	38.76	26.60	16	GNI-ABKT	7.8	3.9	2
8	940603	22:40:21.2	28.74	70.07	33	ABKT-KIV	6.9	3.6	2
9	931026	10:03:57.5	36.74	28.05	33	ANTO-KIV	6.3	2.3	2
10	950103	22:51:45.0	34.94	23.55	33	ANTO-KIV	1.3	0.7	2
11	950203	22:29:13.3	34.42	25.03	49	ANTO-KIV	6.7	3.3	2
12	950330	18:17:15.1	34.40	24.79	7	ANTO-KIV	5.8	2.9	2
13	960422	14:42:32.4	39.17	47.37	29	AFIF-RANI	7.2	6.2	3
14	951123	18:07:17.3	29.33	34.75	10	AFIF-HALM	4.7	4.0	3
15	951124	16:43:45.5	28.94	34.71	10	AFIF-HALM	6.8	5.8	3
16	951125	11:41:35.3	29.12	34.84	10	AFIF-HALM	5.5	4.7	3
17	951207	17:48:16.3	14.56	55.77	0	HALM-AFIF	7.6	6.9	3
						RAYN-BGIO	3.0	1.6	3
18	960328	07:28:28.1	11.92	57.81	10	HALM-AFIF	4.9	4.5	3
19	960402	07:59:26.2	37.83	27.00	33	AFIF-HALM	6.6	6.1	3
						BGIO-HALM	4.6	1.8	3
20	960426	07:01:27.5	36.37	28.04	75	AFIF-HALM	4.8	4.5	3
21	960225	16:14:10.9	35.65	57.07	33	HALM-RANI	5.2	4.7	3
22	960225	17:42:04.5	35.65	57.05	33	HALM-RANI	5.3	4.7	3
23	960331	16:02:09.3	31.83	49.03	0	HALM-SODA	1.8	1.2	3
24	960103	08:42:26.3	39.03	48.72	62	RANI-SODA	7.6	6.6	3
25	960128	08:43:16.2	34.26	46.45	33	RANI-SODA	6.1	5.0	3
26	940610	03:00:44.5	38.56	70.63	33	ABKT-BGIO	3.7	1.3	1,2,3
27	951026	04:23:27.5	39.18	72.07	50	ABKT-BGIO	0.4	0.2	1,2,3

closer of each pair to the epicenter. The rightmost column in Table 2 indicates the region traversed by the surface waves. Earthquakes 26 and 27 are not used for determinations of properties within any region since the waves from those events traverse three regions. They are only used to see if velocities and attenuations along those paths are consistent with our results for the individual regions.

Eighteen earthquakes, listed in Table 3, were analyzed using the single-station method. All but one of the earthquakes (event 13 in Table 3) are less than 44 km in depth. The one exception has a focal depth of 88 km. For some events, records from several stations were available, whereas for some events only a single record was used because of the poor signal/noise ratios at other stations. Depths identified with an exclamation mark were determined by Harvard University and those identified with an asterisk were obtained in this study by matching the spectral holes of the theoretical spectra with those of the observed data. The dip, slip and strike angles were obtained by Harvard University. The number in the rightmost column of Table 3 indicates the region that the surface waves traverse.

Interstation Group and Phase Velocities

Before using the two-station method, we removed the mean ground motion from the records, corrected for geometrical spreading and instrument response, and applied bandpass filtering. Figure 4 is an example which shows the original trace (top) recorded at station SODA for an earthquake that occurred on 28 January, 1996, and the windowed trace of ground displacement (bottom) after correcting for geometrical spreading and instrument response, and bandpass filtering between corner frequencies of 0.01 and 1 Hz. The epicentral distance and the starting times of the traces from the origin are indicated in the boxes. The dashed lines above the top trace represent the window with a 10% cosine taper. In this example the window length is 400 sec, and the starting and ending times correspond to group velocities of about 2.2 to 4.5 km/sec, respectively.

The Multiple Filter Technique (MFT) was then utilized to extract group velocities. Figure 5 shows contoured spectral amplitudes of fundamental- and

Table 3

Earthquakes studied using the single-station method

Event	Date	Origin (h:m:s)	Lat. (°N)	Lon. (°E)	Depth (km)	Dip (°)	Slip (°)	Strike (°)	Recording stations	Region
1	940920	05:51:46.0	32.50	48.77	15*	25	87	103	ABKT, GNI	1
									BGIO	3
2	950422	00:21:48.6	30.89	49.91	8*	32	72	117	GNI	1
3	940329	07:56:53.9	29.10	51.26	6*	40	104	334	ABKT, KIV	1
4	941120	14:31:02.2	35.34	39.56	29	57	176	141	KIV	1
5	910724	09:45:41.8	36.52	44.07	10*	44	121	335	KIV	1
6	940620	09:09:02.9	28.97	52.61	14*	67	− 5	251	ANTO	1
7	941214	20:43:53.7	35.10	58.63	5*	32	144	319	ANTO	1
8	950223	21:03:01.3	35.05	32.28	15!	21	140	239	ANTO	1
9	940701	10:12:41.2	40.23	53.38	44!	30	71	251	GNI, KIV	2
10	940701	19:50:04.3	40.22	53.39	44	34	108	290	GNI, KIV	2
11	910712	10:42:21.2	45.36	21.06	11	78	− 1	279	KIV	2
12	911202	08:49:40.2	45.50	21.12	9	72	− 5	103	KIV	2
13	900531	00:17:47.8	45.81	26.77	88	26	54	90	KIV	2
14	960221	04:59:51.2	28.80	34.78	10	30	−104	132	RANI, SODA	3
15	951123	18:07:17.2	29.33	34.75	26*	77	7	199	AFIF, HALM	3
16	960226	08:08:19.2	28.32	57.09	33	7	125	315	RANI, SODA	3
17	960314	21:47:57.9	14.74	55.74	18*	20	−118	72	AFIF, RANI	3
									HALM, SODA	
18	951205	18:49:30.4	39.44	40.15	29!	70	160	136	BGIO	1,3

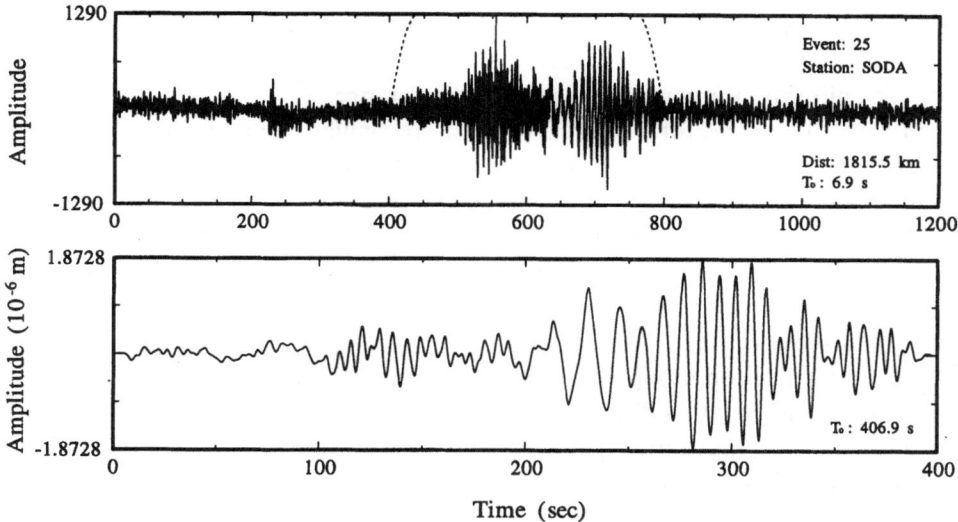

Figure 4

(Top) Original trace (in counts) for the earthquake that occurred on January 28, 1996 as recorded at station SODA. (Bottom) Windowed trace of ground displacement after correction for geometrical spreading and instrument response, and bandpass filtering with corner frequencies of 0.01 and 1 Hz.

higher-mode Rayleigh waves obtained for an earthquake that occurred on 22 April, 1995 and was recorded at station GNI. The windowed ground displacement, after correction for the geometrical spreading and instrument response, and bandpass filtering, is displayed at the rightmost part of the figure. The same trace with a linear group velocity scale is plotted adjacent to the dispersion diagram. This trace can be used to correlate group velocity contour values to the trace itself. The long, smooth curve composed of squares identifies the fundamental-mode group velocity dispersion. The shorter curve with a group velocity of about 3.1 km/s at 5 sec is a superposition of higher modes. The square, circle, triangle, and plus signs correspond to the four largest amplitudes in the time domain when the MFT is applied.

A phase-matched filter technique as proposed by HERRIN and GOFORTH (1977), and coded by HERRMANN (1987) was used to extract the phase-velocity information. The interstation group and phase velocities can then be obtained for the three regions as shown in Figures 6, 7, and 8. In each of the three figures, group velocities appear below the phase velocities.

Different features can be observed for the dispersion curves of the three regions. For Region 1 (Figure 6), the group velocities obtained for different events are similar to one another at periods below 20 sec and display greater scatter at longer periods. The velocities increase rapidly with increasing period at shorter periods, and reach a value of about 2.89 km/s at 10 sec. The velocity values remain almost unchanged in the period range 10–20 sec. Above 20 sec, the velocities increase again, smoothly and monotonically, reaching their maximum values at the maximum observable period. The phase velocity curves for this region, however, increase uniformly throughout the entire period range.

Most of the curves for Region 2 (Fig. 7) are similar to one another, although velocities in a few cases deviate substantially from mean values. This deviation may be due to systematic errors caused by lateral variations of properties in this region. It is comprised of both oceanic (Black and Caspian Seas) and continental portions. Both phase and group velocities, at shorter periods, are lower than those for Region 1. This implies that the shear velocities at shallow depths in Region 2 are lower than those in Region 1.

Fifteen interstation group and phase velocities are obtained for Region 3 (Fig. 8). The largest interstation distance is 1370 km between stations BGIO and RAYN. The distances for the station pairs AFIF-RANI, AFIF-HALM, HALM-RANI, and HALM-SODA are, however, in the range of 178–543 km. Short interstation distances can cause problems when using the two-station method. Any

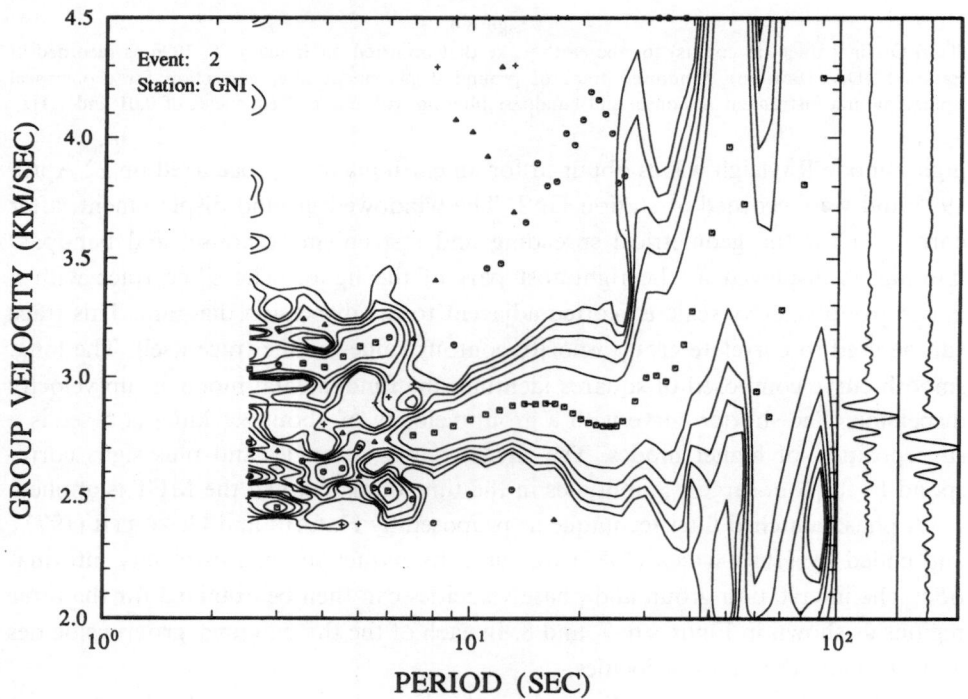

Figure 5

Contoured spectral amplitudes of fundamental- and higher-mode Rayleigh waves obtained for an earthquake that occurred on 22 April, 1995 and recorded at station GNI. The windowed ground displacement, after correction for the geometrical spreading and instrument response, and bandpass filtering, is displayed at the rightmost part of the figure. The same trace with a linear group velocity scale is plotted adjacent to the dispersion diagram.

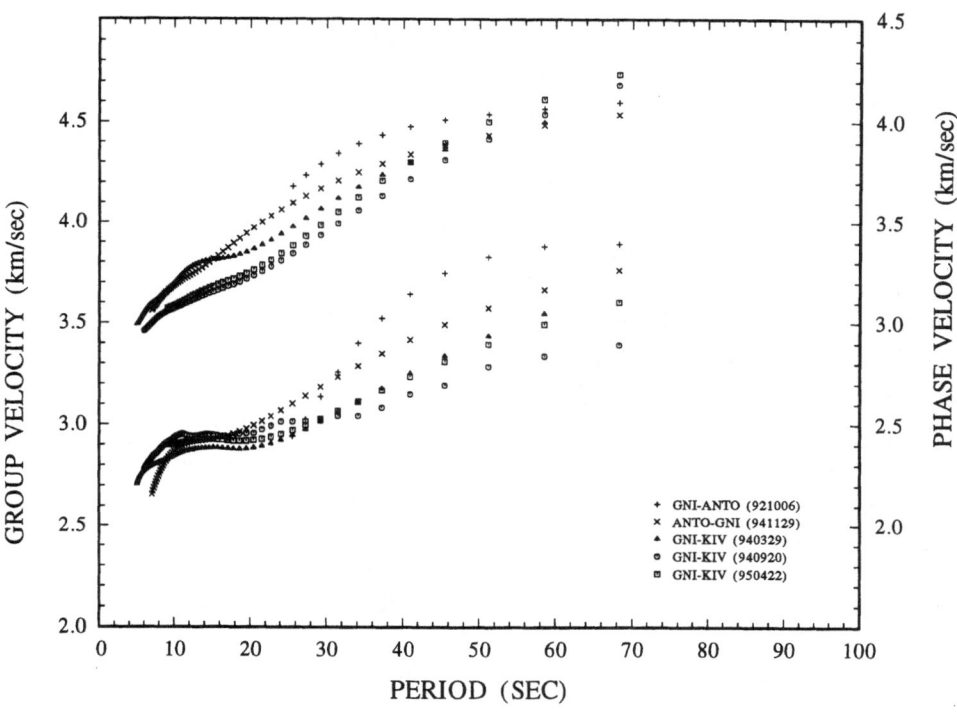

Figure 6
Interstation group (bottom) and phase (top) velocities for the station pairs in Region 1. The caption shows symbols used for various station pairs. Phase velocities are given on the ordinate on the right side of the plots, whereas group velocities are given on the left ordinate on the left.

departures from a great-circle path or measurement errors can result in severe bias in the velocity and attenuation determinations. To control the errors, we tried to collect as many events for each station pair as possible. For example, seven events were used for the station pair AFIF-HALM. If observed values are similar, the standard deviations of the mean values will be reduced. In addition, results for the station pairs with shorter interstation distances will be compared with those with longer interstation distances, and also be compared with the average values to examine the usability of the data. The curves for the station pairs RAYN-BGIO and BGIO-HALM lie in the middle of the group, suggesting that the selected curves for the shorter interstation distances are realistic. The models obtained after inversion will be compared with those reported in the previous studies to check the validity of the interstation results. The mean value of group velocities at 10 sec for Region 3 is higher than that found for either Region 1 or 2, being about 2.98 km/s. This suggests that the near-surface shear velocity is higher here than in the other two regions.

The interstation group and phase velocities for the station pair ABKT-BGIO are plotted in Figure 9. The paths connecting the two stations cross the southern edge of the Caspian Sea, the Turkish and Iranian Plateaus, and the northern Arabian Peninsula as shown in Figure 2. According to our regionalization, the paths cross all of the three regions. Since our goal is to obtain velocity and attenuation structure for individual regions, we did not use the dispersion data for this station pair in our inversion. If our regionalization is appropriate, i.e., the three regions are geologically different with different velocity and attenuation structure, the group and phase velocities, and the attenuation values for this station pair should provide an average for the three regions. The average group velocities at 10 sec are about 2.89, 2.35, and 2.98 km/s for Regions 1, 2, and 3, respectively. The average group velocity at 10 sec for these two events is about 2.70 km/s, which is in the middle of the three values. The velocities increase with period and have a value of about 3.38 km/s at 40 sec, which is also in the middle of the average values of 3.33, 3.30, and 3.55 km/s for Regions 1, 2, and 3 at that period, respectively. The group velocities for these two events are higher than those for Region 2, lower than those for Region 3, and similar to those for Region 1 for the periods greater

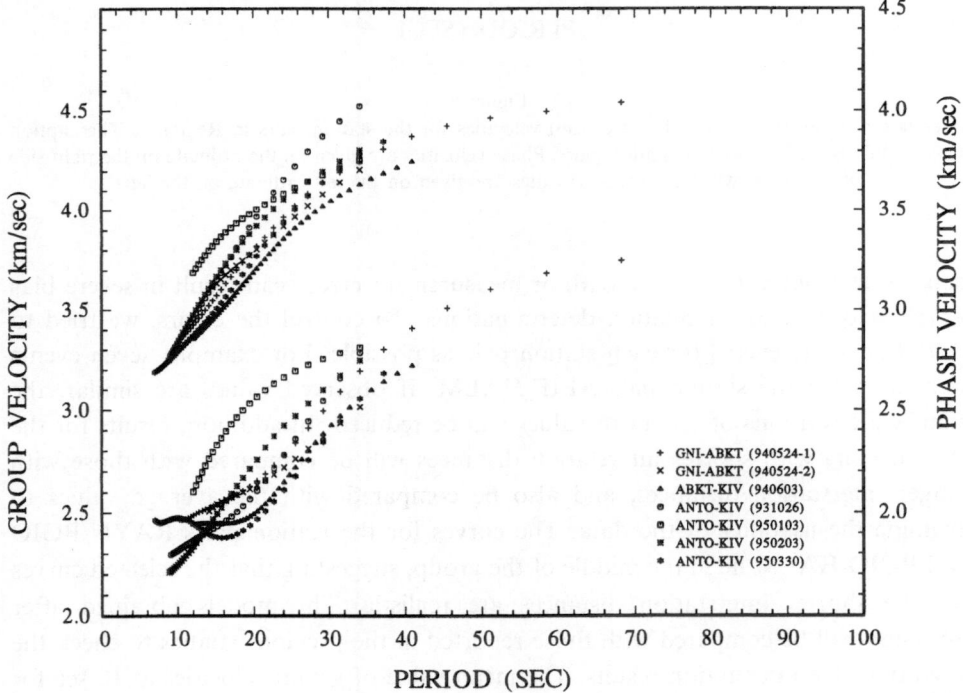

Figure 7
Interstation group (bottom) and phase (top) velocities for the station pairs in Region 2. See the caption for Figure 6.

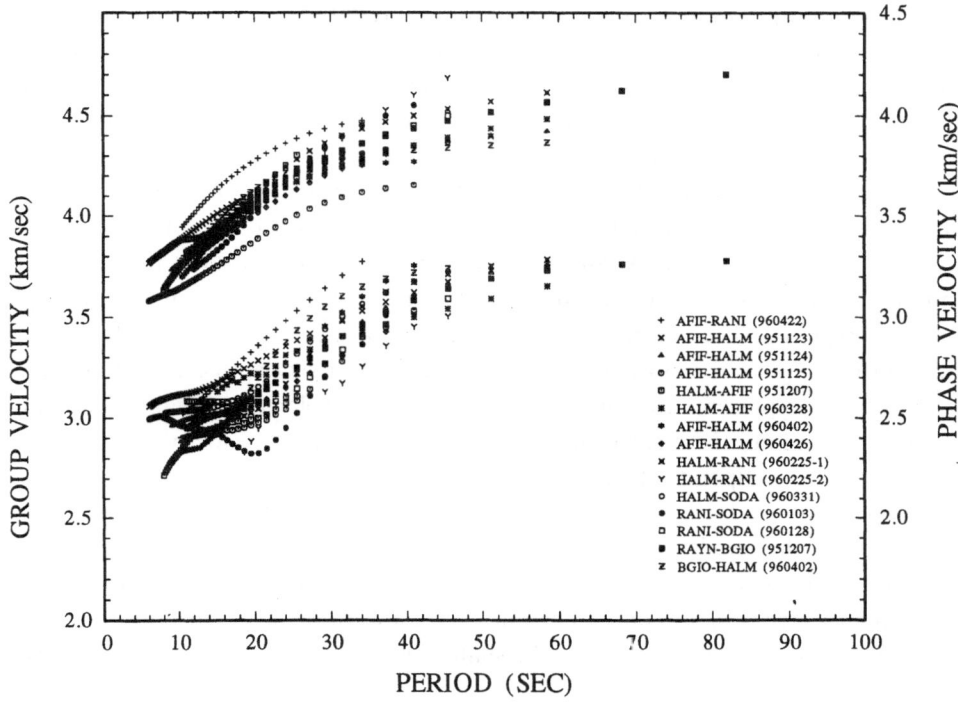

Figure 8
Interstation group (bottom) and phase (top) velocities for the station pairs in Region 3. See the caption for Figure 6.

than 20 sec. The average phase velocity for these two events at 10 sec is about 3.12 km/s. This value is also in the middle of the corresponding values, 3.13, 2.82, and 3.28 for the three regions. From the above comparisons we conclude that our division of the Middle East into three regions is appropriate.

Shear-velocity Models Obtained Using the Two-station Method

Since the dispersion curves in each group exhibit similar features, the average values and the standard errors were calculated for each of the three regions. It should be noted, however, that for some periods only single observations are available and the standard deviations cannot be calculated. In these cases, we compared the measurement errors at these periods with the estimated standard deviations at other periods where multiple observations are available, choosing the largest values as our conservative error estimates. We chose 0.3 km/s as the error estimates at those points.

To save computing time, the average dispersion data were simplified to make a roughly even distribution over a linear period range. The simplified group and phase velocities were then inverted using a differential inverse technique to obtain the shear-velocity models for the three regions. In the inversion process, the thickness of each layer was fixed. The initial values of compressional and shear velocities, and densities are 6.0 km/s, 3.5 km/s, and 3.0 g/cm^3 for all of the layers. These values were subsequently changed to note the effect of initial values on inversion results. Figure 10 is an example which shows the shear velocity models and the resolving kernels obtained by inverting the fundamental-mode Rayleigh data for Region 1. The left panel of the figure displays the velocity model, whereas the right panel gives the normalized resolving kernels at six depths. The dashed line in each figure represents the initial value of the shear velocities, being 3.5 km/sec. The lateral bars are the error estimates of the corresponding velocity values, being plotted at the midpoint of the corresponding layers. Figure 11 illustrates the observed group and phase velocities as well as the theoretical dispersion curves predicted by the model obtained for Region 1. The circles denote the group velocities, whereas the squares represent the phase velocities. The vertical bars are the standard errors of the data. The curves for the resolving kernels on the right panel of Figure 10 demonstrate that the shallow structure is better resolved than the

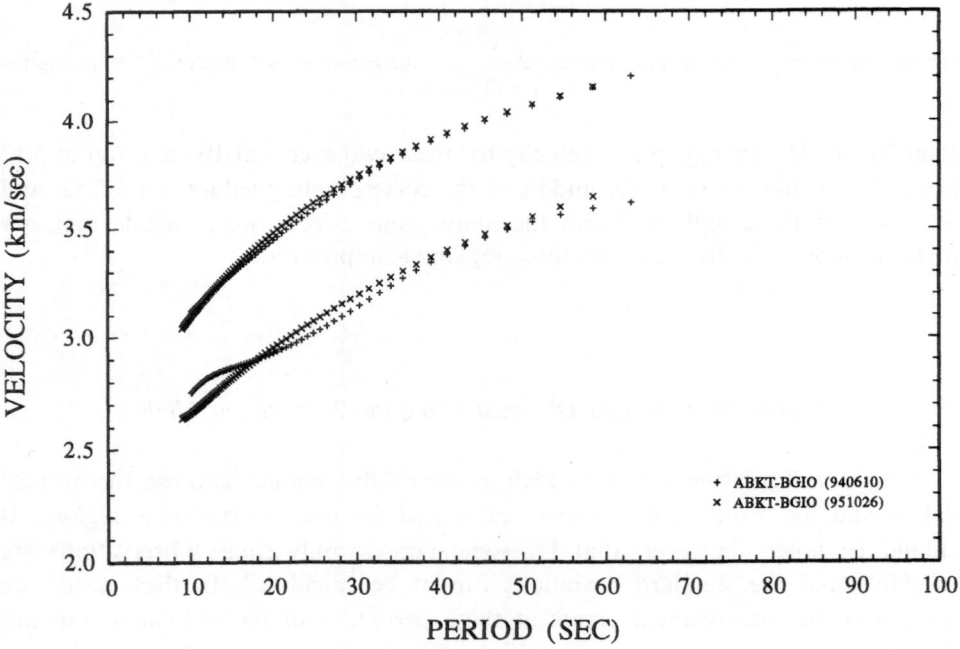

Figure 9
Interstation group (bottom) and phase (top) velocities for the station pair ABKT-BGIO.

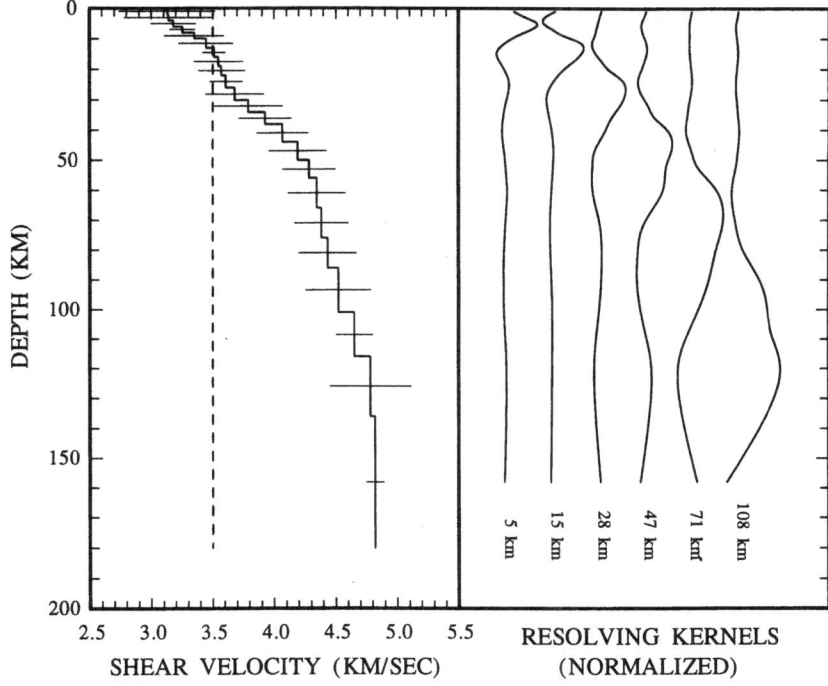

Figure 10

Shear-velocity model and resolving kernels for Region 1. The dashed lines indicate the starting model. The horizontal lines indicate the error estimates for the models. Narrow resolving kernels centered at the given depths indicate good resolution whereas broad kernels indicate poorer resolution.

deeper structure. The shapes of the velocity resolving kernels are similar for all three regions, and are therefore not plotted for the other two regions. The velocity models obtained in this study, from an analysis of the resolving kernels, are restricted to the crust and the uppermost part of the mantle within a depth of about 120 km. The match of the theoretical velocity dispersion curves to the observed data (Fig. 11) is good over the entire period range. Similar results were obtained for the other two regions, and are not plotted.

For comparison, we plot the shear-velocity models for the three regions together in Figure 12. The dotted, dashed, and solid lines are the models for Regions 1, 2, and 3, respectively. The uncertainties of the velocities in each layer are also plotted as horizontal bars, which overlap for the deeper layers. The velocities for the three regions differ greatly in the upper 10 km. Region 3 has the highest velocity values, whereas Region 2 has the lowest velocities. The average velocities for the upper 10 km are about 3.21, 2.85, and 3.39 km/s for Regions 1, 2, and 3, respectively. Below 10 km in the crust, the velocities in Regions 2 and 3 increase gradationally with no obvious gradient change, whereas the velocity gradient in Region 1 changes significantly at depths between about 12 and 30 km. In the uppermost mantle, the velocity gradients decrease for all three regions, and velocity values in the upper-most mantle do not differ much from region to region, especially for Regions 1 and 3. At depths near 120 km, the velocities for the three regions approach a high value in the range 4.64–4.78 km/s. No low-velocity zones are resolved for any of the three regions in the present study. Since Region 2 includes both continental and oceanic portions, the model for this region may not represent either of those portions very well.

Figure 13 compares our velocity models with those of other studies in the Arabian Peninsula. The bold line is the model obtained in this study, whereas the thin lines and dashed or dotted lines represent the results from other studies. The models labeled by "Crust H" and "Crust L" are cited from the study of KNOPOFF and FOUDA (1975). Their study showed a variety of solutions which can explain the data. We selected two typical models, in the depth range 0–180 km, from the Crust H and Crust L models to compare with our model.

MOKHTAR and AL-SAEED (1994), using data recorded by station RYD in the center of the Arabian Peninsula, studied three major propagation paths. Path I connects the station and the events in the Red Sea, Path II is for the events in Gulf of Aden, and Path III is for the events in the Zagros folded zone. The results are plotted in Figure 13 using the lines labeled by Path I, Path II, and Path III,

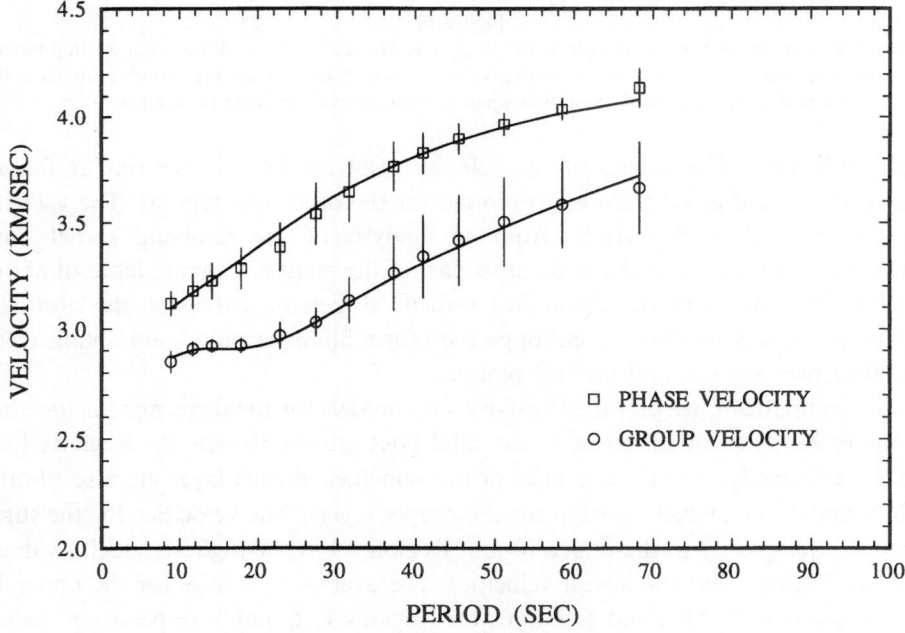

Figure 11

Group and phase velocities, and predicted dispersion curves for Region 1. The vertical lines indicate one standard deviation.

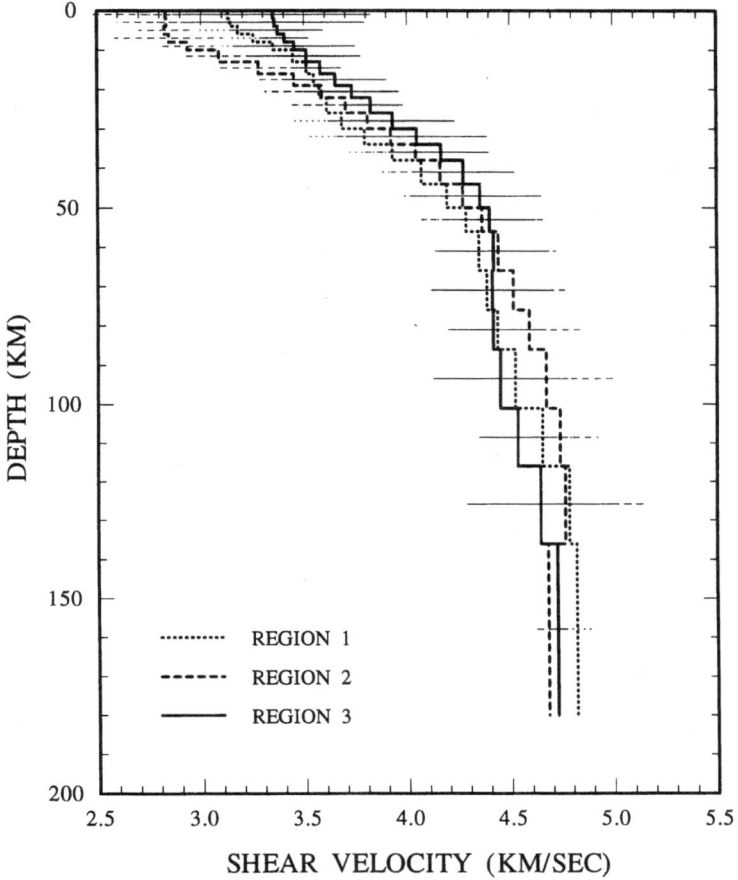

Figure 12

Shear-velocity models for the three regions. The horizontal lines indicate the error estimates for the models.

respectively. Because the period range of the dispersion data is from 4 to 42 sec, the maximum depth for the models is 60 km.

GHALIB (1992) applied data recorded at several stations surrounding the Arabian Peninsula to invert for the velocity structure using a single-station method. Since his maximum period is about 70 sec, the maximum depth he used for the inversion is 100 km. Many paths were included in this study. For purposes of comparison, we select only two typical paths crossing the entire Arabian Peninsula, one trending north-south, another trending east-west. The models for the two paths are plotted in Figure 13 as the lines labeled by T190 and S212f.

Figure 13 shows that the shear-velocity model obtained in this study is in good agreement with models obtained in other studies, especially for the depth range

20–40 km. For the upper 20 km our velocities agree with the average of other studies because we used paths at several azimuths and averaged the dispersion values to obtain an average velocity model of the entire peninsula. For the uppermost mantle, at depths of 40–100 km, there is a wide range of velocities (4.4–4.7 km/s). The model obtained in this study (4.4 km/s) lies in the lower portion of that region.

The velocity model of Region 1 is compared with those of MINDEVALLI and MITCHELL (1989) as shown in Figure 14. In their study the region is divided into two subregions: eastern and western Turkey. The figure shows that the velocity model of this study is in good agreement with their models. In the lower crust and the uppermost mantle (20–80 km) the three models are nearly identical. For the upper crust, the model of this study agrees well with the model for eastern Turkey,

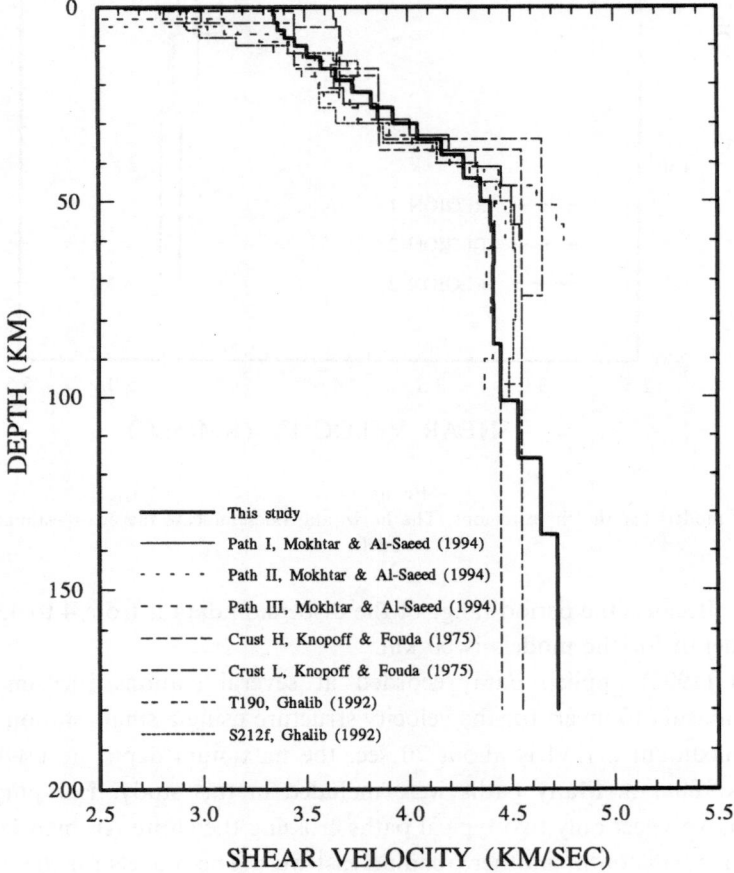

Figure 13
Comparison of the shear-velocity model for the Arabian Peninsula with those from other studies in the same region.

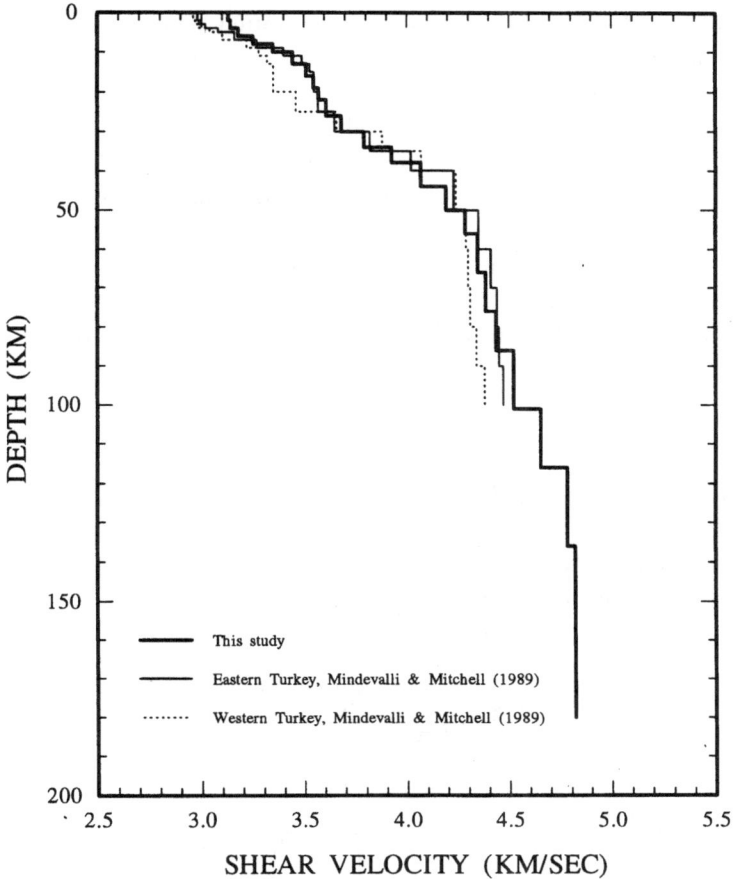

Figure 14
Comparison of the shear-velocity model for the Turkish and Iranian Plateaus with those from other studies in the same region.

and is slightly higher than that for western Turkey. This is not surprising because the path connecting stations ANTO and GNI is located within eastern Turkey. In addition to using data for the station pair ANTO-GNI, we have used data from the station pair GNI-KIV to determine an average velocity model for this region. The model therefore appears to be valid for the mountain area between the Black Sea and Caspian Sea as well as for eastern Turkey.

In the inversion process, we used an initial velocity value of 3.5 km/s for all layers. The velocity models obtained can be affected by the initial model values, especially at depths where the resolving kernels are broad. To investigate the effect of initial values on the inversion results, we performed the inversion two more times, using initial values of 4.0 and 4.5 km/s separately for all layers. The predicted group and phase velocities, and the resolving kernels are almost identical to those previously obtained. They are, therefore, not plotted here. The velocity models obtained using different initial values for the three regions are presented in Figure 15. The bold, thin, and dashed lines in each set of curves correspond to the initial values of 3.5, 4.0, and 4.5 km/s, respectively.

Figure 15 indicates that the velocities at crustal depths are not changed by the variation of the initial values. At mantle depths the changes in velocity are very limited, far less than the error estimates. None of the curves approaches the initial values at great depths. We conclude that the models for the three regions are not affected by the initial values throughout the depth range. Conservatively speaking, the velocities can be resolved to at least 120 km, a depth similar to that suggested by the resolving kernels.

Shear-wave Q Models Obtained Using the Two-station Method

Figure 16 displays the interstation attenuation coefficients (γ) for Region 3. As with velocities, we calculated the average value of γ for each of the three regions. For the periods during which only a single measurement is available, a value of

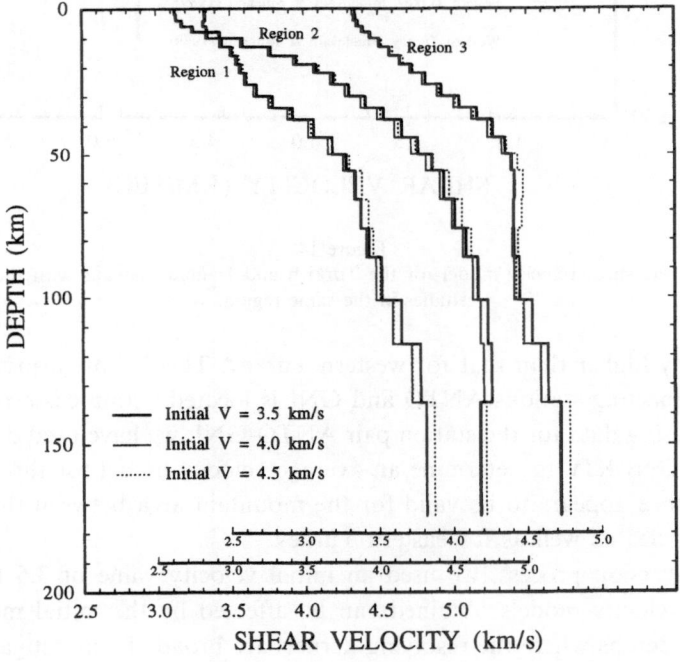

Figure 15
Velocity models for the three regions of this study obtained using different initial values. For Region 2 (middle) and 3 (right) the abscissas are displaced for clarity.

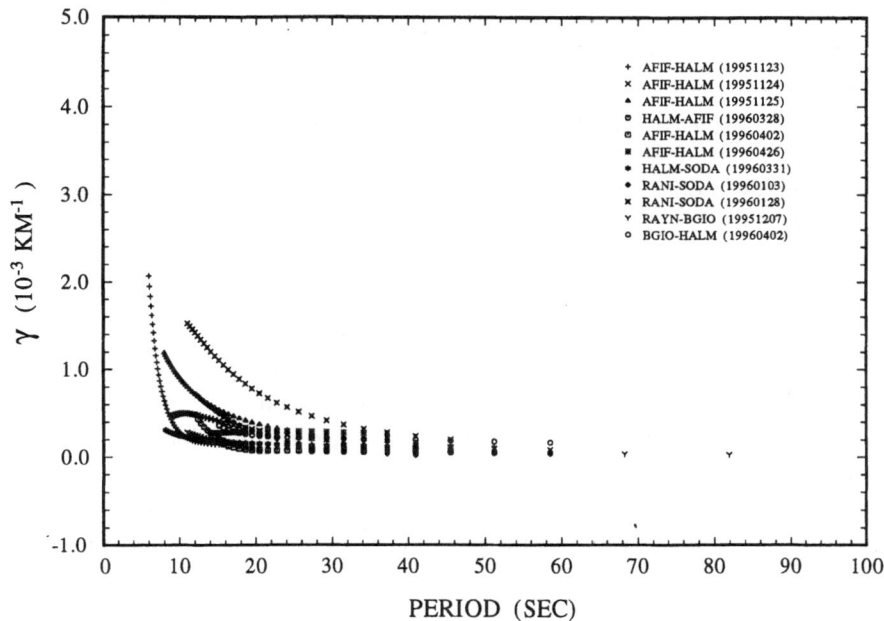

Figure 16
Rayleigh-wave attenuation determined for paths in Region 3. The caption shows symbols used for various station pairs.

0.5×10^{-3} km^{-1} was chosen as a conservative error estimate. The average attenuation data were then simplified to make a roughly even distribution over a linear period range. The differential inverse procedure was used to obtain the Q_μ models for the three regions, taking 0.0 as an initial value of Q^{-1} for all layers. Figure 17 is an example which shows the Q_μ models and the resolving kernels for Region 3. The final Q_μ model with standard deviations is plotted along with the starting model in the left panel. The right panel gives the normalized resolving kernels at six depths. The resolving kernels for Q_μ are similar for all of the three regions, and those for the other two regions are, therefore, not plotted.

Figure 17 shows that the resolving kernels for Q_μ are wider at great depths than those for the shear velocity. This indicates that the applicable depth of our Q_μ models is shallower than that of shear velocity. We restrict the Q_μ models of this study to the uppermost 80 km. The observed attenuation coefficients and theoretical dispersion curves predicted by the model for Region 3 are plotted in Figure 18. The agreement between them is good over the entire period range. Similar results were obtained for the other two regions, and are not plotted.

The Q_μ models obtained for the three regions are presented in Figure 19. The dotted, dashed, ad solid lines are the models for Regions 1, 2, and 3, respectively. The uncertainties in Q_μ values for each layer are also plotted as horizontal bars, which overlap for deeper layers. The Q_μ models exhibit different features from region to region. Q_μ in Region 3 is highest through the upper 60 km, whereas Q_μ in Region 1 is the lowest in the same depth range. For the upper 10 km, the average Q_μ values for Regions 1, 2, and 3 are 63, 71, and 201, respectively. Q_μ values for Regions 1 and 3 decrease with depth in the upper crust, whereas Q_μ for Region 2 increases. The Q_μ model for Region 1 has a minimum at a depth of 20 km. The Q_μ model for Region 3 exhibits a zone of constant Q in the depth range 20–80 km. The Q_μ model for Region 2 fluctuates, establishing a minimum and a maximum at depths of about 12 and 30 km, respectively. This feature may be specious, and is probably due to systematic errors in our one-dimensional inversion in a region where Q_μ varies rapidly in three dimensions. As with the discussion for velocities, we think that details of the Q_μ model for this region should be ignored. At depths greater than 80 km, the Q_μ values for all three regions increase gradually, approaching the initial value.

We originally used a initial value of 0.0 for Q_μ^{-1} in the inversion. We tested the effect of the initial values of Q_μ on the resulting Q_μ models by inverting using initial

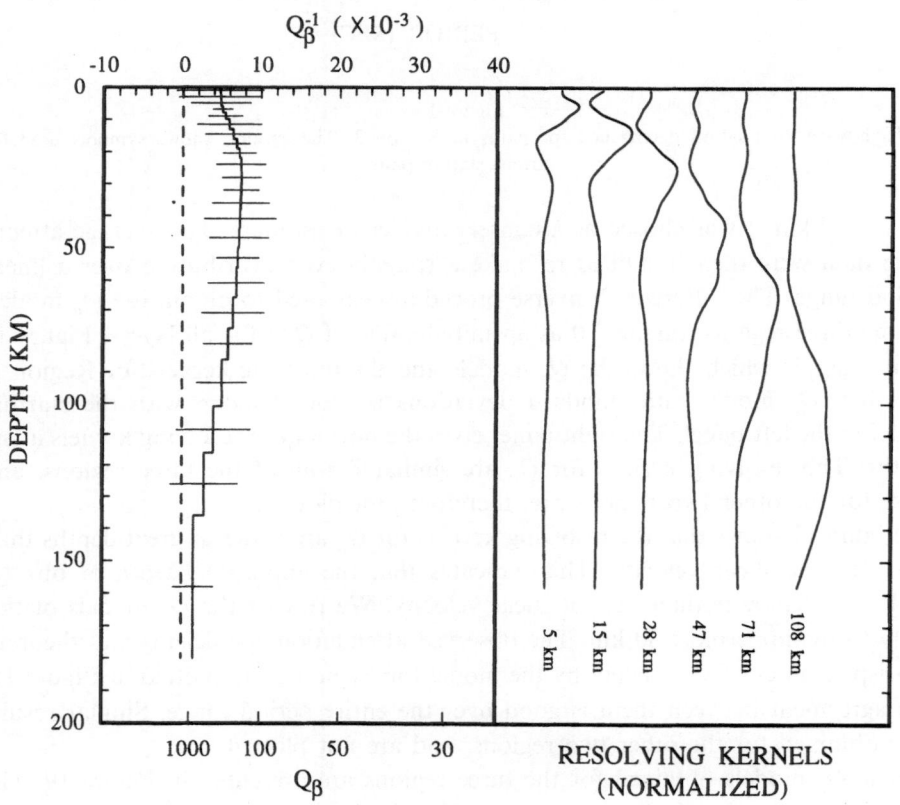

Figure 17
Shear-wave Q model and resolving kernels for Region 3. See the caption for Figure 10.

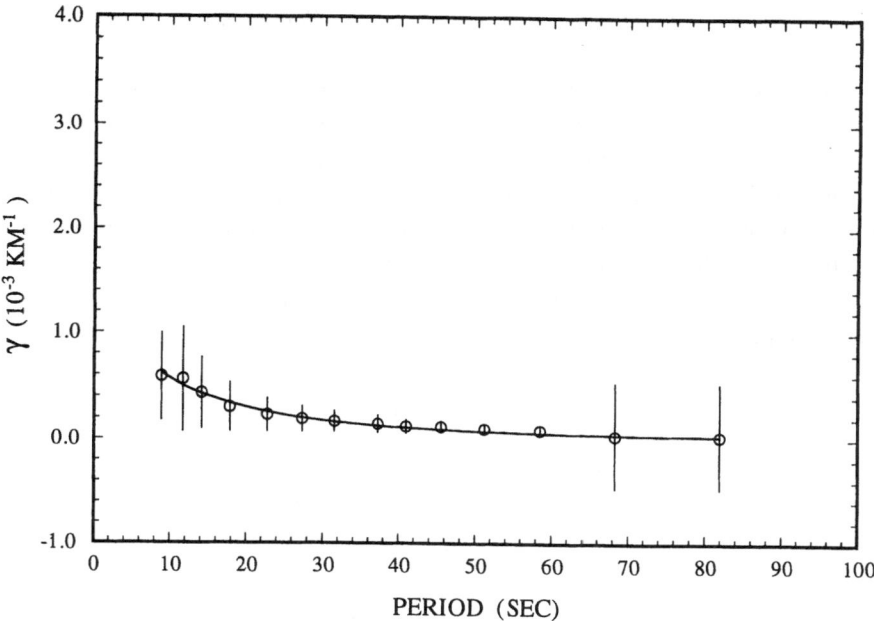

Figure 18

Observed (circles) and predicted (line) attenuation coefficients for Region 3. The vertical lines indicate one standard deviation.

values of 0.005 and 0.01 to Q_μ^{-1} (corresponding to Q_μ values of 200 and 100). The resulting models for the three regions are presented in Figure 20. The Q_μ models are more easily affected by the initial values than are the velocity models, indicating that the resolvable depth is less than that for the velocity. At crustal depths, none of the models for the three regions is greatly affected by the variation of initial values, indicating that resolution is good at those depths. The depths at which the three curves corresponding to the three initial values begin to deviate are about 80, 120, and 90, for Regions 1, 2, and 3, respectively. At greater depths the Q_μ values approach the initial values, indicating poor resolution at those depths. To be conservative, we choose 80 km as the applicable depth of the Q_μ models for all three regions. All of our conclusions concerning the Q_μ models will be restricted to this depth. This is also in agreement with the result obtained by an analysis of the resolving kernels. Our results indicate that the resolvable depth for Q_μ, when inverting Rayleigh wave data, is less than that for shear velocity.

Shear-wave Q Models Obtained Using the Single-station Method

Determination of surface wave attenuation using the two-station method requires that the azimuths between earthquakes and stations are similar to the azimuths of the line connecting the two stations. This greatly reduces the number of usable earthquakes. In addition, if the two stations are located in different geological provinces, only the average attenuation of the different provinces between the two stations can be obtained. For these reasons, a single-station method was developed by CHENG and MITCHELL (1981) for the purpose of improved regional attenuation studies. The importance of the method lies in its ability to

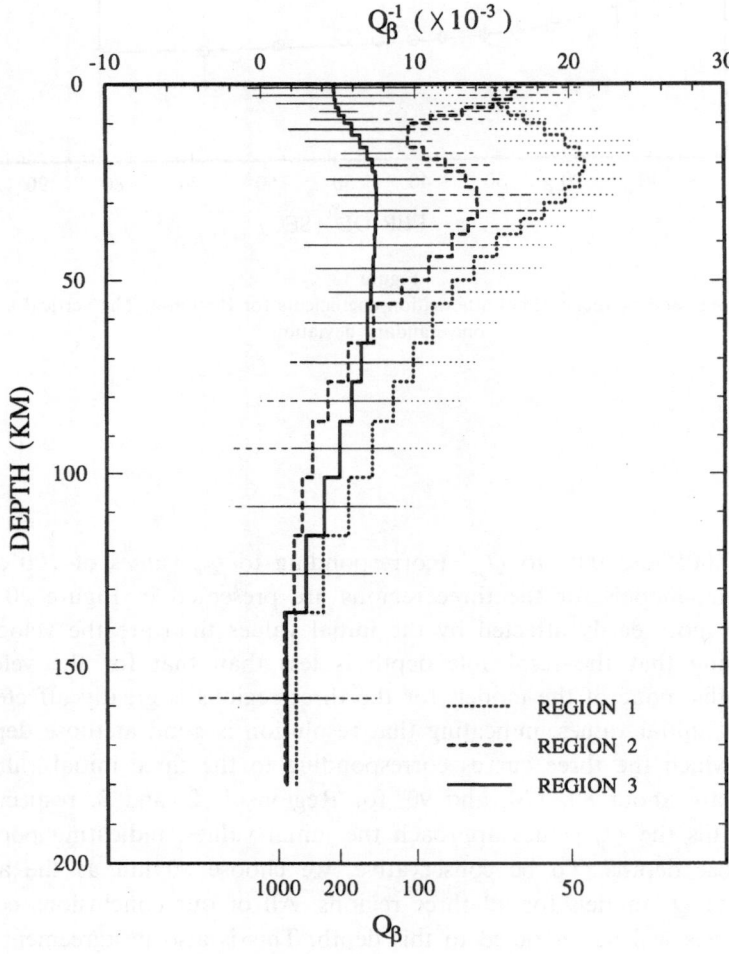

Figure 19
Shear-wave *Q* models for the three regions. The horizontal lines indicate the error estimates for the models.

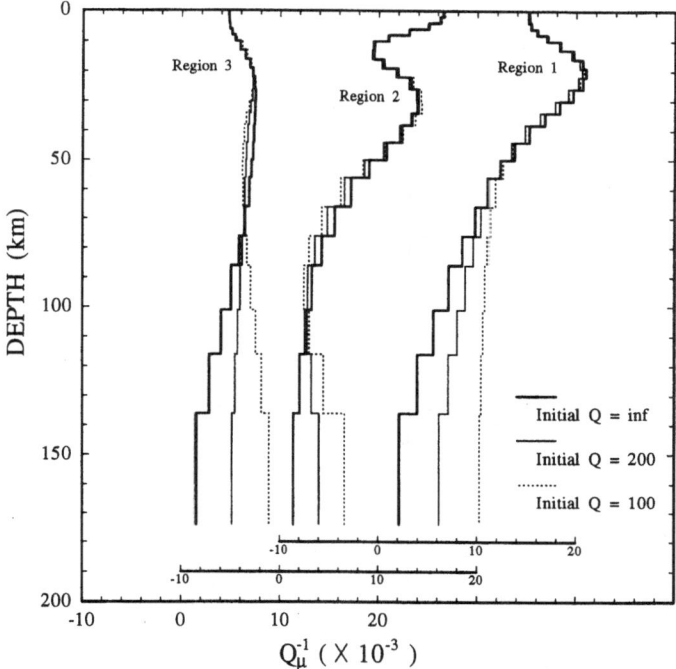

Figure 20

Shear-wave Q models of the three regions of this study obtained using different initial values. See the caption for Figure 15.

utilize relatively short paths between a single source and a single receiver, both of which are preferably located within a single geological province.

The method attempts to match theoretical and observed surface-wave spectra. This is done by a trial-and-error procedure in this study. The method utilizes both fundamental-mode and higher-mode spectra since use of the amplitude ratios for the two spectra obviates the need to know the seismic moment of the earthquake used. To extract the observed spectrum, the multiple-filter technique was applied to the data as described in the previous section. Computations of theoretical surface-wave spectra requires knowledge of the source mechanism of the earthquake. For all earthquakes we used the focal mechanism determined by Harvard University.

Figure 21 is an example which shows the process of Q_μ determination for each layer. The earthquake in this example occurred on March 29, 1994 (event 3 in Table 3) as recorded at two stations KIV and ABKT. Station names are shown to the left of the corresponding spectra. In the tests, three layers were assumed with layer thicknesses of 10, 30, and 40 km, overlying a halfspace. The Q_μ values in the layers are denoted by $Q1$, $Q2$, $Q3$ and $Q4$. In the first test (Fig. 21a), $Q1$ varies between 30 and 130, with a step of 20. $Q2$, $Q3$, and $Q4$ are fixed at 50, 90, and 700, respectively. These values were inferred from the results obtained using the two-sta-

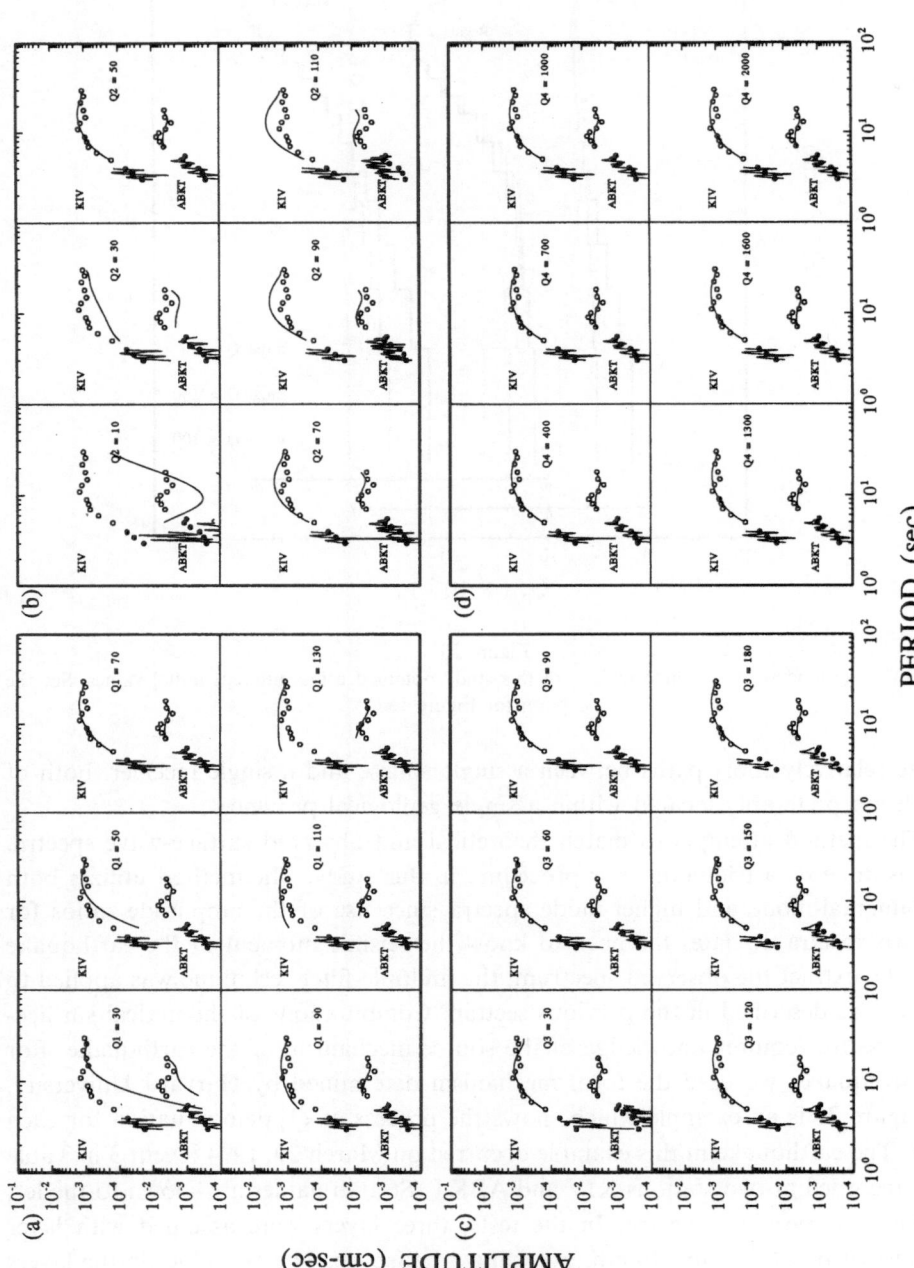

Figure 21

An example showing the determination of shear-wave Q in each layer. (a) Q_μ varies in layer 1. (b) Using the best value (70) for layer 1, Q_μ is varied in layer 2. (c) Using the best values in layer 1 (70) and layer 2 (50), Q_μ is varied in layer 3. (d) Using the best values in layer 1 (70), layer 2 (50), and layer 3 (90), Q_μ is varied in layer 4.

tion method. The Q_μ value in the halfspace, $Q4$, is somewhat arbitrary since we already know that Q_μ cannot be resolved at those depths. As can be seen by comparing the spectra for both stations, the only acceptable solution for $Q1$, among the values shown, is 70. When $Q1$ is 90, the fit for the higher modes is good, but at shorter periods theoretical values of the fundamental mode for station KIV are two high.

Figure 21b displays comparisons of the spectra when $Q2$ varies. In this test $Q1$ takes the value of 70 as obtained in the previous test. $Q3$ and $Q4$ were fixed at 90 and 700, respectively. $Q2$ varies between 10 and 110 with a step of 20. Both fundamental- and higher-mode spectra are sensitive to the variation of $Q2$. When $Q2$ is less than 50 the theoretical curves are lower than observed values for both fundamental and higher modes at the two stations. When $Q2$ is greater than 50, the opposite occurs. The only good match occurs when $Q2$ is 50.

Figure 21c shows the comparison when $Q3$ varies. In this case $Q1$ and $Q2$ are fixed at 70 and 50, respectively, as determined before, and $Q4$ is fixed at 700. $Q3$ varies between 30 and 180 with a step of 30. The theoretical curves for the fundamental mode are far less sensitive to the variation of $Q3$ than are the higher modes. A good match can be detected when $Q3$ takes the values between 60 and 120. A middle value of 90 is therefore chosen as the solution for $Q3$.

In Figure 21d we assign 70, 50 and 90 to $Q1$, $Q2$, and $Q3$, allowing $Q4$ to vary from 400 to 2000. No spectral changes can be detected when $Q4$ varies. This indicates that the spectra are not sensitive to changes in Q_μ at great depths. No sensitivity to Q_μ at depths greater than 80 km was found for any case tested. Therefore, the Q_μ models obtained using the single-station method are restricted to the upper 80 km in this study, the same depth range determined in data processing using the two-station method.

The Q_μ models for the three regions obtained using the single-station method are listed in Table 4, and superposed on those obtained using the two-station method (bold lines) as shown in Figure 22. Comparisons of theoretical spectra predicted by the corresponding models with the observed data for all the earthquakes studied in Regions 1, 2, and 3 are presented in Figures 23, 24, and 25, respectively. The events, the recording stations, and models obtained are indicated in the lower right corners of the panels.

For events 1 and 3 in Region 1 (Figure 23), two seismograms are available. In general, the match is good for both fundamental-mode and higher-mode spectra for all of the events used. For this region, four models were obtained, indicating that the Q_μ models are path dependent. The four models differ only in the crust. Q_μ varies between 50 and 70 for the upper 10 km, and between 50 and 60 for the underlying layer (10–40 km). Although the four models differ, they all lie close to the model obtained by the two-station method (bold line). This indicates that the model obtained using the two-station method for this region is applicable and can be thought of as an average model for the region.

Table 4

Q_μ models obtained using the single-station method

Layer number	Depth (km)	Q_μ (model 1)	Q_μ (model 2)	Q_μ (model 3)	Q_μ (model 4)
		Region 1			
1	0–10	60	70	60	50
2	10–40	50	50	60	50
3	40–80	90	90	90	90
4	>80	700	700	700	700
		Region 2			
1	0–10	60			
2	10–20	100			
3	20–40	70			
4	40–80	120			
5	>80	700			
		Region 3			
1	0–20	180	140	40 (for 0–2 km) 180 (for 2 – 20 km)	
2	20–80	140	140	140	
3	>80	700	700	700	

One of the advantages of using the single-station method is that the applicable range for the models obtained using the two-station method can be broadened. We previously used the station pairs ANTO-GNI and GNI-KIV in Region 1. The results for the velocity and Q_μ are only valid for the Turkish Plateau or, strictly speaking, the region along the interstation paths. Figure 3 shows that some paths lie in the Iranian Plateau and some paths go through both the Turkish and Iranian Plateaus. This leads us to conclude that the velocity and Q_μ models previously obtained for the Turkish Plateau also apply to the Iranian Plateau.

Comparisons of the theoretical and observed spectra for Region 2 are plotted in Figure 24. Five events recorded at stations GNI and KIV were used. Three paths cross the northern portion of the Black Sea, whereas 4 paths cross the Caspian Sea. Events 9 and 10 are recorded at both stations. All paths include both oceanic and continental segments.

The method used for Region 1 is not applicable to Region 2 since no higher-mode spectra were observable for any of the 5 events. In this case, we first simplified the previously obtained model for this region, and computed only the fundamental-mode spectrum. Strictly speaking, the model listed in Table 4 for this region, therefore, is not the one obtained using the single-station method.

Higher-mode energy is probably absent because it cannot propagate through the oceanic crust beneath the Caspian and Black Seas. The model obtained for this region is less unique than models obtained using multimode spectra. Fluctuations in the model may not represent a real feature of the crust, but may be due to systematic errors in the attenuation data caused by the lateral variations of Q_μ and velocity in this region.

Five events recorded at 5 stations were used for Region 3 (Figure 25). Events 14 and 15 are located at the southern end of the Dead Sea fault. Paths connect these two events and stations cross the central and western portions of the peninsula. Event 17 lies in the Gulf of Aden and the path to the station traverses a small portion of the Gulf. Most of the paths are in the central and southern parts of the peninsula. Event 16 is in the Zagros folded zone. The path from that event crosses the shield, the shelf, and the basins.

Panels 1 and 2 are for event 14 recorded at stations RANI and SODA, respectively. For this event, model 1 (180 for the upper 20 km and 140 for depth range 20–80 km) was obtained for both of the two paths. Panels 3 and 4 are for event 15, which is located very close to event 14. For this event, the records of two stations (AFIF and HALM) are available for Q_μ modeling. Since the paths deviate only slightly from that connecting event 14 and RANI, we tried the same model as for event 14. As shown, the theoretical spectrum for the fundamental mode is in

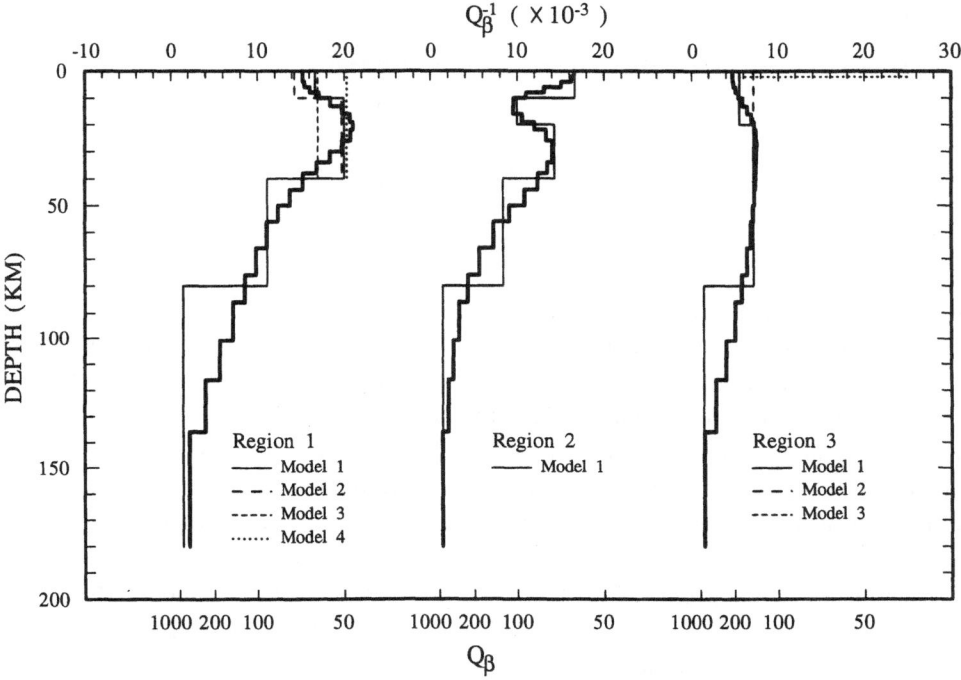

Figure 22

Shear-wave Q models for the three regions obtained using the single-station method (light lines) compared to models obtained usint the two-station method (heavy lines). The four models for Region 1 are for paths indicated in Figure 23, and the three models for Region 3 are for the paths indicated in Figure 25.

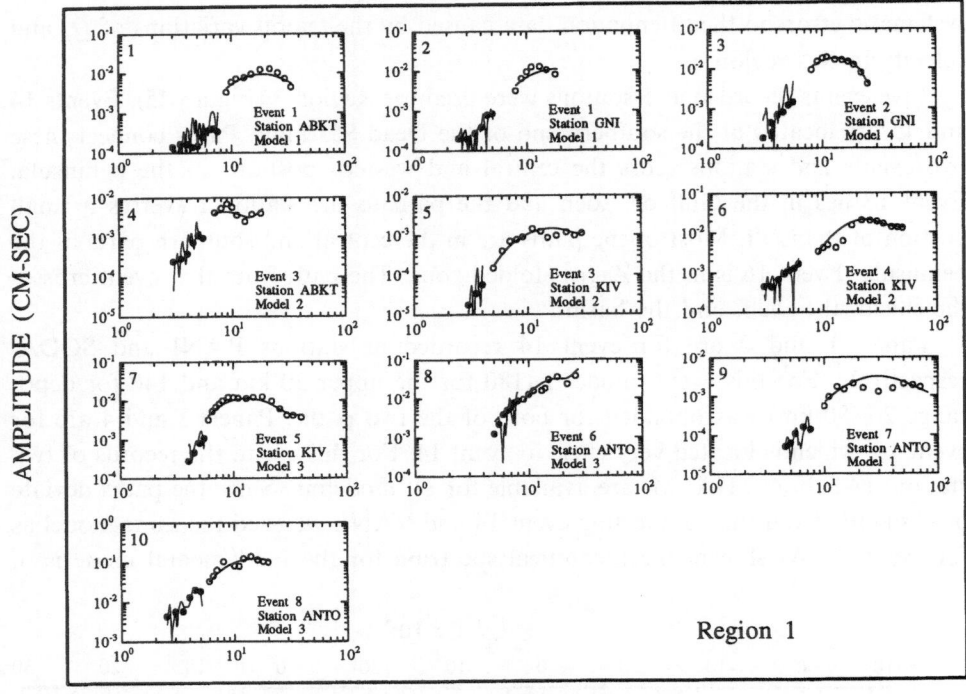

AMPLITUDE (CM-SEC)

PERIOD (SEC)

Figure 23

Comparison of theoretical multi-mode spectra for models of Region 1 with observed spectra. Funda-mental-mode observations are shown with open symbols and higher-mode observations are denoted by closed symbols.

good agreement with the observed data at each station. For higher modes, however, the observed spectra are higher than expected by nearly an order of magnitude. Even using an extremely high Q_μ value of 1000 for all of the layers, the synthetics for the higher modes are still lower than the observed spectra. This suggests that the large amplitudes of Lg in these two records cannot be explained by high Q_μ in the crust. Focusing or site effects that affect high-frequency waves emanating from northwest of the peninsula might be responsible for the unusually high values of the spectra.

The paths from event 16 to the recording stations SODA and RANI cross basins where a thick cover of low-Q sediment exists. Q_μ for those paths (model 2 in Table 4) is 140 for the upper 20 km of the crust. This value is 22% lower than Q_μ for model 1. Panels 7 and 8 in Figure 25 show an optional model (model 3) for the same paths as panels 5 and 6. In this model, Q_μ in the uppermost 2 km is 40 and in the next 18 km is 180 as in model 1. A model with a 2-km-thick layer with a Q_μ of 40 can explain the observed spectra in the eastern portion of the Arabian

Peninsula. Alternative models can also explain the spectra. One possibility is a 3-km-thick layer in which Q_μ is 50 and another is a 4-km-thick layer in which Q_μ is 60. The predicted spectra are nearly identical to those in panels 7 and 8, and are therefore not plotted.

Panels 9 through 12 are for event 17 and the recording stations AFIF, RANI, HALM, and SODA, respectively. For those paths, model 2 provides the best fit to the observed spectra. Since about 20–25% of those paths cross water, we might expect that the best model for the southern portion of the peninsula would be model 1 rather than model 2. The observed spectra, however, exhibit small spectral holes at about 12 sec that are not predicted by the given focal mechanism. This may mean that the reported focal mechanism is incorrect and our Q_μ values at shallow depths may be biased too low for this reason.

In general, for the western, central, and southern portions of the Arabian Peninsula, the crust and uppermost mantle can be modeled as a 20-km-thick layer with Q_μ of 180 and a 60-km-thick layer with Q_μ of 140. For the eastern portion of the peninsula, the low Q sediments must be taken into account. Q_μ for the sediments can be as low as 40–60.

For Regions 1 and 3, the Q_μ models obtained using the two- and single-station methods show good agreement. For Region 2, however, Q_μ models are not verified by the single-station method. Since the higher-mode energy was not detected in the observation, we could not vary the Q_μ value of any layer, but used the simplified

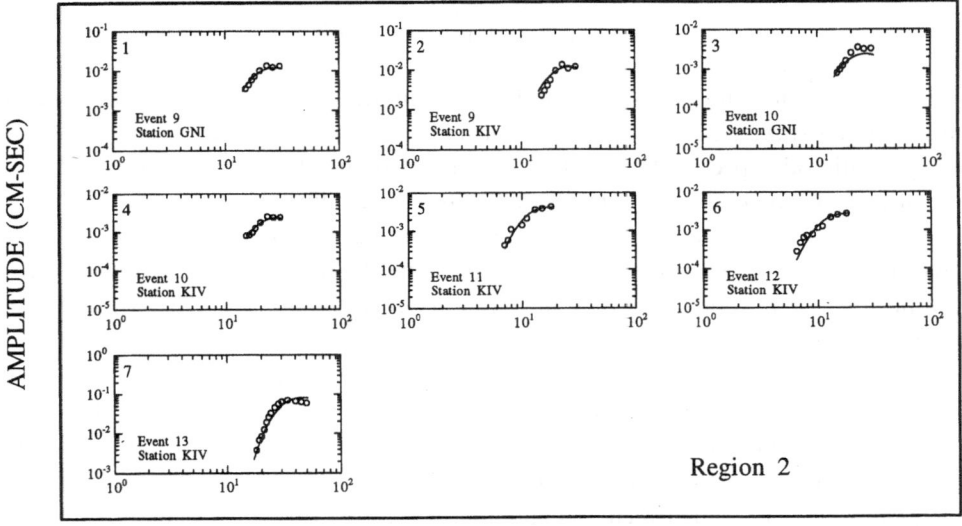

PERIOD (SEC)

Figure 24
Comparison of theoretical fundamental-mode spectra for models of Region 2 with observed spectra.
Higher modes are not observed in this region.

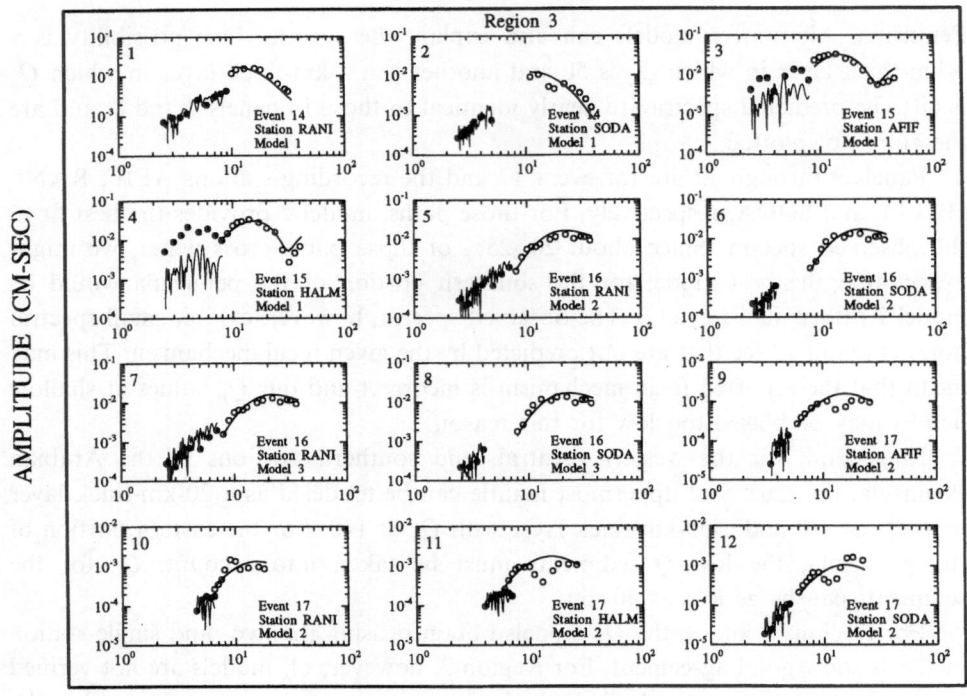

Figure 25
Same as Figure 23, but for Region 3.

model to compute the theoretical fundamental-mode spectrum and compared it to the observed spectrum. The results are probably less unique than models obtained using multimode spectra. For both methods the paths cross both continental and oceanic areas which are characterized by different seismic velocity and Q_μ distributions. We attempted to model this laterally complex region using one-dimension inversion. We think that the fluctuations in the model may be caused by the lateral complexities of Q_μ and velocity, and may not represent the real crustal properties of this region.

Our results indicate that Q_μ is higher in the Arabian Peninsula than in the Turkish and Iranian Plateaus. This is consistent with the results obtained from Lg coda Q measurements (CONG and MITCHELL, this volume). In that study, Lg coda Q is low throughout the plateaus, ranging from 150–300. It is higher for much of the peninsula, ranging between 350 and 450. Results of this study demonstrate that Q_μ is higher in the peninsula than in the plateaus. The average Q_μ values for the uppermost 40 km are 55 and 150 for Regions 1 and 3, respectively. In the Lg coda Q study, Q at 1 Hz varies laterally in the Arabian Peninsula with higher values (about 450) in the shield area, and lower values (about 350) in the platform and basins. In this study we obtained a Q_μ model of 180 for the uppermost 20 km and 140 for the layer 20–80 km form the northern, western, and southern portions of the Arabian Peninsula. If the paths cross the shield, the shelf, and basins, we must include a low-Q layer in the model to explain the data. This is also in agreement with the results from the Lg coda Q study.

Discussion

Previous studies indicate that Q_μ varies regionally, being typically high in stable regions and low in tectonically active regions (MITCHELL, 1995). The northern portion of the Middle East is characterized by severe deformation and young tectonic features. Continental collision is occurring along the Zagros suture zone, resulting in the Zagros folded belt. The left-lateral East Anatolian fault and right-lateral North Anatolian fault became active 5 Ma. As proposed in recent research (MITCHELL, 1995; MITCHELL et al., 1997) Q_μ, and Q of any wave propagating through the crust in any region, is largely governed by the type, severity, and temporal occurrence of tectonic activity there. According to that proposal, enhanced temperature associated with tectonic activity has released water of hydration that resides in cracks and permeable rock of the crust, causing low Q there.

Tectonically active regions are usually characterized by high heat flow. ILKISIK (1995) recently found that the average heat flow value in western Anatolia is about 107 mW m^{-2}, which is about 60% above the world average value. A value as high as 247 mW m^{-2} was even measured at a location of 38°57′N and 29°13.2′E. WYLLIE (1988) described ways in which hydrothermal fluids could be generated by elevated temperatures through broad regions of the mantle. In regions of plate convergence, such as Turkey and Iran, those fluids may originate in a portion of the Arabian plate subducting beneath the Eurasian plate. Upper mantle fluids, once released, propagate slowly upward and eventually reside, predominantly, in the crust. The energy used to move fluids through that permeable crust is lost from waves propagating through it, resulting in higher wave attenuation.

Although the Q_μ values in the Arabian Shield are higher than those in the northern portion of the Middle East, they are lower than those found for other stable regions of the world. Throughout the entire Middle East the lower crust exhibits Q_μ values no greater than about 150. This contrasts sharply with results from other continental regions where Q_μ increases rapidly with depth at mid-crustal depths, usually having values as high as several hundred to a thousand. The Arabian Peninsula has been the site of large-scale uplift and extension with associated volcanic activity. The Afro-Arabian dome began forming in the early Cretaceous and is the result of two or more still active generating systems (ALMOND, 1986). Temperatures beneath the dome are as high as 900°C at a depth of 40 km and these abnormally high temperatures extend as far as 200 km inland from the Red Sea (MCGUIRE and BOHANNON, 1989). The low crustal Q in the Arabian Peninsula may therefore be due to water released from hydrothermal reactions at those elevated temperatures produced by an upper mantle heat source.

Conclusions

Shear-wave velocity models were obtained using a two-station method for three regions of the Middle East: the Turkish and Iranian Plateaus (Region 1), a region surrounding and including the Black and Caspian Seas (Region 2), and the Arabian Peninsula (Region 3). The shear-velocity models of the three regions differ mostly in the uppermost 20 km, velocity being fastest in Region 3, and slowest in Region 2. The velocity models for Regions 1 and 3 agree well with those found in other studies in the Middle East. Analysis of the resolving kernels and the effect of initial values on inversion indicate that our velocity models are valid for at least the upper 120 km of the crust and upper mantle.

Q_μ models obtained using the two-station method show that Q_μ varies both with depth and regionally. Q_μ is higher in the relatively stable Arabian Peninsula than in the tectonically active plateau regions in the north by a factor of 2–3. Q_μ in the Turkish and Iranian Plateaus is 57 in the upper crust, 54 in the lower crust, and 89 in the upper mantle, whereas Q_μ values in the Arabian Peninsula at the same depths are, respectively, 178, 135, and 148. Analysis of the resolving kernels and the effect of initial values on inversion indicate that our Q_μ models are only valid for the uppermost 80 km of the crust and upper mantle.

Single-station studies have allowed us to map regional variations of anelastic properties in greater detail than is possible by two-station studies. For Regions 1 and 3, they support our models obtained using the two-station method. They also show that regional variations in Q_μ occur within both Region 1 and Region 3. The differences occur in the uppermost crust where they vary between 50 and 70 in the upper 10 km, between 50 and 60 in the depth range 10–40 km for the Turkish and Iranian Plateaus, and between 140 and 180 in the upper 20 km of the Arabian Peninsula. These results support the idea (MITCHELL, 1995) that Q_μ in the crust of any region is directly proportional to the length of time that has elapsed since the most recent major tectonic activity there. Our results indicate that zones of plate convergence and regional upper mantle heating are characterized by low Q_μ extending from the surface to upper mantle depths. This contrasts with results in stable regions and regions of extension where Q_μ increases rapidly at mid-crustal depths.

Average Q_μ values for both the upper and lower crust in the three regions are lower than those obtained in other continental regions of the world. Since the northern portion of the Middle East is a region of plate convergence and high upper mantle temperature and uplift occur in Saudi Arabia, fluid-filled cracks may reside at lower, as well as upper crustal depths in both regions. Those fluids were probably released by metamorphic processes that occurred there. With time, we expect that those fluids will be forced to shallower depths and Q_μ at lower crustal depths will become larger.

Acknowledgments

We thank the IRIS Data Management Center and the German Geophone Data Center for providing the data used in this study. This research was supported by the U.S. Department of Energy and was monitored by the Phillips Laboratory under contract F19628-95-K-0004.

REFERENCES

ALMOND, D. C. (1986), *Geological Evolution of the Afro-Arabian Dome*, Tectonophysics *131*, 301–332.

BERBERIAN, M. (1976), *Contribution to the Seismotectonics of Iran, II*, Rep. Feol. Surv. Iran *39*, 516 pp.

BROWN, G. F. (1972), *Tectonic Map of the Arabian Peninsula, Saudi Arabian Peninsula Map AP-2*, Saudi Arabian Dir. Gen. Miner. Resour.

CANITEZ, N., and UCER S. B. (1967), *Computer Determinations for the Fault-plane Solutions in and Near Anatolia*, Tectonophysics *4*, 235–244.

CHENG, C. C., and MITCHELL, B. J. (1981), *Crustal Q Structure in the United States from Multi-mode Surface Waves*, Bull. Seismol. Soc. Am. *71*, 161–181.

COLEMAN, R. G., *Ophiolites. Ancient Oceanic Lithosphere?* (Springer-Verlag, Berlin 1977) 229 pp.

CONG, L., and MITCHELL, B. J., *Lg Coda Q and it Relation to the Geology and Tectonics of the Middle East*, Pure appl. geophys., *153*, 563–585.

GHALIB, H., *Seismic Velocity Structure and Attenuation of the Arabian Plate*, Ph.D. Dissertation, St. Louis University, pp. 314, 1992.

HERRIN, E. H., and GOFORTH, T. (1977), *Phase-matched Filters: Application to the Study of Rayleigh Waves*, Bull. Seismol. Soc. Am. *67*, 1259–1275.

HERRMANN, R. B., *Computer Programs in Seismology*, User's manual, vol. IV (St. Louis University, Misssouri 1987).

ILKISIK, O. M. (1995), *Regional Heat Flow in Western Anatolia Using Silica Temperature Estimates from Thermal Springs*, Tectonophysics *244*, 175–184.

JIH, R. S., and LYNNES, C. S., *Re-examination of regional Lg Q variation in Iranian Plateau*. In *Proc. 14th Ann. PL/DARPA Seismic Research Symposium*, 16–17 September 1992 (eds. Lewkowicz, J. F. and McPhetres, M.) (Phillips Laboratory 1992) pp. 200–206.

KADINSKY-CADE, K., BARAZANGI, M., OLIVER, J., and ISACKS, B. (1981), *Lateral Variations of High-frequency Seismic Wave Propagation at Regional Distances across the Turkish and Iranian Plateaus*, J. Geophys. Res. *86*, 9377–9396.

KETIN, I. (1966), *Tectonic Units of Anatolia*, Bull. Miner. Res. Explo. Inst. Turk. *66*, 23–34.

KNOPOFF, L., and FOUDA, A. A. (1975), *Upper Mantle Structure under the Arabian Peninsula*, Tectonophysics *26*, 121–134.

McGUIRE, A. V., and BOHANNON, R. G. (1989), *Timing of Mantle Upwelling: Evidence for a Passive Origin for the Red Sea Rift*, J. Geophys. Res. *94*, 1677–1682.

McKENZIE, D. P. (1972), *Active Tectonics of the Mediterranean Regions*, Geophys. J. R. Astr. Soc. *30*, 109–185.

MINDEVALLI, O., and MITCHELL, B. J. (1989), *Crustal Structure and Possible Anisotropy in Turkey from Seismic Surface Wave Dispersion*, Geophys. J. *97*, 93–106.

MITCHELL, B. J. (1995), *Anelastic Structure and Evolution of the Continental Crust and Upper Mantle from Seismic Surface Wave Attenuation*, Rev. Geophys. *33*, 441–462.

MITCHELL, B. J., PAN, Y., XIE, J., and CONG, L. (1997), *Lg Coda Q Variation and the Crustal Evolution of Eurasia*, J. Geophys. Res. *102*, 22767–22779.

MOKHTAR, T. A., and AL-SAEED, M. M. (1994), *Shear Wave Velocity Structures of the Arabian Peninsula*, Tectonophysics *230*, 105–125.

NIAZI, M. (1968), *Crustal Thickness in the Central Saudi Arabian Peninsula*, Geophys. J. R. Astr. Soc. *15*, 545–547.

NOWROOZI, A. A. (1972), *Focal Mechanism of Earthquakes in Persia, Turkey, West Pakistan, and Afghanistan and Plate Tectonics of the Middle East*, Bull. Seismol. Soc. Am. *62*, 823–850.

NUTTLI, O. W. (1980), *The Excitation and Attenuation of Seismic Crustal Phases in Iran*, Bull. Seismol. Soc. Am. *70*, 469–485.

POWERS, R. W., RAMIREZ, L. F., REDMOND, C. P., and ELBERG, E. L. (1966), *Geology of the Arabian Peninsula—Sedimentary Geology of Saudi Arabia*, U.S. Geol. Surv., Prof. Pap. 560-D, 147 pp.

SEBER, D., and MITCHELL, B. J. (1992), *Attenuation of Surface Waves across the Arabian Peninsula*, Tectonophysics *204*, 137–150.

SEBER, D., VALLVÉ, M., SANDVOL, E., STEER, D., and BARAZANGI, M. (1997), *Middle East Tectonics: Applications of Geographic Information Systems (GIS)*, GSA Today 7, 1–6.

SENGÖR, A. M. C., and KIDD, W. S. F. (1979), *Post-collisional Tectonics of the Turkish and Iranian Plateaus and a Comparison with Tibet*, Tectonophysics *55*, 361–376.

STOCKLIN, J., *Possible ancient continental margins in Iran*. In *Geology of Continental Margins* (eds. C. Burk and C. Drake), (Springer-Verlag, New York 1974) pp. 873–877.

STONELEY, R. (1981), *The Geology of the Kuh-e Dalneshin Area of Southern Iran, and its Bearing on the Evolution of Southern Tethys*, J. Geol. Soc., London *138*, 509–526.

WYLLIE, P. J. (1988), *Magma Genesis, Plate Tectonics, and Chemical Differentiation of the Earth*, Rev. Geophys. *26*, 370–404.

(Received December 10, 1997, revised April 2, 1998, accepted April 23, 1998)

 To access this journal online:
http://www.birkhauser.ch

Pure appl. geophys. 153 (1998) 539–561
0033–4553/98/040539–23 $ 1.50 + 0.20/0

❚ Pure and Applied Geophysics

Seismic Attenuation Computed from GLIMPCE Reflection Data and Comparison with Refraction Results

MICHAEL P. MATHENEY[1] and ROBERT L. NOWACK[2]

Abstract—Instantaneous frequency matching has been used to compute differential t^* values for seismic reflection data from the Great Lakes International Multidisciplinary Program on Crustal Evolution (GLIMPCE) experiment. The differential attenuation values were converted to apparent Q^{-1} models by a fitting procedure that simultaneously solves for the interval Q^{-1} values using non-negative least squares. The bootstrap method was then used to estimate the variance in the interval Q^{-1} models. The shallow Q^{-1} structure obtained from the seismic reflection data corresponds closely with an attenuation model derived using instantaneous frequency matching on seismic refraction data along the same transect. This suggests that the effects of wave propagation and scattering on the apparent attenuation are similar for the two data sets. The Q^{-1} model from the reflection data was then compared with the structural interpretation of the reflectivity data. The highest interval Q^{-1} values (>0.01) were found near the surface, corresponding to the sedimentary rock sequence of the upper Keweenawan. Low Q^{-1} values (<0.0006) are found beneath the Midcontinent rift's central basin. In addition to structural interpretation, seismic attenuation models derived in this way can be used to correct reflection data for dispersion, frequency and amplitude effects, and allow for improved imaging of the subsurface.

Key words: Seismic attenuation, seismic Q, Midcontinent rift.

Introduction

Seismic attenuation causes a preferential loss of the higher frequencies, as well as phase distortions, in a seismic signal with increasing propagation distance. These effects can cause discrepancies between seismic reflection data and well-log data. Several techniques have recently been developed to incorporate attenuation in the processing of seismic reflection data. For example, inverse-Q filtering (HARGREAVES and CALVERT, 1991; BICKEL, 1993) and dispersion corrections (DUREN and TRANTHAM, 1997) use an attenuation model with depth to correct pre-stack seismic data for frequency loss and phase distortion. Q-deconvolution (BICKEL and NATARAJAN, 1985) has been shown to effectively remove amplitude, frequency, and

[1] Exxon Exploration Company, Houston, TX, 77253. E-mail: matheney@geo.eas.purdue.edu
[2] Purdue University, Dept. of Earth and Atmospheric Science, West Lafayette, IN 47907. E-mail: nowack@geo.eas.purdue.edu

dispersion effects resulting from wave propagation in an attenuating medium. Amplitude statics can be applied to correct for amplitude differences in the reflection data that are due to near-surface, localized low Q areas (BRZOSTOWSKI and MCMECHAN, 1992). In addition, migration algorithms have incorporated attenuation information to provide improved images of the subsurface (SOLLIE and MITTET, 1994).

While there has been increased interest in using attenuation information for the processing of reflection data, most studies have assumed that an attenuation model is known, and relatively few studies have concentrated on the determination of Q models. In this study, an instantaneous frequency matching procedure similar to that developed by MATHENEY and Nowack (1995) is used to determine the differential attenuation between a reference arrival and other arrivals on a seismic reflection trace. The differential attenuation values are then used to obtain the interval Q structure.

The instantaneous frequency matching procedure is first tested using synthetic reflection data. This allows for an analysis of the effects of noise and interference on the derived attenuation values. The method is then applied to seismic reflection data of the 1986 GLIMPCE Lake Superior seismic experiment and used to compute interval Q^{-1} values. The resulting Q^{-1} model is then compared to an independently derived attenuation model obtained from seismic refraction data by MATHENEY et al. (1997) along the same transect. Finally, the derived attenuation structure is compared with the interpreted reflectivity structure from CANNON et al. (1989).

Method of Analysis

Instantaneous Frequency Matching

In order to apply the instantaneous frequency matching procedure of MA-THENEY and NOWACK (1995) to seismic reflection data, a reference pulse is first selected and windowed from a reflection trace. The instantaneous frequency of the reference pulse is then matched with the instantaneous frequency of each reflection on the seismic trace. This is performed by attenuating the reference pulse using a nearly-constant Q attenuator, and comparing the instananeous frequency of the attenuated reference pulse to an observed seismic pulse. The advantage of using a nearly constant Q operator, as opposed to an exactly constant Q operator (KJAR-TANSSON, 1979), is that for the nearly constant Q operator a t^* parameter can be expressed as an integration of the attenuation along the ray. The transfer function for the nearly constant Q operator can then be written as

$$A(\omega) = e^{i\omega[T - (t^*/\pi)\ln(\omega/\omega_r)]} e^{-(\omega/2)t^*}, \tag{1}$$

where $T = \int 1/c(s)\,ds$ is the travel time, $t^* = \int Q^{-1}(s)/c(s)\,ds$ is the integrated attenuation along the ray, $c(s)$ is the velocity at the reference frequency ω_r, $Q^{-1}(s)$ is the seismic quality factor, and s is the path length along the ray (AKI and RICHARDS, 1980). The amount of attenuation for this operator is controlled by adjusting the t^* value. By matching the instantaneous frequency of the attenuated reference pulse at its peak envelope amplitude to the instantaneous frequency of the observed pulse, the differential attenuation between the two arrivals can be obtained.

The instantaneous frequency matching procedure assumes that the locations of the reflections on a seismic trace are known. Due to the large number of traces in a typical seismic reflection survey and the numerous reflections on each trace, it is impractical to manually pick each reflection. Since the instantaneous frequency matching procedure uses the frequency values at peaks of the envelope amplitude, it is sufficient to consider the peaks of the envelope on a seismic trace as possible locations of seismic arrivals. By applying several checks to the current time position along the trace, it can be determined if a given peak of the envelope is a well isolated prominent reflection.

In order to identify the largest, envelope amplitude arrival over a small window on the seismic trace, the current peak envelope location is checked against adjacent peaks of the envelope. The decay of the envelope before and after the current peak envelope location is then computed. If the envelope does not decrease by at least 30% of the maximum on each side of the peak, then it is assumed that the current peak location is contaminated by interfering arrivals and is not used. The interference between arrivals can cause errors in the determination of differential attenuation along a seismic trace (MATHENEY and NOWACK, 1995). A final check to determine if a specific envelope peak is an isolated reflection peak is performed by windowing out the current arrival using a cosine-bell taper and computing its amplitude spectrum. The amplitude spectrum is then compared against the amplitude spectrum from a noise sample obtained prior to the first arrival on the trace. If the amplitude spectrum of the current arrival is above the noise level over a bandwidth of at least 15 Hz, then it is considered as a possible reflection peak. Otherwise, the envelope peak is excluded. The minimum bandwidth of 15 Hz was empirically determined, and is dependent on the frequency content of the reflection arrivals.

For the application of the instantaneous frequency matching procedure, windowed segments of the trace are filtered based on the frequency content of the reflection arrival and the noise level of the trace. The instantaneous frequency of the reference pulse is then matched to the instantaneous frequency of the filtered, reflection arrival at its envelope peak. This process was shown to provide a more accurate estimate of attenuation than spectral ratios, particularly for data with lower signal-to-noise-ratios (MATHENEY and NOWACK, 1995). The matching procedure results in differential attenuation t^* values as a function of time along a seismic reflection trace which can then be used to obtain interval Q^{-1} values.

Estimation of Interval Q^{-1} Values

In order to obtain interval Q^{-1} values from seismic reflection data, the differential t^* values, as well as travel times, between a reference pulse and reflected arrivals on a seismic trace are required. The travel times are obtained from the envelope peaks of the reference pulse and the reflected pulses. The differential attenuation value (Δt_n^*) for a reflection arrival at time t_n is given by

$$\Delta t_n^* = t^*(t_n) - t_{\text{ref}}^*(t_0) = \int_0^{t_n} Q^{-1}(t')\, dt' - \int_0^{t_0} Q^{-1}(t')\, dt'. \tag{2a}$$

Assuming nearly normal incidence, this can be approximated by

$$\Delta t_n^* = t^*(t_n) - t_{\text{ref}}^*(t_0) \cong \int_{t_0}^{t_n} Q^{-1}(t')\, dt' = Q_n^{-1\text{avg}}(t_n - t_0), \tag{2b}$$

where $Q_n^{-1\text{avg}}$ is the average Q^{-1} value between the reference pulse at t_0 and the n-th reflection pulse at t_n. The average attenuation value $Q_n^{-1\text{avg}}$ is a time weighted average of the interval Q^{-1} values for the individual layers in depth.

Since the derived t^* values will have scatter due to interfering arrivals and noise, it is necessary to fit a curve through the t^* data points before computing the interval Q's. For nearly zero-offset reflection data, the differential t^* values between early and later reflected arrivals must increase with time. This can be seen from eqn. (2b) where Δt_n^* at time t_n will be greater than Δt_{n-1}^*, for t_n greater than t_{n-1} and Q^{-1} greater than zero. Thus the attenuation values should increase with time down a zero offset reflection trace. In order to satisfy this constraint, as well as find a best-fitting curve through the data, a non-negative least-squares procedure (LAWSON and HANSEN, 1995) is used to fit all the observed t^* data points simultaneously.

Figure 1a shows an example of t^* values versus time extracted from several observed seismic reflection traces. In order to compute the best-fit curve, the t^* data are broken into a number of constant time segments. For each time segment the Q^{-1} value is assumed to be constant. The slope for each of the time segments is then solved for simultaneously giving the interval Q^{-1} values. In matrix form, the linear set of equations is given by

$$
\begin{array}{l}
\xrightarrow{\hspace{1.5cm}} \\
\text{Interval } \#1 \\
\xrightarrow{\hspace{1.5cm}} \\
\\
\text{Interval } \#2 \\
\\
\xrightarrow{\hspace{1.5cm}} \\
\text{Interval } \#3 \\
\xrightarrow{\hspace{1.5cm}}
\end{array}
\begin{bmatrix}
\Delta t_{1,1} & 0 & 0 & 0 \\
\Delta t_{2,1} & 0 & 0 & 0 \\
\Delta t_1 & \Delta t_{3,2} & 0 & 0 \\
\Delta t_1 & \Delta t_{4,2} & 0 & 0 \\
\Delta t_1 & \Delta t_{5,2} & 0 & 0 \\
\Delta t_1 & \Delta t_2 & \cdot & \Delta t_{m-1,n} \\
\Delta t_1 & \Delta t_2 & \cdot & \Delta t_{m,n}
\end{bmatrix}
\begin{bmatrix}
Q_1^{-1} \\
Q_2^{-1} \\
\cdot \\
\cdot \\
\cdot \\
Q_n^{-1}
\end{bmatrix}
=
\begin{bmatrix}
\Delta t_1^* \\
\Delta t_2^* \\
\cdot \\
\cdot \\
\cdot \\
\\
\Delta t_m^*
\end{bmatrix}, \tag{3}
$$

where Δt_j is the length of time of the j-th time interval, $\Delta t_{i,j}$ is the incremental time between the observed data point (i) and the start of the current time interval (j), Q_j^{-1} is the interval Q^{-1} value for the j-th time interval, Δt_i^* is the i-th data point, m is the total number of t^* data points for all time intervals, and n is the number of time intervals. This problem is then solved using non-negative least squares in which all the Q_j^{-1} values must be non-negative. The resulting solution gives the interval Q^{-1} values for all the layers simultaneously, with a constraint of positive slope and a minimization of the squared errors between the observed data and the line segments. Figure 1b shows the computed best-fit curve for the data shown in Figure 1a. The slope of each of the line segments is the Q^{-1} value for that layer.

Bootstrap Method for Error Estimation

An advantage of using curve fitting for the estimation of the interval Q^{-1} values is that the fitting procedure allows for an estimation of the uncertainties in the Q^{-1}

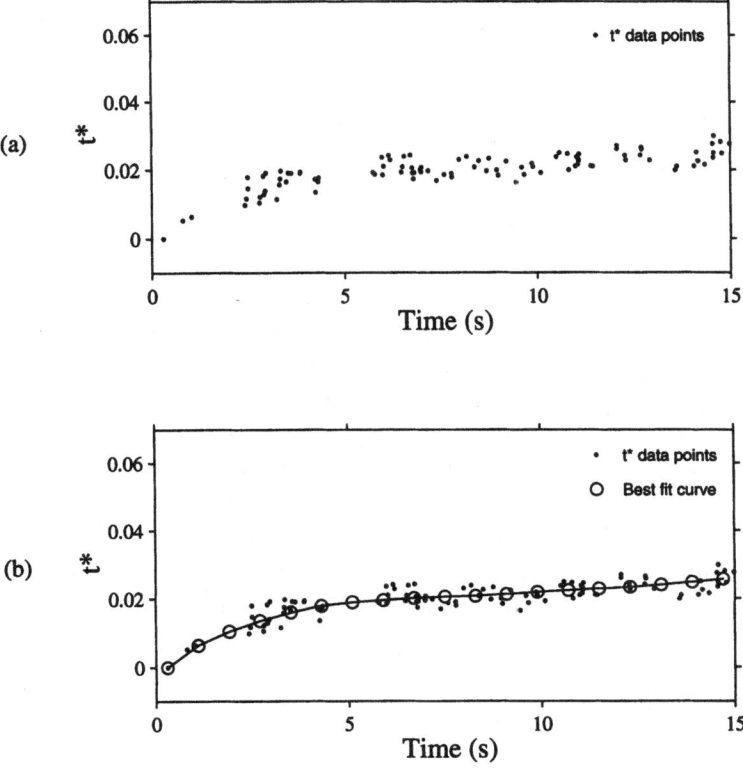

Figure 1
(a) Differential t^* values extracted from several observed seismic traces. (b) Differential t^* values shown in (a) with the computed best-fit curve using non-negative least squares.

model. This can be accomplished by using the bootstrap resampling method. For the bootstrap method, the original $t*$ data set is randomly resampled many times, and the model fit for each of these resampled data sets provides information about the model variance (EFRON and TIBSHIRAMI, 1993). The resampling method used for bootstrapping is the random selection of m data points from the original set of m data. This implies that the resampled data sets can contain a given data point more than once. The bootstrap's estimator of standard deviation ($\hat{\sigma}_{\text{boot}}$) can be determined using a large number of realizations (L) of bootstrap estimations ($\hat{\theta}_i^*$), $i = 1, L$. In our case, the bootstrap estimator ($\hat{\theta}_i^*$) is the Q^{-1} value for each layer. The bootstrap estimate of standard deviation is then

$$\hat{\sigma}_{\text{boot}} = \left[\frac{1}{L-1} \sum_i (\hat{\theta}_i^* - \bar{\theta})^T (\hat{\theta}_i^* - \bar{\theta}) \right]^{1/2}, \tag{4}$$

where

$$\bar{\theta} = \sum_i \hat{\theta}_i^* / L \tag{5}$$

(TICHELAAR and RUFF, 1989). For the error calculations presented below, one hundred resamplings ($L = 100$) have been used for estimating the standard deviation. This value of L was found to give results similar to those obtained from larger resampling values for this data.

Synthetic Data Examples

The method for computing interval Q^{-1} values described above was first tested using synthetic data examples. Figure 2 shows the velocity and attenuation models used to produce synthetic seismic data. The velocity and Q^{-1} models include a thin, low velocity and high Q layer at the surface corresponding to the water layer. Directly beneath this layer both the velocity and the Q^{-1} values increase. At greater depths, the velocity continues to increase and the Q^{-1} values decrease. Figure 3 delineates 60 synthetic seismic traces generated employing ray methods. Amplitudes have been scaled using AGC for display purposes, but true amplitudes are used for the computation of differential $t*$ values. For this computed seismic section, the source-receiver offset for the first trace is 232 meters and the receiver spacing is 25 meters. The seismic section consists of the eleven primary reflections from the layered model shown in Figure 2 and eight water column multiples associated with each primary reflection. The primary effect of the Q^{-1} model is to broaden the pulses. The effect of the attenuation model on the reflection coefficients has not been included for the synthetic examples since this has been shown to have a small effect except in cases of very high Q^{-1} materials or very small velocity contrasts (BOURBIÉ and NUR, 1984).

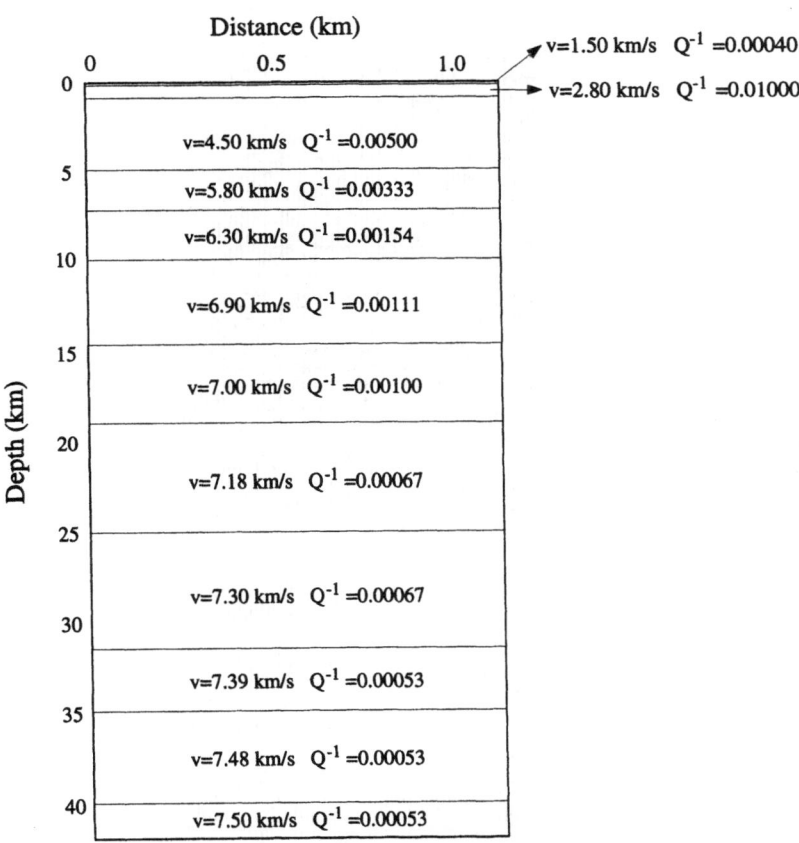

Figure 2
Velocity and attenuation models used to generate synthetic reflection seismograms.

In Figure 4, band-passed Gaussian noise is added to each trace to simulate real seismic data. A mute is applied to the first part of the traces in order to remove as much of the water reverberation as possible. The shot gather is then dip filtered using a slope cutoff of -0.015 s/trace and 0.015 s/trace.

In order to remove the water reverberations from the primary reflections, the seismic gather is deconvolved using a gap deconvolution. For this method, the seismic trace is modeled as a sequence of primary reflections plus water reverberations at the source and receiver. For each primary reflection, this has an impulse response of

$$g(t) = \sum_{n=0}^{\infty} (-1)^n (n+1) R^n \delta(t - n\tau), \qquad (6)$$

Distance (km)

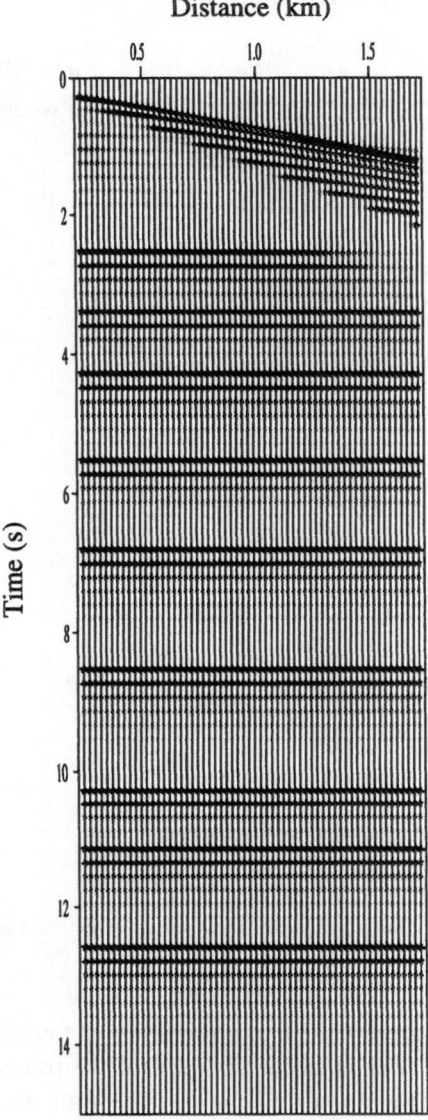

Figure 3
Synthetic seismic shot gather with eleven primary reflections and their associated water reverberations.
AGC has been applied.

where τ is the two-way travel time from the water bottom to the surface and R is
the reflection coefficient at the water bottom. The water bottom multiple portion of
the seismic record can be removed by convolving the traces with an inverse filter
with an impulse response of

Distance (km)

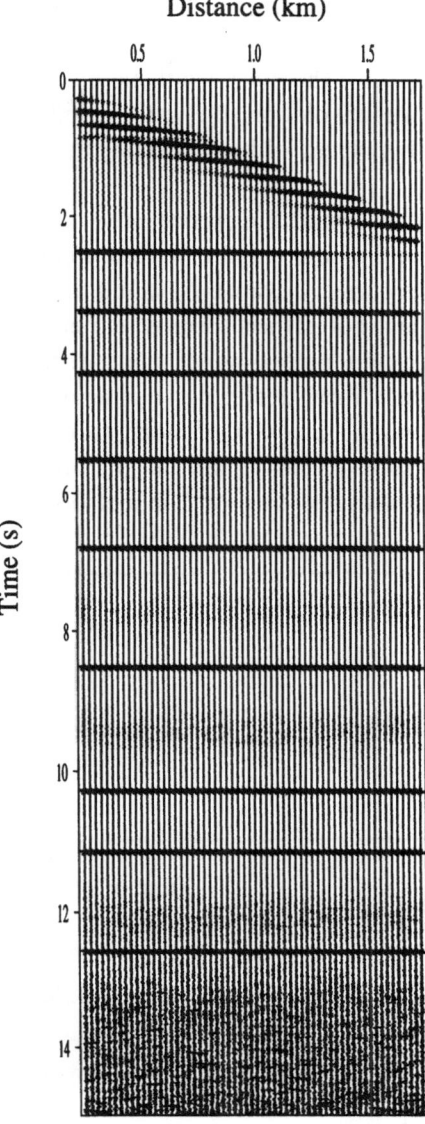

Figure 4
Noisy synthetic seismic shot gather with eleven primary reflections. Dip´ filtering and gap deconvolution
have been applied to remove water bottom multiples. The data is plotted with an AGC of the
amplitudes.

$$h(t) = \delta(t) + 2R\delta(t - \tau) + R^2\delta(t - 2\tau), \tag{7}$$

(BACKUS, 1959). In order to implement this deconvolution operator, the two-way
travel time in the water column and the reflection coefficient of the water bottom

must be known. In the synthetic example of Figure 4 the two-way travel time through the water column is 0.2 s and the water bottom reflection coefficient is 0.3. Using these parameters to deconvolve the noisy synthetic data results in the shot gather shown in Figure 4. After deconvolution, the water reverberations for each primary reflection have been greatly reduced without affecting the pulse shape of the primary reflections.

The instantaneous frequency matching procedure was applied on traces 5 through 20 of Figure 4 to calculate the differential t^* values. Figure 5a shows the extracted t^* values ($\cdot$) along with the correct t^* values ($+$) and the best-fit curve ($\bigcirc$) through the t^* data points. For this shot gather, there is scatter in the computed t^* values associated with the noise and processing of the gather. Also, there are several mispicks early in the seismic traces due to noise generated by the processing of the shot gather. Figure 5b shows the interval Q^{-1} model (solid line) and the two standard deviation uncertainties (short-dashed lines) in the model computed using the bootstrap method, along with the true Q^{-1} model (long-dashed line). Although

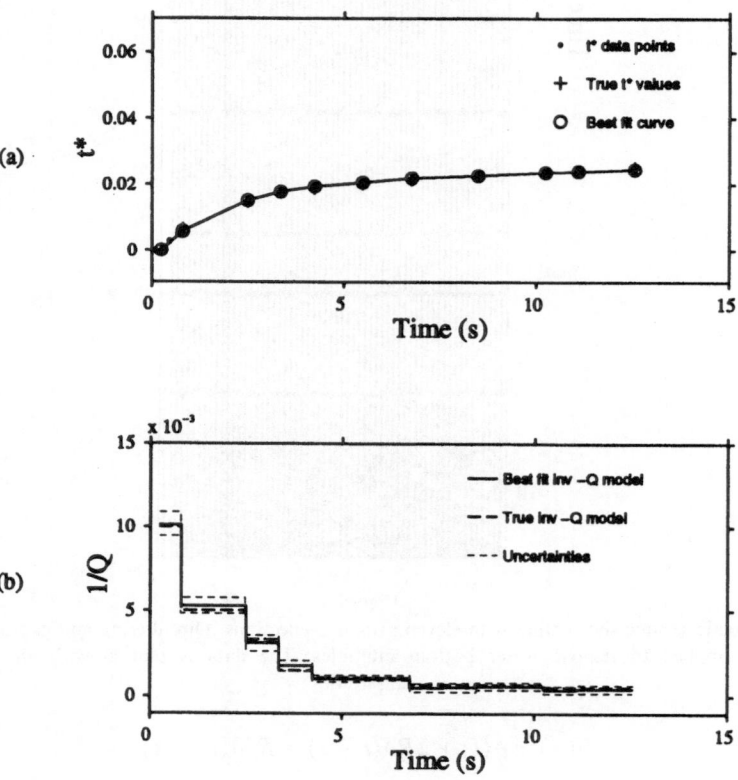

Figure 5
(a) Differential attenuation values with the best-fit curve through the t^* data points. (b) Inverse-Q model computed from the attenuation values in (a).

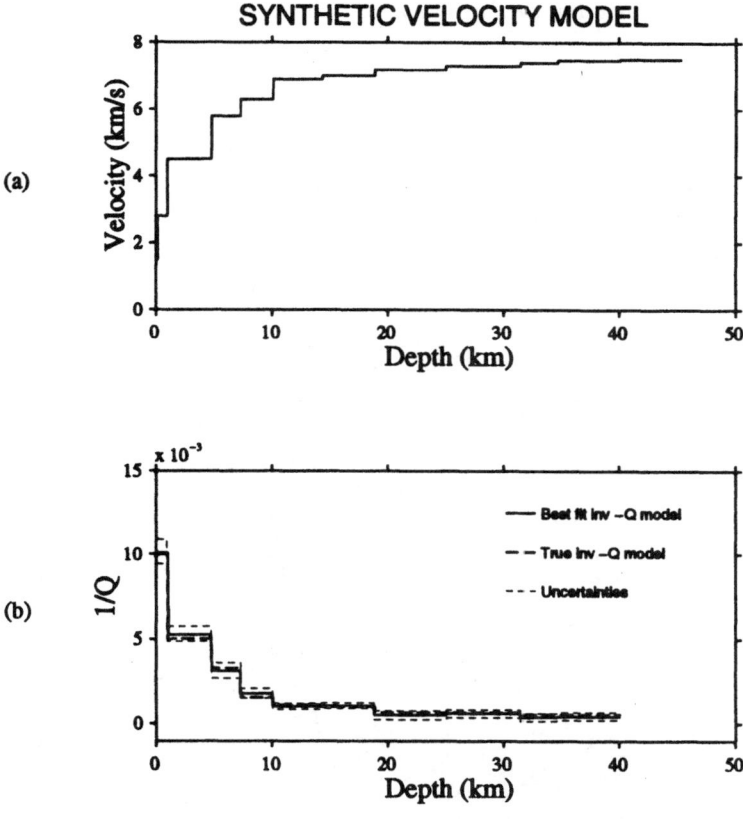

Figure 6
(a) Velocity model used to convert time to depth. (b) Inverse-Q model versus depth.

the uncertainties in the model are slightly larger than in noise-free examples, the estimated and true Q^{-1} models agree to within the error bounds obtained by the bootstrap method. Predictive deconvolution could have been used to remove the reverberations. However, testing with synthetic, as well as real data, demonstrated predictive deconvolution to be more likely to disrupt the frequency content of the primary reflections than the analytic deconvolution operator in eqn. (7). Any change in the frequency of the arrivals will cause erroneous differential t^* calculations.

For a direct comparison of the extracted Q^{-1} model with the original model employed computing the synthetic data (Fig. 2), the Q^{-1} versus time model of Figure 5b can be converted to depth using an estimated velocity model. By calculating the average velocity to different depths, the travel time to any depth can be computed. Figure 6a displays the velocity model used in computing synthetics. In figure 6b the best-fit Q^{-1} model converted to depth is shown by the solid line, the short-dashed line is the uncertainty in the model, and the long-dashed line is the true Q^{-1} model.

Application to the GLIMPCE Lake Superior Data

Seismic Reflection Results

The instantaneous frequency matching procedure described above is now applied to data from the 1986 GLIMPCE Lake Superior reflection experiment. Figure 7 is a map of line A showing the shot number locations across the lake. For this experiment, a tuned airgun array was used as the source, and 120 hydrophones with an initial source-receiver offset of 232 m and a receiver spacing of 25 m were used

LAKE SUPERIOR GLIMPCE
REFLECTION LINE A

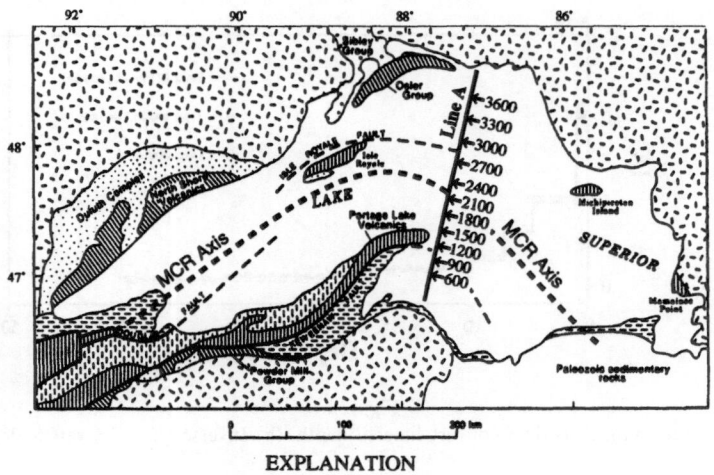

EXPLANATION

☰	Upper Keweenawan sedimentary rocks (Bayfield Group and Jacobsville Sandstone)
▦	Upper Keweenawan sedimentary rocks (Oronto Group)
▨	Middle Keweenawan (Basalt flows and sedimentary rocks/Gabbroic intrusives)
▧	Lower Keweenawan (Basalt flows and underlying sedimentary rocks)
⋰	Lower Keweenawan? (Sibley Group)
⋱	Archean and Lower Proterozoic crystalline rocks

Figure 7

Map of the Lake Superior area showing the probable lithologies and the location of line A of the GLIMPCE experiment. Shot number locations used for computing the attenuation models are specified along line A. The dashed line shows the axis of the Midcontinent rift (MCR) (Geology adapted from CANNON *et al.*, 1989).

DATA PROCESSING FLOW CHART

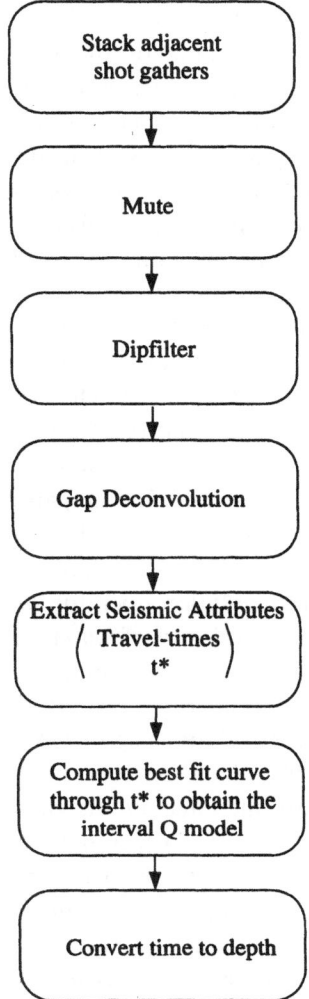

Figure 8
Processing steps used to obtain interval Q values.

to record the data. For shot locations less than 2300 (see Fig. 7) a shot spacing of 50 m was used and for the other shots a spacing of 62.5 m was used. The first fifteen seconds of the seismic reflection records at eleven locations across Lake Superior are used to compute Q^{-1} models. These locations are centered at shot locations 600, 900, 1200, 1500, 1800, 2100, 2400, 2700, 3000, 3300, and 3600 shown in Figure 7.

Figure 8 shows the processing steps that were applied to the reflection data. The first step was to stack traces from adjacent shot gathers to increase the signal-to-noise ratio. Without the stacking of shot gathers, the determination of t^* values below several seconds would have been impossible due to the low signal-to-noise ratio of the reflections. Before stacking of the shot gathers, the water depth variability was checked and shot and receiver locations with water depths that varied by more than 6 meters were not included in a given stack. Large depth variations would cause a misalignment of the reflections from adjacent shots and also make the gap deconvolution ineffective. Shot locations near 1500 and 2700 had too large a variation in the lake bathymetry and were excluded from the differential t^* computations. At the other nine locations, 30 to 45 shot gathers were stacked depending on the lake bathymetry. This corresponds to stacking over a horizontal distance of between 1.5 km and 2.8 km for each location. After stacking, a mute was applied to the start of each shot gather. The time for the mute increased from 0.1 s for the first trace to 2.1 s for trace 60. Each shot gather was then dipfiltered using a filter cutoff of -0.015 s/trace and 0.015 s/trace.

After dipfiltering, the shot gathers were gap deconvolved using the deterministic deconvolution described above. The time for the gap deconvolution operator was estimated from the original experiment logs which gave the water depths. This was checked against the water bottom reflection times for each shot gather. The reflection coefficient for the deconvolution operator was estimated using a velocity of 1450 m/s (PRESS, 1966) for the velocity of water with a density of 1000 kg/m³. The velocity of the subsurface bottom was estimated from the seismic refraction results of MATHENEY and NOWACK (1997). The density of the lake bottom was obtained from the density-velocity relationship of GARDNER et al. (1974). Figure 9 shows shot gather location 3600, which has been fully processed using deconvolution parameters of 0.198 s for the time in the deconvolution operator, and a reflection coefficient of 0.45 for the subsurface bottom.

Although deconvolution was used to remove some of the multiple reverberations from the water column and improve estimates of the differential t^* values, application of the instantaneous frequency matching procedure to the undeconvolved data was found to produce similar results for the attenuation estimates. This is due to the fact that the matching algorithm uses arrivals with the maximum envelope amplitude over each small window and this eliminates some of the multiple arrivals. Also, the water column has considerably lower attenuation.

After processing the shot gathers, the instantaneous frequency matching procedure was applied to trace 5 through 20 for each of the nine shot locations. As an example, Figure 10 shows windowed arrivals from shot location 3600 in Figure 9. In this figure, the arrival at 0.3 s shown in (a) is extracted from the first trace and used as the reference pulse. This arrival corresponds to the water bottom reflection. The computed instantaneous frequency at the envelope peak is shown for each arrival and illustrates the general decrease in the reflected arrival's frequency with increasing time. This decrease in frequency is utilized by the matching procedure to obtain the differential t^* values.

Figure 11 shows the computed differential t^* values for the nine shot locations across Lake Superior. The computed best-fit curves are shown by the circles. A time interval of 0.8 s was used for the curve-fitting procedure for all the shot locations.

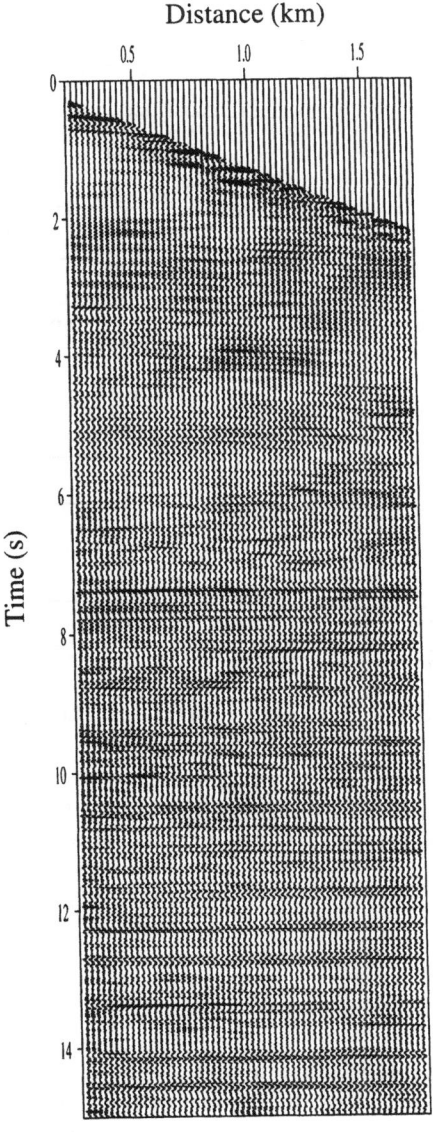

Figure 9

Processed shot gather centered around shot location 3600. The gather has been muted, dipfiltered and gap deconvolved. An AGC has been applied for display purposes.

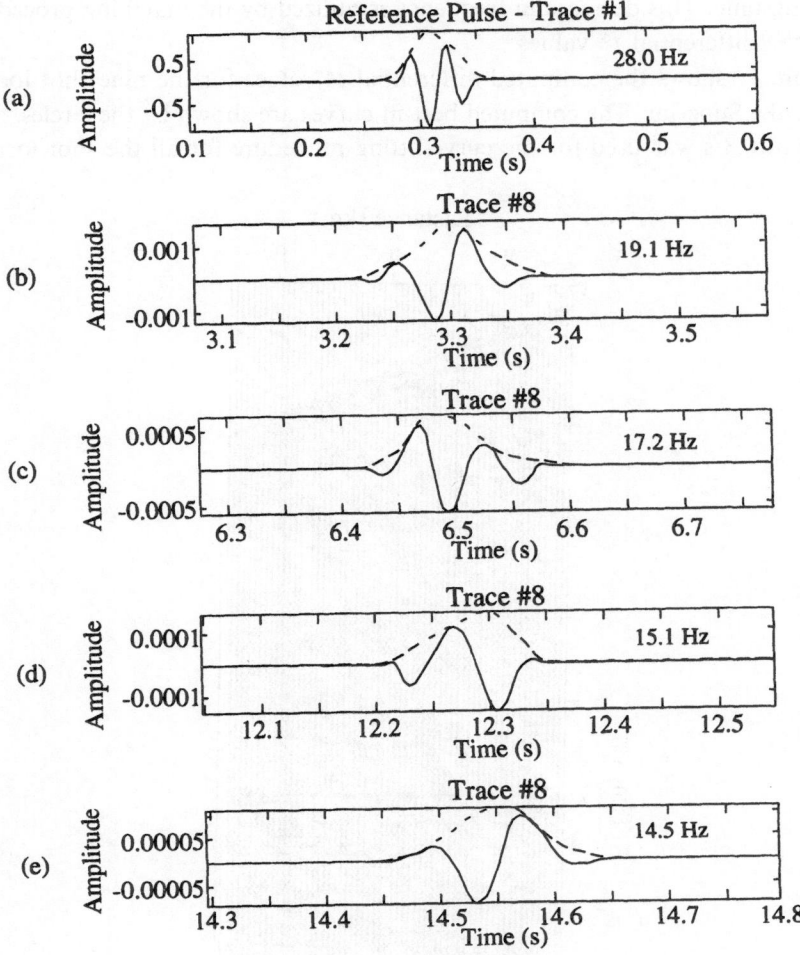

Figure 10
Windowed arrivals from the processed gather at shot location 3600 for trace number 8. The computed envelope amplitudes are also shown along with the instantaneous frequency at the peak of the envelope.

Figure 11 shows a general decrease in the slopes of all the best-fit curves with increasing time. This is similar to that shown in the synthetic example, and results from a decrease in Q^{-1} values with depth. The interval Q^{-1} models determined by the best-fit curve slopes are also shown in Figure 11 along with the uncertainties in the models computed using the bootstrap method.

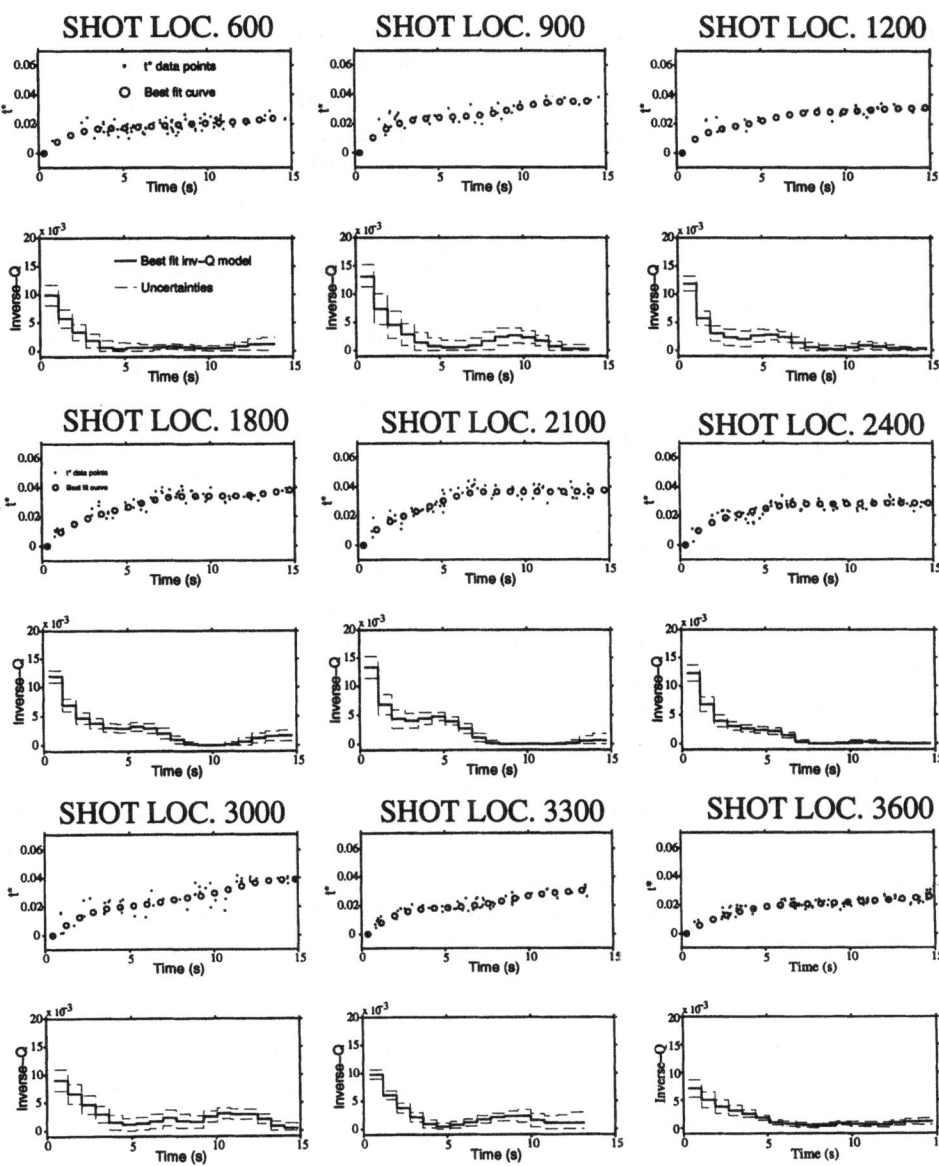

Figure 11
Differential *t** values with the best-fit curve from the nine profiles and the resulting inverse-*Q* models
with computed uncertainties. For each shot location, the *t** values with travel time are shown above and
the estimated inverse-*Q* with time are shown below.

Comparison with Refraction Results

Seismic refraction data as well as reflection data were recorded during the 1986 GLIMPCE Lake Superior experiment. The refraction data along Line A were used in previous studies to determine the velocity structure (TRÉHU *et al.*, 1991; Hamilton and Mereu, 1993; Lutter *et al.*, 1993). In addition, MATHENEY *et al.* (1997) simultaneously inverted for velocity and attenuation along line A, using the method presented by NOWACK and MATHENEY (1997a) for the inversion of multiple seismic attributes.

The coincident refraction and reflection profiles provide an opportunity for the comparison of the attenuation models obtained from two independent data sets. Since the mode of wave propagation is different for refracted and reflection data, the apparent Q^{-1} models derived form the two data sets may not necessarily be equivalent. SCHOENBERGER and LEVIN (1974) demonstrated that thin layering in a rock sequence can cause an increase in the apparent attenuation of reflection data due to interference of arrivals and multiples. Also refractions from interfaces with strong velocity contrasts can effect the apparent attenuation (NOWACK and MATHENEY, 1997b).

In order to compare the reflection results of this study to the refraction results, the Q^{-1} versus time models from the reflection data must be converted to depth. This is accomplished in a similar fashion as in Figure 6 for the synthetic data. A velocity model from the refraction studies has been used at each location to compute an average velocity and the two-way travel time to any depth is then obtained. The inverted, velocity model of MATHENEY *et al.* (1997) was used to convert the reflection data to depth for each of the nine Q^{-1} profiles.

Figure 12 shows a comparison between Q^{-1} profiles obtained from the reflection profiles for the upper 10 km of this study and the Q^{-1} model from the refraction study of MATHENEY *et al.* (1997). The solid lines are Q^{-1} profiles from the reflection data and the long-dashed curves are the refraction results. Uncertainties in the two Q^{-1} models are also shown. For the reflection model, the average Q^{-1} is computed for 0.8 s long time windows which results in Q^{-1} layers which are approximately 2-km-thick. The refraction model is a spline interpolated model which is continuous and not layered. Despite this difference, there is good agreement between the attenuation models derived from two data sets. This implies that the wave propagation and scattering effects on apparent attenuation are similar between the two data sets.

Discussion

The GLIMPCE Lake Superior seismic experiment was a survey designed to provide insight into the Midcontinent rift (MCR), a ≈ 1.1 Ga old paleorift which extends from Kansas up through Lake Superior and into Michigan. Line A of the Lake Superior seismic experiment transverses the MCR (see Fig. 7) and includes

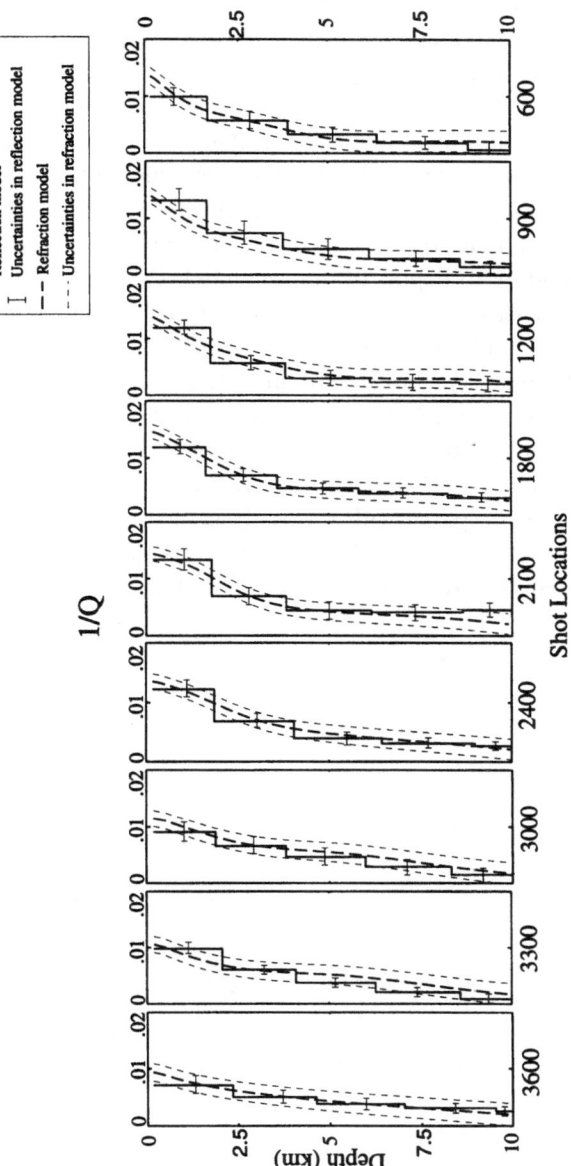

Figure 12

Comparison of the reflection attenuation profiles to the refraction inverse-Q model from MATHENEY *et al.* (1997) for the nine shot locations shown in Figure 7.

coincident refraction and reflection profiles. The seismic refraction and reflection data from this line have been used to invert for the velocity and reflectivity structure of the subsurface (MATHENEY et al., 1997; LUTTER et al., 1993; SHAY and TRÉHU, 1993; HAMILTON and MEREU, 1993; CANNON et al., 1989). These prior results allow the attenuation model presented in this study to be related to the lithology and geologic features across the MCR.

The major shallow (< 10 km) geologic features across the MCR include a large central rift basin between shot locations 1000 to 2600, a smaller northern rift basin extending from shot locations 2900 to 3600, the Isle Royale fault near shot location 2850, and the Keweenaw fault near shot location 1000 (see Fig. 7). Figure 12 shows the shallow Q^{-1} structure from this study. In the central basin the Q^{-1} values for the first layer corresponding to depths of approximately 2 km are greater than 0.01 ($Q < 100$). For the deeper layers between 5 and 10 km in the central basin, the Q^{-1} values are between 0.002 and 0.004. Outside this basin and in the northern basin, the Q^{-1} values for the first layer are less than 0.01 ($Q > 100$) and for the deeper layers Q^{-1} values are on average lower than in the central basin. This is similar to attenuation results obtained from the refraction data by MATHENEY et al. (1997).

For the deeper regions beneath Lake Superior, the Q^{-1} results shown in Figure 11 have been shaded and compared with the analysis of the reflectivity structure by CANNON et al. (1989) in Figure 13. A logarithmic gray-shade scale is used to plot the model in order to emphasize the deeper portions of the model. Overlain on the

INVERSE-Q MODEL

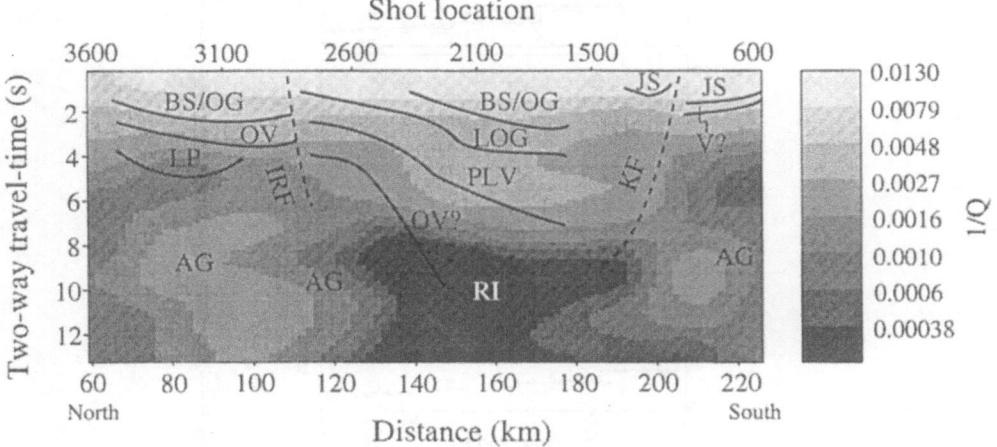

Figure 13
Gray-shade plot of the inverse-Q profiles derived from the seismic reflection data. The solid and dashed lines are adapted from the interpretation of the seismic reflection data from CANNON et al. (1989). See text for an explanation of the abbreviations. Note that the vertical scale is time whereas Figure 12 is depth.

gray-shade map is a line drawing adaptation of the interpretation of CANNON et al. (1989) along with the possible lithologies. The interpretation of CANNON et al. (1989) was derived from the reflectivity structure on the stacked and processed seismic reflection profile. The major structural features described previously, including the Isle Royale and Keweenaw faults, the large central basin and a smaller northern basin, can be seen on the reflection model of CANNON et al. (1989), as well as on the attenuation structure obtained here. The abbreviations for the different formations on Figure 13 are BG for Bayfield Group, OG for Oronto Group, OV for Osler volcanics, LOG for Lower Oronto Group, PLV for Portage Lake Volcanics, LP for Lower Proterozoic, RI for rift related rocks, JS for Jacobsville Sandstone, V for volcanics, and AG for Archean crystalline rocks.

Over the first several seconds of the model there is a layer of high Q^{-1}. This higher Q^{-1} portion of the model corresponds with the upper Keweenawan sedimentary rocks of the Bayfield, Oronto and Jacobsville sandstone. The thickest sequence of the Oronto and Bayfield rocks occurs in the central basin around shotpoint 1600. Beneath these upper sedimentary rocks, the Lower Oronto and Portage Lake Volcanics are present in the central basin, which are primarily basaltic volcanic flows with interflow clastic sedimentary rocks. This sequence of units in the central basin is evident in the attenuation model with generally higher Q^{-1} values than the adjacent Osler and lower Proterozoic volcanics outside the basin. The rocks beneath the central rift basin, noted by RI, have distinctly lower Q^{-1} values and lower attenuation than the rest of the model. Previously obtained velocity models by HAMILTON and MEREU (1993) and SHAY and TRÉHU (1993) also have significantly higher velocities beneath the central rift basin. The surrounding Archean crystalline rocks noted by AG in Figure 13 have Q^{-1} values similar to those of the Osler volcanics which have interbed sedimentary rocks. In the velocity model of TRÉHU et al. (1991), there are correspondingly lower velocities at these depths near shot locations 800 and 3600 for the crystalline rocks. While lithologies seem to be a significant factor in the apparent Q^{-1} values of Figure 13, scattering, interference, and thin bed layering may also have effects on apparent attenuation.

Conclusion

The method of instantaneous frequency matching has been used to compute differential t^* values from the GLIMPCE Lake Superior reflection data. The attenuation values were converted to apparent Q^{-1} models by a fitting procedure that simultaneously solved for all the interval Q's using nonnegative least squares. The bootstrap method was then used to estimate the variance in the interval Q^{-1} models. The Q^{-1} model obtained from the reflection data corresponds closely with a prior attenuation model derived using seismic refraction data along the same profile. This suggests that the effects of wave propagation and scattering on the apparent Q models are similar for the two data sets. The Q^{-1} model from the reflection data was then compared with the interpretation of the seismic reflectivity structure and was found to be in good agreement. Higher interval Q^{-1} values (>0.01) were found near the surface corresponding to the sedimentary rock sequences of the upper Keweenawan. Low Q^{-1} values (< 0.0006) were found beneath the central rift basin. The surrounding crystalline formations were found to have Q^{-1} values typical of those found in the basaltic flows with interflow sedimentary rocks. Finally, attenuation models derived in this way can be used to correct reflection data for dispersion, frequency and amplitude effects and allow for improved imaging of the subsurface.

Acknowledgments

The authors would like to thank Dr. Myung Lee of the United States Geological Survey for providing the seismic reflection data from the GLIMPCE experiment. Partial support was provided by National Science Foundation Grant No. EAR-9614772.

REFERENCES

AKI, K., and RICHARDS, P., *Quantitative Seismology, Theory and Methods* (W. H. Freeman, New York 1989).

BACKUS, M. M. (1959), *Water Reverberations—Their Nature and Elimination*, Geophys. 24, 233–261.

BICKEL, S. H. (1993), *Similarity and the Inverse Q Filter: The Pareto-Levy Stretch*, Geophys. 58, 1629–1633.

BICKEL, S. H., and NATARAJAN, R. R. (1985), *Plane-wave Q Deconvolution*, Geophys. 50, 1426–1439.

BOURBIÉ, T., and NUR, A. (1984), *Effects of Attenuation on Reflections: Experimental Test*, J. Geophys. Res. 89, 6197–6202.

BRZOSTOWSKI, M. A., and MCMECHAN, G. A. (1992), *3-D Tomographic Imaging of Near-surface Seismic Velocity and Attenuation*, Geophys. 57, 396–403.

CANNON, W. F., GREEN, A. G., HUTCHINSON, D. R., LEE, M., MILKEREIT, B., BEHRENDT, J. C., HALLS, H. C., GREEN, J. C., DICKAS, A. B., MOREY, G. B., SUTCLIFFE, R., and SPENCER, C. (1989), *The North American Midcontinent Rift Beneath Lake Superior from GLIMPCE Seismic Reflection Profiling*, Tectonics 8, 305–332.

DUREN, R. E., and TRANTHAM, E. C. (1997), *Sensitivity of the Dispersion Correction to Q Error*, Geophys. 62, 288–290.

EFRON, B., and TIBSHIRAMI, R. J., *An Introduction to the Bootstrap* (Chapman and Hall, New York 1993).

GARDNER, G. H. F., GARDNER, L. W., and GREGORY, A. R. (1974), *Formation Velocity and Density: The Diagnostic Basis for Stratigraphic Traps*, Geophys. 39, 770–780.

HAMILTON, D. A., and MEREU, R. F. (1993), *2-D Tomographic Imaging across the North American Midcontinent Rift System*, Geophys. J. Int. 112, 344–358.

HARGREAVES, N. D., and CALVERT, A. J. (1991), *Inverse Q Filtering by Fourier Transform*, Geophys. 56, 519–527.

KJARTANSSON, E. (1976), *Constant Q-wave Propagation and Attenuation*, J. Geophys. Res. 84, 4737–4748.

LAWSON, C. L., and HANSEN, R. J., *Solving Least-squares Problems* (SIAM Press, Philadelphia 1995).

LUTTER, W. J., TRÉHU, A. M., and NOWACK, R. L. (1993), *Application of 2-D Travel-time Inversion of Seismic Refraction Data to the Mid-continent Rift beneath Lake Superior*, Geophys. Res. Lett. *20*, 615–618.

MATHENEY, M. P., and NOWACK, R. L. (1995), *Seismic Attenuation Values Obtained from Instantaneous Frequency Matching and Spectral Ratios*, Geophys. J. Int. *123*, 1–15.

MATHENEY, M. P., NOWACK, R. L., and TRÉHU, A. M. (1997), *Seismic Attribute Inversion for Velocity and Attenuation Structure Using Data from the GLIMPCE Lake Superior Experiment*, J. Geophys. Res. *102*, 9949–9960.

NOWACK, R. L., and MATHENEY, M. P. (1997a), *Inversion of Seismic Attributes for Velocity and Attenuation Structure*, Geophys. J. Int. *128*, 689–700.

NOWACK, R. L., and MATHENEY, M. P. (1997b), *Extraction of Seismic Attributes from Wide-angle Synthetic Data Derived from Models with Interfaces*, Trans. Am. Geophys. Un. (EOS) *78*.

PRESS, F. In *Handbook of Physical Constants* (Clark, S. P., ed.) (Geological Society of America, Boulder, CO 1966).

SCHOENBERGER, M., and LEVIN, F. K. (1974), *Apparent Attenuation due to Intrabed Multiples*, Geophys. *39*, 278–291.

SHAY, J., and TRÉHU, A. M. (1993), *Crustal Structure of the Central Graben of the Midcontinent Rift beneath Lake Superior*, Tectonophysics *225*, 301–335.

SOLLIE, R., and MITTET, R. (1994), *Prestack depth migration; sensitivity to the macro absorption model*, SEG Annu. Meet. Expanded Tech. Program Abstr. *64*, 1422–1425.

TICHELAAR, B. W., and RUFF, L. J. (1989), *How Good are our Best Models? Jacknifing, Bootstrapping and Earthquake Depth*, Trans. Am. Geophys. Un. (EOS) *70*, 593.

TRÉHU, A., MOREL-Á-L'HUISSIER, P., MEYER, R., HAJNAL, A., KARL, J., MEREU, R., SEXTON, J., SHAY, J., CHAN, W. K., EPILI, D., JEFFERSON, T., SHIH, X. R., WENDLING, S., MILKEREIT, B., GREEN, A., and HUTCHINSON, D. (1991), *Imaging the Midcontinent Rift Beneath Lake Superior Using Large Aperture Seismic Data*, Geophys. Res. Lett *18*, 625–628.

(Received November 10, 1997, revised March 20, 1998, accepted April 23, 1998)

Pure appl. geophys. 153 (1998) 563–585
0033–4553/98/040561–23 $ 1.50 + 0.20/0

❙ Pure and Applied Geophysics

Lg Coda *Q* and its Relation to the Geology and Tectonics of the Middle East

LIANLI CONG[1,2] and B. J. MITCHELL[1]

Abstract—Regional seismograms were collected to image the lateral variations of *Lg* coda *Q* at 1 Hz (Q_0) and its frequency dependence (η) in the Middle East using a back-projection method. The data include 124 vertical-component traces recorded at 10 stations during the period 1986–1996. The resulting images reveal lateral variations in both Q_0 and η. In the Turkish and Iranian Plateaus, a highly deformed and tectonically active region, Q_0 ranges between about 150 and 300, with the lowest values occurring in western Anatolia where extremely high heat flow has been measured. The low Q_0 values found in this region agree with those found in other tectonically active regions of the world. Throughout most of the Arabian Peninsula, a relatively stable region, Q_0 varies between 350 and 450, being highest in the shield area and lowest in the eastern basins. All values are considerably lower than those found in most other stable regions. Low Q values throughout the Middle East may be caused by interstitial fluids that have migrated to the crust from the upper mantle, where they were probably generated by hydrothermal reactions at elevated temperatures known to occur there. Low Q_0 values (about 250) are also found in the Oman folded zone, a region with thick sedimentary deposits. η varies inversely with Q_0 throughout most of the Middle East, with lower values (0.4–0.5) in the Arabian Peninsula and higher values (0.6–0.8) in Iran and Turkey. Q_0 and η are both low in the Oman folded zone and western Anatolia.

Key words: *Lg*, *Lg* coda, *Q*, Middle East, tectonics.

Introduction

Lg coda *Q* tomography is a recently developed tool for investigating lateral variations of the attenuative properties of the earth's crust. It has been applied successfully in Africa and much of Eurasia, where it has provided new insights on the relation of seismic wave attenuation to the tectonic history of continents.

Coda *Q* tomography became possible when determinations of *Lg* coda *Q* on individual traces could be obtained with sufficient stability and precision. A stacked spectral ratio method developed by XIE and NUTTLI (1988) yielded values of *Lg*

[1] Department of Earth and Atmospheric Sciences, St. Louis University, 3507 Laclede Avenue, St. Louis, Missouri 63103, U.S.A.

[2] Now at Department of Earth Sciences, Yunnan University, Kunming, Yunnan, People's Republic of China.

coda Q at 1 Hz (Q_0) and the frequency dependence of Q (η) with the needed stability and precision. XIE and MITCHELL (1990a) applied back-projection tomography to those values to obtain large-scale lateral variations of Lg coda Q in Africa. Later studies obtained tomographic maps of Lg coda Q in Eurasia (MITCHELL et al., 1997), North America (BAQER and MITCHELL, 1998), South America (SOUZA and MITCHELL, 1998), and Australia (MITCHELL et al., 1998).

For the Middle East, however, there are only a few studies of Lg Q and Lg coda Q and those are restricted to average values along paths between sources and stations. Values for Lg Q in Iran were obtained by NUTTLI (1980) and JIH and LINNES (1992). The results of both studies indicate that Lg Q is low in that region, and the latter study showed that it is also quite variable for different paths across the plateau. GHALIB (1992) applied the extended coda-Q technique to short-period Lg coda waves, obtaining a quite uniform distribution of Lg coda Q with an average value of 214 for the Arabian plate.

Previous studies of regional phase propagation (KADINSKY-CADE, 1981; GHALIB, 1992) and of surface waves in the Middle East (SEBER and MITCHELL, 1992) also indicate that wave propagation is poor throughout that region. It is especially poor in the northern portion which is tectonically active and geologically complex, however is also relatively poor in the Arabian Peninsula which is stable and is less complex than the northern regions.

Prior to 1995 poor station coverage throughout the Arabian Peninsula prevented detailed study of that region. A new data set is now available from a temporary station network that was deployed in Saudi Arabia from September 1995 to May 1996. These stations greatly enhanced the likelihood of a detailed Lg coda Q study because they lie in the center of the Arabian Peninsula, providing much improved data coverage for the central and southern portions of the peninsula. In addition, the shorter paths permitted by those stations allowed us to utilize records with a higher signal/noise ratio than had been available earlier. Detailed tomographic maps of Lg coda Q for the entire region of the Middle East will be presented in this study.

Tectonic Framework of the Middle East

The Middle East is a tectonically complex region where several plates (the Arabian, African, Eurasian plates, and the Anatolian block/subplate) meet. The African plate is moving north relative to the Eurasian plate, while the Arabian plate is moving north relative to the African plate. The Arabian plate is therefore moving north relative to the Eurasian plate with a counterclockwise rotation. The Anatolian block/subplate is moving westward relative to the Eurasian plate. Figure 1 is a tectonic map in the Middle East sketching the relative movement of the plates, the major tectonic features, and the boundaries of tectonic provinces within the Arabian plate.

   The Arabian plate is comprised of two major provinces: relatively stable crust in the central and western parts of the plate, and a folded region in the northeastern part of the plate. The first province can be divided into three subprovinces: the shield in the west, the shelf in the middle, and the platform and basins in the east, each representing roughly one third of the peninsula. The Arabian shield consists predominantly of Precambrian gneiss and metamorphosed sedimentary and vol-canic rocks that have been intruded by granites (POWERS *et al.*, 1966; BROWN, 1972). The thin basaltic lava flows that have been emplaced along the western border of the shield since Miocene time reflect the intensity of volcanism in the region (BROWN, 1972; COLEMAN, 1977). East of the shield, the shelf is tectonically stable, consisting of Paleozoic and Mesozoic sedimentary rocks that unconformably

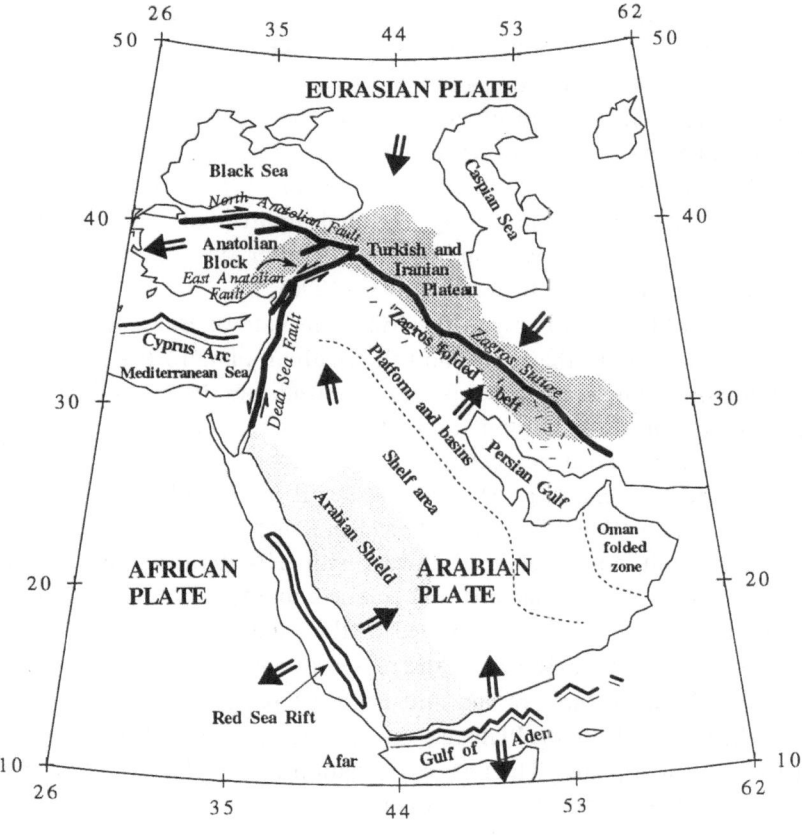

Figure 1

Tectonic map (modified from SEBER *et al.*, 1997) showing the relative movement of the plates in the Middle East area, the major tectonic features, and the tectonic provinces within the Arabian plate. The shaded areas are the Arabian Shield (lighter), and the Turkish and Iranian Plateaus (darker). For the latter region only elevations higher than 1500 m are shaded.

overlap the basement rocks and gently dip to the east towards the Persian Gulf. Thick sediments (up to about 10 km) occupy much of the platform and basins.

Continental collision between the Arabian plate and Eurasian plate is occurring along the Zagros suture zone in western Iran. This collision zone has experienced two major episodes of collisional orogeny during late Mesozoic and Miocene times (STOCKLIN, 1974; SENGÖR and KIDD, 1979; STONELEY, 1981). The collision occurred in the Zagros folded belt (COLMAN-SADD, 1978; STOCKLIN, 1968; FAL-CON, 1974), which is Pliocene to present in age. The total thickness of the Zagros sedimentary cover varies between 5 and 10 km (MORRIS, 1977; NATIONAL IRA-NIAN OIL COMPANY, 1975, 1976).

The left-lateral Dead Sea fault system forms the boundary between the Arabian and African plates, extending approximately 900 km from the Taurus convergence zone to the Red Sea (FREUND, 1965; FREUND et al., 1970; BEN-MENAHEM et al., 1976; NUR and BEN-AVRAHAM, 1978; BEN-AVRAHAM, et al., 1979; GARFUNKEL, 1981; ZAK and FREUND, 1981; WESTAWAY, 1994). The fault system was developed as a result of the mid-Cenozoic breakaway of Arabia from Africa. The fault crosses a continental area that was consolidated during the late Proterozoic Pan-African Orogeny (GARFUNKEL and BEN-AVRAHAM, 1996). The left-lateral transform motion began 20 Ma (EYAL et al., 1981; STEINITZ et al., 1980). The formation of the Dead Sea fault was accompanied by regional uplift and widespread igneous activity. The transform flanks were uplifted mostly after 10 Ma while the transform was active (GARFUNKEL and BEN-AVRAHAM, 1996).

The Anatolian block lies between the Eurasian plate or Black Sea block to the north, the African plate to the south, and the Arabian plate to the east. The active North Anatolian Fault Zone consists of numerous parallel to subparallel fault segments, and is about 1200 km long and a few hundred meters to 40 km wide. The initiation age of the dextral movement along the North Anatolian Fault Zone is some time in the Pliocene (e.g. SENGÖR, 1979; BARKA and GÜLEN, 1988; KO-CYIGIT, 1988a, 1988b, 1989; TOPRAK, 1988; BOZKURT and KOCYIGIT, 1996). The East Anatolian Fault Zone trends southwest with left-lateral movement, allowing the Turkish block to move westward. Kinematic consistency requires that the East and North Anatolian Fault Zones formed at the same time, about 5 Ma (WEST-AWAY, 1994). Before it became active, the region of southeastern Turkey shortened by compression between the Arabian and Eurasian plates.

The Turkish and Iranian Plateaus resulted from the collision between the Arabian and Eurasian plates. The average elevation of the plateau is about 1500 m (SENGÖR and KIDD, 1979). After continental collision during the late Miocene, convergence between Arabia and Eurasia continued until the present, as shown by the Pliocene to recent folding and thrusting in the Zagros, the border folds of southeastern Turkey (KETIN, 1966), and the present diffuse seismicity of the entire Turkish and Iranian Plateaus (CANITEZ and UCER, 1967; MCKENZIE, 1972; NOWROOZI, 1972; BERBERIAN, 1976). The continued convergence across the colli-

Table 1

Station information

Station	Latitude (°N)	Longitude (°E)
ABKT	37.9304	58.1189
ANTO	39.8689	32.7936
BGIO	31.7220	35.0878
GNI	40.0530	44.7240
KIV	43.9562	42.6888
AFIF	23.9310	43.0400
HALM	22.8454	44.3173
RANI	21.3116	42.7761
RAYN	23.5220	45.5008
SODA	18.2921	42.3769

sion site resulted in the shortening of the plateau across strike by thickening and by sideways motion of parts of it (SENGÖR and KIDD, 1979).

Data Acquisition

The Middle East is a tectonically active region in which numerous earthquakes are available for this study. We have used vertical-component broadband seismograms with clear *Lg* and coda recorded at 10 stations. This includes four Incorporated Research Institutions for Seismology (IRIS) stations (ANTO, ABKT, GNI, KIV), one German GEOPHONE station (BGIO), and five stations in Saudi Arabia (RANI, RAYN, HALM, AFIF, SODA), deployed temporarily by the University of California at San Diego and Boise State University. Table 1 lists the station names and their locations. A total of 124 *Lg* coda time series were used in the study. Table 2 lists the dates, the origin times, and locations of the earthquakes, as well as the recording stations used, the epicentral distances, and the back azimuths for the individual traces. The table also lists the Q_0 (*Lg* coda *Q* at 1 Hz) and η (frequency dependence of *Lg* coda *Q*) values determined in this study.

The earthquakes occurred during the period July 1986 through August 1996 and varied in magnitude between 4.0 and 6.1. Earthquakes with smaller magnitudes were avoided to ensure a good signal/noise ratio. The temporary stations in Saudi Arabia are of key importance because they were located in the central and southern portions of the Arabian Peninsula, and provided the first opportunity to study the regional variation of *Q* across that region. Those stations provided 61 seismograms from earthquakes with magnitudes ranging between 4.0 and 5.2. The ability to use smaller events suggests that *Lg* coda *Q* for paths situated in the Arabian Peninsula is higher there than that in the Turkish and Iranian Plateaus. Epicentral distances

Table 2

Earthquakes and corresponding results of single-trace analysis of Lg coda Q

Date	Origin (h:m:s)	Lat. (°N)	Lon. (°E)	m_b	Recording station	Distance (km)	Backazimuth (°)	Q_0	η
930609	17:33:36.0	34.763	53.276	5.0	ABKT	558.9	232.5	207 ± 9	0.62 ± 0.11
931002	01:17:30.4	39.066	69.966	5.0	ABKT	1040.6	79.4	266 ± 14	0.59 ± 0.14
931021	21:52:30.3	30.154	51.235	5.2	ABKT	1070.7	218.4	206 ± 58	0.70 ± 0.58
931116	15:52:48.5	30.798	67.219	5.4	ABKT	1150.6	130.7	359 ± 20	0.47 ± 0.09
931226	08:45:48.6	36.628	68.926	5.0	ABKT	948.4	82.0	321 ± 19	0.42 ± 0.13
940223	08:02:04.7	30.853	60.596	6.1	ABKT	817.3	163.1	349 ± 8	0.51 ± 0.05
940226	02:31:11.0	30.897	60.549	5.8	ABKT	811.4	163.3	362 ± 14	0.52 ± 0.09
940301	03:49:00.8	29.096	52.617	5.8	ABKT	1104.5	209.1	334 ± 37	0.65 ± 0.13
940305	04:03:52.5	36.579	68.659	5.1	ABKT	946.6	95.9	330 ± 10	0.46 ± 0.08
940414	11:03:40.2	28.290	55.340	5.2	ABKT	1099.9	194.4	321 ± 73	0.73 ± 0.33
940525	07:42:58.4	40.232	63.134	5.2	ABKT	503.6	58.0	366 ± 27	0.46 ± 0.16
940620	09:09:02.9	28.968	52.614	5.9	ABKT	1117.4	208.8	342 ± 24	0.68 ± 0.14
940701	19:50:04.3	40.219	53.391	5.6	ABKT	481.6	303.3	358 ± 13	0.61 ± 0.08
940920	05:51:46.0	32.501	48.770	5.0	ABKT	1065.6	237.9	251 ± 76	0.77 ± 0.32
930204	02:22:57.1	38.242	22.668	5.0	ANTO	894.6	261.6	165 ± 43	0.47 ± 0.53
930322	11:03:43.5	34.697	34.402	5.4	ANTO	591.5	165.6	185 ± 21	0.44 ± 0.24
930331	18:20:44.1	39.150	28.017	4.3	ANTO	418.5	260.5	185 ± 15	0.53 ± 0.10
930706	19:48:09.3	37.891	39.315	4.8	ANTO	606.9	109.1	254 ± 56	0.74 ± 0.25
930714	12:31:49.4	38.224	21.756	5.3	ANTO	972.4	262.7	200 ± 18	0.68 ± 0.18
930826	10:03:57.5	36.736	28.051	5.3	ANTO	541.3	231.5	170 ± 21	0.36 ± 0.37
940730	10:37:45.1	37.493	36.189	4.7	ANTO	396.1	130.7	193 ± 28	0.60 ± 0.18
940917	02:24:37.8	37.885	41.584	5.1	ANTO	793.7	103.3	223 ± 28	0.45 ± 0.18
941109	05:09:15.1	36.931	28.999	4.3	ANTO	465.0	226.7	237 ± 72	0.31 ± 0.45
941113	06:56:00.5	36.910	29.060	4.9	ANTO	462.9	226.0	193 ± 14	0.32 ± 0.08
941230	06:56:16.8	38.179	39.670	4.7	ANTO	624.3	105.3	198 ± 22	0.57 ± 0.12
950214	11:13:20.8	37.772	42.724	4.7	ANTO	892.9	102.0	270 ± 53	0.83 ± 0.22
950223	21:03:01.3	35.046	32.279	5.8	ANTO	537.3	185.0	202 ± 11	0.42 ± 0.06
950413	20:23:18.4	37.467	36.197	4.6	ANTO	398.5	130.9	196 ± 9	0.55 ± 0.07

950424	23:17:59.0	37.169	34.893	4.4	ANTO	147.9	351.2	177 ± 11	0.40 ± 0.17
950529	04:58:32.4	35.039	32.246	5.3	ANTO	185.3	538.3	223 ± 22	0.43 ± 0.16
960720	00:00:41.9	36.147	27.103	5.7	ANTO	232.2	648.3	254 ± 30	0.45 ± 0.13
910905	19:23:04.8	38.847	41.417	4.3	GNI	245.9	314.6	180 ± 7	0.56 ± 0.08
940917	02:24:37.8	37.885	41.584	5.1	GNI	229.5	363.3	246 ± 9	0.70 ± 0.08
941011	20:31:17.5	33.558	45.678	4.7	GNI	173.0	725.8	280 ± 79	0.63 ± 0.43
941102	12:31:01.0	38.152	48.315	5.0	GNI	123.1	375.6	185 ± 5	0.53 ± 0.07
941120	14:31:02.2	35.335	39.557	5.1	GNI	222.6	694.0	206 ± 24	0.59 ± 0.26
890413	07:49:20.3	41.764	45.855	4.8	KIV	132.2	355.4	235 ± 14	0.84 ± 0.13
890824	18:55:21.1	41.687	49.273	5.2	KIV	112.8	594.5	256 ± 9	0.47 ± 0.09
900420	23:30:03.4	40.002	40.069	5.0	KIV	207.2	490.1	229 ± 20	0.42 ± 0.13
900527	18:27:57.3	40.869	44.236	4.9	KIV	159.1	365.9	247 ± 8	0.75 ± 0.07
900706	19:34:52.4	36.861	49.303	5.3	KIV	142.3	967.0	319 ± 4	0.83 ± 0.03
910427	03:31:58.5	40.093	43.719	4.2	KIV	168.4	437.6	204 ± 25	0.71 ± 0.18
910521	17:37:38.8	42.867	48.028	5.0	KIV	103.8	449.2	225 ± 11	0.69 ± 0.09
910603	10:22:40.4	40.048	42.859	5.0	KIV	178.1	434.5	260 ± 17	0.76 ± 0.10
911021	11:58:19.3	41.982	46.746	4.5	KIV	122.1	392.7	300 ± 15	0.59 ± 0.11
920311	02:26:06.7	42.575	46.391	4.3	KIV	115.8	337.6	211 ± 15	0.61 ± 0.16
920507	19:15:03.3	38.698	40.143	5.0	KIV	200.9	621.8	287 ± 22	0.72 ± 0.11
920922	14:05:55.0	36.334	52.701	5.2	KIV	131.4	1200.4	303 ± 39	0.38 ± 0.92
921127	21:09:16.6	37.473	59.857	5.1	KIV	110.6	1615.7	229 ± 9	0.50 ± 0.12
960814	01:55:02.6	40.754	35.340	5.3	KIV	242.1	702.1	290 ± 24	0.38 ± 0.12
860712	07:54:26.8	29.962	51.582	5.7	BGIO	92.8	1588.0	309 ± 38	0.15 ± 0.26
860803	01:33:20.3	37.200	37.300	5.0	BGIO	17.9	640.7	259 ± 23	0.53 ± 0.18
940620	09:09:02.9	28.968	52.614	5.9	BGIO	95.8	1710.2	424 ± 38	0.69 ± 0.11
940701	20:02:31.8	38.267	38.824	4.7	BGIO	24.1	802.0	254 ± 34	0.63 ± 0.20
940703	14:26:41.3	28.742	34.573	4.3	BGIO	188.7	333.9	347 ± 14	0.60 ± 0.07
940705	10:01:00.1	28.693	34.607	4.0	BGIO	188.0	338.8	298 ± 43	0.30 ± 0.19
940731	05:15:39.5	32.558	48.369	5.3	BGIO	82.3	1255.5	377 ± 68	0.30 ± 0.22
940806	21:02:14.6	26.991	54.363	5.3	BGIO	100.8	1940.1	409 ± 71	0.42 ± 0.21
940917	02:24:37.8	37.885	41.584	5.1	BGIO	39.2	905.4	300 ± 20	0.70 ± 0.11
940920	05:51:46.0	32.501	48.770	5.0	BGIO	82.6	1293.1	359 ± 24	0.76 ± 0.10
941120	14:31:02.2	35.335	39.557	5.1	BGIO	44.8	576.8	380 ± 16	0.74 ± 0.07
950607	23:09:47.0	32.461	48.737	5.0	BGIO	82.8	1290.0	375 ± 55	0.71 ± 0.18

Table 2 (continued)

Date	Origin (h:m:s)	Lat. (°N)	Lon. (°E)	mb	Recording station	Distance (km)	Backazimuth (°)	Q_0	η
951205	18:49:30.4	39.440	40.153	5.5	BGIO	971.2	26.8	295 ± 26	0.72 ± 0.13
951123	18:07:17.2	29.333	34.749	5.2	AFIF	1018.7	307.7	439 ± 11	0.62 ± 0.04
951124	16:43:45.4	28.941	34.710	4.9	AFIF	998.3	305.5	403 ± 17	0.33 ± 0.09
951125	11:41:35.2	29.120	34.836	4.8	AFIF	998.6	306.9	412 ± 49	0.24 ± 0.20
951202	00:47:22.2	29.308	34.657	4.4	AFIF	1024.6	307.3	392 ± 21	0.37 ± 0.13
951206	17:57:54.7	28.862	34.679	4.0	AFIF	996.3	305.0	442 ± 38	0.50 ± 0.11
951208	04:12:38.9	28.919	34.651	4.4	AFIF	1001.9	305.2	369 ± 22	0.46 ± 0.13
951211	01:32:06.5	28.878	34.690	5.0	AFIF	996.3	305.1	398 ± 12	0.65 ± 0.05
951214	03:53:56.5	28.851	34.635	4.2	AFIF	999.3	304.8	349 ± 27	0.47 ± 0.17
951218	03:45:11.0	30.558	50.644	4.6	AFIF	1050.8	44.0	314 ± 28	0.84 ± 0.16
960104	10:31:41.1	32.104	49.518	4.7	AFIF	1106.3	33.7	414 ± 10	0.63 ± 0.05
960331	15:12:16.8	29.657	50.621	4.3	AFIF	984.3	48.3	277 ± 19	0.49 ± 0.18
960407	05:57:52.8	15.703	42.328	4.1	AFIF	913.2	184.8	361 ± 32	0.46 ± 0.23
960420	18:30:28.2	28.056	51.874	4.0	AFIF	994.7	60.8	434 ± 26	0.74 ± 0.16
960430	17:36:09.0	15.296	42.063	4.1	AFIF	960.6	186.3	325 ± 50	0.24 ± 0.22
960501	11:45:20.6	15.448	41.987	4.3	AFIF	944.8	186.9	463 ± 19	0.53 ± 0.07
951123	18:07:17.2	29.333	34.749	5.2	HALM	1195.6	308.9	389 ± 13	0.80 ± 0.07
951124	16:43:45.4	28.941	34.710	4.9	HALM	1174.6	307.1	372 ± 18	0.67 ± 0.07
951125	11:41:35.2	29.120	34.836	4.8	HALM	1175.3	308.2	341 ± 11	0.72 ± 0.05
951202	00:47:22.2	29.308	34.657	4.4	HALM	1201.4	308.6	356 ± 17	0.61 ± 0.09
951204	19:35:36.1	37.833	54.883	4.8	HALM	1197.3	60.4	314 ± 21	0.49 ± 0.09
951206	17:57:54.7	28.862	34.679	4.0	HALM	1172.4	306.6	351 ± 11	0.58 ± 0.06
951208	04:12:38.9	28.919	34.651	4.4	HALM	1178.1	306.8	389 ± 14	0.62 ± 0.06
951211	01:32:06.5	28.878	34.690	5.0	HALM	1172.4	306.7	364 ± 9	0.65 ± 0.04
951214	03:53:56.5	28.851	34.635	4.2	HALM	1175.4	306.5	384 ± 20	0.51 ± 0.09
951218	03:45:11.0	30.558	50.644	4.6	HALM	1060.4	35.0	447 ± 23	0.51 ± 0.09
951231	11:56:39.5	29.385	52.442	4.7	HALM	1087.5	46.6	432 ± 31	0.80 ± 0.13
960104	10:31:41.1	32.104	49.518	4.7	HALM	1146.6	25.5	417 ± 15	0.59 ± 0.07
960126	13:11:14.0	29.388	50.976	4.5	HALM	983.4	41.2	399 ± 25	0.78 ± 0.11

960128	08:43:16.2	34.265	46.453	4.8	HALM	1282.2	8.9	393 ± 23	0.54 ± 0.13
960207	12:27:09.3	38.510	39.271	4.1	HALM	1801.4	345.7	259 ± 31	0.53 ± 0.27
960225	16:14:10.9	35.648	57.067	4.8	HALM	1879.8	38.2	314 ± 37	0.57 ± 0.14
960407	05:57:52.8	15.703	42.328	4.1	HALM	817.1	195.2	393 ± 30	0.66 ± 0.18
960420	18:30:28.2	28.056	51.874	4.0	HALM	953.5	51.2	421 ± 41	0.37 ± 0.24
960430	17:36:09.0	15.296	42.063	4.1	HALM	867.9	196.2	382 ± 19	0.57 ± 0.11
960501	11:45:20.6	15.448	41.987	4.3	HALM	854.0	197.1	445 ± 22	0.51 ± 0.09
951206	17:57:54.7	28.862	34.679	4.0	RANI	1167.7	317.3	443 ± 33	0.61 ± 0.16
951208	04:12:38.9	28.919	34.651	4.4	RANI	1174.0	317.5	391 ± 23	0.54 ± 0.10
951211	01:32:06.5	28.878	34.690	5.0	RANI	1168.2	317.4	439 ± 19	0.66 ± 0.08
951214	03:53:56.5	28.851	34.635	4.2	RANI	1169.9	317.1	428 ± 46	0.33 ± 0.20
951218	03:45:11.0	30.558	50.644	4.6	RANI	1290.9	35.9	350 ± 32	0.53 ± 0.24
960104	10:31:41.1	32.104	49.518	4.7	RANI	1369.6	27.8	421 ± 17	0.30 ± 0.10
960126	13:11:14.0	29.388	50.976	4.5	RANI	1215.8	41.0	399 ± 32	0.70 ± 0.15
960128	08:43:16.2	34.265	46.453	4.8	RANI	1479.5	13.3	411 ± 28	0.55 ± 0.16
960207	12:27:09.3	38.510	39.271	4.1	RANI	1935.3	350.8	279 ± 88	0.38 ± 1.18
960407	05:57:52.8	15.703	42.328	4.1	RANI	622.1	184.4	494 ± 30	0.55 ± 0.13
960430	17:36:09.0	15.296	42.063	4.1	RANI	669.5	186.6	464 ± 34	0.62 ± 0.16
960501	11:45:20.6	15.448	41.987	4.3	RANI	653.8	187.5	499 ± 34	0.57 ± 0.17
951204	19:35:36.1	27.833	54.883	4.8	RAYN	1055.0	61.1	290 ± 30	0.46 ± 0.17
951206	17:57:54.7	28.862	34.679	4.0	RAYN	1231.5	301.0	350 ± 15	0.49 ± 0.09
951208	04:12:38.9	28.919	34.651	4.4	RAYN	1236.7	301.2	326 ± 14	0.68 ± 0.08
951211	01:32:06.5	28.878	34.690	5.0	RAYN	1231.3	301.1	355 ± 12	0.82 ± 0.05
951214	03:53:56.5	28.851	34.635	4.2	RAYN	1234.8	300.8	363 ± 33	0.66 ± 0.15
951218	03:45:11.0	32.104	49.518	4.6	RAYN	931.0	32.1	340 ± 32	0.53 ± 0.24
951206	17:57:54.7	28.862	34.651	4.0	SODA	1408.1	327.6	518 ± 68	0.51 ± 0.33
951208	04:12:38.9	28.919	34.690	4.4	SODA	1414.8	327.6	514 ± 39	0.42 ± 0.14
951211	01:32:06.5	28.878	34.635	5.0	SODA	1408.9	327.7	419 ± 14	0.76 ± 0.05
951214	03:53:56.5	28.851	34.651	4.2	SODA	1409.6	327.4	505 ± 75	0.36 ± 0.36
960104	10:31:41.1	32.104	49.518	4.7	SODA	1688.7	23.7	356 ± 23	0.60 ± 0.16
960126	13:11:14.0	29.388	50.976	4.5	SODA	1507.0	33.8	345 ± 24	0.35 ± 0.16
960128	08:43:16.2	34.265	46.453	4.8	SODA	1814.5	12.1	379 ± 55	0.49 ± 0.39
960331	15:12:16.8	29.657	50.621	4.3	SODA	1510.6	32.1	326 ± 21	0.49 ± 0.16

for the entire region of study range between 315 and 1940 km. We ruled out all earthquakes with epicentral distance less than 300 km to avoid confusion of Lg coda with local coda.

Three typical traces recorded at stations SODA, HALM, and ANTO are presented in Figure 2. The traces are plotted so that the arrivals corresponding to the group velocity of 3.5 km/s are aligned at 200 sec. The top trace (station SODA) and middle trace (station HALM) are from the same earthquake (951211). The Lg coda amplitudes decay more rapidly at HALM, at a distance of 543 km northeast of the station SODA. The difference in amplitude decay for the two traces indicates that the Arabian Peninsula cannot be modeled uniformly, and that Lg coda Q is smaller in the northeastern part of the peninsula than it is in the southwestern part. Figure 2 also indicates that high-frequency energy dies off far more rapidly for the path of ANTO in the Turkish Plateau than along the other two paths. This indicates that Lg coda Q in the northern portion of the Middle East is substantially lower than in the peninsula. Lg travels much faster in the peninsula (top and middle plots of Fig. 2) than in the northern portion of the Middle East (bottom of Fig. 2).

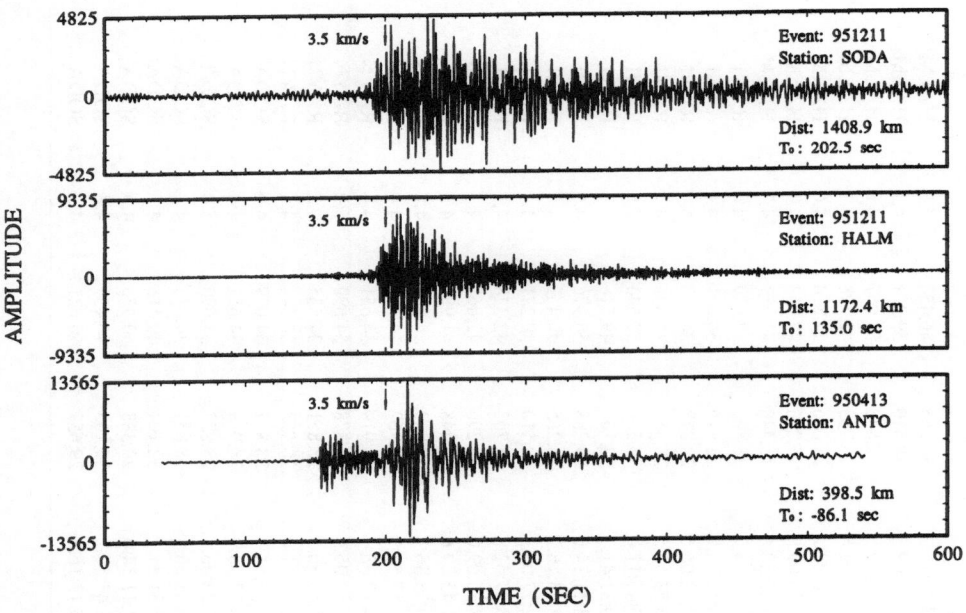

Figure 2
Comparison of seismograms recorded for paths through high-Q (top and mediate) and low-Q (bottom) regions. The vertical axis is amplitude in counts. The data of each event, the station name, the epicentral distance, and the time of the time-scale origin measured from the event origin time are indicated in each box.

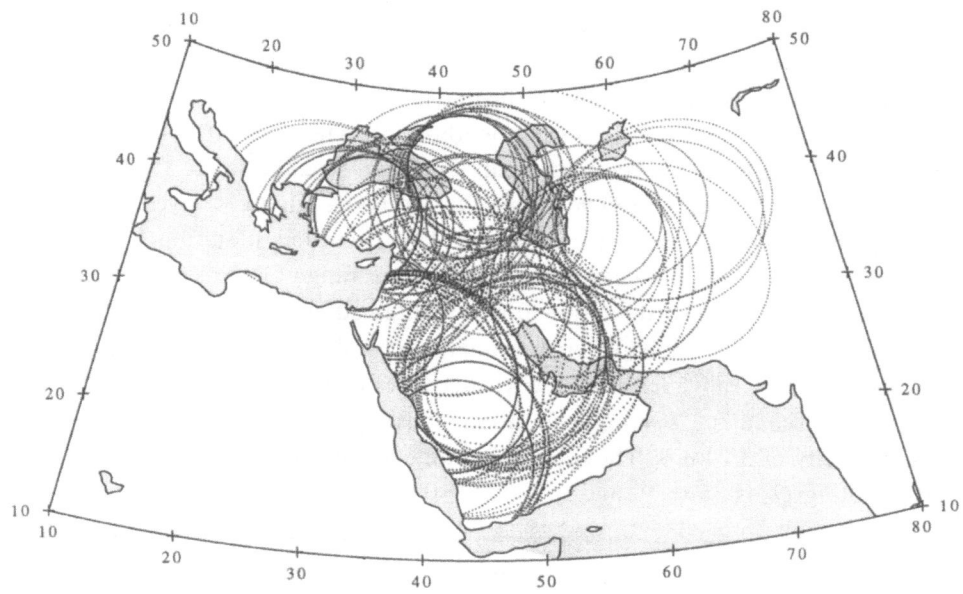

Figure 3

Sampling pattern of *Lg* coda. Each record of *Lg* coda is assumed to sample an elliptical area corresponding to the maximum lapse time used in the analysis. When a continental boundary is encountered we assumed it is a barrier to *Lg* propagation and the ellipses are truncated at those boundaries.

To map this variation in detail we have determined *Lg* coda *Q* for a large number of seismograms covering the Middle East. Figure 3 shows the sampling pattern of *Lg* coda. Each record of *Lg* coda is assumed to sample an elliptical area corresponding to the maximum lapse time used in the single trace analysis. The event and station lie at the two foci of each ellipse. The event and station configurations provide excellent data coverage for the Middle East.

Single Trace Analysis

We take *Lg* coda to start at a group velocity of about 3.15 km/s. The starting time of the coda was adjusted for each trace so that a clear amplitude decay can be detected. Each of the coda time series was divided into numerous windows with the same window length, and a Fourier transform was performed for each window. The spectra of the windowed signals were then compared with those of noise spectra taken from a time interval prior to the *P* arrival to determine a usable portion of the coda with good signal/noise ratio. The lengths of the usable portion of the coda vary from trace to trace, depending on the event magnitude, the epicentral distance, and the attenuation along the path. In general, the usable length is in the range 200–500 sec. The effect of noise was empirically reduced by subtracting the reference power spectrum of noise from that of the signal.

The stacked spectral ratio (SSR) method of XIE and NUTTLI (1988) was employed for single-trace Q measurements. Fitting a straight line to the SSR values by least-squares yields Q_0 and η values for the coda. Figure 4 gives an example showing the SSR analysis for earthquake 951211 recorded at station HALM. The upper right portion of the figure displays the time range of the selected coda, the sampling interval of the data, the points contained in the coda, the epicentral distance, the value of Q_0 and its error estimate, the value of η and its error estimate, and the frequency range used for the least-squares determination of Q_0 and η. For this event the coda is 326.4 sec in duration, starting at a time corresponding to the group velocity of 3.1 km/s. The frequency range for fitting the SSR is 0.3 to 3.0 Hz. The resulting Q_0 is 364 ± 9, and η is 0.65 ± 0.04.

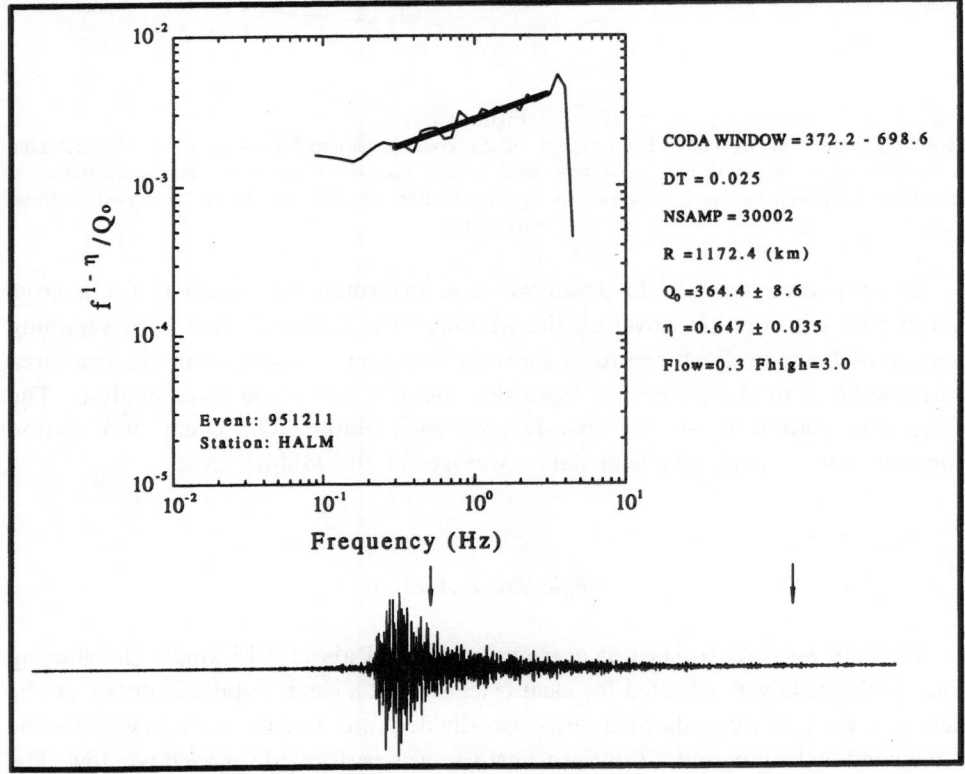

Figure 4

An example of linear regression determination of Lg coda Q and its frequency dependence at 1 Hz for earthquake 951211 recorded at station HALM. Arrows above the seismogram indicate the beginning and end times for the portion of coda used in the analysis.

In the analysis, we treated the Red Sea and Mediterranean Sea as oceanic regions, that is we assumed that neither *Lg* nor *Lg* coda propagated there. When such a boundary between oceanic and continental crust is encountered by the *Lg* wave coming from the continental side, part of the energy is scattered back into the continent. We did not make that assumption for either the Black Sea or Caspian Sea although it is known that portions of those seas overlie oceanic type crust. We assumed that the oceanic portions of the Caspian and Black Seas were small enough to ignore and *Lg* coda would be scattered around these regions. For that reason, the scattering ellipses in Figure 3 were not truncated at those margins. Our results could be incorrect in and near the seas if our assumption regarding them is incorrect.

Lateral Variation of Q_0

The back-projection method (XIE and MITCHELL, 1990a) was then employed to map the *Lg* coda *Q* images, using cells that are 2° by 2° in area. The resulting Q_0 values are shown in Figure 5. The Q_0 varies smoothly across the Middle East in the range between 150 and 450.

In the Arabian Peninsula Q_0 varies between about 450 and 250 with most values in the range of 350–450. The highest Q_0 values of 400–450 occur in the shield area of the Arabian Peninsula where the crust is relatively stable. Q_0 values decrease smoothly and slowly eastward, northward and northeastward. They drop to 400 or less in the shelf area, and 350 in the platform and basin region. The Oman folded zone, in the southeastern portion of the peninsula, exhibits Q_0 values of about 250, reflecting the effect of the thick sediments there. The results for the Arabian Peninsula are higher than the average value of 214 for the same region reported by GHALIB (1992). The difference in Q_0 values can be attributed to the improved station distribution and hence the higher signal/noise ratios available in this study.

In Iran Q_0 varies between 250 and 350 with higher values in the south and lower values in the north. These values are slightly higher than the Q_0 estimate of approximately 200, obtained by NUTTLI (1980). Q_0 decreases gradationally to about 200 or less in the northern Caspian Sea and surrounding regions.

The Turkish Plateau and the Black Sea exhibit low Q_0 values, ranging between about 150 and 250, with higher values in the east and lower values in the west. This agrees with the results of GHALIB (1992). The results obtained by GHALIB (1992) show low *Q* values of about 200 for several paths crossing the plateau. There is a small region centered at 46°N and 42°E where Q_0 is slightly higher than in surrounding regions, roughly 250 to 300. This might be real or might be caused by inadequate data coverage in this area.

In summary, Q_0 values are relatively high in the Arabian Peninsula, especially in the old, relatively stable shield region, and lower in the Zagros fault zone, the

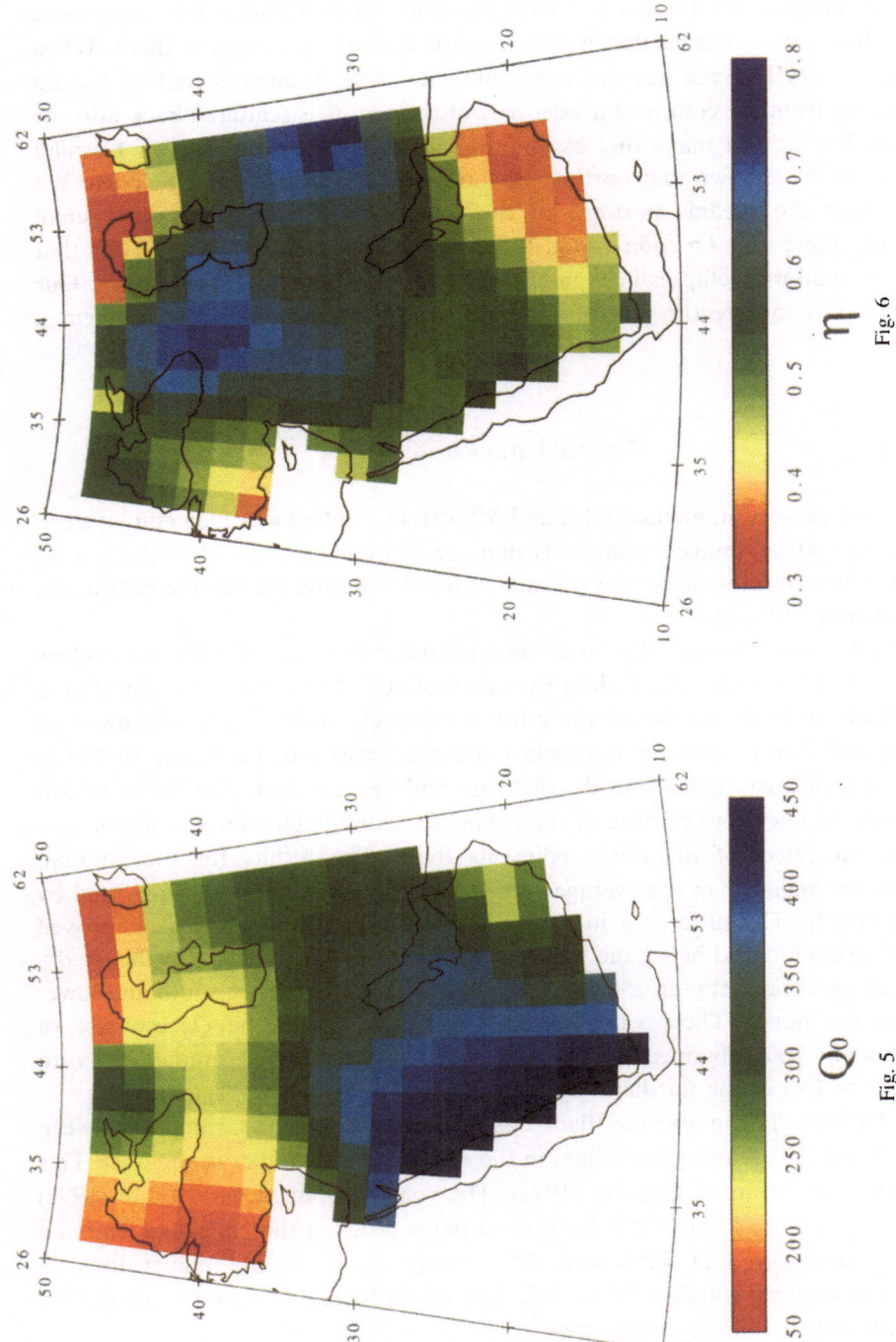

Figure 5
Tomographic map of *Lg* coda *Q* at 1 Hz for the Middle East. Each cell is 2° by 2° in area.

Figure 6
Tomographic map of the frequency dependence of *Lg* coda *Q* at 1 Hz. Each cell is 2° by 2° in area.

Turkish and the Iranian Plateaus, and the Black and Caspian Seas. The lowest values were found in western Anatolia. The southern and northern portions of the Middle East can be characterized by distinctly different Q_0, with values registering between 350 and 450 in most of the Arabian Peninsula and between 150 and 300 in the northern portion which is more seismically active.

Lateral Variation of η

Following the procedures of XIE and MITCHELL (1990a), we extrapolated Q to 3 Hz for each trace using Q_0 and η determined at 1 Hz. The calculated value represents the areal average of Q at 3 Hz within each elliptical area. The Q values at 3 Hz for all of the cells can then be calculated, and the areal distribution of η can be obtained. Figure 6 illustrates the resulting lateral variation of η in the Middle East. η varies between 0.3 and 0.8 across the region. η is quite uniform in the central and western portions of the peninsula, being about 0.5. It decreases southeastward to a value of 0.4 in the Oman folded zone and increases gradationally northward and eastward from the central part of the peninsula. η is highest in the Iranian Plateau, where it ranges 0.6–0.8, and the eastern Turkish Plateau, where it is between 0.7 and 0.8. In western Anatolia η decreases to between 0.4 and 0.5. In the northeastern corner of the region studied η is between 0.3 and 0.4. Errors may, however, be large in this region because of poorer data coverage there. If we ignore the southeastern margin and the western Turkish Plateau, our images of lateral variation of η and Q_0 support the suggestion by NUTTLI (1988) that η correlates inversely with Q_0. The southern portion of the area under study exhibits a higher Q_0 (greater than 350) and lower η (less than 0.5); the northern portion has a lower Q_0 (less than 300) and higher η (greater than 0.6). The two exceptions are the Oman folded zone and western Anatolia, where Q_0 and η are both low. These low values may indicate anomalous properties in these regions or it is also possible that Q_0 and η determinations are affected systematically by continent-ocean boundaries in those regions.

Resolution and Error

To estimate the resolution we use the "point-spreading function" (*psf*) as employed by HUMPHREYS and CLAYTON (1988), and XIE and MITCHELL (1990a) as an approximation of the resolution kernel. To obtain the *psf*, a unit value of Q^{-1} is assigned to the cell of interest and the Q^{-1} values in all other cells are set to zero. Synthetic data using this model are then computed and inverted. The breadth of the resulting image will then provide a measure of resolution.

We have calculated the *psf* for 8 locations to estimate the resolving power. The results appear in Figure 7. Four plots are included in the figure, each containing two *psf* patterns. The center locations of the cells are 43°N, 45°E and 21°N, 45°E (Fig. 7a); 37°N, 43°E and 21°N, 51°E (Fig. 7b); 43°N, 57°E and 33°N,

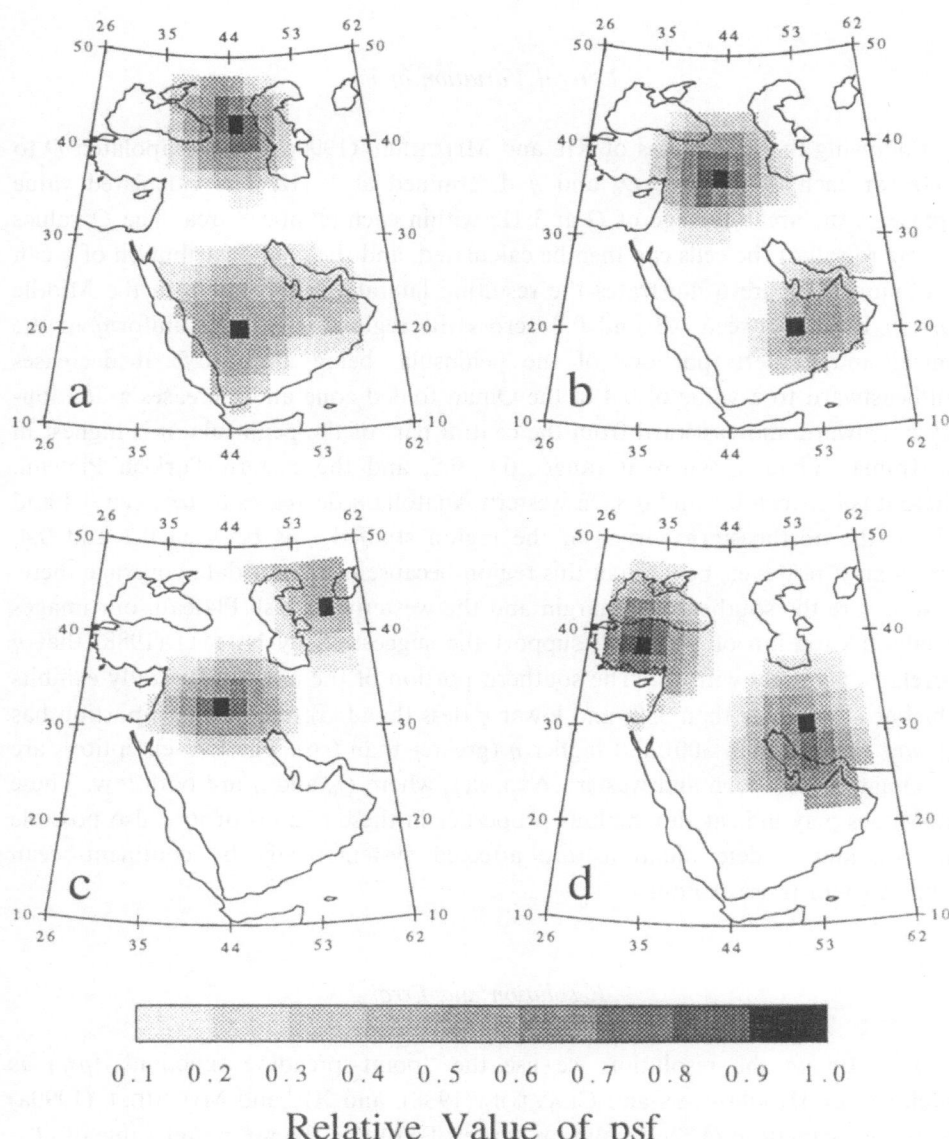

Figure 7
Point-spreading function computed at eight locations.

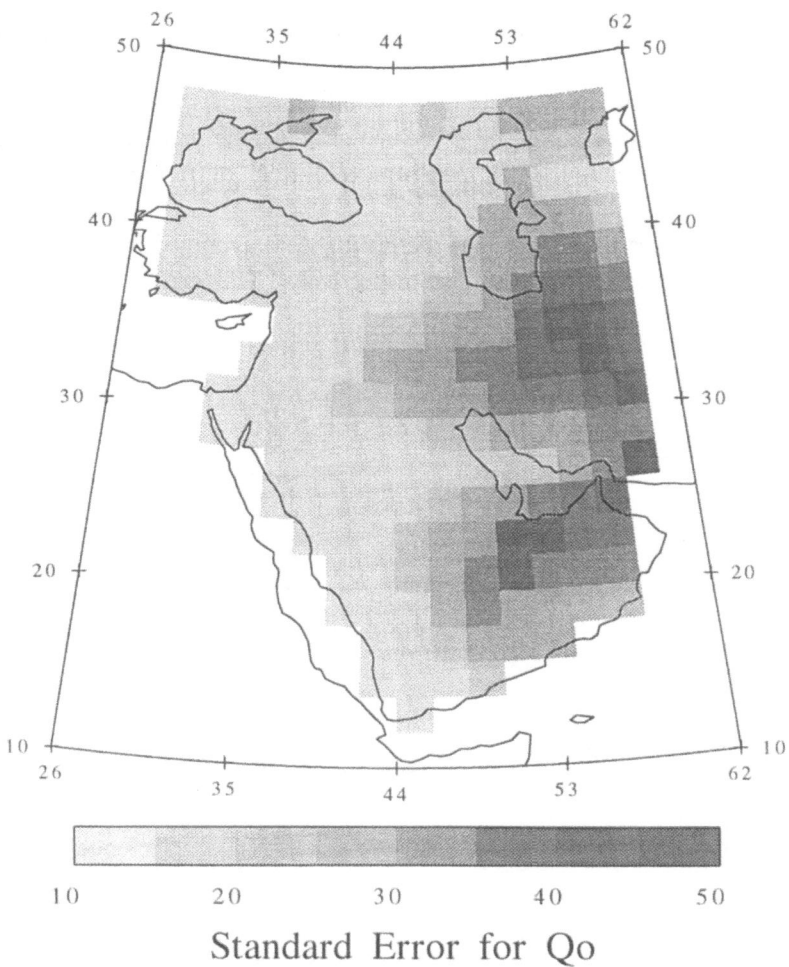

Figure 8
Error distribution for the image of *Lg* coda *Q*.

43°E (Fig. 7c); 39°N, 33°E, and 31°N, 53°E (Fig. 7d). The *psf* values are normalized so that the maximum values, occurring at the center cells, are unity. In general, the spatial resolution is good for most of the region under study, ranging from 6° or less to about 10°. The best resolution occurs west of the Persian Gulf and in the center of the Arabian Peninsula, while poorer resolution characterizes Turkey, southern Iran, and the region between the Caspian and Black Seas. Because of the excellent data coverage, the resolution obtained in this study is considerably better than that of XIE and MITCHELL (1990a) for Africa.

To estimate the errors of Q_0 and η, we first constructed an error series, which consists of the errors in single-trace measurements. We then gave each element a randomly chosen sign to form a new error series. A new Q_0 series was thus formed and inverted to obtain a new Q model. The difference between the new image of Q values and the original image of Q values gave us the error estimates of our Q values. This process empirically measures the effect of random noise on the Q image.

In practice, we ran ten tests, and the results were averaged. Figures 8–9 present the estimated errors in Q_0 and η, respectively. The errors in Q_0 are minor

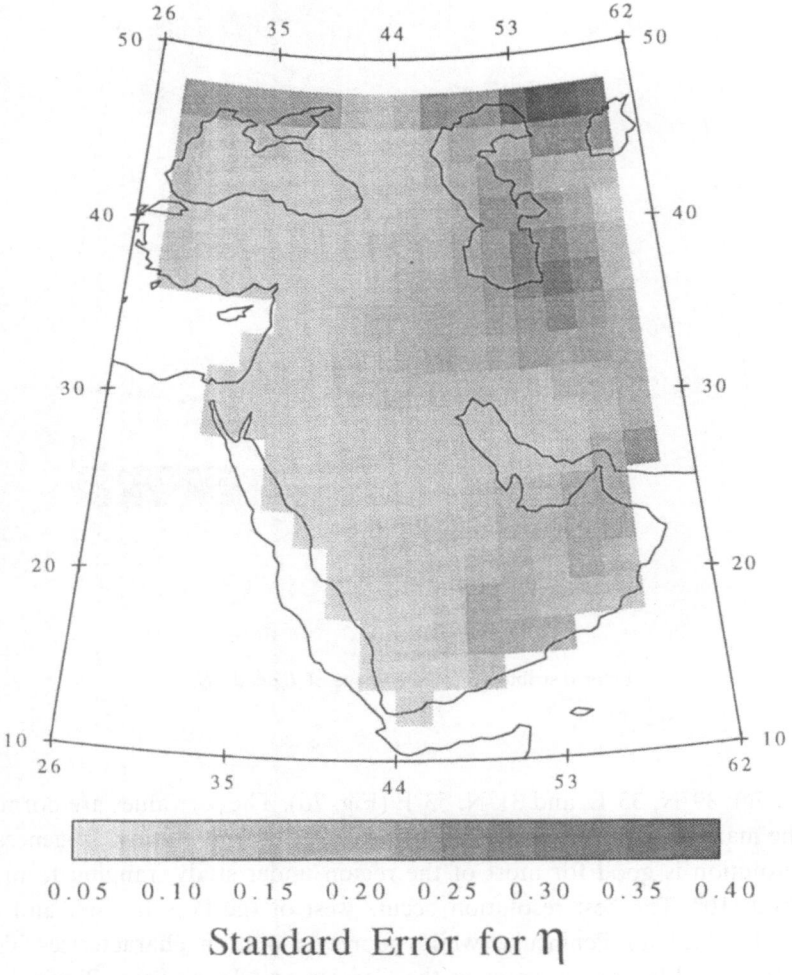

Standard Error for η

Figure 9
Error distribution for the image of the frequency dependence of Lg coda Q.

throughout the region, being less than 50. In the Arabian shield the errors are less than 20; the relative errors being less than 6%. In the shelf and basins the errors become larger, as high as 44 in the southeastern margin, a value equivalent to a relative error of 14%. The lowest errors occur in the Turkish Plateau and the Black Sea, where they are less than 15, a value equivalent to a relative error of about 10%. The highest errors are in the Iranian Plateau, being between 25 and 50, producing a relative error there as high as 15%.

The estimated error in η is in the range between 0.05 and 0.2 for most of the Middle East. It is interesting that the distributions of η and Q_0 are similar in that errors in the western portion of the Middle East are much lower than those in the eastern portion. Errors in the peninsula are as low as 0.1 except near the southeastern margin, where errors increase to about 0.2. Like the errors Q_0, the errors in η in the Turkish Plateau and the Black Sea are low (less than 0.1). East of this region η errors increase. In Iran and the Caspian Sea errors in η range between 0.1 and 0.25. North of the Caspian and Black Seas, where data coverage is poor, the errors can be as high as 0.4.

Discussion

Previous studies have revealed that Q_0 is usually low in the tectonically active regions. Examples are the East Africa Rift (GUMPER and POMEROY, 1970; XIE and MITCHELL, 1990a), and the western United States (SINGH and HERRMANN, 1983; XIE and MITCHELL, 1990b; BAQER and MITCHELL, 1998). The northern portion of the Middle East is a tectonically active region. High seismicity and deformation occur along the Zagros folded belt, the East and North Anatolian Fault Zones, the Dead Sea fault system, and the Turkish and Iranian Plateaus. Q_0 varies between 150 and 300, comparable to the results for the western United States and other tectonically active regions. The lowest Q_0 value of about 150 is in the western Anatolian block. This region is dominated by extensional deformation and a horst-graben system (SENGÖR, 1987). High levels of seismicity can be found there. More than 1/3 of the total number of earthquakes in the Middle East occur in this region. High heat flow in this region was measured by ILKISIK (1995). The average heat flow there is about 107 mW m^{-2} and the highest value at 29°13.2'E is 247 mW m^{-2}.

MITCHELL (1995) proposed that both *Lg* coda *Q* and shear-wave *Q* in the upper crust correlate with the time elapsed since the most recent tectonic activity. Since the northern portion of the Middle East is currently active, the low *Lg* coda *Q* values found there were expected. According to MITCHELL (1995), the heating caused by subduction in tectonically active regions triggers hydrothermal reactions that release fluids. Those fluids gradually move to the surface through cracks and permeable rock. As seismic waves travel through those regions they lose energy that is used to move fluids short distances through the permeable rock.

Previous studies indicate that Q_0 is typically high in stable regions. For example, Q_0 is about 900 in the Siberian shield (MITCHELL *et al.*, 1997), 700 or more in the central United States (BAQER and MITCHELL, 1998), and 800–1000 in the cratons of Africa (XIE and MITCHELL, 1990a). The Arabian Peninsula is also relatively stable. Nevertheless, although Q_0 values are higher there than the northern portion of the Middle East, they are much lower (300–450) than Q_0 observed in other shield areas.

The Arabian Peninsula has been the site of large-scale uplift and extension with associated volcanic activity. The Afro-Arabian dome began forming in the early Cretaceous and is the result of two or more generating systems that are still active (ALMOND, 1986). Temperatures beneath the dome are as high as 900°C at the depth of 40 km and these abnormally high temperatures extend as far as 200 km inland from the Red Sea (MCGUIRE and BOHANNON, 1989). The low crustal Q may be due to water released from hydrothermal reactions at those elevated temperatures or may result from high temperatures at depth which may drive the circulation of fluids beneath the surface (WYLLIE, 1988). Some of the reduced Q_0 values may, however, be caused by thick young sediments. The Q_0 values decrease gradually to 350 in the Basins and 250 in the Oman folded zone. Average crustal shear-wave Q is also lower in this region than it is in the western portion of the Arabian Peninsula (CONG and MITCHELL, 1998). A profile (SEBER *et al.*, 1997), trending roughly east-west, shows that the thickness of the sediments increases gradually across the shelf area, more rapidly to the east, and reaches about 10 km in the platform and basins.

Conclusions

Q_0 varies laterally across the Middle East ranging between 150 and 450. On a gross scale the Middle East can be divided into two regions characterized by different ranges of *Lg* coda Q values. The northern portion consisting of the Turkish and Iranian Plateaus and the Black Caspian Seas, is characterized by lower Q_0 values, between 150 and 300. The southern portion consisting of most of the Arabian Peninsula is characterized by higher Q_0 values in the range 350–450. In the Oman folded zone, however, a lower Q_0 of about 250 was found. The low Q_0 values in the northern portion of the Middle East can be explained by recent tectonic activity, severe deformation, and very high heat flow values. Deviation from the typical high Q_0 values in the peninsula can be explained by extension and doming with recent volcanism there. The Q_0 distribution in the Arabian Peninsula correlates well with geology. The highest Q_0 (450) occurs in the shield area of western Saudi Arabia. Q_0 decreases gradually eastward to 350 in the basins and 250 near the southeastern margin where thick sediments occur.

Most of the Middle East exhibits an inverse relation between η and Q_0. The Arabian Peninsula with relatively high Q_0 of 350–450 exhibits relatively low η of 0.4–0.5. The Turkish and Iranian Plateaus with relatively low Q_0 values of 150–300 show relatively high η of 0.6–0.8. The western Anatolia and Oman are exceptions, where both Q_0 and η are low.

Acknowledgments

We thank the IRIS Data Management Center and the German GEOPHONE Data Center for providing the data used in this study. This research was supported by the U.S. Department of Energy and was monitored by the Phillips Laboratory under contract F19628–95–K–0004.

REFERENCES

ALMOND, D. C. (1986), *Geological Evolution of the Afro-Arabian Dome*, Tectonophysics *131*, 301–332.

BAQER, S. A., and MITCHELL, B. J. (1998), *Regional Variation of Lg Coda Q in the Continental United States and it Relation to Crustal Structure and Evolution*, Pure and appl. geophys., *153*, 613–638.

BARKA, A. A., and GÜLEN, L. (1988), *New Constraints on Age and Total Offset of the North Anatolian Fault Zone: Implications for Tectonics of the Eastern Mediterranean Region, Middle East Technical University*, J. Pure Appl. Sci. *21*, 39–63.

BEN-AVRAHAM, Z., GARFUNKEL, Z., ALMAGRO, G., and HALL, J. K. (1979), *Continental Breakup by Leaky Transform: The Gulf of Elat (Aqaba)*, Science *206*, 214–216.

BEN-MENAHEM, A., AUR, A., and VERED, M. (1976), *Tectonics, Seismicity and Structure of the Afro-Eurasian Junction—The Breaking of an Incoherent Plate*, Phys. Earth Planet. Int. *12*, 1–50.

BERBERIAN, M. (1976), *Contribution to the Seismotectonics of Iran, II*, Rep. Feol. Surv. Iran *39*, pp. 516.

BOZKURT, E., and KOCYIGIT, A. (1996), *The Kazova Basin: An Active Negative Flower Structure on the Almus Fault Zone, a Splay Fault System of the North Anatolian Fault Zone, Turkey*, Tectonophysics *265*, 239–254.

BROWN, G. F. (1972), *Tectonic Map of the Arabian Peninsula, Saudi Arabian Peninsula Map AP-2*. Saudi Arabian Dir. Gen. Miner. Resour.

CANITEZ, N., and UCER, S. B. (1967), *Computer Determinations for the Fault-plane Solutions in and near Anatolia*, Tectonophysics *4*, 235–244.

COLEMAN, R. G., *Ophiolites. Ancient Oceanic Lithosphere?* (Springer-Verlag, Berlin, 1977) 229 pp.

COLMAN-SADD, S. (1978), *Fold Development in Zagros Simply Folded Belt, Southwest Iran*, Am. Assoc. Pet. Geol. Bull. *62*, 984–1003.

CONG, L., and MITCHELL, B. J. (1998), *Seismic Velocity and Q Structure of the Middle Eastern Crust and Upper Mantle from Surface-wave Dispersion and Attenuation*, Pure and appl. geophys., *153*, 503–538.

EYAL, M., EYAL, Y., BARTOV, Y., and STEINITZ, G. (1981), *The Tectonic Development of the Western Margin of the Gulf of Elat (Aqaba) Rift*, Tectonophysics *80*, 39–66.

FALCON, N., *Southern Iran: Zagros Mountains, Mesozoic-Cenozoic Orogenic Belts* (ed. A. Spencer), vol. 4 (Spec. Publ. Geol. Soc. London 1974) pp. 199–211.

FREUND, R. (1965), *A Model of the Structural Development of Israel and Adjacent Areas since Upper Cretaceous Times*, Geol. Mag. *102*, 189–205.

FREUND, R., GARFUNKEL, Z., ZAK, I., GOLDBERG, M., WEISSBROD, T., and DERIN, B. (1970), *The Shear along the Dead Sea Rift*, Phil. Trans. Roy. Soc. London *A267*, 107–130.

GARFUNKEL, Z. (1981), *Internal Structure of the Dead Sea Leaky Transform (Rift) in Relation to Plate Kinematics*, Tectonophysics *70*, 71–107.

GARFUNKEL, Z., and BEN-AVRAHAM, Z. (1996), *The Structure of the Dead Sea Basin*, Tectonophysics *266*, 155–176.

GHALIB, H. (1992), *Seismic Velocity Structure and Attenuation of the Arabian Plate*, Ph.D. Dissertation, St. Louis University, 314 pp.

GUMPER, F., and POMEROY, P. W. (1970), *Seismic Wave Velocities and Earth Structure on the African Continent*, Bull. Seismol. Soc. Am. *60*, 651–668.

HUMPHREYS, E., and CLAYTON, R. W. (1988), *Adaptation of Back-projection Tomography to Seismic Travel-time Problems*, J. Geophys. Res. *93*, 1073–1086.

ILKISIK, O. M. (1995), *Regional Heat Flow in Western Anatolia Using Silica Temperature Estimates from Thermal Springs*, Tectonophysics *244*, 175–184.

JIH, R. S., and LYNNES, C. S., *Re-examination of regional Lg Q variation in Iranian Plateau. In* Proc. 14th Ann. PL/DARPA Seismic Research Symposium, 16–17 September 1992 (eds. Lewkowicz, J. F. and McPhetres, M.), (Phillips Laboratory, 1992) pp. 200–206.

KADINSKI-CADE, K., BARANZANGI, M., OLIVER, J., and ISACKS, B. (1981), *Lateral Variations of High-frequency Seismic Wave Propagation at Regional Distances across the Turkish and Iranian Plateaus*, J. Geophys. Res. *86*, 9377–9396.

KETIN, I. (1966), *Tectonic Units of Anatolia*, Bull. Miner. Res. Explo. Inst. Turk. *66*, 23–34.

KOCYIGIT, A. (1988a), *Tectonic Setting of the Geyve Basin: Age and Total Displacement of the Geyve Fault Zone*, Middle East Tech. Univ., J. Pure Appl. Sci. *21*, 81–104.

KOCYIGIT, A. (1988b), *Basic Geological Characteristics and Total Offset of North Anatolian Fault Zone in Susehri Area, NE Turkey*, Middle East Tech. Univ., J. Pure Appl. Sci. *22*, 43–68.

KOCYIGIT, A. (1989), *Susehri Basin: An Active Fault-wedge Basin on the North Anatolian Fault Zone, Turkey*, Tectonophysics *167*, 13–29.

McGUIRE, A. V., and BOHANNON, R. G. (1989), *Timing of Mantle Upwelling: Evidence for a Passive Origin for the Red Sea Rift*, J. Geophys. Res. *94*, 1677–1682.

McKENZIE, D. P. (1972), *Active Tectonics of the Mediterranean Regions*, Geophys. J. R. Astr. Soc. *30*, 109–185.

MITCHELL, B. J. (1995), *Anelastic Structure and Evolution of the Continental Crust and Upper Mantle from Seismic Surface Wave Attenuation*, Rev. Geophys. *33*, 441–462.

MITCHELL, B. J., PAN, Y., XIE, J., and CONG, L. (1997), *Lg coda Q Variation across Eurasia and its Relation to Evolution*, J. Geophys. Res *102*, 22767–22779.

MITCHELL, B. J., BAGER, S., AKINCI, A., and CONG, L. (1998), *Lg code Q in Australia and its Relation to Crustal Structure and Evolution*, Pure and appl. geophys., *153*, 639–657.

MORRIS, P. (1977), *Basement Structure as Suggested by Aeromagnetic Surveys in S. W. Iran*, Internal Rep., Oil Serv. Co. of Iran, Tehran, Iran.

NATIONAL IRANIAN OIL COMPANY (1975), *Geological Cross Sections South-central Iran, Scale 1:500,000*, Explor. and Prod., Tehran, Iran.

NATIONAL IRANIAN OIL COMPANY (1976), *Geological Cross Sections Southwest Iran and Northern Persian Gulf, Scale 1:500,000*, Explor. and Prod., Tehran, Iran.

NOWROOZI, A. A. (1972), *Focal Mechanism of Earthquakes in Persia, Turkey, West Pakistan, and Afghanistan and Plate Tectonics of the Middle East*, Bull. Seismol. Soc. Am. *62*, 823–850.

NUTTLI, O. W. (1980), *The Excitation and Attenuation of Seismic Crustal Phases in Iran*, Bull. Seismol. Soc. Am. *70*, 469–485.

NUTTLI, O. W. (1988), *Lg Magnitudes and Yield Estimates for Underground Novaya Zemlya Nuclear Explosions*, Bull. Seismol. Soc. Am. *78*, 873–884.

POWERS, R. W., RAMIREZ, L. F., REDMOND, C. P., and ELBERG, E. L. (1966), *Geology of the Arabian Peninsula—Sedimentary Geology of Saudi Arabia*, U.S. Geol. Surv., Prof. Pap. *560-D*, 147 pp.

SEBER, D., and MITCHELL, B. J. (1992), *Attenuation of Surface Waves across the Arabian Peninsula*, Tectonophysics *204*, 137–150.

SEBER, D., VALLVÉ, M., SANDVOL, E., STEER, D., and BARAZANGI, M. (1997), *Middle East Tectonics: Applications of Geographic Information Systems (GIS)*, GSA Today *7*, 1–6.

SENGÖR, A. M. C., *Cross-faults and differential stretching of hanging walls in regions of low-angle normal faulting: examples from western Turkey. In* Continental Extensional Tectonics (eds. Coward, M. P., Dewey, J. F., and Hancock, P. L.) (Geol. Soc. London, Spec. Publ. *28*, 1987) pp. 575–589.

SENGÖR, A. M. C., and KIDD W. S. F. (1979), *Post-collisional Tectonics of the Turkish and Iranian Plateaus and a Comparison with Tibet*, Tectonophysics *55*, 361–376.

SINGH, S. K., and HERRMANN, R. B. (1983), *Regionalization of Crustal Coda Q in the Continental United States*, J. Geophys. Res. *88*, 527–538.

SOUZA, J. L., and MITCHELL, B. J. (1998), *Lg Coda Q Variations across South America and their Relation to Crustal Evolution*, Pure and appl. geophys., *153* 587–612.

STEINITZ, G., EYAL, Y., EYAL, M., and EYAL, Y. (1980), *K-Ar Age Determination of Tertiary Magmatism along the Western Margin of the Gulf of Elat*, Geol. Surv. Isr. Curr. Res, 27–29.

STOCKLIN, J. (1968), *Structural History and Tectonic Iran: A Review*, Am. Assoc. Pet. Geol. Bull. *52*, 1229–1258.

STOCKLIN, J., *Possible ancient continental margins in Iran. In* Geology of Continental Margins (eds. Burk, C. and Drake, C.) (Springer-Verlag, New York, 1974) pp. 873–877.

STONELEY, R. (1981), *The Geology of the Kuh-e Dalneshin Area of Southern Iran, and it Bearing on the Evolution of Southern Tethys*, J. Geol. Soc. London *138*, 509–526.

TOPRAK, V. (1988), *Neotectonic Characteristics of the North Anatolian Fault Zone between Koyul-hisar and Susehri (NE Turkey)*, Middle East Tech. Univ., J. Pure Appl. Sci. *21*, 155–168.

WESTAWAY, R. (1994), *Present-day Kinematics of the Middle East and Eastern Mediterranean*, J. Geophys. Res. *99*, 12071–12090.

WYLLIE, P. J. (1988), *Magma Genesis, Plate Tectonics, and Chemical Differentiation of the Earth*, Rev. Geophys. *26*, 370–404.

XIE, J., and NUTTLI, O. W. (1988), *Interpretation of High-frequency Coda at Large Distances: Stochastic Modeling and Method of Inversion*, Geophysical Journal *95*, 579–595.

XIE, J., and MITCHELL, B. J. (1990a), *A Back-projection Method for Imaging Large-scale Lateral Variations of Lg coda Q with Application to Continental Africa*, Geophys. J. Int. *100*, 161–181.

XIE, J., and MITCHELL, B. J. (1990b), *Attenuation of Multiphase Surface Waves in the Basin and Range Province, Part I: Lg and Lg Coda*, Geophys. J. Int. *102*, 121–137.

ZAK. I., and FREUND, R. (1981), *Asymmetry and Basin Migration in the Dead Sea Rift*, Tectonophysics *80*, 27–38.

(Received October 30, 1997, revised May 6, 1998, accepted May 27, 1998)

To access this journal online:
http://www.birkhauser.ch

Pure appl. geophys. 153 (1998) 587–612
0033–4553/98/040587–26 $ 1.50 + 0.20/0

❙Pure and Applied Geophysics

Lg Coda *Q* Variations across South America and their Relation to Crustal Evolution

J. L. DE SOUZA[1] and B. J. MITCHELL[2]

Abstract—Nine broadband seismograph stations in South America have provided 389 recordings of *Lg* coda with paths that cover most of the continent. *Lg* coda *Q* (Q_0) and frequency dependence (η) values at 1 Hz, obtained from these records, were inverted using back-projection tomography to obtain regionalized maps of Q_0 and η. The entire western margin of the continent (the active Andean mountain belt) is typified by low Q_0 (250–450), whereas broad regions of high Q_0 (700–1100) span the central Brazilian shield and contiguous regions to the north and south. Intermediate Q_0 (450–700) characterizes the northern Patagonia platform and most of the Atlantic shield. Reduced Q_0 in the Atlantic shield may be related to tectonic or igneous activity that occurred during the breakup of Gondwanaland during the Jurassic period. This *Q* distribution is generally consistent with earlier studies where Q_0 was found to be directly proportional to the time that has elapsed since the most recent episode of major tectonic or orogenic activity in any region. Reduced Q_0 in the Patagonian platform may, however, be due to young sediments there.

Q_0 is slightly higher in two portions of the Andean belt (between latitudes 2.0°N and 10.0°S, and between latitudes 24.0°S and 34.0°S) than in other portions of the belt. These variations are consistent with results of earlier studies of body-wave attenuation and heat flow in the Andean mountain belt.

Spatial variations of η generally vary inversely with Q_0, being low (0.0–0.2) throughout a broad region centered in the central Brazil shield and extending to the northeastern coast. All surrounding regions except that to the northeast exhibit intermediate to high (0.4–0.8 and possibly higher) η values. Possible biasing of *Lg* coda *Q* measurements by proximity to the transition between the South American and Pacific plates was examined using records from a station near that boundary and was found to be small.

Key words: *Lg*, coda, *Q*, South America, crust, attenuation.

[1] Observatório Nacional, Departamento de Geofísica, R: Gal. José Cristino, 77, São Cristovão, Rio de Janeiro, 20921-400, Brazil.
[2] Department of Earth and Atmospheric Sciences, Saint Louis University, 3507 Laclede Avenue, St. Louis, Missouri, U.S.A.

1. Introduction

South America is a geologically diverse continent with a complex tectonic history. In shield regions the oldest rocks are more than 3.5 By in age while the Andean mountain belt is evolving and being deformed today by interaction with the subducting Nazca plate.

The seismological parameter, Q, which is inversely proportional to seismic wave attenuation, is strongly affected by the tectonic evolutionary history of any region (MITCHELL, 1995). For that reason it is an interesting parameter to study in South America. Study of seismic Q also has practical benefits, since it is important for magnitude determinations using regional phases (NUTTLI, 1973).

Studies of seismic wave attenuation have, for several reasons, been limited in South America. These include the small number of available stations, their inadequate distribution, and because recording systems of older stations were often not well calibrated or were not sensitive to ground motion in frequency ranges of interest. Those shortcomings have been mitigated in recent years with the installation of several high-quality broadband, digital seismic stations in different parts of South America.

The objective of the present study is to use those modern stations and recently developed methods of data analysis and tomographic inversion to image the lateral variations of Lg coda Q (Q_0) and its frequency dependence (η) at 1 Hz for as much of South America as possible. Lg coda Q has been found, in many earlier studies, to be primarily sensitive to anelastic properties of the crust. We will relate variations of these crustal anelastic properties to aspects of the tectonic evolution and geology of South America.

2. Geotectonic Framework of South America

The principal geological features of South America and their relation to the plate tectonic framework of that region have become increasingly well understood over the past three decades. Of the three major tectonic provinces (the South American platform, the Patagonian platform, and the Andean Cordillera), two are stable and one is tectonically active (Fig. 1). Several authors (e.g., JAMES, 1971; GANSSER, 1973; CORDANI et al., 1973; ALMEIDA et al., 1981; FORSYTHE, 1982; DALZIEL, 1986; BRIO NEVES and CORDANI, 1991) have described aspects of the geological and tectonic development of these provinces.

The largest of the provinces, the South American platform, is also the oldest portion of the South American plate. It is bounded on the west by the tectonically active Andean belt and on the east by the Atlantic Ocean. The platform consists essentially of Precambrian basement, much of which is covered by sedimentary and volcanic rocks of Phanerozoic age. It includes three large exposed shield areas: (1)

the Guiana shield, consisting of granulites and gneisses with ages varying from 3700 to 3400 Ma, (2) the central Brazilian shield, where rocks with ages as great as 2500 Ma have been found, and (3) the Atlantic shield where ages are as great as 3400 Ma but were extensively overprinted at about 600 Ma.

Most of the sediments within the South American platform lie in four great intracratonic basins (Fig. 1). These basins, the Amazonian, Parnaíba, Paraná and Chaco are discussed by ALMEIDA *et al.* (1981) and we will follow his discussion in our descriptions of the basins given below.

The Amazonian basin extends across the northern portion of the South American platform and is located between the Guiana and central Brazilian shield areas. It is composed of Paleozoic marine deposits from the Early Silurian or Late Ordovician.

The Parnaíba basin is located entirely in northern Brazil. It is almost circular in shape and its elevation is in most areas less than 600 m. Subsidence of the province began in the Early Silurian or Late Ordovician, after erosion of Precambrian basement rocks. The region was affected by several episodes of marine and

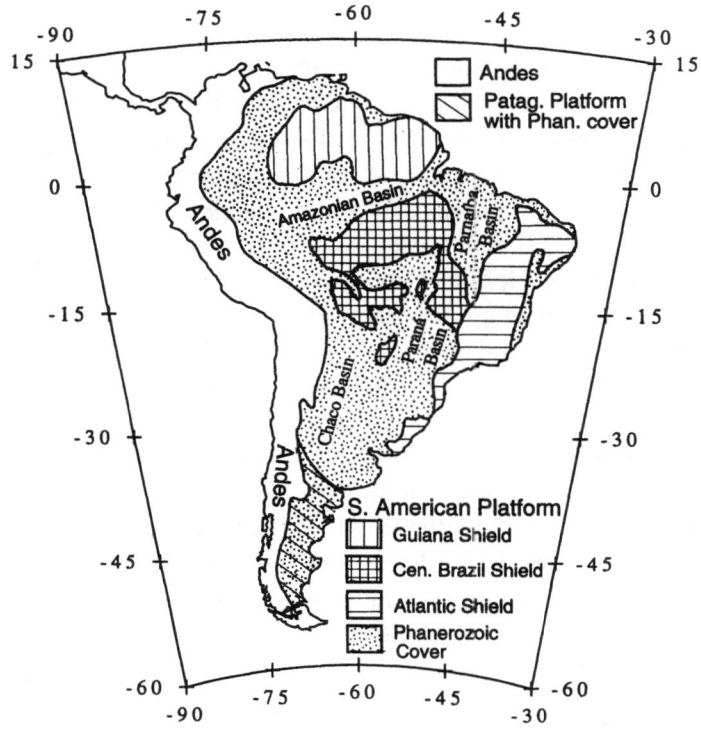

Figure 1
Map showing the three main tectonic provinces of South America and subdivisions within the South American platform (from HASUI and ALMEIDA, 1985).

continental sedimentation with ages ranging from Early Devonian to Early Permian.

The Paraná basin is situated mainly in southern Brazil, although it also extends into Paraguay, Uruguay and Argentina (Fig. 1). The oldest sediments are of Silurian age and are found in Paraguay. The evolution of the basin is marked by episodes of marine (Devonian and Carboniferous) and continental (Permian) sedimentation. 4.2 km of sediments were deposited during the later episodes. During the separation of South America and Africa in the Late Jurassic, the province was extensively covered by continental sediments of 0.4 km thickness as well as by basalt flows which can reach 1.5 km thickness in the north-central part of the province.

The Chaco basin is located in the southern and southwestern parts of the South American platform (Fig. 1). Its shape and dimensions are not very well defined, but it is most clearly developed in Argentina. This province is mostly covered by a few hundred meters of Cenozoic sediments. In its deeper parts the basin reaches 3.2 km in thickness. The oldest sediments in the Chaco basin are probably from the Early Silurian or Early Devonian and the youngest are of Quaternary age.

The Patagonian platform is a second major tectonic province of South America. It is much smaller than the South American platform and lies at the southern tip of the continent where it is bounded on the west by the southern portion of the Andes belt and on the east by the Atlantic Ocean. Basement rocks of this region are covered by sediments and volcanics of the Carboniferous and Jurassic age.

A third major tectonic province of South America, the Andean mountain belt, formed, according to DALZIEL (1986), by the uplift of Precambrian and Paleozoic rocks during the Mesozoic and Cenozoic eras. It is a complex and tectonically active mobile belt extending the entire length of the western margin of the continent. The main central part of this spectacular mountain chain has been interpreted as resulting from the subduction of oceanic lithosphere (the Nazca plate) beneath a continental plate (the South American plate) over the last 200 Ma (JAMES, 1971). Formation of the Andean Cordillera began in the Middle to Late Triassic in the north and the Middle to Late Jurassic in the south. Mesozoic sedimentation took place along the length of the mountain chain with subaerial to shallow marine deposits intercalated with volcanic rocks. Jurassic and Early Cretaceous subsidence occurred throughout the Andean mountain chain. During the Mid-Cretaceous, deposits located along western South America were uplifted and deformed by tectonic compression (DALZIEL, 1986). Between the Late Cretaceous and Cenozoic the locus of Andean magmatic activity migrated eastward. Changes in both magmatic history and tectonic style occurred along strike, and deformation during that period was essentially confined to the continental side of the Cordillera. Further details of the geology and tectonics of the Andean mountain belt can be found in a detailed review by GANSSER (1973).

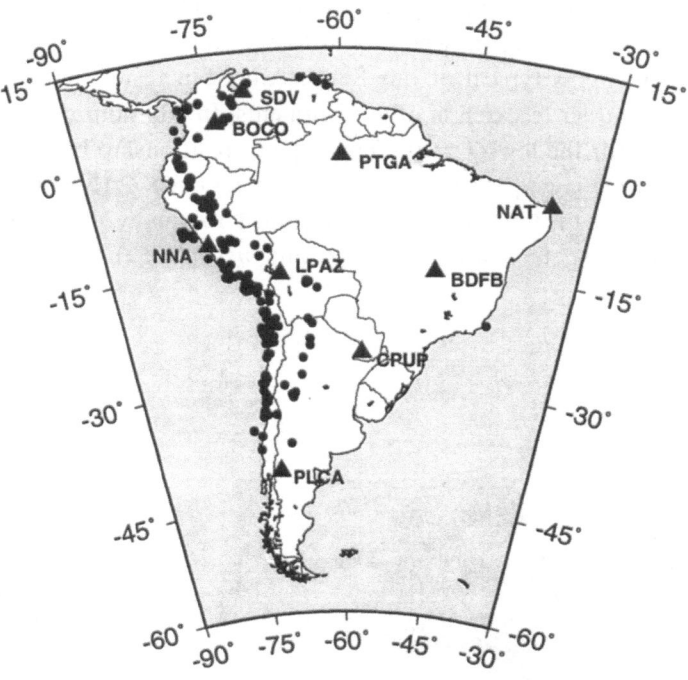

Figure 2
Map of South America showing the locations of earthquakes and seismic stations used in this study.

3. *Lg Coda Q Determination*

The stacked spectral ratio (SSR) method for determining *Lg* coda *Q* and its frequency dependence was developed by XIE and NUTTLI (1988) and is described by those authors and by XIE and MITCHELL (1990a). In this volume, the method is summarized by CONG and MITCHELL (1998). We applied the SSR technique to a large set of broadband vertical-component recordings of *Lg* coda in South America. The locations of events and stations used in our study appear in Figure 2 and are listed in Tables 1 and 2. Figure 3 is a map of assumed *Lg* coda scattering ellipses corresponding to those event and station locations. Those ellipses show that the southernmost portion of the continent lacks any data coverage and that coverage for the region east of 45°W is poor. As discussed in earlier studies, we assume that the continental boundaries form barriers to the propagation of *Lg* coda and truncate the scattering ellipses that intersect those boundaries.

Figures 4 displays a sample seismogram from one station used in this study, along with an SSR plot and Q_0 and η values obtained from that plot. The Q_0 value (381) in Figure 4 obtained at station PLCA for a path near the eastern margin of the Andes is lower than values (550 and higher) found throughout most of the South American platform. This pattern, where Q_0 is low in tectonically active regions and high in stable regions, is typical of our South American results and is consistent with the results of earlier research in other continents. In addition, η is lower for the high-Q paths than for the low-Q paths. This type of relationship between Q_0 and η has been reported in single-path studies of Lg and Lg coda Q (NUTTLI, 1988). By applying this method to 389 seismograms obtained for many different paths, we have determined the spatial variation of Q_0 and η (Table 2) for most of South America.

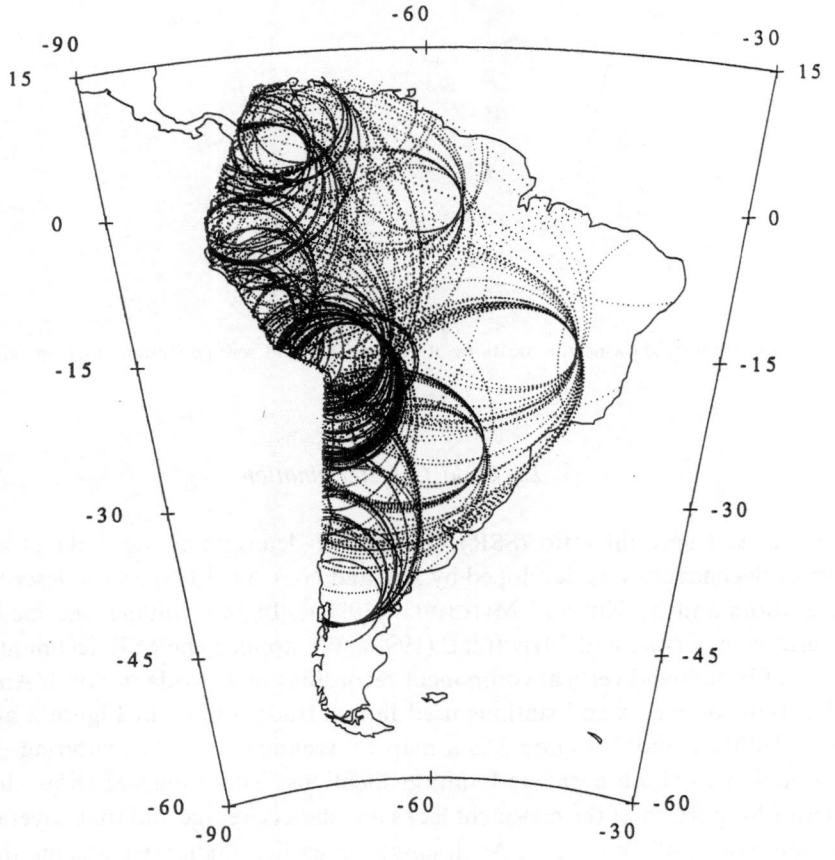

Figure 3

Plots of scattering ellipses obtained assuming single scattering for Lg coda where an earthquake source lies at one focus of each ellipse and the recording station lies at the other focus. The spatial coverage of these ellipses corresponds to that for Lg coda near the maximum lapse time of records used in this study. Ellipses that intersect a continent-ocean boundary are truncated because we assume that that boundary forms a barrier to Lg coda propagation.

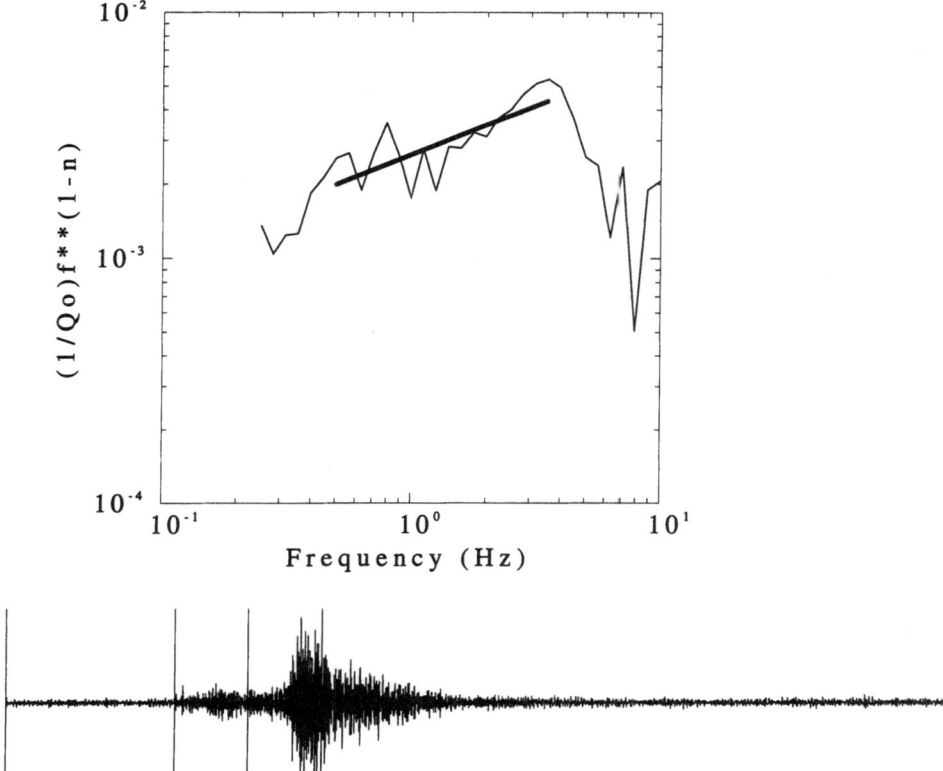

Figure 4

Bottom—Seismogram generated by the m_b 5.0 earthquake of 1 August 1995, which occurred in western Argentina and was recorded by station PLCA at a distance of 1103 km from the epicenter. The ray path between the epicenter and station lies near the eastern margin of the Andes. Vertical lines indicate the expected arrival time of the *P* wave, the *S* wave, and the point in *Lg* coda with a group velocity at 3.15 km/s. Measurements of *Lg* coda begin at a lapse time corresponding to a group velocity of 3.15 km/s and continue until the coda signal is lost in background noise. Durations of seismograms used in this study vary between about 300 and 450 seconds. Top—Stacked Spectral Ratio (SSR) for the seismogram below. Q_0 for this path along the eastern edge of the Andes is 381 ± 24 and η is 0.60 ± 0.11.

4. *Lg Coda Q Tomography—Results, Resolution, and Error Analysis*

Regionalization of *Lg* coda *Q* is performed using the tomographic method of XIE and MITCHELL (1990a). In the present volume this method is also discussed in CONG and MITCHELL (1998). We divided the South American continent into 2° by 2° cells, and inverted the observed *Lg* coda *Q* data to obtain maps of Q_0 and η (Figs. 5 and 6). Those results are discussed in the ensuing paragraphs of this section and discussed in terms of the geology and tectonics of South America in the following section.

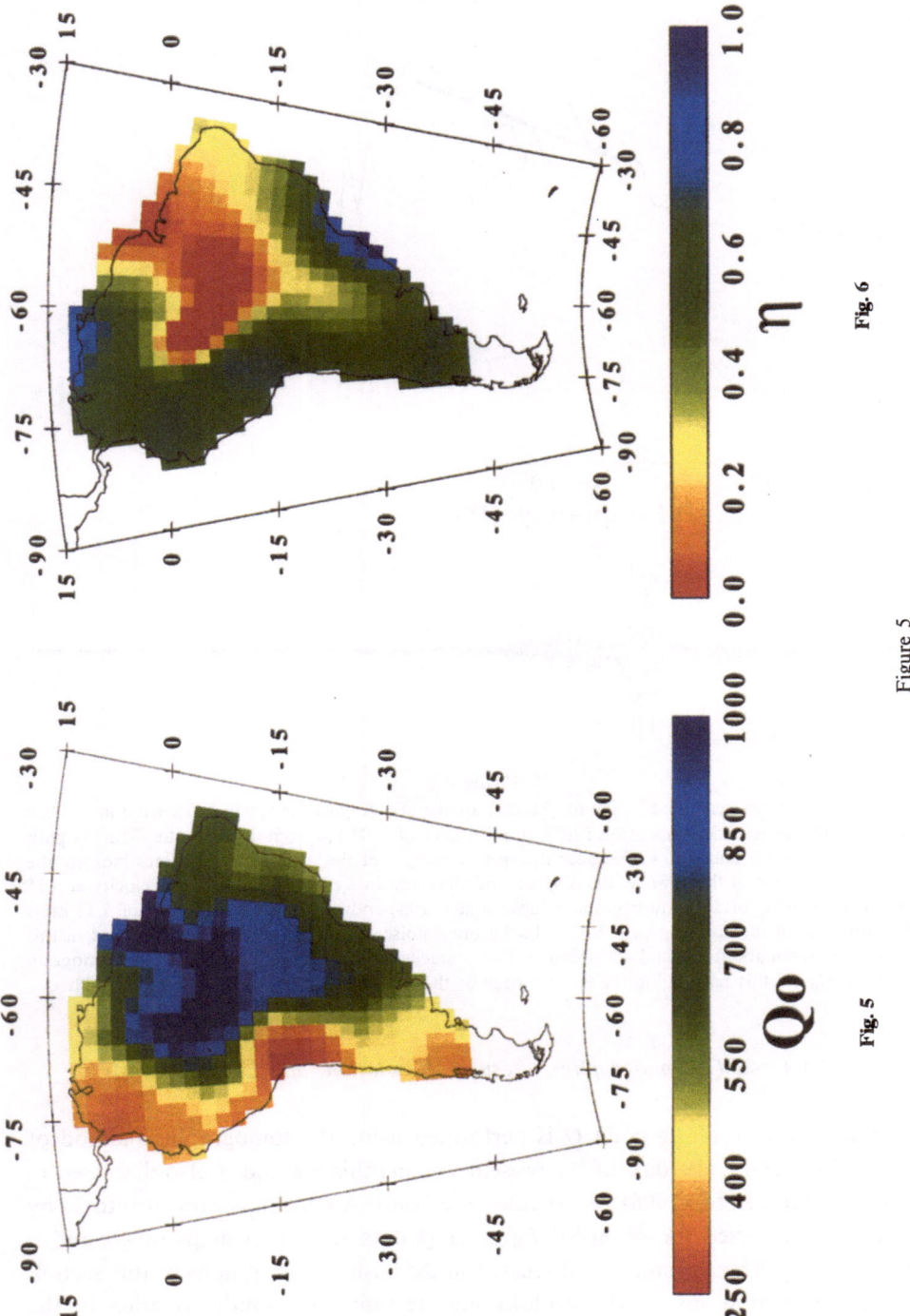

Figure 5
Tomographic map of Lg coda Q at 1 Hz (Q_0) for most of South America.

Figure 6
Tomographic map of the frequency dependence (η) of Lg coda Q at 1 Hz for most of South America.

Q_0 varies between about 1100 and 200. All values greater than 600, except possibly some in the southernmost portion of the map, lie in the South American platform. Values greater than 800 are centered in the central Brazilian shield and extend northward into the Amazonian basin, southward into the Chaco basin, and possibly northeastward into the Parnaíba basin to the northeastern coast. Values less than about 400 all appear to be related to the Andean mountain belt. There is variation along the belt, with two regions of somewhat higher Q_0 being centered at latitudes of about 7°S and 32°S. Intermediate values of Q_0 (about 700) occur along most of the eastern coast of South America. Somewhat higher values occur at

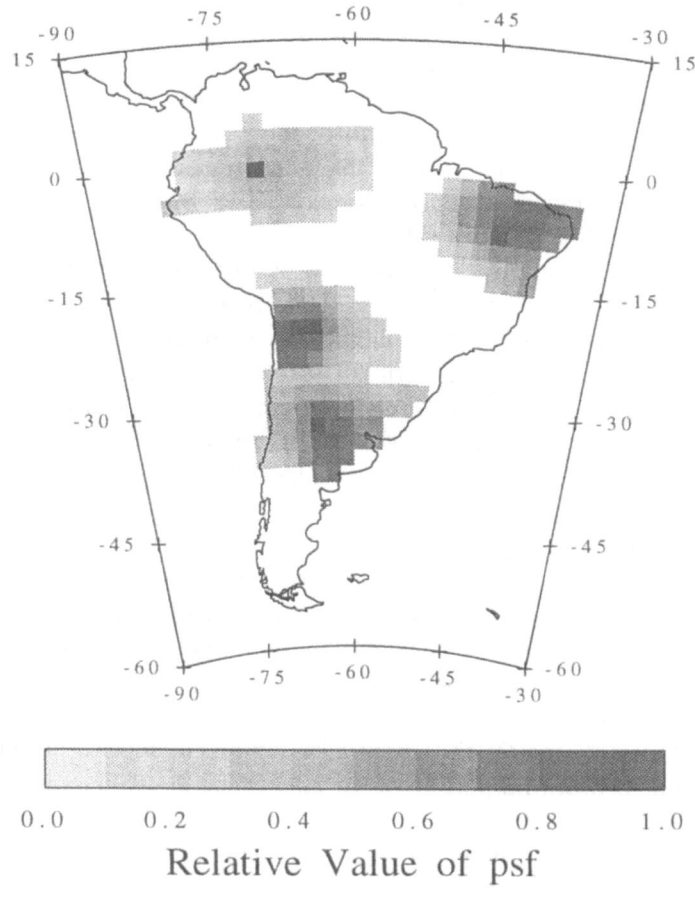

Figure 7

Point-spreading functions (*psf*) for points centered at 33.0°S and 65.0°W, at 21.0°S and 65.0°W, at 9.0°S and 43.0°W, and at 1.0°S and 71.0°W. All *psf*'s are normalized to unity at their centers. Sharp *psf*'s indicate good resolution and broad *psf*'s indicate poor resolution.

about 3°N latitude and lower values occur in the southernmost portion of the eastern coast for which we have data ($\sim 40°S$). Data coverage, however, is poor in those regions as well as other parts of the eastern coast; therefore these values may not be reliable.

The pattern of frequency dependence variation (Fig. 6) contains one broad region of low η (0.0–0.3) surrounded on three sides by high values. The low η region roughly coincides with the high-Q region of Figure 5, although it is somewhat broader in the northeastern region. Considerably higher values of η (0.5–0.8) surround that region on its northwestern, southwestern, and southeastern sides. At one point along the eastern coast ($\sim 30°S$) η is about 1.0, but since this is only one small region on the periphery of the map the high value might be spurious and could result from poor data coverage.

The tomographic method allows the determination of resolution using the point-spreading function (*psf*) method and of standard errors for Q_0 and η (see CONG and MITCHELL, this volume). Four sample *psf* plots appear in Figure 7. They indicate that resolution is good throughout most of the continent. *psf* plots for the Chaco basin (33°S, 65°W) and Bolivian Altiplano (21°S, 65°W) indicate that Lg coda Q and its frequency dependence can be resolved to areas with dimensions as small as about 400 km on a side. In northeastern Peru (1.0°S, 71.0°W) the *psf* plot indicates that Q_0 and η can be resolved to dimensions as small as 200 km. In northeastern Brazil the *psf* plot indicates that resolution is much poorer than in the other regions. This result is consistent with the scattering ellipses shown in Figure 3 where coverage in the northeastern part of South America is much poorer than in other regions. Resolution of Lg coda Q variation found in this study is better than that obtained for Africa by XIE and MITCHELL (1990a). This improvement is due to the greater amount of data available (389 records in the present study compared to 253 for Africa) enabling us to use smaller cell dimensions.

The error maps (Figs. 8 and 9) show that the random errors in both Q_0 and η are generally small throughout South America. In western South America, where the spatial coverage is good, the errors in Q_0 and η are generally smaller than 40 and 0.1, respectively, and percentage errors are about 10% for Q_0 and 15% for η. In eastern South America, where the spatial coverage is poorer, the errors in Q_0 and η assume values as high as 250 and 0.27, respectively. These translate to percentage errors of about 22% for Q_0 and about 80% for η. The highest absolute Q_0 errors occur in the high-Q central portion of the map where Q_0 assumes its highest values; however, percentagewise errors are highest along the eastern coast where data coverage is poorest. Absolute and percentagewise errors for η exhibit the same patterns as do errors for Q_0.

The error analysis discussed above, while providing quantitative information on random errors in determinations of Q_0 and η, does not address the possibility of systematic errors affecting those determinations. Systematic errors might arise from features in the earth, such as major reflective boundaries that affect all of the data

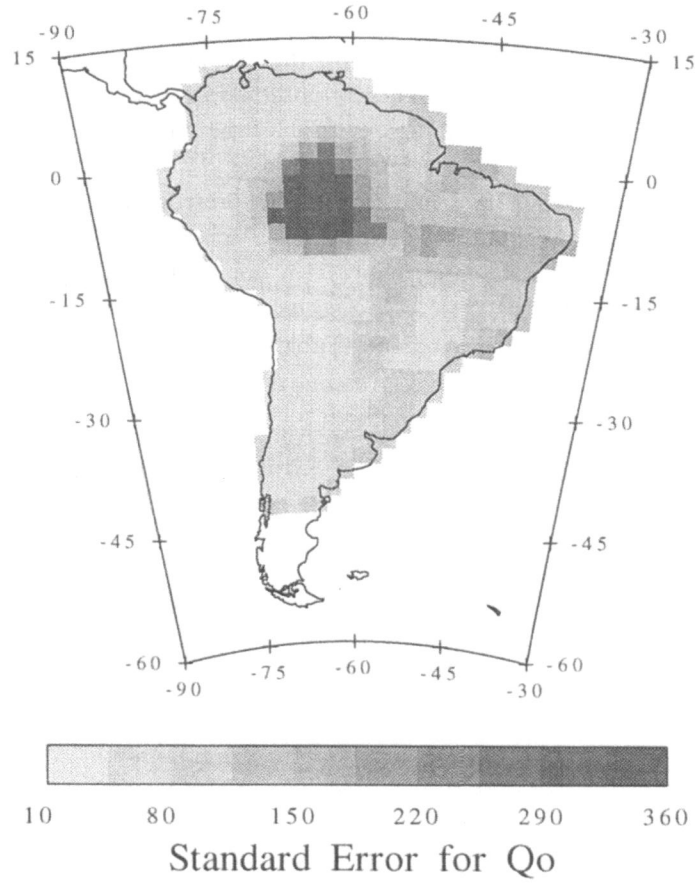

Standard Error for Qo

Figure 8
Map of standard errors for the tomographic image Q_0 for South America.

in the same way. Those systematic errors might be misinterpreted as reductions or increases of Q in the region of interest.

XIE and MITCHELL (1990b) addressed the possibility that interaction of *Lg* coda with the coast of California and the Pacific lithosphere to the west might systematically affect Q_0 and η determinations for southern California and the Basin and Range province. They used a group of earthquakes with epicenters along the California coast that were recorded by stations of the Lawrence Livermore Laboratory Network in the Basin and Range province to the east. They divided the earthquakes into two groups where plots of scattering ellipses for one group indicate that more than one third of the ellipse overlies oceanic crust while the ellipses for the other group are mostly confined to continental crust. The SSR method yielded values of Q_0 for the first group that were significantly higher (and exhibited greater scatter) than those for the second group. η values for both groups also show scatter, although the scatter for the first group was slightly greater than that for the second group. Based on these observations, XIE and MITCHELL (1990b) concluded that data for which a large portion of the scattering ellipse lies in oceanic areas should not be used to calculate Lg coda Q.

Many of the earthquakes utilized in the present study occurred near the boundary between the South American and Pacific plates. For that reason we investigated possible systematic errors that might be introduced by Lg coda propagation in the Pacific plate. We selected six single-trace records of Lg coda recorded at station PLCA in Argentia (Table 1) on which to perform the analysis.

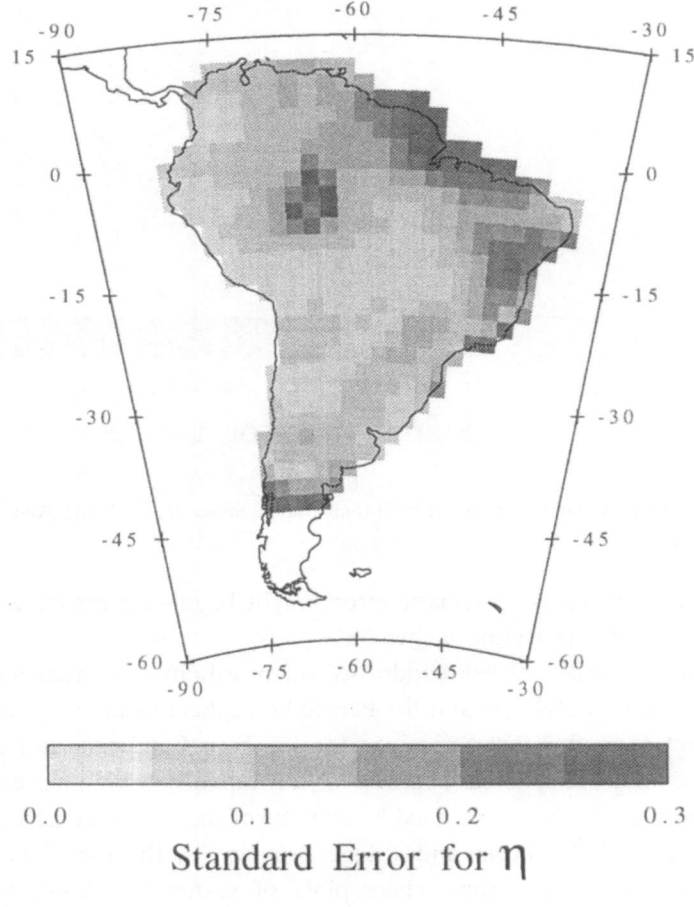

Standard Error for η

Figure 9
Map of standard errors for the tomographic image of η for South America.

Table 1

Seismic station information

Station Code	Location	Coordinates		Data Period
		Latitude	Longitude	
Broadband Stations				
BDFB	Brasília, Brazil	15.6440° S	48.0141° W	1993–1996
BOCO	Bogota, Colombia	4.5869 N	74.0432 W	1994–1996
CPUP	Vila Florida, Paraguay	26.3306 S	57.3292 W	1994–1996
LPAZ	La Paz, Bolivia	16.2879 S	68.1307 W	1993–1996
NNA	Naña, Peru	11.9875 S	76.8422 W	1992–1996
PLCA	Paso Flores, Argentina	40.7306 S	70.5500 W	1994–1996
PTGA	Pitinga, Brazil	0.7308 S	59.9668 W	1995–1996
SDV	Santo Domingo, Venezuela	8.8790 N	70.6330 W	1994–1996
Short-period Stations				
BDF	Brasília, Brazil	15.6639 S	47.9033 W	1988–1993
BOCO	Bogota, Colombia	4.5869 N	74.0432 W	1978–1979
NAT	Natal, Brazil	5.1167 S	35.0333 W	1969–1974
ZOBO	Zongo Valley, Bolivia	16.2700 S	68.1250 W	1976–1993

The scattering ellipses for two of the events are restricted mostly to the continent, while significant portions of the ellipses for four events lie in oceanic crust.

The mean Q_0 value and standard deviation for the set of six events obtained by applying the SSR method to the data is 401 ± 35 while for η it is 0.60 ± 0.11. These relatively small standard deviations suggest that the coastal effect noted by XIE and MITCHELL (1990b) for California has only a slight effect on Q_0 and η determination in South America.

5. Comparison with Previous Studies

Several studies of seismic wave attenuation, employing various phases, have been conducted in South America. The following paragraphs compare our *Lg* coda *Q* determinations with results of those studies. In addition, since elevated temperature in the crust or upper mantle is a commonly proposed mechanism for reducing *Q*, we compare our *Q* results with spatial variations of measured heat flow in South America wherever those measurements have been made. High temperature may play a direct role in reducing *Q* or may generate hydrothermal fluids that contribute to the dissipation of seismic wave energy. Energy may be lost either because it goes into moving the fluids through permeable media or because it is scattered by fluid-filled zones with an enhanced reflection coefficient (MITCHELL, 1995).

Table 2

Hypocentral parameters of the earthquakes used in this study, station abbreviations, epicentral distances, and coda attenuation parameters (Q_0 and η) with standard deviations

Date Y/M/D	Origin time h min. sec.	Depth km	Mag. m_b	Recording station	Ep. dist. km	$Q_0 \pm \Delta Q_0$	$\eta \pm \Delta \eta$
94 07 08	10 36 44.7	47	4.8	BOCO	393.63	263 ± 38	0.682 ± 0.213
94 07 07	22 54 10.3	09	5.0	BOCO	789.48	424 ± 33	0.623 ± 0.207
94 08 04	02 50 49.2	10	4.4	BOCO	306.90	318 ± 16	0.609 ± 0.049
94 08 11	08 39 06.4	33	4.5	BOCO	437.00	716 ± 51	0.436 ± 0.076
94 08 12	03 37 00.4	33	4.5	BOCO	406.97	198 ± 28	0.516 ± 0.299
94 09 13	10 01 34.8	33	5.8	BOCO	403.64	228 ± 9	0.565 ± 0.046
94 09 13	11 45 22.4	33	4.6	BOCO	414.76	264 ± 29	0.541 ± 0.176
94 10 10	18 49 13.4	33	5.0	BOCO	1019.66	301 ± 29	0.470 × 0.149
94 10 24	20 37 17.9	24	4.0	BOCO	1408.84	335 ± 77	0.566 ± 0.336
94 08 14	10 58 19.1	36	5.5	CPUP	1452.21	540 ± 30	0.489 ± 0.084
94 09 17	12 22 14.8	28	5.5	CPUP	1535.60	420 ± 33	0.579 ± 0.118
93 09 20	19 38 37.0	33	4.3	LPAZ	383.62	377 ± 17	0.658 ± 0.029
93 09 25	21 06 31.8	10	5.1	LPAZ	831.74	310 ± 27	0.474 ± 0.226
93 10 09	11 47 02.5	10	4.3	LPAZ	425.88	266 ± 12	0.723 ± 0.079
93 10 27	17 51 36.8	52	4.8	LPAZ	815.06	442 ± 69	0.602 ± 0.176
93 12 01	13 23 56.4	33	5.3	LPAZ	817.28	372 ± 33	0.713 ± 0.136
93 12 12	13 58 24.5	33	4.4	LPAZ	597.12	300 ± 21	0.751 ± 0.107
94 02 21	22 17 03.6	22	5.4	LPAZ	2788.77	544 ± 39	0.808 ± 0.108
94 03 13	16 39 58.9	33	5.1	LPAZ	567.09	297 ± 31	0.682 ± 0.160
94 04 19	08 28 38.0	49	5.2	LPAZ	741.67	386 ± 22	0.552 ± 0.106
94 05 10	01 49 03.0	53	5.9	LPAZ	421.43	310 ± 17	0.758 ± 0.089
94 05 11	11 53 28.0	19	5.2	LPAZ	2184.98	589 ± 35	0.751 ± 0.161
94 05 31	17 41 55.8	13	6.3	LPAZ	2656.45	817 ± 27	0.600 ± 0.069
94 05 31	20 45 55.0	33	5.6	LPAZ	2648.66	829 ± 21	0.548 ± 0.062
94 06 06	20 47 39.8	09	6.8	LPAZ	2292.84	689 ± 102	0.769 ± 0.218
94 07 07	22 54 10.3	09	5.9	LPAZ	2019.30	470 ± 168	0.817 ± 0.346
94 07 08	05 52 46.1	48	4.6	LPAZ	830.63	477 ± 299	0.425 ± 0.486
94 07 14	22 09 33.9	37	4.8	LPAZ	565.98	288 ± 29	0.433 ± 0.119
94 09 13	10 01 34.8	33	5.8	LPAZ	2748.74	555 ± 190	0.580 ± 0.453
94 09 14	05 02 12.2	38	5.2	LPAZ	1692.39	253 ± 129	0.339 ± 0.605
94 09 22	14 24 34.1	27	5.5	LPAZ	1802.47	399 ± 181	0.381 ± 0.402
94 12 10	03 39 31.4	37	6.2	PLCA	1901.43	394 ± 28	0.603 ± 0.180
94 09 13	10 01 34.8	33	5.8	SDV	698.30	303 ± 18	0.792 ± 0.136
94 09 13	11 45 22.4	33	4.6	SDV	708.31	332 ± 133	0.779 ± 0.475
95 02 12	01 02 07.0	22	5.7	BOCO	1169.77	322 ± 31	0.746 ± 0.168
95 02 16	09 35 17.9	18	5.3	BOCO	1167.05	377 ± 22	0.829 ± 0.146
95 03 22	20 05 55.2	40	5.2	CPUP	1844.73	456 ± 32	0.359 ± 0.267
95 01 19	15 05 03.6	18	6.4	LPAZ	2418.49	792 ± 36	0.579 ± 0.114
95 01 21	09 18 17.9	33	5.3	LPAZ	2412.93	516 ± 105	0.612 ± 0.132
95 01 22	10 41 27.7	22	5.6	LPAZ	2425.166	560 ± 84	0.636 ± 0.165
95 02 09	12 59 25.1	33	4.8	LPAZ	733.89	203 ± 53	0.899 ± 0.200
95 02 12	01 02 07.0	22	5.7	LPAZ	1451.09	450 ± 57	0.858 ± 0.640
95 02 16	09 35 17.9	18	5.3	LPAZ	1452.21	457 ± 85	0.712 ± 0.221
95 05 04	18 47 48.6	33	5.0	LPAZ	875.10	295 ± 82	0.413 ± 0.256
94 10 01	07 33 52.6	33	4.8	NNA	689.41	216 ± 18	0.812 ± 0.121
94 10 10	18 49 13.4	33	5.0	NNA	914.02	219 ± 6	0.791 ± 0.078
94 10 27	23 46 29.9	45	5.1	NNA	371.39	193 ± 7	0.637 ± 0.068

Table 2 (Continued)

Date Y/M/D	Origin time h min. sec.	Depth km	Mag. m_b	Recording station	Ep. dist. km	$Q_0 \pm \Delta Q_0$	$\eta \pm \Delta\eta$
94 11 09	05 02 45.4	33	4.5	NNA	465.91	219 ± 4	0.655 ± 0.023
95 01 19	15 05 03.6	18	6.4	NNA	1933.68	449 ± 88	0.576 ± 0.275
95 01 22	10 41 27.7	22	5.6	NNA	1938.13	492 ± 105	0.512 ± 0.197
95 02 12	01 02 07.0	22	5.7	NNA	688.30	300 ± 46	0.559 ± 0.169
95 02 16	09 35 17.9	18	5.3	NNA	690.52	298 ± 12	0.786 ± 0.092
95 03 25	01 13 01.2	33	5.0	NNA	490.37	253 ± 8	0.532 ± 0.057
95 03 14	03 32 33.5	33	5.2	PLCA	395.85	286 ± 14	0.679 ± 0.118
95 03 22	20 05 55.2	40	5.2	PLCA	538.18	302 ± 11	0.779 ± 0.068
95 03 23	09 18 21.2	43	5.2	PLCA	538.18	325 ± 10	0.790 ± 0.057
95 05 04	03 11 54.1	33	4.5	PLCA	416.98	322 ± 10	0.736 ± 0.061
95 01 19	15 05 03.6	18	6.4	SDV	490.37	272 ± 35	0.515 ± 0.280
95 01 19	17 34 54.3	33	4.9	SDV	499.27	330 ± 22	0.385 ± 0.151
95 01 20	13 59 20.2	33	5.2	SDV	475.91	369 ± 26	0.664 ± 0.119
95 01 21	09 18 19.9	33	5.3	SDV	498.15	373 ± 18	0.525 ± 0.125
95 01 22	10 41 27.7	22	5.6	SDV	488.15	362 ± 42	0.496 ± 0.215
95 01 23	00 36 51.4	33	4.6	SDV	479.25	377 ± 65	0.501 ± 0.253
95 01 23	08 00 58.8	33	4.7	SDV	497.04	307 ± 29	0.371 ± 0.218
95 02 12	01 02 07.0	22	5.7	SDV	1732.42	606 ± 37	0.759 ± 0.101
95 03 19	17 04 07.5	33	4.9	SDV	690.52	224 ± 37	0.500 ± 0.365
95 03 23	02 08 34.5	10	5.1	SDV	687.19	217 ± 22	0.393 ± 0.296
95 03 29	16 10 52.0	40	4.8	SDV	716.10	290 ± 28	0.623 ± 0.213
95 04 12	18 22 32.6	41	5.2	SDV	1088.60	331 ± 38	0.795 ± 0.201
95 04 24	00 29 41.5	33	4.7	SDV	447.00	325 ± 24	0.782 ± 0.077
89 07 21	02 45 02.9	29	4.8	BDF	2843.26	886 ± 61	0.545 ± 0.216
89 09 14	04 42 9.8	33	5.3	BDF	2636.43	1109 ± 176	0.488 ± 0.286
89 12 30	12 38 11.2	32	5.3	BDF	2617.53	871 ± 93	0.661 ± 0.395
90 08 02	05 24 08.5	36	5.5	BDF	2988.92	594 ± 132	0.585 ± 0.147
90 08 13	22 36 30.9	22	4.9	BDF	2376.24	667 ± 38	0.459 ± 0.093
91 02 27	08 34 35.1	35	5.4	BDF	2668.68	740 ± 33	0.743 ± 0.103
91 04 05	15 50 47.3	50	5.7	BDF	2970.02	899 ± 44	0.525 ± 0.124
91 12 21	08 01 18.3	33	5.0	BDF	1879.20	668 ± 22	0.487 ± 0.077
92 07 30	23 00 41.7	05	5.5	BDF	2639.77	563 ± 38	0.773 ± 0.214
92 07 31	13 03 55.9	05	5.2	BDF	2676.02	610 ± 34	0.727 ± 0.174
92 09 02	22 37 28.8	28	5.1	BDF	2516.34	826 ± 33	0.560 ± 0.100
92 10 29	14 39 17.4	46	5.5	BDF	2819.91	640 ± 76	0.359 ± 0.194
92 11 04	21 32 33.9	19	5.8	BDF	2974.47	483 ± 47	0.483 ± 0.144
92 11 06	20 51 37.8	14	5.3	BDF	2978.91	893 ± 79	0.325 ± 0.222
93 05 09	03 59 18.6	39	4.5	BDF	1769.11	1010 ± 83	0.423 ± 0.180
94 03 10	08 02 41.2	51	4.8	BDFB	2032.64	632 ± 20	0.486 ± 0.081
94 04 16	09 55 20.7	35	5.5	BDFB	2952.23	613 ± 53	0.330 ± 0.381
94 05 24	02 05 57.4	37	5.5	BDFB	2864.38	990 ± 260	0.285 ± 0.821
94 07 16	17 51 39.3	33	4.8	BDFB	2905.53	904 ± 193	0.410 ± 0.224
94 08 14	10 58 19.1	36	5.5	BDFB	2872.17	604 ± 82	0.296 ± 0.155
94 09 12	06 29 56.3	53	6.0	BDFB	2948.89	520 ± 14	0.585 ± 0.052
94 09 14	05 02 12.2	38	5.2	BDFB	2953.34	614 ± 57	0.575 ± 0.154
94 10 11	01 37 20.8	51	5.8	BDFB	2988.92	524 ± 44	0.517 ± 0.124
78 03 24	00 31 22.1	42	5.3	BOCO	1096.38	352 ± 66	0.667 ± 0.347
78 05 20	13 10 31.1	50	5.6	BOCO	1710.18	407 ± 34	0.532 ± 0.210
78 06 07	08 00 34.2	12	4.9	BOCO	751.68	479 ± 22	0.429 ± 0.129

Table 2 (Continued)

Date Y/M/D	Origin time h min. sec.	Depth km	Mag. m_b	Recording station	Ep. dist. km	$Q_0 \pm \Delta Q_0$	$\eta \pm \Delta\eta$
78 11 22	06 24 17.6	33	4.9	BOCO	1333.23	373 ± 42	0.653 ± 0.166
79 01 05	09 29 38.4	28	5.1	BOCO	2268.38	694 ± 153	0.593 ± 0.250
79 01 06	01 31 47.6	33	5.7	BOCO	1500.02	590 ± 47	0.455 ± 0.101
79 09 13	04 40 58.7	34	5.0	BOCO	1164.21	501 ± 20	0.715 ± 0.102
94 02 21	22 17 03.6	22	5.4	LPAZ	2788.77	715 ± 29	0.632 ± 0.103
94 04 03	09 01 03.3	30	5.3	LPAZ	386.96	348 ± 39	0.605 ± 0.091
94 05 11	11 53 28.0	19	5.2	LPAZ	2184.98	553 ± 75	0.875 ± 0.089
94 05 31	20 45 05.0	33	5.6	LPAZ	2648.66	670 ± 51	0.596 ± 0.155
94 07 07	22 54 10.3	09	5.0	LPAZ	2019.30	587 ± 89	0.650 ± 0.119
94 07 08	05 52 46.1	48	4.6	LPAZ	830.63	422 ± 120	0.626 ± 0.320
94 08 11	08 39 06.4	33	4.5	LPAZ	2714.27	754 ± 41	0.349 ± 0.142
76 10 05	16 36 29.1	33	5.2	ZOBO	2083.79	389 ± 147	0.679 ± 0.390
76 10 06	09 12 38.9	33	5.7	ZOBO	2074.90	227 ± 72	0.328 ± 0.306
76 10 25	22 19 04.9	37	5.3	ZOBO	366.94	216 ± 7	0.863 ± 0.054
76 11 26	12 46 25.9	52	4.8	ZOBO	637.15	243 ± 29	0.784 ± 0.304
79 06 25	20 41 59.5	28	5.0	ZOBO	391.41	198 ± 24	0.505 ± 0.183
79 07 21	02 38 02.5	43	5.3	ZOBO	737.22	337 ± 38	0.724 ± 0.133
79 10 04	11 18 30.9	28	5.1	ZOBO	811.72	294 ± 36	0.457 ± 0.181
79 10 28	05 13 16.7	33	4.7	ZOBO	756.13	309 ± 44	0.809 ± 0.218
79 11 06	15 49 09.3	51	5.0	ZOBO	324.69	245 ± 20	0.678 ± 0.149
79 11 15	13 36 29.7	33	5.4	ZOBO	713.87	284 ± 35	0.712 ± 0.140
88 01 20	02 54 26.0	33	4.8	ZOBO	592.67	212 ± 12	0.870 ± 0.071
88 02 01	13 32 51.9	30	5.2	ZOBO	959.61	272 ± 48	0.648 ± 0.260
88 03 07	11 20 04.0	43	5.1	ZOBO	818.40	387 ± 95	0.655 ± 0.228
88 04 13	01 59 48.5	33	4.8	ZOBO	505.94	293 ± 30	0.624 ± 0.156
88 04 13	02 16 01.8	33	4.9	ZOBO	454.76	201 ± 13	0.639 ± 0.102
88 04 13	06 22 31.5	37	5.2	ZOBO	485.92	214 ± 21	0.608 ± 0.146
88 04 13	10 43 40.5	45	4.9	ZOBO	507.05	320 ± 43	0.780 ± 0.166
88 04 14	08 09 49.6	48	4.7	ZOBO	520.39	255 ± 20	0.605 ± 0.122
88 04 21	08 48 17.7	52	5.0	ZOBO	469.24	228 ± 18	0.644 ± 0.119
88 07 02	05 27 40.8	33	4.7	ZOBO	584.89	280 ± 18	0.771 ± 0.097
88 07 18	02 30 29.3	19	5.3	ZOBO	669.39	224 ± 17	0.654 ± 0.119
88 08 03	07 52 25.7	10	5.4	ZOBO	492.52	219 ± 27	0.887 ± 0.162
88 10 22	05 41 20.8	10	4.4	ZOBO	443.67	308 ± 17	0.839 ± 0.085
88 10 25	19 01 07.8	38	4.6	ZOBO	542.63	278 ± 36	0.717 ± 0.209
88 12 06	00 18 33.3	51	5.1	ZOBO	558.20	287 ± 45	0.736 ± 0.207
89 02 17	15 12 52.6	46	5.2	ZOBO	493.71	216 ± 15	0.676 ± 0.107
89 02 19	04 23 44.6	48	5.2	ZOBO	1123.07	306 ± 41	0.726 ± 0.295
89 04 18	04 21 58.4	52	4.4	ZOBO	664.95	227 ± 22	0.718 ± 0.149
89 05 04	10 30 06.9	33	5.5	ZOBO	1353.24	364 ± 42	0.894 ± 0.121
89 05 29	22 22 30.7	32	5.5	ZOBO	875.10	230 ± 16	0.734 ± 0.105
89 06 01	20 56 00.2	44	5.2	ZOBO	871.70	212 ± 23	0.552 ± 0.132
89 11 23	15 34 22.2	33	4.5	ZOBO	372.50	297 ± 10	0.721 ± 0.032
93 01 19	11 17 56.1	33	4.8	ZOBO	481.47	258 ± 23	0.829 ± 0.163
93 02 01	00 27 51.1	10	4.1	ZOBO	806.16	257 ± 17	0.718 ± 0.099
93 03 19	16 01 10.2	20	5.4	ZOBO	1187.56	263 ± 41	0.766 ± 0.228
93 04 12	19 48 43.4	33	4.5	ZOBO	502.60	201 ± 21	0.845 ± 0.125
93 05 09	03 59 18.6	39	4.5	ZOBO	418.09	340 ± 26	0.764 ± 0.116
93 07 07	11 24 36.1	37	4.6	ZOBO	522.62	223 ± 9	0.495 ± 0.061

Table 2 (Continued)

Date Y/M/D	Origin time h min. sec.	Depth km	Mag. m_b	Recording station	Ep. dist. km	$Q_0 \pm \Delta Q_0$	$\eta \pm \Delta\eta$
69 07 18	23 17 10.6	19	5.6	NAT	3397.01	690 ± 55	0.294 ± 0.213
72 10 24	15 36 35.9	33	5.3	NAT	1930.35	539 ± 107	0.484 ± 0.177
73 11 19	16 56 21.5	33	5.3	NAT	3824.00	906 ± 172	0.432 ± 0.195
74 07 01	16 51 51.5	13	5.5	NAT	3707.24	588 ± 111	0.406 ± 0.221
75 04 15	09 47 43.6	47	5.4	NAT	3339.19	854 ± 45	0.189 ± 0.229
74 01 07	16 35 57.1	33	5.8	NAT	4049.70	507 ± 69	0.399 ± 0.140
95 11 17	02 12 23.6	33	5.3	PTGA	1568.96	681 ± 36	0.349 ± 0.082
95 12 06	15 38 23.9	33	4.8	PTGA	1826.93	744 ± 109	0.361 ± 0.365
95 06 12	03 35 48.8	34	5.7	BDFB	3137.92	782 ± 31	0.479 ± 0.173
95 10 03	12 44 58.0	17	6.0	BDFB	3562.69	764 ± 41	0.302 ± 0.088
95 10 03	16 08 17.8	37	5.5	BDFB	2951.12	513 ± 46	0.315 ± 0.146
95 10 07	21 28 03.1	12	5.8	BDFB	3560.46	702 ± 65	0.443 ± 0.123
95 10 08	10 27 39.1	33	5.3	BDFB	3570.47	666 ± 59	0.345 ± 0.073
95 11 17	02 12 23.3	33	5.1	BDFB	3623.84	962 ± 76	0.493 ± 0.128
95 12 29	13 01 40.9	33	5.4	BDFB	3736.15	707 ± 29	0.402 ± 0.052
96 02 22	13 40 53.5	44	5.9	BDFB	3105.68	626 ± 48	0.489 ± 0.087
96 03 10	08 56 22.3	33	5.8	BDFB	2326.20	820 ± 31	0.394 ± 0.095
96 05 09	00 50 38.6	45	4.7	BDFB	3552.68	705 ± 82	0.422 ± 0.144
95 06 12	03 35 48.8	34	5.7	BOCO	1438.36	482 ± 46	0.674 ± 0.253
95 06 25	09 11 32.3	33	4.7	BOCO	391.41	268 ± 16	0.623 ± 0.090
95 09 01	04 54 25.1	42	4.6	BOCO	514.83	233 ± 15	0.371 ± 0.241
95 10 03	01 51 23.9	24	6.5	BOCO	916.25	257 ± 29	0.599 ± 0.075
95 10 03	04 58 19.8	33	4.6	BOCO	855.11	249 ± 20	0.378 ± 0.276
95 10 03	05 20 26.3	33	5.0	BOCO	919.58	297 ± 11	0.722 ± 0.067
95 10 03	09 30 52.3	33	4.7	BOCO	945.16	313 ± 16	0.714 ± 0.072
95 10 03	10 06 59.0	27	4.7	BOCO	918.47	340 ± 37	0.771 ± 0.175
95 10 03	11 12 21.6	33	4.6	BOCO	888.45	223 ± 16	0.864 ± 0.277
95 10 03	11 40 16.0	33	4.9	BOCO	902.90	342 ± 12	0.472 ± 0.093
95 10 03	12 44 58.0	17	6.0	BOCO	917.36	368 ± 20	0.882 ± 0.073
95 10 03	14 29 27.9	33	4.7	BOCO	934.04	394 ± 94	0.735 ± 0.388
95 10 03	17 01 00.9	33	4.9	BOCO	929.59	310 ± 14	0.709 ± 0.061
95 10 03	18 04 27.6	46	4.7	BOCO	930.70	274 ± 36	0.810 ± 0.340
95 10 03	30 08 58.0	33	4.6	BOCO	904.02	296 ± 29	0.778 ± 0.158
95 10 03	01 36 55.6	33	5.1	BOCO	912.91	272 ± 34	0.677 ± 0.100
95 10 05	02 45 18.2	33	4.9	BOCO	920.69	271 ± 23	0.805 ± 0.141
95 10 05	12 04 28.4	33	4.6	BOCO	924.03	161 ± 24	0.597 ± 0.274
95 10 06	00 14 45.5	33	4.6	BOCO	936.26	295 ± 41	0.727 ± 0.364
95 10 06	18 05 54.9	33	4.8	BOCO	910.69	187 ± 42	0.677 ± 0.332
95 10 07	05 53 52.6	33	4.5	BOCO	921.81	235 ± 25	0.700 ± 0.275
95 10 07	09 15 24.9	33	4.7	BOCO	917.36	281 ± 9	0.832 ± 0.057
95 10 07	09 58 53.7	33	4.5	BOCO	920.69	310 ± 35	0.676 ± 0.170
95 10 07	21 19 49.5	33	4.8	BOCO	914.02	234 ± 44	0.788 ± 0.274
95 10 07	21 20 10.5	33	4.8	BOCO	920.30	215 ± 33	0.520 ± 0.225
95 10 07	21 28 03.1	12	5.8	BOCO	915.13	599 ± 70	0.996 ± 0.171
95 10 08	10 27 39.1	33	5.3	BOCO	897.34	283 ± 19	0.813 ± 0.111
95 10 29	05 28 48.5	33	4.9	BOCO	908.46	275 ± 42	0.676 ± 0.227
95 10 31	05 24 04.4	49	4.8	BOCO	381.40	229 ± 9	0.670 ± 0.067
96 01 07	07 47 07.6	33	4.7	BOCO	2028.20	345 ± 44	0.564 ± 0.248
96 01 19	19 01 58.3	36	5.6	BOCO	1739.09	430 ± 118	0.589 ± 0.402

Table 2 (Continued)

Date Y/M/D	Origin time h min. sec.	Depth km	Mag. m_b	Recording station	Ep. dist. km	$Q_0 \pm \Delta Q_0$	$\eta \pm \Delta\eta$
96 02 23	04 07 16.0	33	4.5	BOCO	509.27	211 ± 24	0.460 ± 0.498
96 03 06	07 42 20.0	33	5.0	BOCO	1257.62	321 ± 99	0.769 ± 0.451
96 03 10	08 56 22.3	33	5.8	BOCO	2004.85	348 ± 99	0.419 ± 0.377
96 03 28	23 03 49.8	33	5.8	BOCO	810.61	379 ± 38	0.817 ± 0.198
96 05 19	03 25 37.0	17	4.6	BOCO	1395.50	3122 ± 61	0.865 ± 0.285
96 05 23	01 57 22.9	33	5.5	BOCO	418.09	219 ± 20	0.559 ± 0.104
95 06 04	12 18 58.9	33	4.9	CPUP	1626.89	456 ± 59	0.396 ± 0.190
96 06 29	01 27 02.8	41	5.1	CPUP	1667.93	439 ± 63	0.433 ± 0.095
95 07 01	09 14 53.9	43	4.9	CPUP	1672.37	539 ± 44	0.472 ± 0.126
95 07 30	10 35 39.2	11	5.8	CPUP	1363.25	596 ± 27	0.496 ± 0.099
95 07 30	16 19 24.4	33	5.3	CPUP	1363.24	529 ± 118	0.595 ± 0.362
95 07 30	18 02 45.8	33	5.0	CPUP	1374.37	709 ± 82	0.705 ± 0.173
95 07 30	21 05 47.7	14	5.6	CPUP	1381.04	698 ± 46	0.398 ± 0.108
95 08 09	08 23 01.1	35	5.3	CPUP	1344.35	800 ± 48	0.435 ± 0.129
95 08 10	18 10 37.2	33	5.4	CPUP	1364.36	650 ± 53	0.630 ± 0.137
96 03 07	08 38 57.5	31	5.3	CPUP	1352.13	766 ± 47	0.453 ± 0.092
96 03 10	08 56 22.3	33	5.8	CPUP	1944.80	762 ± 37	0.483 ± 0.074
95 06 12	03 35 48.8	34	5.7	LPAZ	1220.92	429 ± 34	0.629 ± 0.200
95 06 14	16 12 59.1	49	5.4	LPAZ	965.17	243 ± 55	0.575 ± 0.240
95 07 30	10 16 09.8	33	4.5	LPAZ	905.13	260 ± 24	0.426 ± 0.068
95 07 30	10 56 13.0	33	5.4	LPAZ	799.49	272 ± 20	0.652 ± 0.111
95 07 30	13 27 19.7	33	4.6	LPAZ	897.34	241 ± 42	0.798 ± 0.379
95 07 30	13 45 56.9	33	4.6	LPAZ	795.04	134 ± 8	0.454 ± 0.207
95 07 30	14 44 46.0	33	4.8	LPAZ	882.89	236 ± 35	0.708 ± 0.231
95 07 30	15 40 51.7	33	4.5	LPAZ	908.46	188 ± 28	0.437 ± 0.208
95 07 30	16 16 16.4	33	4.6	LPAZ	851.75	199 ± 48	0.444 ± 0.184
95 07 30	18 02 45.8	33	5.0	LPAZ	897.34	282 ± 49	0.617 ± 0.254
95 07 30	18 26 31.9	33	4.7	LPAZ	858.43	239 ± 57	0.496 ± 0.351
95 07 30	21 05 47.7	14	5.6	LPAZ	822.84	267 ± 19	0.696 ± 0.153
95 07 30	22 27 55.0	33	4.9	LPAZ	798.38	225 ± 22	0.420 ± 0.198
95 07 30	23 06 44.0	33	4.7	LPAZ	847.31	246 ± 24	0.562 ± 0.243
95 07 31	00 14 47.3	33	5.0	LPAZ	894.01	269 ± 39	0.763 ± 0.496
95 07 31	02 07 12.8	33	4.8	LPAZ	792.82	202 ± 20	0.491 ± 0.126
95 07 31	06 08 35.2	33	4.6	LPAZ	843.97	182 ± 20	0.822 ± 0.239
95 07 31	09 25 32.9	44	4.8	LPAZ	850.64	286 ± 22	0.841 ± 0.241
95 08 01	02 44 46.5	31	4.8	LPAZ	809.50	218 ± 22	0.638 ± 0.214
95 08 01	05 10 57.4	30	5.2	LPAZ	997.42	283 ± 30	0.468 ± 0.224
95 08 01	06 00 37.4	33	4.7	LPAZ	920.69	266 ± 77	0.665 ± 0.346
95 08 01	12 37 19.9	23	4.8	LPAZ	880.66	278 ± 41	0.721 ± 0.169
95 08 01	21 55 43.9	33	4.6	LPAZ	827.29	257 ± 26	0.605 ± 0.299
95 08 02	00 14 09.4	33	5.4	LPAZ	812.94	260 ± 28	0.827 ± 0.162
95 08 02	05 22 21.5	33	4.8	LPAZ	796.16	241 ± 30	0.852 ± 0.389
95 08 02	11 05 38.9	33	5.2	LPAZ	791.71	294 ± 33	0.480 ± 0.188
95 08 02	20 20 13.9	33	5.1	LPAZ	781.70	253 ± 34	0.400 ± 0.167
95 08 03	01 57 19.9	17	5.4	LPAZ	792.82	257 ± 36	0.557 ± 0.176
95 08 03	04 43 04.5	34	4.7	LPAZ	798.38	244 ± 38	0.511 ± 0.177
95 08 03	12 00 27.7	33	5.0	LPAZ	788.37	280 ± 16	0.768 ± 0.185
95 08 03	14 03 05.8	33	4.6	LPAZ	789.48	228 ± 40	0.612 ± 0.283
95 08 03	19 07 45.2	34	4.6	LPAZ	897.37	274 ± 27	0.810 ± 0.313

Table 2 (Continued)

Date Y/M/D	Origin time h min. sec.	Depth km	Mag. m_b	Recording station	Ep. dist. km	$Q_0 \pm \Delta Q_0$	$\eta \pm \Delta\eta$
95 08 03	19 50 37.5	45	4.9	LPAZ	366.21	215 ± 25	0.735 ± 0.175
95 08 03	23 07 12.1	33	4.5	LPAZ	322.89	287 ± 34	0.815 ± 0.176
95 08 04	11 28 50.3	32	4.5	LPAZ	387.34	262 ± 46	0.452 ± 0.288
95 08 05	00 32 46.6	38	4.6	LPAZ	310.61	284 ± 28	0.712 ± 0.163
95 08 05	01 50 13.4	33	4.7	LPAZ	797.27	297 ± 31	0.179 ± 0.170
95 08 05	08 21 06.1	33	4.8	LPAZ	792.82	247 ± 32	0.653 ± 0.210
95 08 06	22 38 32.9	33	4.6	LPAZ	860.65	298 ± 29	0.636 ± 0.214
95 08 09	07 20 37.4	34	5.0	LPAZ	930.70	256 ± 19	0.857 ± 0.120
95 08 09	08 23 01.1	35	5.3	LPAZ	782.81	291 ± 28	0.791 ± 0.159
95 08 10	18 10 37.2	33	5.4	LPAZ	866.21	281 ± 43	0.599 ± 0.174
95 08 18	19 44 49.9	33	4.5	LPAZ	878.44	294 ± 27	0.707 ± 0.164
95 08 20	03 09 05.1	33	4.6	LPAZ	808.39	294 ± 36	0.674 ± 0.202
95 08 31	04 16 34.2	39	4.5	LPAZ	689.41	294 ± 54	0.761 ± 0.558
95 09 12	08 40 43.1	33	4.5	LPAZ	881.78	268 ± 23	0.509 ± 0.271
95 10 03	11 40 16.0	33	4.9	LPAZ	1838.05	428 ± 80	0.709 ± 0.220
95 10 03	14 29 27.9	33	4.7	LPAZ	1822.49	529 ± 247	0.853 ± 0.549
95 10 03	17 01 00.9	33	4.9	LPAZ	1826.93	523 ± 110	0.703 ± 0.261
95 10 05	01 36 55.6	33	5.1	LPAZ	1840.28	563 ± 163	0.634 ± 0.329
95 10 05	02 45 18.2	33	4.9	LPAZ	1829.16	651 ± 99	0.785 ± 0.190
95 10 07	03 32 56.8	33	4.8	LPAZ	802.83	310 ± 29	0.750 ± 0.156
95 10 07	09 15 24.9	33	4.7	LPAZ	1830.27	373 ± 79	0.781 ± 0.240
95 10 07	09 58 53.7	33	4.5	LPAZ	1832.49	612 ± 91	0.743 ± 0.465
95 10 07	21 28 03.1	12	5.8	LPAZ	1831.38	395 ± 136	0.452 ± 0.320
95 10 08	10 27 39.1	33	5.3	LPAZ	1849.17	472 ± 129	0.655 ± 0.331
95 10 12	22 57 09.9	32	5.3	LPAZ	782.81	310 ± 46	0.537 ± 0.168
95 10 16	16 36 19.4	28	5.4	LPAZ	809.50	285 ± 24	0.657 ± 0.138
95 10 16	16 44 20.1	30	5.0	LPAZ	827.29	208 ± 42	0.730 ± 0.588
95 10 18	02 25 35.8	33	4.7	LPAZ	1833.61	61 ± 95	0.611 ± 0.245
95 10 19	09 13 40.9	33	4.6	LPAZ	1821.37	652 ± 70	0.928 ± 0.268
95 10 20	20 18 56.0	33	4.9	LPAZ	1826.93	340 ± 82	0.845 ± 0.352
95 10 23	00 42 10.8	33	4.8	LPAZ	1826.93	514 ± 51	0.898 ± 0.226
95 10 25	21 26 55.0	33	4.5	LPAZ	836.19	282 ± 61	0.703 ± 0.679
95 10 28	18 05 35.9	33	5.0	LPAZ	992.97	274 ± 62	0.657 ± 0.369
95 10 29	05 28 48.5	33	4.9	LPAZ	1839.37	475 ± 83	0.705 ± 0.184
95 10 31	01 55 57.4	33	5.2	LPAZ	1438.36	347 ± 90	0.674 ± 0.294
95 11 01	00 35 32.3	20	6.3	LPAZ	1444.42	367 ± 66	0.592 ± 0.263
95 11 17	02 12 23.3	33	5.1	LPAZ	2599.74	848 ± 37	0.391 ± 0.139
95 12 29	13 01 40.9	33	5.4	LPAZ	2916.64	551 ± 19	0.729 ± 0.090
95 12 31	15 12 33.3	40	5.1	LPAZ	2921.09	693 ± 25	0.646 ± 0.090
96 03 01	02 27 24.1	27	5.1	LPAZ	828.40	289 ± 30	0.725 ± 0.171
96 03 07	08 38 57.5	31	5.3	LPAZ	803.94	281 ± 25	0.517 ± 0.110
96 03 10	08 56 22.3	33	5.8	LPAZ	392.52	391 ± 48	0.784 ± 0.140
96 03 15	12 45 44.1	29	5.0	LPAZ	1310.99	412 ± 88	0.404 ± 0.358
96 03 18	05 53 21.2	19	4.5	LPAZ	1309.88	533 ± 29	0.449 ± 0.173
96 03 29	15 35 47.8	33	4.5	LPAZ	957.39	203 ± 39	0.629 ± 0.419
96 04 10	12 43 40.6	29	4.6	LPAZ	640.48	323 ± 52	0.894 ± 0.189
96 04 14	13 47 05.8	33	4.6	LPAZ	499.27	285 ± 27	0.851 ± 0.141
96 05 09	00 50 38.6	45	4.7	LPAZ	2374.01	766 ± 139	0.768 ± 0.206
96 05 09	16 44 26.6	33	4.7	LPAZ	713.87	338 ± 90	0.562 ± 0.303

Table 2 (Continued)

Date Y/M/D	Origin time h min. sec.	Depth km	Mag. m_b	Recording station	Ep. dist. km	$Q_0 \pm \Delta Q_0$	$\eta \pm \Delta \eta$
96 05 19	03 25 37.0	17	4.6	LPAZ	3018.94	498 ± 66	0.553 ± 0.399
95 08 17	12 11 28.6	26	5.1	NNA	372.50	243 ± 8	0.703 ± 0.083
95 08 31	04 16 34.2	39	4.5	NNA	402.53	210 ± 13	0.630 ± 0.071
95 10 03	01 51 23.9	24	6.5	NNA	1027.44	308 ± 49	0.332 ± 0.197
95 10 03	10 06 59.0	27	4.7	NNA	1019.66	317 ± 62	0.440 ± 0.351
95 10 03	11 40 16.0	33	4.9	NNA	1034.11	396 ± 105	0.420 ± 0.329
95 10 03	12 44 58.0	17	6.0	NNA	1022.99	339 ± 11	0.895 ± 0.083
95 10 05	01 36 55.6	33	5.1	NNA	1029.67	231 ± 47	0.325 ± 0.254
95 10 07	21 28 03.1	12	5.8	NNA	1022.99	346 ± 34	0.787 ± 0.231
95 10 08	10 27 39.1	33	5.3	NNA	1045.23	303 ± 39	0.779 ± 0.159
95 10 18	02 25 35.8	33	4.7	NNA	1017.43	335 ± 29	0.610 ± 0.181
95 10 19	09 13 40.9	33	4.6	NNA	1008.54	266 ± 27	0.492 ± 0.122
95 10 23	00 42 10.8	33	4.8	NNA	1019.66	368 ± 46	0.674 ± 0.256
95 10 25	21 26 55.0	33	4.5	NNA	309.12	271 ± 21	0.804 ± 0.103
95 11 14	06 25 45.2	26	4.8	NNA	1005.20	269 ± 15	0.456 ± 0.115
96 01 19	19 01 58.4	36	5.6	NNA	272.43	268 ± 15	0.493 ± 0.093
96 02 21	12 51 04.3	33	5.8	NNA	395.85	206 ± 18	0.554 ± 0.143
96 02 21	13 47 17.0	33	4.8	NNA	403.64	237 ± 12	0.747 ± 0.132
96 02 21	14 47 24.7	33	4.8	NNA	418.09	215 ± 18	0.578 ± 0.193
96 02 22	09 20 19.2	33	4.6	NNA	290.29	220 ± 5	0.549 ± 0.067
96 02 22	15 08 56.1	48	4.5	NNA	398.08	200 ± 7	0.633 ± 0.094
96 03 14	07 29 03.4	33	4.5	NNA	319.13	212 ± 7	0.500 ± 0.088
96 03 15	12 45 44.1	29	5.0	NNA	343.59	233 ± 11	0.690 ± 0.127
96 03 18	05 53 21.2	19	4.5	NNA	593.78	361 ± 26	0.428 ± 0.100
96 03 28	23 03 49.8	33	5.8	NNA	1227.59	368 ± 37	0.566 ± 0.114
96 04 10	12 43 40.6	29	4.6	NNA	553.75	223 ± 10	0.580 ± 0.113
96 04 14	13 47 05.8	33	4.6	NNA	709.42	248 ± 18	0.494 ± 0.125
96 04 21	20 47 12.0	32	4.5	NNA	230.17	213 ± 12	0.590 ± 0.103
96 05 06	18 30 57.3	23	4.5	NNA	691.63	329 ± 35	0.635 ± 0.199
96 05 11	15 57 40.8	33	4.6	NNA	365.83	228 ± 28	0.640 ± 0.203
95 06 04	12 18 58.9	33	4.9	PLCA	836.19	467 ± 58	0.713 ± 0.167
95 06 07	22 49 18.1	25	5.0	PLCA	853.98	596 ± 87	0.585 ± 0.237
95 06 12	17 47 16.4	9	5.0	PLCA	773.92	445 ± 22	0.858 ± 0.124
95 06 14	16 12 59.1	49	5.4	PLCA	1764.66	342 ± 36	0.329 ± 0.172
95 06 29	01 27 02.8	41	5.1	PLCA	772.81	363 ± 71	0.539 ± 0.320
95 07 09	13 47 26.7	43	4.7	PLCA	1071.92	388 ± 22	0.400 ± 0.143
95 07 30	10 35 39.2	11	5.8	PLCA	1815.81	329 ± 11	0.452 ± 0.086
95 07 30	10 56 13.0	33	5.4	PLCA	1959.26	457 ± 75	0.644 ± 0.133
95 07 30	18 02 45.8	33	5.0	PLCA	1853.62	353 ± 42	0.372 ± 0.310
95 07 30	21 05 47.7	14	5.6	PLCA	1927.01	514 ± 47	0.787 ± 0.231
95 07 31	00 14 43.3	33	5.0	PLCA	1841.39	327 ± 27	0.916 ± 0.257
95 08 01	05 10 57.4	30	5.2	PLCA	1754.66	405 ± 42	0.660 ± 0.170
95 08 01	06 00 37.4	33	4.7	PLCA	1832.49	360 ± 21	0.519 ± 0.146
95 08 01	13 29 42.9	25	5.0	PLCA	1103.05	380 ± 24	0.600 ± 0.106
95 08 02	00 14 09.4	33	5.4	PLCA	1940.35	434 ± 33	0.621 ± 0.193
95 08 03	01 57 19.9	17	5.4	PLCA	1959.26	401 ± 44	0.708 ± 0.274
95 08 10	18 10 37.2	33	5.4	PLCA	1880.31	420 ± 52	0.511 ± 0.205
95 09 17	07 25 26.9	8	5.9	PLCA	656.05	303 ± 22	0.525 ± 0.123
95 10 31	01 55 57.4	33	5.2	PLCA	1309.88	411 ± 28	0.619 ± 0.174

Table 2 (Continued)

Date Y/M/D	Origin time h min. sec.	Depth km	Mag. m_b	Recording station	Ep. dist. km	$Q_0 \pm \Delta Q_0$	$\eta \pm \Delta\eta$
95 11 01	00 35 32.3	30	6.3	PLCA	1308.77	472 ± 20	0.714 ± 0.107
95 11 21	18 17 04.5	40	5.1	PLCA	838.41	542 ± 43	0.815 ± 0.151
95 01 03	09 33 10.7	46	4.9	PLCA	1318.77	435 ± 26	0.670 ± 0.151
96 01 03	19 36 25.4	33	4.8	PLCA	1329.89	401 ± 30	0.383 ± 0.198
96 02 22	13 40 53.5	44	5.9	PLCA	787.26	477 ± 18	0.821 ± 0.051
96 03 01	02 27 24.1	27	5.1	PLCA	1912.55	470 ± 112	0.760 ± 0.494
96 03 07	08 38 57.5	31	5.3	PLCA	1935.90	340 ± 30	0.520 ± 0.236
96 04 19	00 19 31.1	50	6.0	PLCA	1816.40	392 ± 17	0.726 ± 0.109
96 01 17	19 46 01.2	24	4.8	PTGA	2113.82	607 ± 37	0.511 ± 0.153
96 01 19	19 01 58.3	36	5.6	PTGA	2338.43	625 ± 30	0.675 ± 0.073
96 02 21	12 51 04.3	33	5.8	PTGA	2380.68	662 ± 78	0.555 ± 0.133
96 02 21	13 47 17.0	33	4.8	PTGA	2411.82	639 ± 91	0.517 ± 0.178
96 02 21	14 47 24.7	33	4.8	PTGA	2411.82	537 ± 47	0.683 ± 0.278
96 02 22	09 20 19.2	33	4.6	PTGA	2392.92	612 ± 171	0.681 ± 0.380
96 02 22	15 08 56.1	48	4.5	PTGA	2381.80	481 ± 110	0.616 ± 0.363
96 03 06	07 42 20.0	33	5.0	PTGA	1335.45	596 ± 48	0.586 ± 0.132
96 03 10	08 56 22.3	33	5.0	PTGA	1706.84	778 ± 30	0.380 ± 0.072
96 03 15	12 45 44.1	29	5.0	PTGA	2161.63	502 ± 26	0.668 ± 0.081
96 03 18	05 53 21.2	19	4.5	PTGA	1834.72	651 ± 52	0.567 ± 0.141
96 03 28	23 03 49.8	33	5.8	PTGA	2087.13	602 ± 61	0.649 ± 0.168
96 04 19	00 19 31.1	50	6.0	PTGA	2788.77	804 ± 143	0.309 ± 0.229
96 05 09	00 50 38.6	45	4.7	PTGA	1570.07	591 ± 49	0.654 ± 0.153
96 05 13	04 53 47.7	28	5.1	PTGA	2078.23	578 ± 53	0.550 ± 0.139
96 05 19	03 25 27.2	17	4.6	PTGA	1284.30	548 ± 26	0.643 ± 0.088
96 05 23	01 57 22.9	33	5.5	PTGA	2088.24	543 ± 73	0.701 ± 0.198
96 06 02	00 50 37.0	33	5.3	PTGA	2385.13	577 ± 58	0.651 ± 0.157
95 06 25	09 11 32.3	33	4.7	SDV	553.75	246 ± 9	0.733 ± 0.094
95 08 03	01 57 19.9	27	5.4	SDV	3529.33	676 ± 138	0.496 ± 0.143
95 08 17	12 11 28.6	26	5.1	SDV	2311.74	597 ± 84	0.546 ± 0.308
95 09 01	04 54 25.1	42	4.6	SDV	823.95	230 ± 15	0.810 ± 0.098
95 10 03	01 51 23.9	24	6.5	SDV	1515.59	335 ± 63	0.518 ± 0.278
95 10 03	05 20 26.3	33	5.0	SDV	1517.81	390 ± 47	0.375 ± 0.281
95 10 03	09 30 52.3	33	4.7	SDV	1544.50	415 ± 88	0.553 ± 0.360
95 10 03	11 12 21.6	33	4.6	SDV	1486.68	329 ± 56	0.896 ± 0.300
95 10 03	11 40 16.0	33	4.9	SDV	1502.24	292 ± 81	0.625 ± 0.270
95 10 03	12 44 58.0	17	6.0	SDV	1515.59	397 ± 110	0.417 ± 0.339
95 10 03	14 29 27.9	33	4.7	SDV	1532.27	389 ± 97	0.515 ± 0.241
95 10 03	18 04 27.6	46	4.9	SDV	1527.82	297 ± 85	0.521 ± 0.325
95 10 03	20 08 59.0	33	4.6	SDV	1503.36	373 ± 131	0.839 ± 0.339
95 10 07	09 15 24.9	33	4.7	SDV	1516.70	359 ± 66	0.663 ± 0.208
95 10 07	21 28 03.1	12	5.8	SDV	1514.48	374 ± 31	0.876 ± 0.135
95 10 08	10 27 39.1	33	5.3	SDV	1496.68	338 ± 68	0.408 ± 0.294
95 10 19	08 51 17.1	33	4.5	SDV	1493.35	385 ± 73	0.465 ± 0.347
95 11 17	02 12 23.3	33	5.1	SDV	252.41	231 ± 24	0.881 ± 0.245
96 01 19	19 01 58.3	36	5.6	SDV	2315.08	455 ± 93	0.668 ± 0.198
96 06 02	00 50 37.0	33	5.3	SDV	2277.27	589 ± 174	0.380 ± 0.334
96 08 14	21 23 37.4	33	4.9	PLCA	323.58	219 ± 18	0.515 ± 0.134
96 09 05	21 22 21.0	20	4.6	PLCA	567.06	245 ± 30	0.541 ± 0.225

In the first study of South American attenuation which covered most of the continent, RAOOF and NUTTLI (1985) studied the attenuation of both direct Lg waves and Lg coda at frequencies near 1 Hz. Owing to the low density of seismological stations in eastern South America, their results still, however, were weighted toward the western part of the continent. They found that western South America is dominated by high attenuation and high frequency dependence of Q. Q_0 values vary from about 130 to 350, and frequency dependence ranges between 0.4 and 0.7. The low Q_0 band coincides with the tectonically active Andean belt. RAOOF and NUTTLI (1985) found Q_0 to be between 580 and 980 and frequency dependence to be between 0.0 and 0.2 in the eastern part of the continent. They found intermediate values of both Q_0 (420 to 580) and frequency dependence (0.2 to 0.3) in northern and northeastern Argentina. Their Q_0 determinations, using direct Lg and Lg coda are generally similar. There are, however, some differences, mainly in the eastern part of the continent, where Q_0 obtained for direct Lg is slightly lower than that obtained for Lg coda.

Our tomographic images of Q_0 and η are similar in their large-scale features to the results of RAOOF and NUTTLI (1985). Differences in results between the two studies can be explained to some extent by differences in methodology, although mainly by improved data coverage in central and eastern South America in the present study. All other studies of attenuation in South America are principally restricted to the western portion of the continent, mainly along the Andean mountain belt. The ensuing paragraphs compare our results with those of earlier studies in the Andean belt.

In the northern Andes, Q_0 for Lg coda is low (300 to 400) over much of the distance between the Pacific coast at 2.5°N and the Guyana-Venezuela border (about 13°N, 60°W). MOLNAR and OLIVER (1969) reported inefficient propagation of Sn phases in the eastern part of this region and estimated that Q is less than about 300. High heat flow (>100 mW/m²) has been reported for the eastern portion of this region in northern Venezuela (HAMZA and MUÑOZ (1996).

Q_0 is slightly higher between 2.0°N (southern Colombia) and 8.0°S (northern Peru) than it is further north. Sn Q in this region is also higher, with values of 300 or more (MOLNAR and OLIVER, 1969), and high-frequency shear waves propagate efficiently (CHINN et al., 1980). HAMZA and MUÑOZ (1996) report heat flows of 40–60 mW/m² for Peru in this region, which are substantially lower than those further north and are consistent with the higher Q values in this region.

Q_0 is again lower (about 250) between latitudes 10.0°S (central Peru) and 24.0°S (northern Chile). Within this region MOLNAR and OLIVER (1969) report inefficient propagation for Sn, and CHINN et al. (1980) refer to a mix between efficient and inefficient Sn propagation with inefficient propagation predominating beneath the Altiplano and Puna. Melting of the mantle beneath the Altiplano and eastern Cordillera, possibly due to convective removal of the base of the lithosphere (HOKE *et al.*, 1994), and high heat flow (>100 mW/m²), in both the Altiplano and Puna (HAMZA and MUÑOZ, 1996), suggest higher than normal temperatures beneath this region.

Between latitudes 24.0°S and 34.0°S (central Chile and western Argentina), our Q_0 map shows values of about 450. This is significantly higher than in either the Altiplano to the north or in southern Chile, between 34.0°S and 42.0°S. This region exhibits efficient propagation for high-frequency shear waves from shallow sub-Andean events (CHINN *et al.*, 1980). Heat flow determinations are limited to the region between latitudes 30.0°S and 33.0°S but indicate moderate values of 60–80 mW/m² (HAMZA and MUÑOZ, 1996).

In the Patagonian platform (Fig. 1) our determinations are limited only to the northern portion of the province (Fig. 5), where Q_0 varies between some 400 and 500. No other attenuation measurements are available for this region; however, moderate heat flow (60–80 mWm²) has been inferred from a few measurements there (HAMZA and MUÑOZ, 1996).

BARAZANGI *et al.* (1975) made many measurements of *Q* for p*P* phases along western South America and found low attenuation in some regions and high attenuation in others in the upper 300 km of the mantle. SACKS and OKADA (1974) also found *Q* to be high for direct *P* waves throughout much of the upper mantle beneath portions of that region. In addition, WHITMAN *et al.* (1992) estimated *Q* for *P* and *S* waves beneath the central Andean Plateau employing recordings at distances that are primarily sensitive to properties in the uppermost mantle (about 100–250 km depth). They found predominantly high *Q* ($Q_P > 500$ and $Q_S > 350$) for paths north of latitude 22°S and predominantly low *Q* ($Q_P < 350$ and $Q_S < 200$) for paths to the south of latitude 22°S. The results of these three mantle studies indicate that even though crustal attenuation in the Andean belt (as determined from Lg coda) is high, upper mantle attenuation (as determined from p*P* and *P* and *S* phases) there can be either high or low.

6. *The Relation of Q_0 to the Tectonics and Geology of South America*

The most obvious feature of the Q_0 map of South America is the large difference between the tectonically active Andean mountain belt, where Q_0 is almost everywhere less than 450 and the rest of the continent where it is 550 or more. The Andean mountain belt is currently tectonically active, undergoing uplift and deformation by interaction of the South American plate with the subducting Nazca and Pacific plates. MITCHELL (1995) proposed that crustal *Q* in any region is directly proportional to the length of time that has elapsed since the most recent episode of tectonic activity there (see also MITCHELL and CONG, this volume). In western South America the low *Q* is probably caused by the same mechanism that produces the low *Q* in the Tethysides belt of southern Europe and Asia. In that region it is likely that fluids are generated by hydrothermal reactions in subducting slabs beneath Eurasia and they slowly percolate upward into the crust (MITCHELL *et al.*, 1997). The low Q observed for the Andean mountain belt is consistent with that idea.

As discussed in the preceding section, although Q_0 is low along the entire Andean belt, there are some variations within the belt. In the two regions of lowest Q_0, between latitudes 2.5° and 13°N and between 10° and 24°S, upper mantle temperatures are higher than in adjacent regions. This suggests either that temperature has an effect on crustal Q in these regions or that the elevated temperatures have generated more hydrothermal fluids there.

The region of highest Q_0 is centered in the central Brazilian shield and extends northward into the Amazonian basin and Guiana shield and southward into the Pananá basin. It might be expected that the sediments of the two basins possibly affect observed Q_0. The sediments in those basins are, however, Phanerozoic in age. MITCHELL and HWANG (1987), studying the effect of sediment on *Lg Q* in North America, showed that although Mesozoic and younger sandstones and shales significantly reduced Q for *Lg* waves, sediments of greater age did not. It is likely that the sediments of the Amazonian and Paraná basins are either largely limestones or dolomites or that they have lost fluids with time through the process of lithification. Reduced Q_0 in southern portion of the Chaco basin, however, could possibly be caused by accumulations of sediments with ages as young as Quaternary, as described in section 2.

Q_0 along most of the eastern coast of South America is about 650–700. Although this region is a shield and contains little sedimentary cover over the majority of its length, it is significantly more attenuating than is the central high-Q region. It is interesting that Q_0 is also about 700 along portions of the eastern coast of the United States (SINGH and HERRMANN, 1983; BAQER and MITCHELL, 1998), and along parts of the western coasts of Europe (MITCHELL *et al.*, 1997) and Africa (XIE and MITCHELL, 1990a). This may indicate that Q_0 in those regions is related to tectonic processes that occurred during the breakup of Gondwanaland in Jurassic time.

7. Conclusions

Q_0 varies between about 250 and 1100 across South America with the lowest values being located in the Andean mountain belt, a long currently tectonically active zone along the western coast, and the highest values centered in the central Brazilian shield and extending northward into the Amazonian basin and Guiana shield and southward into the Paraná and Chaco basins. Q_0 varies along the Andean belt and is lowest where heat flow is highest. Most of the spatial variations in Q_0 in South America are consistent with the proposal (MITCHELL, 1995) that the value of that parameter in any region is directly proportional to the length of time that has elapsed since the last major episode of tectonic activity there. One exception is the northern portion of the Chaco basin where enhanced attenuation may be due to accumulations of young sediments.

Reduced Q_0 (about 650–700) values extend along most of the eastern coast of South America. These reduced values are not associated with any recent tectonic activity or sedimentary accumulations and may be related to tectonic processes that occurred during the breakup of Gondwanaland in the Jurassic period.

Acknowledgments

We thank several colleagues for their various contributions to the study. The program which determined *Lg* coda *Q* was written by Jack Xie and modified by R. B. Herrmann. Jack Xie also wrote the programs used in *Lg* coda tomography. Fernando Mancini provided several references pertaining to the geology of South America and Valiya Hanza and Miguel MuÑoz provided a preprint of their paper. One of us (JLS) thanks the Brazilian government (Fundação Coordenação de Aperfeiçoamento de Pessoal de Nível Superior—CAPES) for financial support while he was a scholar at Saint Louis University. He also thanks the Department of Earth and Atmospheric Sciences at Saint Louis University for excellent facilities with which to conduct this research.

REFERENCES

ALMEIDA, F. F. M., HASUI, Y., DE BRITO NEVES, B. B., and FUCK, R. A. (1981), *Brazilian Structural Provinces: An Introduction*, Earth Science Rev. *17*, 1–29.

BAQER, S., and MITCHELL, B. J. (1988), *Regional Variation of Lg Coda Q in the Continental United States and its Relation to Crustal Structure and Evolution*, Pure and appl. geophys., *153*, 613–638.

BARAZANGI, M., PENNINGTON, W., and ISACKS, B. (1975), *Global Study of Seismic Wave Attenuation in the Upper Mantle behind Island Arcs Using pP Waves*, J. Geophys. Res. *80*, 1079–1092.

BRITO NEVES, B. B. DE, and CORDANI, U. G. (1991), *Tectonic Evolution of South America during the Late Proterozoic*, Precambrian Res. *53*, 23–40.

CHINN, D. S., ISACKS, B. L., and BARAZANGI, M. (1980), *High Frequency Seismic Wave Propagation in Western South America along the Continental Margin, in the Nazca Plate and across the Altiplano*, Geophys. J. R. Astr. Soc. *60*, 209–244.

CONG, L., and MITCHELL, B. J. (1998), *Lg Coda Q and its Relation to the Geology and Tectonics of the Middle East*, Pure and appl. geophys., *153*, 563–585.

CORDANI, U. G., AMARAL, G., and KAWASHITA, K. (1973), *The Precambrian Evolution of South America*, Geologishe Rundschau *62*, 309–317.

DALZIEL, I. W. D. (1986), *Collision and cordilleran orogenesis: An Andean perspective*. In *Collision Tectonics*, Geol. Soc. Spec. Publ. *19*, 389–404.

FORSYTHE, T. (1982), *The Late Paleozoic to Early Mesozoic Evolution of Southern South America: A Plate Tectonic Interpretation*, J. Geol. Soc. London *139*, 671–682.

GANSSER, A. (1973), *Facts and Theories on the Andes*, J. Geol. Soc. London *129*, 93–131.

HOKE, L., HILTON, D. R., LAMB, S. H., HAMMERSCHMIDT, K., and FRIEDRICHSEN, H. (1994), *^{3}He Evidence for a Wide Zone of Active Mantle Melting beneath the Central Andes*, Earth Planet. Sci. Lett. *128*, 341–355.

HAMZA, V. M., and MUÑOZ, M. (1996), *Heat Flow Map of South America*, Geothermic 25, 599–646.

HASUI, Y., and ALMEIDA, F. F. M. (1985), *The Central Brazil Shield Reviewed*, Episode 8, 29–37.

JAMES, D. E. (1971), *Plate Tectonic Model for the Evolution of the Central Andes*, Geol. Soc. Am. Bull. 82, 3325–3346.

MITCHELL, B. J. (1995), *Anelastic Structure and Evolution of the Continental Crust and Upper Mantle from Seismic Surface Wave Attenuation*, Rev. Geophys. 33, 441–462.

MITCHELL, B. J., and HWANG, H. J. (1987), *Effect of Low Q Sediments on Lg Attenuation in the United States*, Bull. Seismol. Soc. Am. 77, 1197–1210.

MITCHELL, B. J., and CONG, L. (1998), *Lg Coda Q and its Relation to the Structure and Evolution of Continents: A Global Perspective*, Pure and appl. geophys., 153, 655–663.

MITCHELL, B. J., PAN, Y., XIE, J., and CONG, L. (1997), *The Variation of Lg Coda Q across Eurasia and its Relation to Crustal Evolution*, J. Geophys. Res. 102, 22767–22779.

MOLNAR, P., and OLIVER, J. (1969), *Lateral Variations of Attenuation in the Upper Mantle and Discontinuities in the Lithosphere*, J. Geophys. Res. 74, 2648–2682.

NUTTLI, O. W. (1973), *Seismic Wave Attenuation and Magnitude Relations for Eastern North America*, J. Geophys. Res. 78, 876–885.

NUTTLI, O. W. (1988), *Lg Magnitudes and Yield Estimates for Underground Novaya Zemlya Nuclear Explosions*, Bull. Seismol. Soc. Am. 78, 873–884.

RAOOF, M. M., and NUTTLI, O. W. (1985), *Attenuation of High-frequency Earthquake Waves in South America*, Pure and appl. geophys. 122, 619–644.

SACKS, I. S., and OKADA, H. (1974), *A Comparison of the Anelasticity Structure beneath Western South America and Japan*, Phys. Earth Planet. Int. 9, 211–219.

SINGH, S., and HERRMANN, R. B. (1983), *Regionalization of Crustal Coda Q in the Continental United States*, J. Geophys. Res. 88, 527–538.

WHITMAN, D., ISACKS, B. L., CHATELAIN, CHIU, J.-M., and PEREZ, A. (1992), *Attenuation of High-frequency Seismic Waves beneath the Central Andean Plateau*, J. Geophys. Res. 97, 19929–19947.

XIE, J., and NUTTLI, O. W. (1988), *Interpretation of High-frequency Coda at Large Distances: Stochastic Modeling and Method of Inversion*, Geophys. J. 95, 579–595.

XIE, J., and MITCHELL, B. J. (1990a), *A Back-projection Method for Imaging Large-scale Lateral Variations of Lg coda Q with Application to Continental Africa*, Geophys. J. Int. 100, 161–181.

XIE, J., and MITCHELL, B. J. (1990b), *Attenuation of Multiphase Surface Waves in the Basin and Range Province, I, Lg and Lg Coda*, Geophys. J. Int. 102, 121–137.

(Received October 22, 1997, revised April 21, 1998, accepted June 1, 1998)

To access this journal online:
http://www.birkhauser.ch

Pure appl. geophys. 153 (1998) 613–638
0033–4553/98/040613–26 $ 1.50 + 0.20/0

Pure and Applied Geophysics

Regional Variation of *Lg* Coda *Q* in the Continental United States and its Relation to Crustal Structure and Evolution

SAADIA BAQER[1,2] and B. J. MITCHELL[1]

Abstract—Records from broadband digital stations have allowed us to map regional variations of *Lg* coda *Q* across almost the entire United States. Using a stacked ratio method we obtained estimates of Q_0 (*Lg* coda *Q* at 1 Hz) and its frequency dependence, η, for 218 event-station pairs. Those sets of estimates were inverted using a back-projection method to obtain tomographic images showing regional variations of Q_0 and η. Q_0 is lowest (250–300) in the California coastal regions and the western part of the Basin and Range province, and highest (650–750) in the northern Appalachians and a portion of the Central Lowlands. Intermediate values occur in the Colorado Plateau (300–500), the Columbia Plateau (300–400), the Rocky Mountains (450–550), the Great Plains (500–650), the Gulf Coastal Plain and the southern portion of Atlantic Coastal Plain (400–500), and the portions of the Central Lowlands surrounding the high-*Q* region (500–550). The pattern of Q_0 variations suggests that the United States can be divided into two large *Q* provinces. One province spans the area from the Rocky Mountains to the Atlantic coast, is tectonically stable, and exhibits relatively high Q_0. The other extends westward from the approximate western margin of the Rocky Mountains to the Pacific coast, is tectonically active, and exhibits low Q_0. The transition from high to low *Lg* coda *Q* in the western United States lies further to the west than does an upper mantle transition for *Q* and electrical resistivity found in earlier studies. The difference in Q_0 between the western and eastern United States can be attributed to a greater amount of interstitial crustal fluids in the west. Regions of moderately reduced *Q* within the stable platform often occur where there are accumulations of Mesozoic and younger sediments. Reduced Q_0 in the southeastern United States may not be due to anelasticity but may rather be explained by a gradational velocity increase at the crust-mantle boundary that causes shear energy to leak into the mantle.

Key words: *Lg*, *Lg* coda, *Q*, United States, crust, crustal evolution.

Introduction

Various geophysical studies have shown that many crustal and upper mantle properties differ greatly between the eastern and western United States, with a major boundary coinciding roughly with the eastern front of the Rocky Mountains.

[1] Department of Earth and Atmospheric Sciences, Saint Louis University, 3507 Laclede Avenue, St. Louis, Missouri 63103, U.S.A. Fax: (314) 977-3117. E-mail: mitchell@eas.slu.edu
[2] Now at Department of Earth and Planetary Sciences, Washington University, St. Louis, Missouri, U.S.A.

These studies include determinations of P_n velocity (HERRIN and TAGGART, 1962; BRAILE et al., 1989), upper mantle shear velocity (ALSINA et al., 1996), and upper mantle electrical conductivity (PORATH, 1971; KELLER, 1989). Studies of seismic Q using intermediate-period surface waves (MITCHELL, 1975), direct Lg waves (SUTTON et al., 1967), and Lg coda (SINGH and HERRMANN, 1983) also suggest a major lateral change in Q in the crust somewhere near the Rocky Mountains. Studies of teleseismic body waves, at both intermediate periods (SOLOMON and TOKSÖZ, 1970) and short periods (DER et al., 1975), indicate that a similar change in Q occurs in the upper mantle beneath the United States.

Both of these two major tectonic divisions of the United States can be further divided into smaller tectonic provinces that differ in detail in their rock properties, temperatures, and levels of tectonic activity. The purpose of this study is to map variations of Lg coda Q (Q_{Lg}^c) and its frequency dependence across the United States. Whereas Lg Q (Q_{Lg}) is a measure of the efficiency of propagation of the direct Lg phase, Q_{Lg}^c measures the efficiency of propagation of the oscillatory motion following Lg. Lg coda is continuous with Lg and can remain prominent for up to several minutes after the onset of that phase. Direct Lg travels at a group velocity of 3.2–3.6 km/s, typically being high in stable regions and low in tectonically active regions. It is comprised of higher-mode surface waves and is usually observed at frequencies between about 0.1 and several Hz.

We will use the stacked spectral ratio (SSR) method (XIE and NUTTLI, 1988) and Lg coda Q tomography (XIE and MITCHELL, 1990) to obtain detailed maps of Q_{Lg}^c at 1 Hz (Q_0) and its frequency dependence (η) for the United States. We will then relate, as far as possible, those quantities to the tectonic history and structure of the crust. We will show that some variations are related to the tectonic evolution of the United States while others are related to structural features.

Pertinent Aspects of the Geology and Tectonics of the United States

In earlier studies, we found that seismic Q is influenced by the time that has elapsed since the most recent major episode of tectonic activity and, to a lesser extent, by thick accumulations of young sediments. For that reason we present a simplified discussion of the tectonics and sedimentary accumulations in various physiographic provinces of the United States.

The United States can be divided into several geomorphological units as shown in Figure 1. The region between the Atlantic Coast and the Rocky Mountains is a large, relatively stable, province that consists of several of those units, including the New England province, the Appalachian Mountains, the Central Lowlands, the Great Plains, and the Atlantic and Gulf Coastal Plains. Most of the sediments that cover this region are of Paleozoic and Mesozoic age, while younger sediments are mainly restricted to river valleys and glacial deposits. Precambrian basement crops out in a few places, including the St. Francois Mountains in Missouri, New England, and the central and southern Appalachians. Most of this large region has been stable since the Paleozoic, but the Atlantic and Gulf coastal regions underwent extensional deformation when the North American plate separated from the African plate during Jurassic time.

The Appalachian and Ouachita Mountains are orogens that developed along the eastern and southern continental margins of North America during the Paleozoic era. Geologists have documented considerable diversity in stratigraphic, metamorphic, deformational, and plutonic events in those regions.

The second major province of the United States, extending from the Rocky Mountains to the Pacific Coast, is a structurally complex region called the western United States Cordillera. In contrast to the central and eastern United States, portions of this region have undergone periods of severe tectonic activity from early in the Cenozoic era to the present time. Much of the Cordillera consists largely of oceanic and continental fragments that have been accreted during a complex evolutionary history. It includes various sedimentary and volcanic landforms of diverse rock composition, and tectonic elements that vary in age and geographic distribution.

MILLER *et al.* (1992) divide the Cordillera into regions with distinctive tectonic, sedimentary, and magmatic characteristics. The Rocky Mountains constitute the eastern portion of the Cordillera and were first formed by compressional deforma-

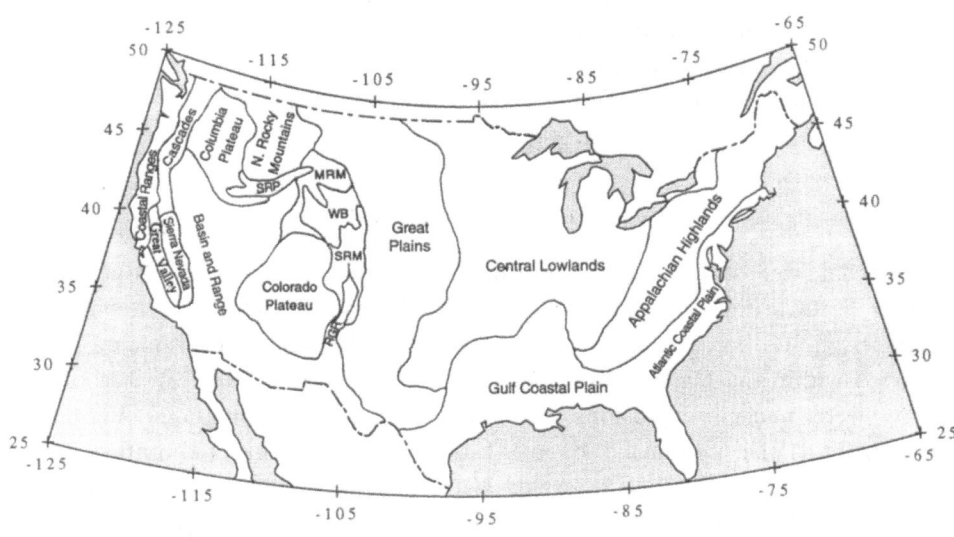

Figure 1
Map of physiographic provinces of the continental United States (adapted from ZOBACK and ZOBACK, 1989).

tion involving folds, thrusts and block uplifts in late Mesozoic and early Tertiary time. They were exposed to a complex sequence of extensional tectonic events that began in the Eocene in the northern Rockies and spread to the southern Rockies in the middle and late Tertiary (PRODEHL and LIPMAN, 1989).

The Colorado Plateau, the Basin and Range and the Snake River Plain form contrasting units within a region termed the intermontane system (ANDERSON, 1989). While the Colorado Plateau has escaped strong deformation and magmatism since Precambrian time, the Basin and Range has undergone repeated orogenesis, and the lava provinces underwent a single major pulse of magmatism and rifting that produced crust of Cenozoic and younger age. The Columbia Plateau, to the west and north of the intermontane system, is a volcanic plateau covered by thick layers of Miocene flood basalts and thick sedimentary basins of Pliocene and Quaternary age (MOONEY and WEAVER, 1989). The adjacent Cascade range has resulted from repeated episodes of volcanism that have occurred along the western margin of North America since late Paleozoic time (McBIRNEY, 1978).

The westernmost part of the United States to be considered in the present study is that along and near the Pacific Coast. The highest elevations in this region are those for the Sierra Nevada. This plutonic mountain range evolved from a Paleozoic margin and was modified by Mesozoic and Cenozoic subduction and late Cenozoic tectonics involving Basin and Range extension (SALEEBY, 1981). The Great Valley of California is a sedimentary basin situated between the granitic and metamorphic terranes of the Sierra Nevada and Coast ranges. These sediments include Jurassic to Late Cretaceous units with overlying Cenozoic deposits. Mountain ranges extend along the entire western coast of the United States. Different portions of these mountains consist of various amounts of marine metasedimentary rocks, plutonic terrains and Tertiary sediments and volcanics.

Data

The data set of this study was obtained by several stations throughout the United States (Fig. 2 and Table 1) and consists of 218 vertical-component, short-period and broadband *Lg* coda records from 108 regional seismic events that occurred within the United States between 1981 and 1996 (Table 2). Ten of these events were nuclear explosions that were part of the Integrated Verification Experiment (IVE), conducted by the Los Alamos National Laboratory Source Region Program. The other 98 events were earthquakes that occurred within, or just outside, the continental boundaries of the United States. *Lg* coda from selected seismic events of this study were all recorded for time durations between 300 and 600 seconds. Frequency components span the range between 0.2 and 3.0 Hz with frequencies in low-Q regions usually being concentrated toward the low-frequency

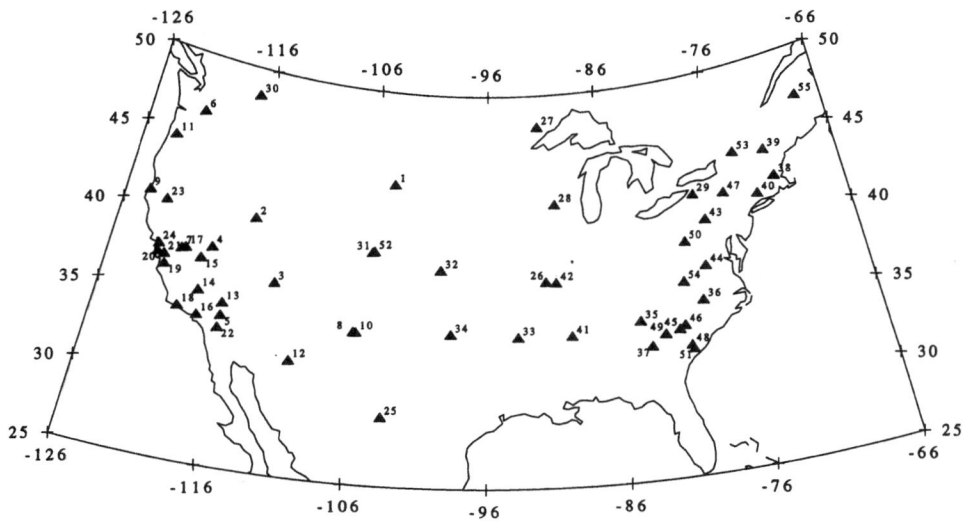

Figure 2
Map of seismic stations that recorded data used in this study. The numbers identify stations listed in Table 1.

end. Sampling rates were either 20 or 25 samples per second, depending upon the station.

Example seismograms of *Lg* and its coda appear in Figure 3 for three different paths, one (recorded at station SCP) in the eastern United States, one (recorded at station CCH) in the central United States, and one (recorded at station PAS) in the western United States. The seismogram for the path in the western United States is depleted in high frequencies relative those in the eastern and central United States, even though the path length in that region is shorter than the others. This depletion of high frequencies suggests that seismic energy has been lost by either intrinsic absorption or scattering along the path of travel for *Lg*.

We applied the stacked spectral ration (SSR) technique (XIE and NUTTLI, 1988) to all records to obtain single-trace determinations of Q_0 and η. Resulting values for Q_0 and η appear along with event information in Table 2. Example SSRs for the three seismograms of Figure 3 appear in Figure 4. The depletion of high-frequency energy in the western United States (station PAS) is apparent in the SSR.

Tomographic Maps of Q_0 and η

Our goal is to map Q_0 and η across the United States and to relate variations in those parameters to crustal and geological properties. To achieve that goal we

Table 1

Station information

Station number	Station name	Latitude (°)	Longitude (°)
1	RSSD	44.0710	−104.0210
2	ELK	40.7440	−115.2400
3	KNB	37.0160	−112.8220
4	MNV	38.4320	−118.1540
5	LAC	34.3900	−116.4110
6	LON	46.4500	−121.4840
7	JAS	37.9470	−120.4380
8	ALQ	34.5630	−106.2730
9	ARC	40.8770	−124.0700
10	ANMO	34.5640	−106.2730
11	COR	44.5860	−123.3030
12	TUC	32.3100	−110.7850
13	GSC	35.1800	−116.4810
14	ISA	35.6650	−118.4700
15	MLA	37.6310	−118.8300
16	PAS	34.1480	−118.1700
17	CMB	38.0350	−120.3900
18	SBC	34.4420	−119.7100
19	SAO	36.7650	−121.4400
20	STAN	37.4040	−122.1700
21	MHC	37.3420	−121.6400
22	PFO	33.6090	−116.4600
23	WCK	40.5800	−122.5400
24	BKS	37.8770	−122.2400
25	LTX	29.3340	−103.6607
26	CCM	38.0557	−91.2440
27	EYMN	47.9462	−91.4950
28	JFWS	42.9149	−90.2480
29	YSNY	42.4758	−78.5370
30	NEW	48.2633	−117.1200
31	GOL	39.7000	−105.3701
32	CBKS	38.8140	−99.7370
33	MIAR	34.5457	−93.5730
34	WMOK	34.7379	−98.7810
35	MYNK	35.0739	−84.1279
36	CEH	35.8908	−79.0928
37	GOGA	33.4112	−83.4666
38	HRV	42.5064	−71.5583
39	LBNH	44.2401	−71.9259
40	LSCT	41.6784˙	−73.2244
41	OXF	34.5118	−89.4092
42	FVM	37.9840	−90.4260
43	SCP	40.7950	−77.8650
44	CVL	37.9814	−78.4608
45	JSC	34.2789	−81.2581
46	LHS	34.4792	−80.8083

Table 1 (continued)

Station number	Station name	Latitude (°)	Longitude (°)
47	BINY	42.1993	−75.9861
48	SGS	33.1927	−80.5118
49	PRM	34.0833	−82.3633
50	MCWV	39.6581	−79.8456
51	HBF	32.9331	−80.3777
52	GLD	39.7506	−105.2210
53	RSNY	44.5483	−74.5300
54	BLA	37.2110	−80.4210
55	CBM	46.9330	−68.1210

inverted Q^c_{Lg} data using a back-projection tomography method which was developed by XIE and MITCHELL (1990a). Some aspects of the back-projection method are described briefly in the present volume by CONG and MITCHELL (1998). Figure 5 shows the coverage achieved for the United States with the available data. The area of each ellipse is assumed to correspond to the surface projection of the volume of the crust that is sampled by *Lg* coda energy that is singly scattered at a time late in the coda of each record. A seismic event lies at one focus of each ellipse and a recording station lies at the other.

Application of the method requires that the area of study be divided into a grid of many approximately rectangular cells. We performed several tests to determine the optimum size for the cells. Trying cells with dimensions of $1.5° \times 1.5°$, $2.0° \times 2.0°$, $2.5° \times 2.5°$, and $3.0° \times 3.0°$, we found that $2.0° \times 2.0°$ cells produced sufficient detail in Q^c_{Lg} variations yet were large enough to avoid spurious fluctuations. This selection resulted in a division of the United States into 420 cells spanning the area between latitudes 25.0°N and 50.0°N and between longitudes 65.0°W and 125.0°W. Q_0 values were initially set at 500 (the approximate mean Q_0 for all measurements) for every cell. Our inversions yielded maps of Q_0 and η for almost the entire United States (Figs. 6 and 7).

The Q_0 map exhibits large differences in Q^c_{Lg} between regions in the central/eastern United States and regions west of the Rocky Mountains, with values ranging between about 250 and 750. Lowest Q_0 (250–300) occurs in southern California and the western part of the Basin and Range province and highest values (700–750) in the northeastern United States. Intermediate values occur in the Gulf coastal plain and southern part of the Atlantic coastal plain (400–500), the Central lowlands (650–670), the Great Plains (450–600), the Rocky Mountains (400–550), the Colorado Plateau (300–500), and the Columbia Plateau (300–500).

The frequency dependence values in Figure 7 range between 0.4 and 0.9 with the highest values occurring in the northern Great Plains while low values occupy three broad regions, the westernmost United States, the northeastern United States, and

Table 2

Events information and Lg Coda Q results

Date	Origin (h:m:s)	Lat. (°)	Lon. (°)	Recording station	Distance (km)	Q_0	η
890627	15:30:00.1	37.28	−116.35	ELK	397.0	225 ± 31	0.46 ± 0.03
890622	21:15:00.0	37.28	−116.41	ELK	397.5	234 ± 62	0.30 ± 0.08
891031	59:30:00.0	37.26	−116.49	ELK	401.4	221 ± 23	0.44 ± 0.03
890309	14:05:00.1	37.14	−116.07	ELK	401.7	235 ± 34	0.51 ± 0.05
890210	20:06:00.0	37.08	−116.00	ELK	412.9	246 ± 29	0.45 ± 0.04
900725	15:00:00.0	37.21	−116.21	ELK	401.7	224 ± 42	0.31 ± 0.06
900613	16:00:00.0	37.26	−116.42	ELK	400.0	239 ± 20	0.38 ± 0.03
890627	15:30:00.1	37.28	−116.35	KNB	314.7	280 ± 38	0.58 ± 0.07
890622	21:15:00.0	37.28	−116.41	KNB	320.1	297 ± 58	0.51 ± 0.06
891031	15:30:00.0	37.26	−116.49	KNB	327.0	254 ± 29	0.60 ± 0.06
890309	14:05:00.1	37.14	−116.07	KNB	288.4	242 ± 35	0.26 ± 0.08
890210	20:06:00.0	37.08	−116.00	KNB	282.4	283 ± 56	0.42 ± 0.06
900613	16:00:00.0	37.26	−116.42	KNB	320.8	287 ± 48	0.45 ± 0.05
900310	16:00:00.0	37.11	−116.06	KNB	288.1	259 ± 45	0.46 ± 0.05
900725	15:00:00.0	37.21	−116.21	KNB	301.8	275 ± 33	0.41 ± 0.03
910404	18:59:59.9	37.29	−116.31	KNB	311.3	276 ± 38	0.51 ± 0.04
920326	16:30:00.0	37.27	−116.36	KNB	315.5	266 ± 24	0.38 ± 0.03
890627	15:30:00.1	37.28	−116.35	MNV	204.2	200 ± 18	0.30 ± 0.04
890622	21:15:00.0	37.28	−116.41	MNV	199.6	268 ± 48	0.46 ± 0.05
891031	59:30:00.0	37.26	−116.49	MNV	195.7	230 ± 23	0.40 ± 0.04
890309	14:05:00.1	37.14	−116.07	MNV	226.9	243 ± 39	0.56 ± 0.05
900725	15:00:00.0	37.21	−116.21	MNV	208.6	233 ± 42	0.35 ± 0.05
900613	16:00:00.0	37.26	−116.42	MNV	200.5	236 ± 26	0.40 ± 0.03
900310	16:00:00.0	37.11	−116.06	MNV	235.0	228 ± 32	0.40 ± 0.04
910404	19:00:00.0	37.29	−116.31	MNV	205.5	224 ± 27	0.35 ± 0.04
891031	59:29:59.9	37.26	−116.49	LAC	318.9	241 ± 34	0.50 ± 0.05
900613	16:00:00.0	37.26	−116.42	LAC	318.7	257 ± 31	0.59 ± 0.04
900310	16:00:00.0	37.11	−116.06	LAC	303.8	265 ± 30	0.38 ± 0.03
900725	15:00:00.0	37.21	−116.21	LAC	313.1	245 ± 45	0.53 ± 0.06
910404	18:59:59.9	37.29	−116.31	LAC	322.6	231 ± 30	0.58 ± 0.04
920326	16:30:00.0	37.27	−116.36	LAC	319.8	235 ± 31	0.59 ± 0.04
831030	01:24:51.1	44.08	−113.97	LON	679.0	370 ± 30	0.30 ± 0.02
830317	07:25:56.6	47.53	−112.70	LON	696.0	444 ± 70	0.60 ± 0.09
831212	04:55:36.3	44.43	−114.10	LON	654.6	456 ± 15	0.50 ± 0.10
831030	01:59:01.4	44.22	−114.19	LON	658.1	370 ± 30	0.76 ± 0.04
831028	19:51:24.3	44.07	−113.91	LON	686.4	370 ± 30	0.67 ± 0.04
840822	09:46:30.2	44.47	−114.01	LON	659.1	377 ± 55	0.51 ± 0.04
840216	11:14:57.5	39.93	−117.76	LON	825.9	383 ± 12	0.41 ± 0.10
840324	00:07:47.8	44.74	−114.43	LON	616.3	384 ± 71	0.47 ± 0.00
840908	13:56:37.6	44.42	−114.15	LON	651.5	265 ± 67	0.36 ± 0.09
840908	06:16:40.2	44.42	−114.15	LON	650.5	382 ± 47	0.65 ± 0.04
850821	18:05:38.2	43.17	−110.78	LON	956.3	450 ± 10	0.60 ± 0.10
850206	19:34:19.5	44.55	−114.18	LON	641.5	314 ± 87	0.60 ± 0.07
850401	09:13:14.2	47.28	−113.23	LON	655.2	390 ± 42	0.70 ± 0.04
850317	06:56:17.0	44.55	−114.18	LON	643.1	372 ± 71	0.52 ± 0.06
821014	04:10:24.3	42.59	−111.43	JAS	922.9	431 ± 124	0.80 ± 0.10

Table 2 (continued)

Date	Origin (h:m:s)	Lat. (°)	Lon. (°)	Recording station	Distance (km)	Q_0	η
821110	11:21:25.6	34.05	−116.67	JAS	550.0	521 ± 16	0.31 ± 0.10
831008	11:57:54.2	40.75	−111.99	JAS	791.7	444 ± 63	0.36 ± 0.10
830713	21:16:48.2	33.20	−115.53	JAS	689.2	281 ± 86	0.60 ± 0.07
831030	01:59:01.6	44.08	−113.97	RSSD	812.9	346 ± 71	0.75 ± 0.08
830317	07:25:56.7	47.53	−112.70	RSSD	696.6	480 ± 84	0.80 ± 0.07
831212	04:55:36.4	44.43	−114.10	RSSD	804.1	546 ± 66	0.70 ± 0.03
831028	19:51:24.4	44.07	−113.91	RSSD	790.7	297 ± 54	1.08 ± 0.09
840822	09:46:30.1	44.47	−114.01	RSSD	796.9	667 ± 46	0.50 ± 0.04
841103	09:30:80.5	42.49	−108.85	RSSD	430.6	485 ± 30	0.52 ± 0.03
850401	09:13:14.2	47.28	−113.23	RSSD	797.3	677 ± 45	0.50 ± 0.03
850317	06:57:45.5	44.55	−114.18	RSSD	810.3	403 ± 95	0.78 ± 0.07
811001	19:00:00.0	37.08	−116.01	ALQ	892.7	524 ± 15	0.40 ± 0.10
810425	07:03:13.5	33.12	−115.65	ALQ	872.4	494 ± 14	0.40 ± 0.10
810606	18:00:00.0	37.30	−116.33	ALQ	925.8	422 ± 66	0.70 ± 0.05
820415	21:52:08.7	38.04	−118.50	ALQ	1064.0	604 ± 14	0.80 ± 0.10
820212	41:55:00.1	37.22	−116.46	ALQ	935.6	465 ± 10	0.80 ± 0.07
811112	15:00:00.2	37.11	−116.05	ANMO	869.8	386 ± 62	0.20 ± 0.09
811001	19:00:00.0	37.08	−116.01	ANMO	892.7	360 ± 93	0.10 ± 0.04
810606	18:00:00.0	37.30	−116.33	ANMO	925.7	397 ± 67	0.50 ± 0.08
810425	07:03:13.7	33.12	−115.65	ANMO	872.5	414 ± 13	0.60 ± 0.10
820524	12:13:26.9	38.71	−112.04	ANMO	646.7	324 ± 78	0.40 ± 0.07
821014	04:10:00.3	42.59	−111.43	ANMO	952.0	257 ± 97	0.50 ± 0.10
820212	14:55:00.2	37.22	−116.46	ANMO	935.5	564 ± 10	0.70 ± 0.10
820425	18:04:59.9	37.26	−116.44	ANMO	935.2	399 ± 65	0.40 ± 0.10
820706	02:10:43.4	34.95	−106.46	ANMO	829.4	218 ± 41	0.30 ± 0.09
830703	18:40:08.3	37.54	−118.86	ANMO	1110.5	300 ± 92	0.90 ± 0.09
831209	08:58:41.4	38.58	−112.58	ANMO	549.8	275 ± 66	0.41 ± 0.08
910817	19:29:39.9	40.26	−124.30	ANMO	1712.9	466 ± 12	1.00 ± 0.10
920629	10:14:22.2	36.71	−116.29	ANMO	937.7	386 ± 74	0.61 ± 0.09
921204	02:08:57.4	34.37	−116.90	ANMO	976.2	359 ± 64	0.50 ± 0.06
940204	02:42:11.9	42.71	−111.03	ANMO	994.1	504 ± 18	0.90 ± 0.10
910817	19:29:39.9	40.26	−124.30	COR	487.6	231 ± 52	0.40 ± 0.10
920308	03:43:04.6	40.24	−123.99	COR	486.6	208 ± 52	0.21 ± 0.10
930325	13:34:00.0	45.04	−122.61	COR	073.3	186 ± 19	0.30 ± 0.03
940119	22:27:31.2	42.30	−121.96	COR	276.3	303 ± 88	0.60 ± 0.10
920624	07:31:20.8	38.89	−111.46	TUC	763.3	359 ± 97	0.72 ± 0.11
921204	02:08:57.4	34.37	−116.90	TUC	613.2	325 ± 57	0.39 ± 0.06
930429	08:21:00.9	35.58	−112.12	TUC	402.3	330 ± 56	0.52 ± 0.06
930528	04:47:40.5	35.09	−119.15	TUC	834.4	293 ± 69	0.54 ± 0.09
930517	23:20:49.1	37.10	−117.80	TUC	833.5	385 ± 80	0.70 ± 0.09
930528	04:47:38.9	35.09	−119.15	TUC	833.2	207 ± 73	0.28 ± 0.18
930520	20:14:14.0	36.02	−117.64	TUC	753.7	253 ± 40	0.30 ± 0.08
930518	01:03:06.4	37.10	−117.75	TUC	830.0	302 ± 80	0.46 ± 0.10
940211	14:59:49.5	42.76	−110.99	TUC	1160.5	363 ± 16	0.86 ± 0.15
940117	12:30:55.3	34.21	−118.54	TUC	752.3	304 ± 47	0.54 ± 0.05
940204	02:42:11.9	42.71	−111.03	TUC	1154.5	330 ± 88	0.51 ± 0.10
910817	19:29:39.9	40.26	−124.30	GSC	890.0	304 ± 57	0.51 ± 0.09
930528	04:47:40.5	35.09	−119.15	GSC	243.6	234 ± 23	0.30 ± 0.04
930517	23:20:49.0	37.10	−117.80	GSC	243.9	379 ± 10	0.65 ± 0.10

Table 2 (continued)

Date	Origin (h:m:s)	Lat. (°)	Lon. (°)	Recording station	Distance (km)	Q_0 .	η
930210	21:48:34.9	40.43	−119.60	GSC	644.5	238 ± 42	0.32 ± 0.08
930116	06:29:34.8	36.90	−121.60	GSC	461.9	235 ± 56	0.40 ± 0.09
940117	12:30:55.3	34.21	−118.54	GSC	216.0	320 ± 62	0.69 ± 0.09
940204	14:59:50.5	42.71	−111.03	GSC	966.6	395 ± 17	0.70 ± 0.20
921204	02:08:57.4	34.37	−116.90	ISA	202.4	238 ± 66	0.49 ± 0.01
920629	10:14:22.2	36.71	−116.29	ISA	227.4	279 ± 49	0.41 ± 0.09
930116	06:29:35.8	36.90	−121.60	ISA	307.1	225 ± 38	0.30 ± 0.07
930210	21:48:35.1	40.43	−119.60	ISA	538.8	207 ± 41	0.30 ± 0.07
921204	02:09:03.1	34.37	−116.90	MLA	402.3	217 ± 56	0.37 ± 0.10
930528	04:47:40.6	35.09	−119.15	MLA	283.1	207 ± 32	0.30 ± 0.07
901024	06:15:20.7	38.05	−119.16	PAS	441.6	221 ± 05	0.55 ± 0.07
910817	19:29:39.9	40.26	−124.30	PAS	869.2	299 ± 66	0.54 ± 0.09
920629	10:14:22.1	36.71	−116.29	PAS	733.3	290 ± 14	0.63 ± 0.33
930517	23:20:49.2	37.10	−117.80	PAS	329.2	279 ± 48	0.60 ± 0.06
930517	23:20:49.1	37.10	−117.80	CMB	251.3	286 ± 74	0.31 ± 0.08
930528	04:47:40.5	35.09	−119.15	CMB	344.6	218 ± 35	0.23 ± 0.05
930116	06:29:34.9	36.90	−121.60	CMB	173.3	218 ± 39	0.25 ± 0.05
910817	19:29:39.9	40.26	−124.30	SBC	762.9	291 ± 74	0.45 ± 0.12
921204	02:08:57.5	34.37	−116.90	SBC	258.0	245 ± 48	0.30 ± 0.08
920629	10:14:22.1	36.71	−116.29	MHC	481.0	287 ± 46	0.31 ± 0.08
930528	04:47:40.6	35.09	−119.15	PFO	297.3	237 ± 45	0.20 ± 0.07
930116	06:29:34.9	36.90	−121.60	ARC	500.3	265 ± 70	0.50 ± 0.17
920629	10:14:22.9	36.71	−116.29	STAN	528.7	348 ± 48	0.49 ± 0.06
920629	10:14:22.2	36.71	−116.29	SAO	459.8	335 ± 68	0.51 ± 0.08
910817	19:29:40.1	40.26	−124.30	WCK	576.4	404 ± 45	0.71 ± 0.05
920629	10:14:22.2	36.71	−116.29	BKS	543.1	304 ± 49	0.66 ± 0.07
920821	16:31:35.1	33.00	−079.10	BLA	462.0	561 ± 124	0.43 ± 0.07
921117	03:58:00.9	45.80	−074.90	BLA	1057.0	451 ± 84	0.61 ± 0.10
940116	01:49:16.0	40.30	−076.00	BLA	514.8	604 ± 86	0.74 ± 0.06
940116	00:42:43.2	40.33	−076.05	BLA	516.5	455 ± 88	0.70 ± 0.11
940925	00:53:28.0	47.78	−069.95	BLA	780.4	677 ± 232	0.57 ± 0.09
950311	08:15:52.0	37.00	−083.20	BLA	243.9	523 ± 52	0.70 ± 0.09
950417	13:45:57.8	32.90	−080.10	BLA	474.2	510 ± 170	0.59 ± 0.10
950527	19:51:10.3	36.20	−089.40	BLA	813.2	619 ± 150	0.57 ± 0.20
960325	14:15:50.4	32.10	−088.70	BLA	946.7	442 ± 78	0.52 ± 0.10
940913	06:01:23.0	38.16	−107.98	CBKS	722.5	427 ± 109	0.96 ± 0.06
950203	15:26:10.7	41.50	−109.60	CBKS	895.2	432 ± 98	0.50 ± 0.09
950414	00:32:54.2	30.20	−103.30	CBKS	1005.8	425 ± 97	0.69 ± 0.06
960206	15:10:27.6	42.50	−097.50	CBKS	451.0	413 ± 51	0.76 ± 0.20
960220	14:40:41.9	27.55	−111.68	CBKS	1670.0	553 ± 72	0.85 ± 0.10
960516	15:41:01.9	44.12	−104.06	CBKS	1054.8	438 ± 46	0.67 ± 0.10
921117	03:58:00.9	45.80	−074.90	CBM	535.1	1102 ± 147	0.25 ± 0.04
900129	13:16:10.6	34.46	−106.88	CCM	1458.6	603 ± 60	0.58 ± 0.01
930409	00:59:49.5	28.81	−098.12	CCM	1207.5	749 ± 89	0.17 ± 0.01
931205	00:58:20.2	27.83	−102.74	CCM	1559.5	512 ± 34	0.24 ± 0.09
940512	00:22:23.0	24.95	−109.28	CCM	2230.4	582 ± 168	1.00 ± 0.20
940925	00:53:28.0	47.78	−069.95	CCM	2036.4	713 ± 162	0.74 ± 0.10
950203	15:26:10.7	41.50	−109.60	CCM	1619.0	511 ± 104	0.42 ± 0.07

Table 2 (continued)

Date	Origin (h:m:s)	Lat. (°)	Lon. (°)	Recording station	Distance (km)	Q_0	η
950311	08:15:52.0	37.00	−083.20	CCM	725.3	697 ± 127	0.74 ± 0.10
950414	00:32:54.2	30.20	−103.30	CCM	1409.2	432 ± 91	0.78 ± 0.06
950417	13:45:57.8	32.90	−080.10	CCM	1160.5	876 ± 128	0.51 ± 0.10
950527	19:51:10.3	36.20	−089.40	CCM	264.2	765 ± 311	0.60 ± 0.10
920102	11:45:35.8	32.30	−103.10	CCM	1250.9	632 ± 29	0.34 ± 0.06
920821	16:31:55.1	33.00	−080.10	CEH	328.8	557 ± 78	0.58 ± 0.04
920822	12:20:32.6	39.10	−070.30	CEH	853.6	636 ± 128	0.54 ± 0.10
940116	01:49:16.0	40.30	−076.00	CEH	560.9	745 ± 71	0.47 ± 0.05
950417	13:45:57.8	32.90	−080.10	CEH	338.6	643 ± 174	0.43 ± 0.07
950527	19:51:10.3	36.20	−089.40	CEH	931.8	384 ± 91	0.49 ± 0.10
950417	13:45:57.8	32.90	−080.10	CVL	577.3	711 ± 153	0.10 ± 0.09
950414	00:32:54.2	30.20	−103.30	EYMN	2210.3	426 ± 71	0.74 ± 0.07
960816	04:56:49.2	49.20	−082.90	EYMN	648.3	490 ± 43	0.59 ± 0.09
950527	19:51:10.3	36.20	−089.40	FVM	220.0	725 ± 180	0.81 ± 0.10
960325	14:15:50.4	32.10	−088.70	FVM	671.4	455 ± 47	0.54 ± 0.10
940116	01:49:16.0	40.30	−076.00	GOGA	1013.4	412 ± 121	0.80 ± 0.07
950527	19:51:10.3	36.20	−089.40	GOGA	625.5	683 ± 190	0.44 ± 0.10
960325	14:15:50.4	32.10	−088.70	GOGA	511.4	469 ± 48	0.43 ± 0.10
950414	00:32:54.2	30.20	−103.30	GLD	1068.7	462 ± 112	0.70 ± 0.08
960516	15:41:01.9	44.10	−104.04	GLD	592.5	316 ± 41	0.60 ± 0.10
931205	00:58:20.2	27.80	−102.70	GOL	1338.6	744 ± 96	0.58 ± 0.30
940913	06:01:23.0	38.20	−107.98	GOL	283.9	427 ± 111	0.64 ± 0.06
950203	15:26:10.7	41.50	−109.60	GOL	414.3	640 ± 117	0.86 ± 0.10
960107	14:32:52.1	35.80	−117.65	GOL	1166.0	528 ± 39	0.52 ± 0.10
950527	19:51:10.3	36.20	−089.40	HBF	904.6	458 ± 184	0.62 ± 0.10
900926	13:18:51.3	37.20	−089.58	HRV	1648.9	534 ± 111	0.54 ± 0.20
901019	07:01:57.4	46.47	−075.59	HRV	545.0	711 ± 50	0.31 ± 0.07
920102	11:45:35.8	32.33	−103.10	HRV	2990.4	443 ± 33	0.80 ± 0.10
940116	01:49:16.0	40.30	−076.00	HRV	445.7	655 ± 130	0.56 ± 0.07
950527	19:51:10.4	36.20	−089.40	JFWS	752.3	788 ± 190	0.67 ± 0.10
940925	00:53:28.0	47.78	−069.95	LBNH	421.1	723 ± 192	0.54 ± 0.05
940116	01:49:16.0	40.35	−076.05	LBNH	551.1	1305 ± 170	0.20 ± 0.06
950417	13:45:57.8	32.9	−80.100	LBNH	1444.9	539 ± 100	0.95 ± 0.09
940116	01:49:16.0	40.30	−076.00	LSCT	280.1	642 ± 84	0.52 ± 0.04
940116	00:42:43.2	40.35	−076.04	MCWV	278.2	682 ± 192	0.80 ± 0.10
940116	01:49:16.0	40.30	−076.00	MCWV	333.8	729 ± 60	0.40 ± 0.03
931205	00:58:20.2	27.83	−102.74	MIAR	1146.2	466 ± 28	0.51 ± 0.09
940913	06:01:23.0	38.16	−107.98	MIAR	1351.8	600 ± 96	0.60 ± 0.06
950311	08:15:52.0	37.00	−083.20	MIAR	980.0	519 ± 187	0.60 ± 0.10
950414	00:32:54.2	30.20	−103.30	MIAR	1033.3	412 ± 49	0.20 ± 0.10
950527	19:51:10.3	36.20	−089.40	MIAR	417.4	618 ± 181	0.79 ± 0.10
960325	14:15:50.4	32.10	−088.70	MYNK	537.4	460 ± 26	0.53 ± 0.06
940116	01:49:16.0	40.35	−076.04	MYNC	921.0	773 ± 150	0.30 ± 0.09
950311	08:15:52.0	37.00	−083.20	OXF	629.9	541 ± 104	0.45 ± 0.10
931205	00:58:20.0	27.83	−102.70	OXF	1468.5	594 ± 65	0.66 ± 0.10
950417	13:45:57.8	32.90	−080.10	PRM	247.7	650 ± 131	0.39 ± 0.07
940116	01:49:16.0	40.35	−076.04	RSNY	484.6	765 ± 103	0.43 ± 0.05
940925	06:01:23.0	47.78	−069.95	RSNY	503.0	730 ± 162	0.69 ± 0.09
950311	08:15:52.0	37.00	−083.20	RSNY	1110.6	637 ± 150	0.25 ± 0.10

Table 2 (continued)

Date	Origin (h:m:s)	Lat. (°)	Lon. (°)	Recording station	Distance (km)	Q_0	η
950527	19:15:52.0	36.20	−089.40	RSNY	1563.0	566 ± 100	0.49 ± 0.09
950527	19:51:10.3	36.20	−089.40	SGS	881.0	402 ± 195	0.88 ± 0.10
930409	00:59:49.5	28.81	−098.12	WMOK	660.0	505 ± 33	0.15 ± 0.01
931205	00:58:20.2	27.83	−102.74	WMOK	853.0	282 ± 07	0.79 ± 0.03
940913	06:01:23.0	38.16	−107.98	WMOK	906.7	554 ± 138	0.69 ± 0.10
950414	00:32:54.2	30.20	−103.30	WMOK	656.0	422 ± 54	0.59 ± 0.03
950527	19:51:10.3	36.20	−089.40	WMOK	863.3	727 ± 175	0.50 ± 0.10
960319	15:31:36.4	25.00	−109.30	WMOK	1481.6	463 ± 18	0.78 ± 0.04
960220	14:40:41.9	27.55	−111.68	WMOK	1463.0	349 ± 40	0.68 ± 0.10
960107	14:32:53.1	35.77	−117.65	WMOK	1718.3	897 ± 50	0.80 ± 0.06
940116	01:49:16.0	40.30	−076.00	YSNY	317.1	947 ± 223	0.30 ± 0.10
960325	14:15:50.4	32.10	−088.70	JSC	734.5	313 ± 14	0.80 ± 0.05
960325	14:15:50.4	32.10	−088.70	LHS	780.7	310 ± 31	0.76 ± 0.10
920102	11:45:35.8	32.34	−103.10	SCP	2435.4	470 ± 36	0.40 ± 0.10
900926	13:18:51.3	37.20	−089.58	SCP	1091.1	647 ± 25	0.63 ± 0.06
901019	07:01:57.4	47.58	−072.56	SCP	656.9	831 ± 17	0.31 ± 0.03
901231	03:53:58.3	47.58	−072.56	SCP	865.0	620 ± 39	0.47 ± 0.09
931205	00:58:20.2	27.83	−102.74	RSSD	1811.2	425 ± 24	0.79 ± 0.09
940913	06:01:23.0	38.16	−107.98	RSSD	740.9	430 ± 84	0.73 ± 0.06
960107	14:32:53.1	35.77	−117.65	RSSD	1485.0	436 ± 85	0.90 ± 0.10
960516	15:41:01.9	44.12	−104.04	RSSD	605.5	412 ± 27	0.70 ± 0.10
960516	15:41:01.9	44.12	−104.04	NEW	782.6	439 ± 44	0.64 ± 0.10
930602	02:08:32.7	25.95	−109.99	LTX	727.7	288 ± 16	0.49 ± 0.07
940512	00:22:23.0	24.90	−109.28	LTX	730.4	317 ± 75	0.74 ± 0.07
940512	01:14:07.7	24.68	−109.10	LTX	730.0	305 ± 65	0.50 ± 0.07
940925	00:53:28.0	47.78	−069.95	BINY	781.7	677 ± 100	0.57 ± 0.09

The origin times and locations are from the U.S. Geological Survey's Preliminary Determination of Epicenters; The Q_0 and η values with standard errors for each path are obtained in this study.

a region extending inland from the coasts of Texas and Louisiana. There is no obviously consistent relation between Q_0 and η. The region of low η in the western United States correlates with low Q_0 but the other two low-η regions occupy regions where Q_0 is 700 or higher in the northeastern United States and where it is 450–550 in the south-central United States.

Resolution and Error

We follow XIE and MITCHELL (1990a) in utilizing the point-spreading function (*psf*) to estimate resolution inherent in our results. Figure 8 displays *psfs* at three locations, centered on different cells, in the United States. A *psf* should fall off rapidly with distance if the resolving power of the data used is good and slowly if resolving power is poor. The *psfs* of Figure 8 indicate that resolving power does not

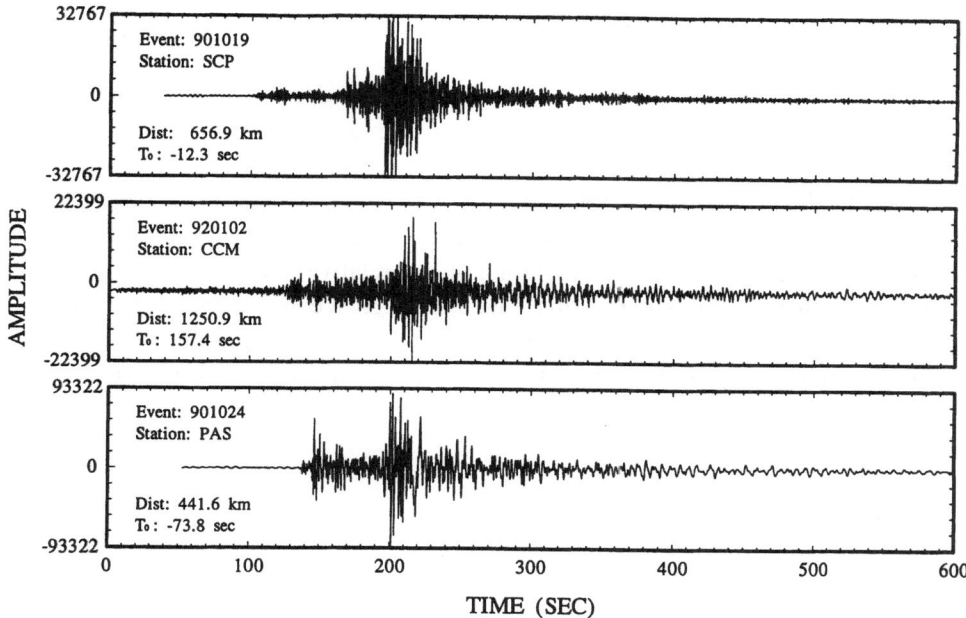

Figure 3

Examples of the *Lg* phase and its coda recorded at three stations in the United States. The top trace is from a path in the eastern United States, the middle trace for a path in the central United States, and the bottom trace is for a path in the western United States. *Lg* onsets are aligned at an arrival time of 200 s and T_0 denotes the time-scale origin measured from the event origin time for each trace.

vary greatly for the three selected regions. If we take the half-width of each *psf* as the minimum dimension of a feature which will be resolved, then resolution can be estimated to be between about 6° and 10° depending on position and azimuthal direction of the feature.

Standard errors were also calculated using the method of XIE and MITCHELL (1990a). Figures 9 and 10 show those errors for Q_0 and η, respectively. The error for Q_0 throughout most of the United States is about 50. The lower mapped errors correlate with good regional data coverage and/or with regions of low Q_0. Errors in Q_0 tend to increase (to about 100) when Q_0 increases in the northeastern United States and in the portion of the western Great Plains and Rocky Mountains in Colorado. Percentage errors, however, show smaller variations.

The estimated error for η is small (between 0.0 and 0.08) throughout most of the western and central United States. It varies between 0.10 and 0.15 in the Great Plains and around Lake Michigan and reaches about 0.20 at some points around the margins of the map. As with the errors for Q_0, those for η correlate inversely with the density of coda sampling.

Possible Biases in Q_0 and η

It is well known that *Lg* does not propagate well in oceanic crust. In the present study we assume that the continent-ocean transition acts a barrier to *Lg* waves and their coda and we therefore truncate the scattering ellipses for data used in our tomographic inversion at that boundary. If, however, *Lg* energy leaks through that boundary, our measured Q_0 could be biased to lower values and η could be biased to either higher or lower values, depending on the relative effect of the boundary on different frequency components that make up *Lg*. We have, therefore, tried to minimize the portions of the scattering ellipses that overlie oceanic regions. It is possible that Q_0 may be biased to lower values along the Pacific Coast, but the

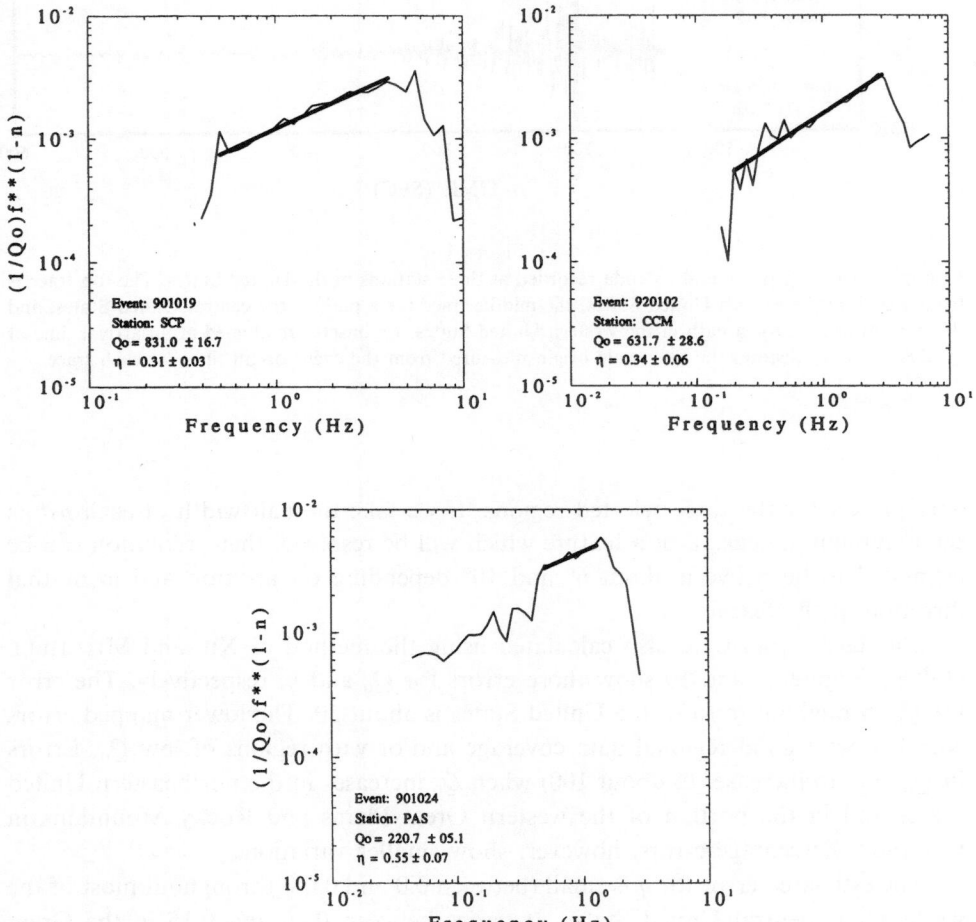

Figure 4
Stacked spectral ratios (SSRs) for the three seismograms in Figure 2.

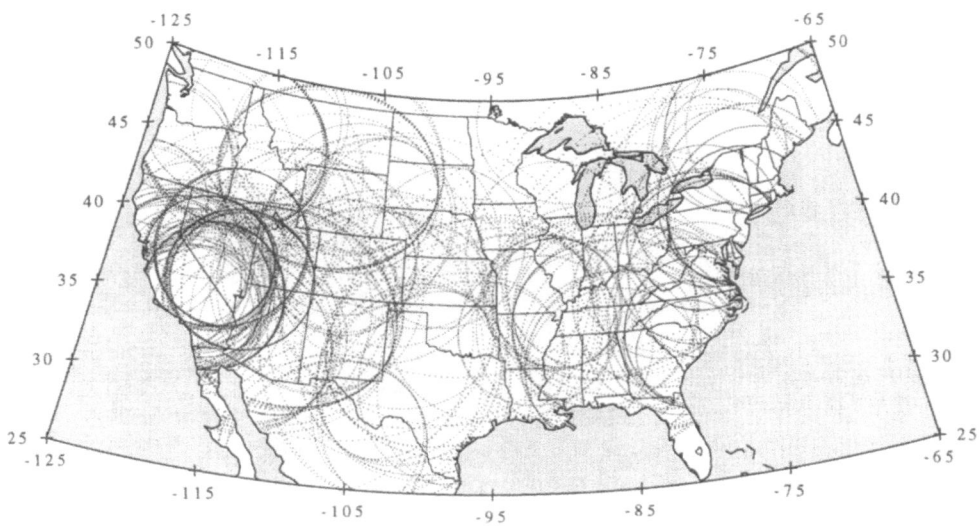

Figure 5

Sampling patterns of *Lg* coda across the United States. The foci of each ellipse correspond to the locations of the source and recording station for each event in Table 2. Each record of *Lg* coda is assumed to sample an elliptical area corresponding to the maximum lapse time used in the analysis. When a continental boundary is encountered it assumed to be a barrier to *Lg* propagation and the ellipses are truncated at those boundaries.

agreement of tomographically derived Q_0 in those regions with directly measured Q_0 for ellipses that are nearby, but do not intersect the coast, suggest that large biases are unlikely.

Comparison with Earlier Studies of Q_{Lg}, Q_{Lg}^c and Q_μ

There have been several studies of attenuation of *Lg* and *Lg* coda for the United States and various subregions of it. Several studies, in various regions of the world (DER *et al.*, 1984; NUTTLI, 1988; XIE and NUTTLI, 1988; XIE *et al.*, 1996), have found that *Q* determined for direct *Lg* and *Lg* coda, where both have been determined, have similar values. That similarity allows us to compare Q_{Lg} and Q_{Lg}^c results if both are obtained at similar frequencies. An early study (SUTTON *et al.*, 1967), using single-station paths, found large regional variations in Q_{Lg} (240–1000) across the United States. They found, with one exception, values in the central and eastern United States to be higher than those in the west. Their results for the western United States (250–500) agree well with the Q_{Lg}^c values of the present study. Their only determination for the eastern United States was 600 which is close to the average of about 650 found in the present study. In the north-central and

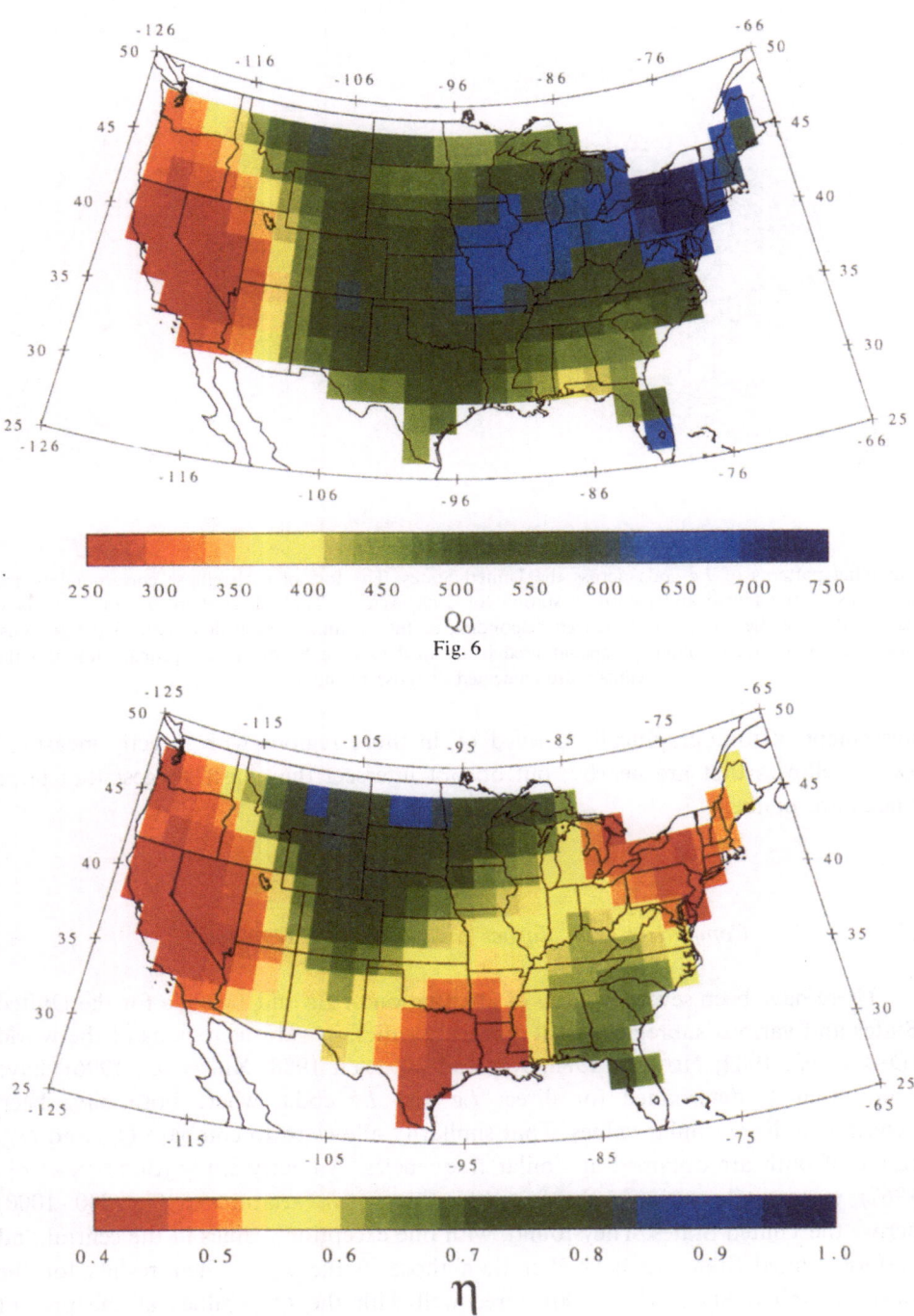

Fig. 6

Fig. 7

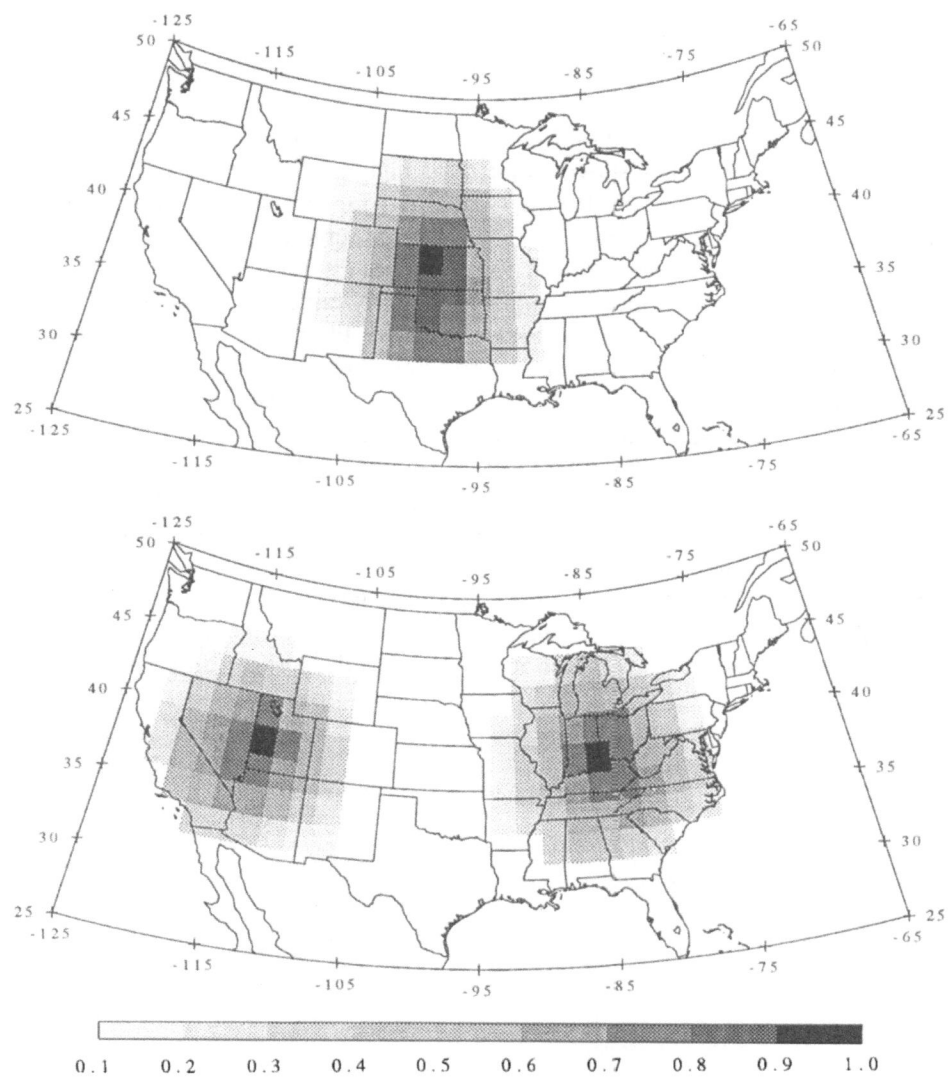

Figure 8

Point spreading functions (psf) determined for regions centered on cells at 38.5°N, 98.5°W (upper diagram) at 38.5°N, 112°W, and at 38.5°N, 85.5°W.

Figure 6

Tomographic map of *Lg* coda *Q* at 1 Hz for the continental United States. Each cell is 2.0° by 2.0° in area.

Figure 7

Tomographic map of the frequency dependence of *Lg* coda *Q* at 1 Hz. Each cell is 2.0° by 2.0° in area.

south-central United States SUTTON *et al.* found 250 and 550, respectively, for Q_{Lg}, while we obtained 450 and 550 in the present study. In addition, the value of 500 obtained by SUTTON *et al.* for the westernmost Great Plains is similar to our average for that region.

SINGH and HERRMANN (1983) constructed a Q_{Lg}^c map for the conterminous United States in which they showed that Q_0 is high east of the Rocky Mountains, much lower from there to the Pacific Coast, and moderately low along the Gulf Coast. From the western margin of the Rocky Mountains westward our map is quite similar to theirs, except that our lowest values (∼ 250) near the Pacific Coast of California are not as low as those (∼ 200) obtained in their study. Large differences, however, occur throughout the Great Plains and Central Lowlands where our values reach a maximum of 650–700 whereas SINGH and HERRMANN (1983) found values as high as 1300. Our results are also significantly lower in the southeastern United States where we find Q_0 to be betwen 400 and 550, whereas SINGH and HERRMANN (1983) find values, in the northern part of that region between 800 and 1000. Our lowest values in the southeastern United States lay in regions not well sampled by *Lg* coda in the Singh and Herrmann study.

Some of the differences between Q_0 values found by SINGH and HERRMANN (1983) and those of the present study are relatively large and may be due to several factors. First, the methodology is different. Singh and Herrmann obtained results for individual paths where our tomographic method averages and smoothes our

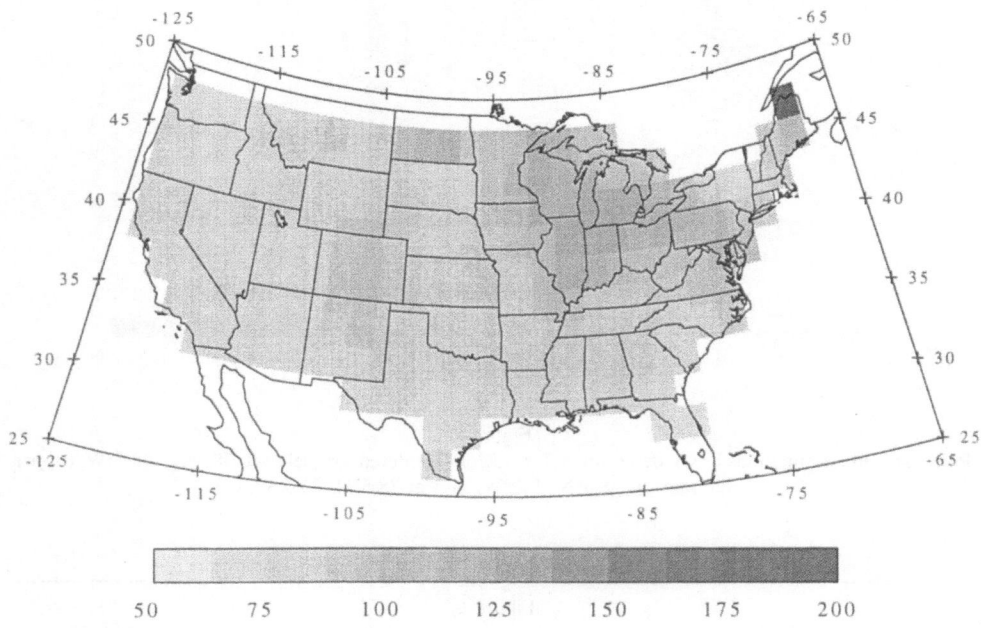

Figure 9
Standard error distribution for the image of Q_0.

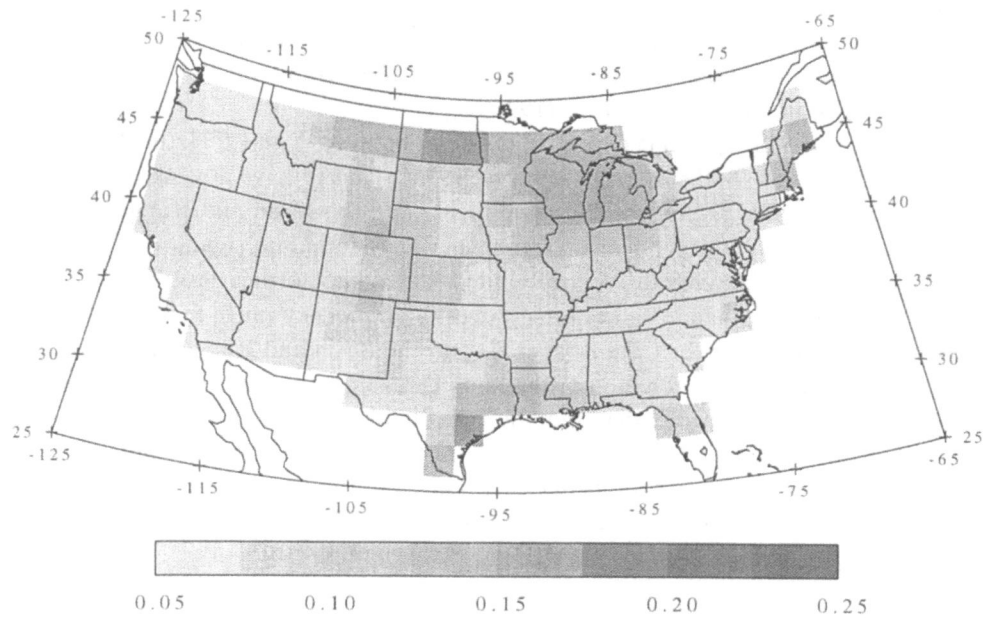

Figure 10
Standard error distribution for the image of η.

plotted values during the inversion process. For that reason, single measurements which obtain high or low values will be smoothed when the means are obtained for different cells. Second, path coverage between the two studies is different. Our study includes better coverage for the southeastern and north-central parts of the United States, a situation that may partly explain large differences between the two studies for those regions. In addition, the analysis of SINGH and HERRMANN (1983) has been found to underestimate Q_0 by as much as 30% because the effect of the bandwidth of the instrument response was not taken into account in that study (R. B. Herrmann, personal communication). This effect will cause the largest overestimates of Q_0 in regions of high Q_0 and can explain why the Q_0 determinations of SINGH and HERRMANN (1983) are substantially higher than those of the present study in many parts of the central and eastern United States.

CHAVEZ and PRIESTLY (1986) determined Q_{Lg} in a broad portion of the Basin and Range province and found Q_0 to be 214 ± 15. XIE and MITCHELL (1990b) found Q_0 for *Lg* to be 267 ± 56 and for *Lg* coda to be 275 ± 26 along a path in the western part of that province. Our values are very close to those of XIE and MITCHELL (1990b) for this region.

SHI *et al.* (1996) obtained average Q_{Lg} for five subregions of the northeastern United States. They obtained 784 in the Adirondack Mountains, 519 in the Erie-Ontario Lowland, 578 for the Appalachian Plateau and folded zone, 560 in the coastal basins and highlands, and 576 in northern New England. Although our Q_{Lg}^c

determinations differ from the Q_{Lg} values of SHI *et al.* (1996) for specific regions, the overall agreement is excellent. Some of the differences probably occur because our tomographic inversion process smoothes over some of the smaller scale fluctuations observed by SHI *et al.* (1996).

BENZ *et al.* (1997) found Q_{Lg} at frequencies between 1 and 5 Hz in the eastern United States and southeastern Canada to be high and to have a weak dependence on frequency, while Q_{Lg} in southern California and the Basin and Range province is low and is strongly dependent on frequency. They found no significant difference in *Lg* attenuation between the central United States, the northeastern United States and southeastern Canada. Values for Q_0 over the frequency range form 1 to several Hz were slightly lower for California (178) and the Basin and Range province (235) and much higher (1291) for the northeastern United States than values obtained in this study. The two sets of determinations probably differ because ours are for a narrow frequency near 1 Hz while those of BENZ *et al.* (1997) extend to higher frequencies. Our values for frequency dependence in the Basin and Range province are similar to those of both XIE and MITCHELL (1990b) and BENZ *et al.* (1997).

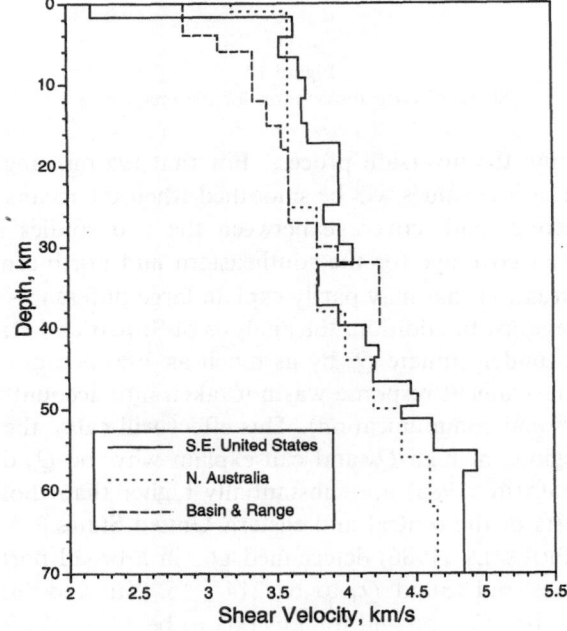

Figure 11

Shear-velocity distributions for the upper 60 km beneath (1) the southeastern United States, (2) the North Australian Craton, (3) the Basin and Range province. The model for (1) was obtained using a receiver function method (OWENS *et al.*, 1987), the model for (2) is taken from BOWMAN and KENNETT (1991), and the model for (3) is a variation of a model for the Basin and Range province (MITCHELL and XIE, 1994).

The patterns of Q_{Lg} and Q_{Lg}^c variation across the United States correlate well with variations of shear-wave Q (Q_μ) in the upper crust obtained from the inversion of Rayleigh wave attenuation coefficients. MITCHELL (1975) reported that average Q_μ in the upper crust, determined at Rayleigh wave periods of 5–40 s, in the western United States is about half as much (125) as that in the central United States (250). CHENG and MITCHELL (1981), using a similar period range, found that differences occurred within the western Cordillera with the Colorado Plateau exhibiting an upper crustal Q_μ of 160 while Q_μ in the upper crust of the Basin and Range province is about 85. This correlation of Q_{Lg} and Q_{Lg}^c variations with upper crustal Q_μ variations suggests that much of the regional variation of Q_{Lg} and Q_{Lg}^c throughout the United States is controlled by intrinsic absorption in the upper crust.

Correlation of Q_{Lg}^c with Crustal Properties and Structure

As indicated earlier, Q_0 in portions of the central and eastern United States is much higher than that observed across most of the western United States. The differences do not appear to be explainable by variations in crustal thickness. Although the lowest values occur in the Basin and Range province where the crust is thin, other regions of relatively thin crust, such as continental shelf regions are marked by higher Q_0. Highest Q_0 occurs in a portion of the northeastern United States near the Great Lakes. Lower values occur in parts of both the western Great Plains and the Rocky Mountains, where the crust is thicker, and beneath the Basin and Range province where the crust is thinner. Similarly, there appears to be no correlation with elevation. Both the Rocky Mountains, with high elevations, and portions of the central lowlands and southeastern United States, with low elevations, exhibit similar Q_0 values.

Variations of Q_0 appear to correlate directly, to some extent, with average crustal and upper mantle P_n velocities. Contour maps of both these parameters (BRAILE et al., 1989) exhibit a rapid decrease going from east to west across the Rocky Mountains in a pattern similar to that in Figure 6. There are other regional fluctuations in both those velocities, however, they show no correlation, or an inverse correlation, with our Q_0 map.

Seismic-wave attenuation is known to increase with increasing temperature. For that reason, it is reasonable to expect that Q_0 will vary inversely with heat flow across the United States. Heat flow maps of the entire United States have been presented by several authors. Those maps (e.g., MORGAN and GOSNOLD, 1989) show that heat flow is much higher, on average, west of the Rocky Mountains than to the east of them. The large difference in Q_0 between the eastern and western United States appears to partly support the idea that Q_0 is lower in regions of high heat flow. Many features of the Q_0 map in Figure 5, however, do not correlate with variations in heat flow. For instance, relatively low heat flows occur in parts of the Colorado Plateau and Rocky Mountains, as well as along parts of the Pacific Coast, regions characterized by low Q_0.

MITCHELL and HWANG (1987) showed that Q_{Lg} in regions of high to moderate crustal Q, can be considerably reduced by thick accumulations of Mesozoic or younger sandstones and shales characterized by low intrinsic Q. Low-Q sediments were deposited extensively in the northern Great Plains and the Gulf Coastal Plain through the Mesozoic era, especially during the Cretaceous period, as well as through some intervals of the Cenozoic era. These deposits may partly explain why Q^c_{Lg} throughout the Great Plains, between longitudes 106°W and 95°W, are lower than expected for stable regions and why Q^c_{Lg} is relatively low though a broad band along the Gulf Coast and the southern portion of the Atlantic Coastal Plain.

BOWMAN and KENNETT (1991) showed that shear-wave energy which would be trapped in a crustal wave guide with a sharp lower boundary can leak out if that boundary is replaced by a velocity gradient. They found that a velocity gradient zone between depths of 30 and 55 km dramatically reduced Q_{Lg} in the North Australian Craton. In another study, MITCHELL and XIE (1994) investigated the effects of a velocity increase of 0.9 km/s over a depth interval of 5 km at the crust-mantle transition for a model of the Basin and Range province and found that the gradient reduced Q_{Lg} by about 8%.

Receiver function studies of teleseismic P waves at several stations in North America (OWENS et al., 1987) have revealed that a significant positive velocity gradient exists at the crust-mantle transition beneath the Regional Seismic Test Network (RSTN) station RSCP in central Tennessee in the southeastern United States. Figure 11 compares the model derived by those authors with the North Australian Craton and Basin and Range models described above. The figure indicates that the velocity gradient at the crust-mantle transition beneath station RSCP is similar to that beneath northern Australia. MITCHELL et al. (1988) investigated the extent to which the velocity gradient at the crust-mantle transition beneath the North Australian Craton could reduce Q_{Lg} and found it be about 20%. Assuming that Q^c_{Lg} is similar in value to Q_{Lg}, we infer that Q^c_{Lg} in the southeastern United States would be about 600, rather than about 480 as indicated by Figure 6, if the crust-mantle transition there were sharp. Q^c_{Lg} of about 600 characterizes much of the eastern United States.

In summary, there appear to be two factors, other than intrinsic absorption, that can reduce Q^c_{Lg} in the central and eastern United States. These are (1) thick accumulations of Mesozoic and younger sediments at the surface and (2) severe velocity gradients over a broad depth range in the crust-mantle transition region. In the western United States where upper crustal Q_μ has been found to be low (MITCHELL, 1975; MITCHELL and XIE, 1994), the sediments do not have a significant effect on Q_{Lg} values (MITCHELL and HWANG, 1987), and we are aware of no instances of major velocity gradients at the crust-mantle transition. For those reasons we attribute the large reductions of Q^c_{Lg} in that region, compared to the central and eastern United States, to enhanced intrinsic absorption of shear energy (or reduced Q_μ) in the crust.

Intrinsic Q_μ and its Relation to Crustal Evolution

The Q_0 map of Figure 6 suggests that we can, to first order, divide the United States into two large provinces within each of which the crust exhibits relatively similar attenuative properties. The larger of these includes all of that region from the western margin of the Rocky Mountains to the Atlantic Coast where Q_{Lg}^c is relatively high (450–750). Variations in Q_{Lg}^c within that region may possibly be explained by variations in thickness of low-Q sediments in the near-surface portions of the crust and/or variations in the severity of velocity gradients in the crust-mantle transition region. The other province is that low-Q region (250–450) lying between the Rocky Mountains and the Pacific Coast. The low Q_{Lg}^c in this region is most likely due to the presence of fluids in the upper crust (MITCHELL, 1995; 1997). Seismic waves lose energy because that energy goes into moving fluids in permeable crust during propagation. Variations of Q_{Lg}^c within that region probably reflect variations in the volume of fluids in faults, cracks and permeable rock in the upper crust. Fluids are likely to be generated by hydrothermal reactions during mantle and crustal heating (e.g., WYLLIE, 1988) and may then be lost by seepage to the surface or absorbed by retrograde metamorphism. This process causes Q_μ, and hence Q_{Lg} and Q_{Lg}^c to increase with time following the most recent episode of tectonic activity when the fluid formed.

An interesting result of this study is that the transition from high to low Q_{Lg}^c going from east to west in the western United States lies further to the west than the transition of upper mantle Q obtained from body waves (DER *et al.*, 1975, SOLOMON and TOKSÖZ, 1970), or electrical resistivity (PORATH, 1971; KELLER, 1989) from high to low values. If Q reductions are related to the presence of fluids as suggested by MITCHELL (1995), this suggests that crustal fluids in the crust of the Rocky Mountains are much smaller in volume than they are further to the west, but that upper mantle fluids are similar in abundance there. An analogous result was found in a recent tomographic study of shear-velocity variations beneath the United States (ALSINA *et al.*, 1996) where maps indicate that the transition from high to low velocities lies further west at depths between 25 and 100 km than at greater depths.

Conclusions

The Q_{Lg}^c distribution in the United States defines two large provinces, one from the Rocky Mountains to the Atlantic Coast and the other from the western margin of the Rocky Mountains to the Pacific Coast. The eastern, and larger province, consists of relatively stable crust in which Q_{Lg}^c ranges between about 450 and 750. The western province, the western United States Cordillera, is tectonically active and Q_{Lg}^c ranges between about 250 and 450. Variations of Q_{Lg}^c within the eastern province may be partly explained by variations in thickness and composition of Mesozoic and younger sediments and, in at least some regions, by variations in the thickness and severity of velocity gradients in the transition region between the crust and upper mantle. The transition from high to low Q in the western United States lies further to the west than the transition from high to low Q and electrical resistivity found in earlier studies and correlates well with the location of the transition from high to low shear velocities found in recent work using surface waves. Unexpectedly low Q_{Lg}^c in the southeastern United States occurs in a region where a thick velocity gradient exists beneath the Cumberland Plateau in Tennessee. Low Q_{Lg}^c in the western United States coincides with crust that has experienced tectonic activity beginning in the Mesozoic and continues to the present time. The low Q values are consistent with the idea that fluids were emplaced in the crust by hydrothermal fluid release during past heating (MITCHELL, 1995) and seismic waves suffer loss of energy during propagation because they expend it to move those fluids through permeable material.

Acknowledgments

Work on this paper and its preparation benefited greatly from discussions with Lianli Cong. The Lg coda Q determinations utilized computer code developed by Jiakang Xie and modified by Robert Herrmann. This research was supported by the U. S. Department of Energy and was monitored by the Phillips Laboratory under contract F19628–95–K–0004.

REFERENCES

ALSINA, D., WOODWARD, R. L., and SNIEDER, R. K. (1996), *Shear Wave Velocity Structure in North America from Large-scale Waveform Inversions of Surface Waves*, J. Geophys. Res. *101*, 15969–15986.

ANDERSON, R. E., *Tectonic evolution of the Intermontane system; Basin and Range, Colorado Plateau, and High Lava Plains. In Geophysical Framework of the Continental United States, Mem. 172* (ed's. Pakiser, L. C., and Mooney, W. D.) (Geol. Soc. Am., Boulder, Colorado)

BENZ, H. M., FRANK, A., pp. 163–176. and BOORE, D. M. (1997), *Regional Lg Attenuation for Continental United States*, Bull. Seismol. Soc. Am. *87*, 606–619.

BOWMAN, J. R., and KENNETT, B. L. N. (1991), *Propagation of Lg Waves in the North Australian Craton: Influence of Crustal Velocity Gradients*, Bull. Seismol. Soc. Am. *81*, 592–610.

BRAILE L. W., HINZE, W. J., VON FRESE, R. R. B., and KELLER, G. R. *Seismic properties of the crust and uppermost mantle of the conterminous United States and adjacent Canada. In Geophysical Framework of the Continental United States, Mem. 172* (eds. Pakiser, L. C., and Mooney, W. D.) (Geol. Soc. Am., Boulder, Colorado 1989) pp. 655–680.

CHÁVEZ, D. E., and PRIESTLEY (1986), *Measurement of Frequency Dependent Lg Attenuation in the Great Basin*, Geophys. Res. Lett. *13*, 551–554.

CHENG, C. C., and MITCHELL, B. J. (1981), *Crustal Q Structure in the United States from Multi-mode Surface Waves*, Bull. Seismol. Soc. Am. *71*, 161–181.

CONG, L., and MITCHELL, B. J. (1998), *Lg Coda Q and its Relation to the Geology and Tectonics of the Middle East*, Pure and appl. geophys. *153*, 563–585.

DER, Z. A., MASSE, R. P., and GURSKI, J. P. (1975), *Regional Attenuation of Short-period P and S Waves in the United States*, Geophys. J. R. Astr. Soc. *40*, 85–106.

DER Z. A., MARSHALL, M. E., O'DONNELL, A., and MCELFRESH, T. W. (1984), *Spatial Coherence Structure and Attenuation of the Lg Phase, Site Effects, and the Interpretation of the Lg Coda*, Bull. Seismol. Soc. Am. *74*, 1125–1147.

HERRIN, E., and TAGGART, J. (1962), *Regional Variations in P_n Velocity and their Effect's on the Location of Epicenters*, Bull. Seismol. Soc. Am. *52*, 1037–1046.

KANE, M. F., and GODSON, R. H. (1989), *A crust/mantle structural framework of the conterminous United States based on gravity and magnetic trends*. In *Geophysical Framework of the Continental United States, Mem. 172* (eds. Pakiser, L. C., and Mooney, W. D.) (Geol. Soc. Am., Boulder, Colorado 1989) pp. 383–403.

KELLER, G. V. (1989), *Electrical structure of the crust and upper mantle beneath the United States; Part 2, Survey of data and interpretation*. In *Geophysical Framework of the Continental United States, Mem. 172* (ed's. Pakiser, L. C., and Mooney, W. D.) (Geol. Soc. Am., Boulder, Colorado 1989) pp. 425–446.

MCBIRNEY, A. R. (1978), *Volcanic Evolution of the Cascade Range*, Earth and Planetary Sciences Ann. Rev. *6*, 437–456.

MILLER, E. L. MILLER, M. M., STEVENS, C. H., WRIGHT, J. E., and MADRID, R., *Late Paleozoic paleogeographic and tectonic evolution of the western U.S. Cordillera*. In *The Cordilleran Orogen: Conterminous U.S., The Geology of North America, G-3* (eds. Burchfiel, B. C., Lipman, P. W., and Zoback, M. L.) (Geol. Soc. Am., Boulder, Colorado 1992) pp. 57–106.

MITCHELL, B. J. (1975), *Regional Rayleigh Wave Attenuation in North America*, J. Geophys. Res. *80*, 4904–4916.

MITCHELL, B. J. (1995), *Anelastic Structure and Evolution of the Continental Crust and Upper Mantle from Seismic Surface Wave Attenuation*, Rev. Geophys. *33*, 441–462.

MITCHELL, B. J. (1997), *Lg Coda Q Variation across Eurasia and its Relation to Crustal Evolution*, J. Geophys. Res. *102*, 22767–22779.

MITCHELL B. J., and HWANG, H. J. (1987), *Effect of Low Q Sediments and Crustal Q and Lg Attenuation in the United States*, Bull. Seismol. Soc. Am. *77*, 1197–1210.

MITCHELL, B. J., and XIE, J. (1994), *Attenuation of Multiphase Surface Waves in the Basin and Range Province—Part III, Inversion for Crustal Anelasticity*, Geophys. J. Int. *116*, 468–484.

MOONEY, W. D., and WEAVER, S. C. (1989), *Regional crustal structure and tectonics of the Pacific coastal states; California, Oregon, and Washington*. In *Geophysical Framework of the Continental United States, Mem. 172* (ed's. Pakiser, L. C., and Mooney, W. D.) (Geol. Soc. Am., Boulder, Colorado 1989) pp. 129–161.

MORGAN, P., and GOSNOLD, W. D. (1989), *Heat flow and thermal regimes in the continental United States*. In *Geophysical Framework of the Continental United States, Mem. 172* (eds Pakiser, L. C., and Mooney, W. D.) (Geol. Soc. Am., Boulder, Colorado 1989) pp. 493–522.

NUTTLI, O. W. (1988), *Lg Magnitudes and Yield Estimates for Underground Novaya Zemlya Nuclear Explosions*, Bull. Seismol. Soc. Am. *78*, 873–884.

OWENS, T. J. TAYLOR, S. R., and ZANDT, G. (1987), *Crustal Structure at Regional Seismic Test Network Stations Determined from Inversion of Broadband Teleseismic P Waveforms*, Bull. Seismol. Soc. Am. *77*, 631–662.

PORATH, H. (1971), *Magnetic Variation Anomalies and Seismic Low-velocity Zone in the Western United States*, J. Geophys. Res. *76*, 2643–2648.

PRODEHL, C., and LIPMAN, W. P. (1989), *Crustal structure of the Rocky Mountain region*. In *Geophysical Framework of the Continental United States, Mem. 172* (eds. Pakiser, L. C., and Mooney, W. D.) (Geol. Soc. Am., Boulder, Colorado 1989) pp. 249–284.

SALEEBY, J. C., *Ocean floor accretion and volcano-plutonic arc evolution of the Mesozoic Sierra Nevada, California*. In *The Geotectonic Development of California* (ed. Ernst, W. G.) (Prentice-Hall, Englewood Cliffs, New Jersey 1981) pp. 130–181.

SHI, J., KIM, W., and RICHARDS, P. G. (1996), *Variability of Crustal Attenuation in the Northeastern United States from Lg Waves*, J. Geophys. Res. *101*, 25231–25242.

SINGH, S. K., and HERRMANN, R. B. (1983), *Regionalization of Crustal Coda Q in the Continental United States*, J. Geophys. Res. *88*, 527–538.

SOLOMON, S. C., and TOKSÖZ, M. N. (1970), *Lateral Variation of Attenuation of P and S Waves beneath the United States*, Bull. Seismol. Soc. Am. *60*, 819–838.

SUTTON, G. H., MITRONOVAS, W., and POMEROY, P. W. (1967), *Short-period Seismic Energy Radiation Patterns from Underground Nuclear Explosions and Small-magnitude Earthquakes*, Bull. Seismol. Soc. Am. *57*, 249–267.

WYLLIE, P. J. (1988), *Magma Genesis, Plate Tectonics, and Chemical Differentiation of the Earth*, Rev. Geophys. *26*, 370–404.

XIE, J., and NUTTLI, O. W. (1988), *Interpretation of High-frequency coda at Large Distances: Stochastic Modeling and Method of Inversion*, Geophys. J. Int. *95*, 579–595.

XIE, J., and MITCHELL, B. J. (1990a), *A Back-projection Method for Imaging Large-scale Lateral Variations of Lg Coda Q with Application to Continental Africa*, Geophys. J. Int. *100*, 161–181.

XIE, J., and MITCHELL, B. J. (1990b), *Attenuation of Multiphase Surface Waves in the Basin and Range Province, Part I: Lg and Lg Coda*, Geophys. J. Int. *102*, 121–137.

XIE, J., CONG, L., and MITCHELL, B. J. (1996), *Spectral Characteristics of the Excitation and Propagation of Lg from Underground Nuclear Explosions in Central Asia*, J. Geophys. Res. *101*, 5813–5822.

ZOBACK, M. L., and ZOBACK, M. D. (1989), *Tectonic stress field of the continental United States*. In *Geophysical Framework of the Continental United States*, Mem. *172* (eds. Pakiser, L. C., and Mooney, W. D.) (Geol. Soc. Am., Boulder, Colorado 1989) pp. 523–539.

(Received November 10, 1997, revised May 21, 1998, accepted June 1, 1998)

To access this journal online:
http://www.birkhauser.ch

Pure appl. geophys. 153 (1998) 639–653
0033–4553/98/040639–15 $ 1.50 + 0.20/0

❙Pure and Applied Geophysics

Lg Coda Q in Australia and its Relation to Crustal Structure and Evolution

B. J. Mitchell,[1] S. Baqer,[1] A. Akinci[1] and L. Cong[2]

Abstract—We have determined *Lg* Coda *Q* (Q_{Lg}^c) from ground motion recorded at seven broadband stations in Australia, using a stacked spectral ratio method. In spite of the relatively small number of events and less than optimum station coverage, we were able to use those data to obtain a tomographic map Q_{Lg}^c and its frequency dependence, at 1 Hz for almost the entire island continent.

Q_{Lg}^c at 1 Hz in Australia varies between about 330 and 600. The lowest values (330–400) characterize the Tasman Fold Belt in eastern Australia; these may be associated with fluids produced by orogenic activity that occurred during the Devonian and Carboniferous periods or by sedimentation that occurred in Jurassic and Triassic times. Smaller reductions of Q_{Lg}^c, relative to maximum values, in central and western Australia, may be associated with sedimentation that occurred over a long-time interval between late Precambrian time and the Carboniferous period or with deformation that occurred in the Central Australian Mobile Belts during the Carboniferous.

Q_{Lg}^c throughout most of Australia is 30 to 60% lower than it is in most other stable continental regions. Assuming that Q_{Lg}^c varies in the same proportion as Q_{Lg}, ground motion computations for one-dimensional models of the Australian crust and upper mantle indicate that up to 20% reductions in Q_{Lg}^c can be produced if a velocity gradient, rather than a sharp boundary, resides at the crust-mantle transition. The remaining portion of the Q_{Lg}^c reduction may be caused by lateral variability in the thickness, depth, and severity of the velocity gradient which causes additional Lg energy to leak into the mantle.

Key words: *Lg*, coda, *Q*, crust, Australia, attenuation.

Introduction

Australia is an island continent, the western two-thirds of which last underwent large-scale orogenic and tectonic activities in Precambrian time. Orogenic activity occurred in the eastern part of the continent in the late Precambrian as well as during Devonian and Permian times. Continental regions where the last orogenic or tectonic activity was so ancient are normally characterized by low rates of seismic

[1] Department of Earth and Atmospheric Sciences, Saint Louis University, 3507 Laclede Avenue, St. Louis, Missouri 63103, U.S.A.

[2] Now at Department of Earth Sciences, Yunnan University, Kunming, Yunnan, People's Republic of China.

wave attenuation, or high Q (MITCHELL, 1995). Australia, however, exhibits lower values of Q for Lg waves (Q_{Lg}) than those found in most stable cratonic regions. An important question is whether the unexpectedly low Q_{Lg} in Australia is due to energy that goes into moving fluids in permeable crust (MITCHELL, 1995) or to a scattering of energy into the mantle. BOWMAN and KENNETT (1991) have proposed a possible mechanism of the latter type. They found, in modeling a region of north-central Australia using synthetic seismograms, that they could achieve a reduction of Q_{Lg} by placing a thick velocity gradient at the crust-mantle transition. That velocity gradient causes a portion of the S-wave energy comprising Lg to leak into the upper mantle. They used crust-upper mantle models obtained for that region (FINLAYSON, 1982) in which the crust-mantle transition exhibits a gradational increase in velocity over a depth range of about 25 km.

This study, like those of other studies of Lg coda Q (Q_{Lg}^c) in this volume, will investigate the variation in attenuation of that phase over the entire continent. As in those studies, we will investigate possible causes for lateral variations of Q_{Lg}^c. Because of the earlier finding that a velocity gradient at the crust-mantle transition can reduce Q_{Lg} (and by inference Q_{Lg}^c), we will emphasize this aspect of our interpretation more than we did in the other studies.

A Tectonic Sketch of Australia

RUTLAND (1976) recognized three main basement provinces in Australia. From west to east, and with decreasing age, these are the Pilbara-Yilgarn province (including the Pilbara and Yilgard subprovinces), the Arunta-Gawler province (including the Arunta and Gawler subprovinces), and the Tasman province (including the Lachlan and New England subprovinces). Rutland assigned, respectively, Archean, Early and Middle Proterozoic, and Late Precambrian to Phanerozoic ages to the three provinces.

The map in Figure 1 shows those provinces, but identifies some of them by names used in more recent literature on Australian crustal structure. The map follows the nomenclature of COLLINS (1991) by renaming the Arunta province of Rutland as the North Australian Craton, and by dividing Rutland's Gawler subprovince into the Gawler craton and the Nullarbor block. Further, we have followed PLUMB (1979) whose map designates the Tasman Province of Rutland as the Tasman Fold Belt. That province includes the Lachlan subprovince which comprises the greatest share of the province and the smaller New England subprovince situated along the eastern coast.

No major tectonic or orogenic activity has occurred in the Pilbara-Yilgarn province since the end of the Precambrian, but some deformation occurred in two small areas in the central part of the country during the Carboniferous. The map in Figure 1 shows those areas in the Central Australian Mobile Belts.

The Tasman Fold Belt has undergone three phases of orogeny: (1) the Penguin orogeny in the Precambrian, (2) the Tabberabberan orogeny in the western portion of the fold belt (Lachlan subprovince) during the Devonian, and (3) the Hunter-Bowen orogeny in the eastern part of the belt (New England subprovince) during the Permian period. As indicated in later sections, Q_{Lg}^c is relatively uniform throughout the western two-thirds of Australia. For that reason, we treat Australia as consisting of only two provinces, the Tasman Fold Belt where there has been orogenic activity as late as the Permian and the rest of Australia where activity was largely restricted to Precambrian time.

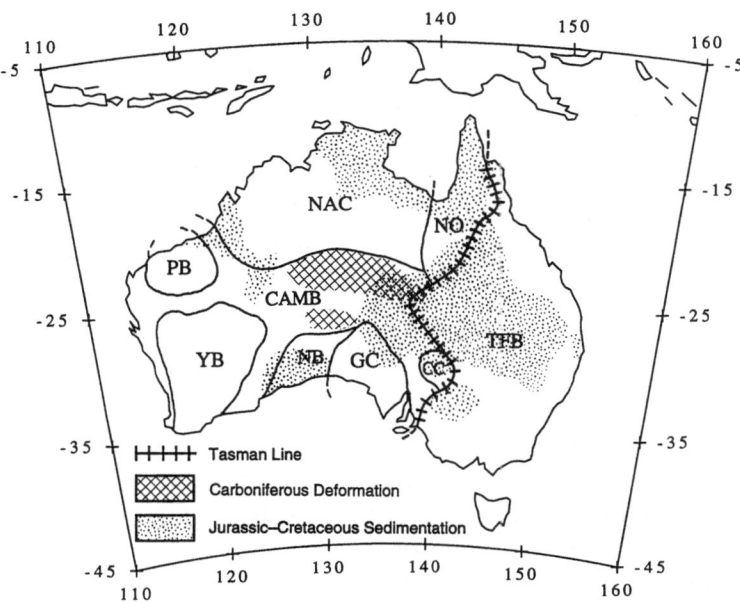

Figure 1

Map of Australia showing major crustal subdivisions, sites of Carboniferous deformation (cross-hatching), and regions of exposed Jurassic-Cretaceous sedimentation (stippling) as shown on maps by PLUMB (1979). Subdivision abbreviations denote the following: PB—Pilbara Block, YB—Yilgarn Block, NAC—North Australian Craton, CAMB—Central Australian Mobile Belts, NB—Nullarbor Block, GC—Gawler Craton, NO—Northeast Orogens, TFB Tasman Fold Belt. The location of the Tasman line separating differing eastern and western terrains follows VEEVERS (1984). The western terrain contains exposed Precambrian blocks and fold belts while the eastern terrain consists of exposed Phanerozoic fold belts. Phanerozoic sediments overlie both terrains, but are younger in the eastern region. Some reductions in *Lg* coda *Q* that appear in Figure 3 correlate with regions of Devonian or younger deformations and/or Mesozoic or younger sedimentation.

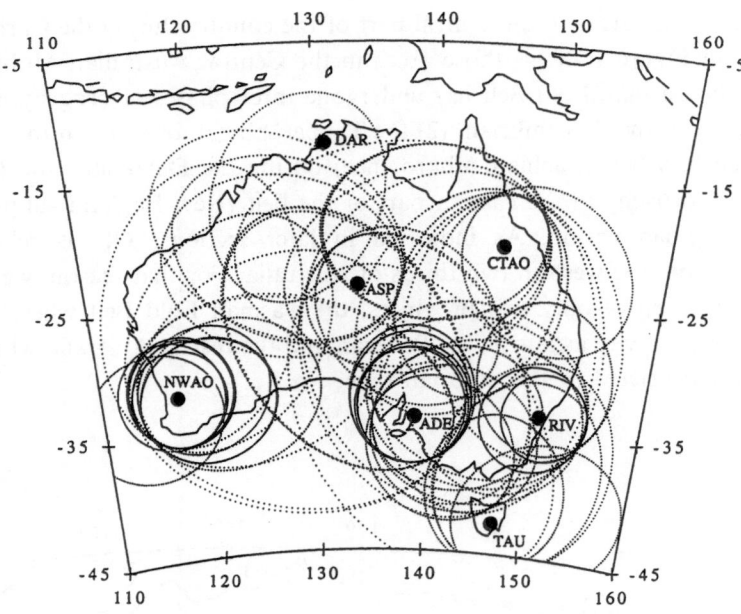

Figure 2

Sampling patterns of *Lg* coda waves across Australia are approximated by the mapped scattering ellipses. One focus of each ellipse corresponds to a source location and the other to a station location. Since Australia is surrounded by a broad shelf on all but the eastern coast, we have not truncated the ellipses as in other *Lg* coda *Q* studies in this volume.

Data

Australia has fewer earthquakes than other continents in which have applied our methods to obtain estimates of Q_{Lg}^c. However those that have occurred show a good distribution across the continent. In addition, the distribution of modern digital stations, although not dense, is fairly uniform (Fig. 2). The largest gap in station coverage lies in the northwestern portion of the continent in a region centered on about latitude 22°S and longitude 125°E. Table 1 lists the stations and their locations. Earthquake magnitudes vary between 4.2 and 6.1 with all but five events being of magnitude 5.8 or smaller. Table 2 gives the epicentral parameters for all events, the stations used for each event, and the values of Q_{Lg}^c at 1 Hz (Q_0), as well as the frequency dependence (η) of Q_{Lg}^c at 1 Hz, obtained from recordings of *Lg* coda at each station.

We applied the stacked spectral ratio (SSR) method of XIE and NUTTLI (1988) for all Q_0 and η determinations in Australia (Table 2). An application of the method in the present volume appears in the paper by CONG and MITCHELL (1998a). Example seismograms and SSR plots appear in BAQER and MITCHELL (1998) and DE SOUZA and MITCHELL (1998) for studies in the United States and

South America, respectively. We follow the same procedure in the present study, thus we will not make similar plots here.

As discussed in the other studies of Q^c_{Lg} in this volume, we represent the sampling pattern of *Lg* coda by an ellipse with the source at one focus and the recording station at the other (Fig. 2). The ellipse approximates the surface projection of the volume over which *Lg* energy is scattered if we assume a single-scattering process. Using this areal representation allows us to obtain better coverage than we could achieve if we had assumed that energy travelled in a straight line. For this reason, even though a relatively small number of earthquakes and stations are used, we can still obtain fairly good coverage of the Australian continent.

Lg Coda Q Tomography

The back-projection tomographic method of XIE and MITCHELL (1990) was employed to map images of Q_0 (Fig. 3) and η (Fig. 4) for all of Australia. After several tests, we divided Australia into cells that are 1.5° by 1.5° in area. Our map indicates that Q_0 varies between about 250 and 600. Consideration of the resolution achievable in this study and the computed standard errors, leads us, however, to ignore the low values that appear in places along the fringes of the map. These lie in the northernmost portions of the North Australian Craton and the Northeast Orogens, as well as along the western coast near the Pilbara block. Our map of data coverage (Fig. 2) reveals that coverage is scant in these regions. A discussion of resolution and error for this study appears in the following section.

Neglecting the low values in the poorly sampled regions, we find that Q_0 varies between about 300 and 600 across Australia. The highest values lie in the Yilgarn block, a small region in the southeastern part of the North Australian Craton, and in the Gawler Craton/Nullarbor block. The smallest values lie in the western part

Table 1

Station information

Station	Latitude (°S)	Longitude (°E)
ADE	34.967	138.709
ASP	23.683	133.897
CTAO	20.088	146.254
DAR	12.408	130.818
NWAO	32.927	117.233
RIV	33.829	151.158
TAU	42.910	147.320

Table 2

Event information and Lg coda Q results

Date ymd	Origin time h:m:s	Lat. deg.	Lon. deg.	Station	Q_0	η
650125	20:22:56.3	32.200S	138.600E	ADE	518	0.88
650302	15:18:53.2	30.500S	138.400E	ADE	514	1.05
				RIV	394	0.35
650603	21:59:58.5	28.100S	150.100E	RIV	425	0.88
650828	00:26:38.1	32.300S	138.100E	ADE	636	1.32
				RIV	372	0.92
700310	17:15:08.7	31.010S	116.540E	ADE	682	0.69
700404	14:09:45.5	21.809S	126.604E	ADE	550	1.12
700720	16:01:47.9	21.738S	126.815E	ADE	468	0.90
710716	08:00:04.9	22.100S	126.500E	DAR	323	0.87
750724	22:23:46.8	21.342S	120.448E	ASP	362	0.92
751003	11:51:09.4	22.126S	126.721E	ADE	681	0.91
770704	20:05:20.3	34.634S	148.850E	RIV	415	0.62
771202	13:32:37.5	37.662S	144.396E	ADE	518	0.75
				ASP	329	0.88
790423	05:45:10.1	16.535S	120.175E	ASP	916	0.92
800315	07:09:45.1	18.832S	121.760E	ASP	404	0.54
801208	00:12:18.8	31.211S	115.651E	NWAO	449	0.90
801210	04:35:00.9	30.594S	117.399E	NWAO	824	0.71
810616	21:33:57.8	38.836S	144.268E	ADE	602	0.65
				ASP	493	0.91
				CTAO	317	0.43
811115	16:58:11.0	34.252S	150.930E	ADE	374	0.61
				CTAO	371	0.79
				RIV	447	0.74
				TAU	870	0.85
820206	15:24:37.8	30.851S	117.357E	NWAO	627	0.90
820206	15:30:35.9	30.851S	117.357E	NWAO	432	0.73
820213	08:26:57.0	22.046S	126.637E	CTAO	399	0.83
				NWAO	426	0.94
820715	01:44:17.6	30.976S	138.224E	ADE	394	1.88
821121	11:34:19.1	37.152S	146.760E	ADE	536	0.89
				CTAO	502	0.56
				TAU	321	0.46
821126	00:11:16.3	33.876S	147.133E	ADE	479	0.99
830126	06:16:12.8	30.656S	117.241E	ADE	509	0.31
				NWAO	955	1.46
830408	19:33:13.6	29.615S	142.150E	ASP	260	0.45
				CTAO	577	0.73
831125	19:56:07.8	40.451S	155.507E	TAU	654	1.11
831229	17:41:58.4	30.762S	137.257E	CTAO	462	1.00
850208	08:23:42.6	25.211S	153.436E	CTAO	318	0.80
850728	07:39:43.9	32.223S	122.420E	NWAO	546	1.02
850728	10:39:27.2	32.294S	122.282E	NWAO	522	0.80
851127	23:18:17.1	30.654S	117.392E	NWAO	679	0.75
860102	21:33:42.3	34.310S	112.097E	NWAO	379	0.93
860330	08:53:53.2	26.194S	132.767E	CTAO	545	0.84
				NWAO	389	0.73
				TAU	542	1.04

Table 2 (continued)

Date ymd	Origin time h:m:s	Lat. deg.	Lon. deg.	Station	Q_0	η
880106	03:42:04.6	31.059S	117.864E	NWAO	640	0.82
880122	00:35:58.0	19.847S	133.803E	CTAO	549	0.92
880122	03:57:25.2	19.798S	133.910E	CTAO	621	1.03
880122	20:54:03.0	19.857S	133.991E	CTAO	494	0.94
880615	09:13:41.1	31.479S	126.440E	NWAO	314	1.00
890528	02:55:19.6	25.053S	130.781E	CTAO	413	0.85
891013	09:59:11.9	17.609S	122.404E	NWAO	341	0.82
891227	23:26:57.0	32.967S	151.619E	TAU	442	0.68
900117	06:38:05.7	31.673S	116.995E	NWAO	594	0.99
910309	01:36:27.7	30.585S	121.589E	NWAO	427	1.04

of the Tasman Fold Belt (in the New England subprovince). Most of the continent is characterized by Q_0 values between about 400 and 500. Much of the Tasman Fold Belt, however, is somewhat lower, between 320 and 380. A zone of relatively low Q_0 coincides with a projection of the Tasman Fold Belt to the west. Somewhat low values (370–400) extend even further to the west into a region of Carboniferous deformation within the Central Australian Mobile Belts shown in Figure 1 (PLUMB, 1979). This suggests that the slightly low Q for that region may be associated with that period of deformation (MITCHELL, 1995). That region of deformation is, however, partially coincident with thick sedimentary basins that formed over a long period of time between late Precambrian time and the Carboniferous period.

Some sediments, especially young sandstone and shale, have been found to significantly reduce Q_{Lg}, especially when they overlie high-Q crust (MITCHELL and HWANG, 1987). It is difficult to assess the effect of these particular sediments on Lg propagation, so we cannot, at this time, determine if the moderately low Q_{Lg}^c in central Australia is caused by fluids associated with past deformation or by low-Q sediments.

Figure 1 shows regions where sedimentation occurred during the Jurassic and Cretaceous periods (PLUMB, 1979). It is apparent that some of these sediments coincide with the low Q mapped in the Tasman Fold Belt and in two portions of the North Australian Craton (northeastern and westernmost regions). Other regions of sedimentation shown in Figure 1, such as those in the Nullarbor block and the Curnamona Craton, are not characterized by reduced Q_{Lg}^c. It is also significant that the region of lowest Q_{Lg} in the New England subprovince in easternmost Australia is devoid of those sediments. This lack of correlation between Q_{Lg}^c and sediment accumulation seems to favor Carboniferous deformation as the causative agent for producing reduced Q_{Lg}^c both in the New England subprovince and in the Central Australian Mobile Belts but, as mentioned above, we cannot be sure of this.

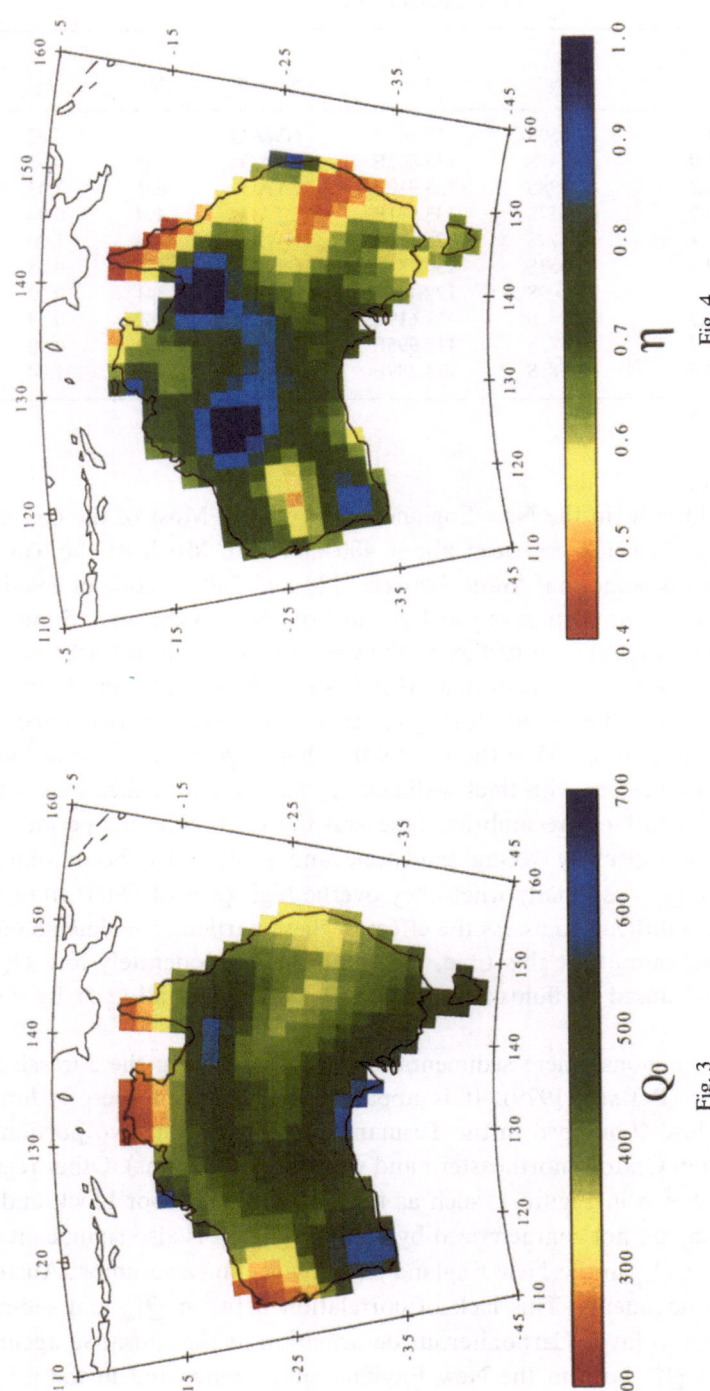

Figure 3

Tomographic map of *Lg* coda *Q* at 1 Hz for all of Australia. Each cell is 1.5° by 1.5° in area. Low *Lg* coda *Q* values along two portions of the northern coast and a portion of the western coast are probably due to poor data coverage in those regions (see Figure 2).

Figure 4

Tomographic map of the frequency dependence of *Lg* coda *Q* at 1 Hz. Each cell is 1.5° by 1.5° in area.

Most values of Q_0 in Australia are lower than those measured in other stable regions. Q_0 ranges between 600 and 1000 in stable portions of Eurasia (MITCHELL *et al.*, 1998), between some 550 and 1000 in stable portions of South America (DE SOUZA and MITCHELL, 1998), between about 650 and 1000 in stable portions of Africa (XIE and MITCHELL, 1990), and between about 500 and 750 in stable portions of the United States (BAQER and MITCHELL, 1998). The only stable region where Q_0 is observed to be lower than in Australia is the Arabian Peninsula (CONG and MITCHELL, 1998a) where Q_0 ranges between about 300 and 450. The Arabian Peninsula, however, overlies an upper mantle with very high temperatures. We have postulated that those elevated temperatures generated hydrothermal fluids that have migrated to the crust and produce the low Q_0 (CONG and MITCHELL, 1998b).

The regional variation of the frequency dependence of Q_{Lg}^c at 1 Hz (η) in Australia appears in Figure 4. The only clear correlation with η with Q_0 occurs in the Tasman Fold Belt of eastern Australia where the lowest values for η correlate with moderately low values for Q_0. All other low values of η are in regions where

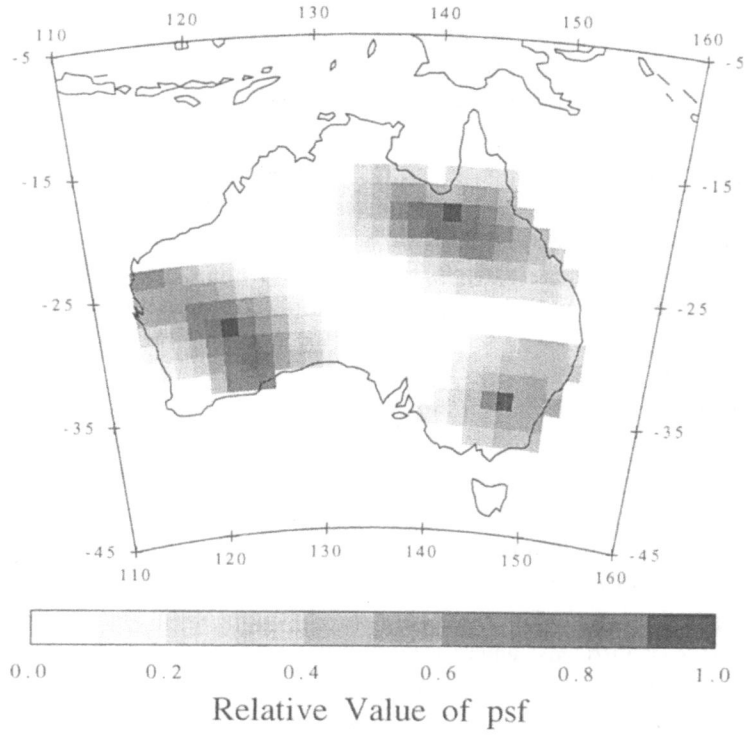

Figure 5
Point-spreading functions (psf) determined for regions centered on cells at 28.25°S, 121.75°E, at 18.75°S, 141.50°E, and at 18.75°S, 141.25°E.

resolution is poor or standard errors are large. η throughout most of Australia lies in the range 0.65 to 0.75, however there are two regions of very high η (about 1.0) in the northern part of the continent. As discussed in the following section, the standard errors are not large in these regions and, although resolution is only moderate, it is difficult to explain these high values as being due to random error. It is, however, possible that these high values are due to systematic errors that cannot be formally evaluated. We obtained η numerically in a process that involves subtraction of Q_{Lg}^c values obtained at 1 Hz from values extrapolated to 3 Hz using measured η obtained from individual records (XIE and NUTTLI, 1988). Systematic errors associated with either measured Q_0 or η could contribute to those high values. Although we cannot disprove the reality of the two regions of high η, we think it is more likely that they result from systematic error. If so, then η for all but the easternmost part of Australia lies between 0.65 and 0.75.

Resolution and Error

XIE and MITCHELL (1990) utilized the point-spreading function (*psf*), as presented by HUMPHREYS and CLAYTON (1988) to estimate resolution. In this volume that method is briefly discussed by CONG and MITCHELL (1998a). For the present study we will just state that the width of the area covered by the *psf* is an indicator of resolution; sharp *psf*s indicate good resolution and broad *psf*s indicate poor resolution. The dimensions that can be resolved can be estimated by determining the distance from the center cell of each *psf* at which it falls to a small value.

Figure 5 shows *psf*s for three locations in Australia. The *psf* in the southeastern part of the continent indicates relatively good resolution (300–500 km) whereas those in northeastern and western Australia are broader and suggest that resolvable features may be as large as 900 km or more.

Figures 6 and 7 show standard errors for Q_0 and η, respectively. These were also obtained following XIE and MITCHELL (1990) and the process is briefly described in the present volume by CONG and MITCHELL (1998a). The standard errors are, as expected, largest along the fringes of the continent where data coverage is poor (compare with Fig. 2). Numerical values on these maps also suggest that they are rough estimates rather than precise values of standard error. An example is the relatively small estimate for standard error that appears in the northern extremity of the Northeast Orogens, where there is little or no data coverage. A *psf* for that region would, however, manifest extremely poor resolution. The largest standard errors for η occur in about the same places where the standard errors for Q_0 are high. As indicated earlier, standard errors are not large in two regions of central Australia where η seems unreasonably large. We infer that systematic errors that cannot be evaluated, produce those high values.

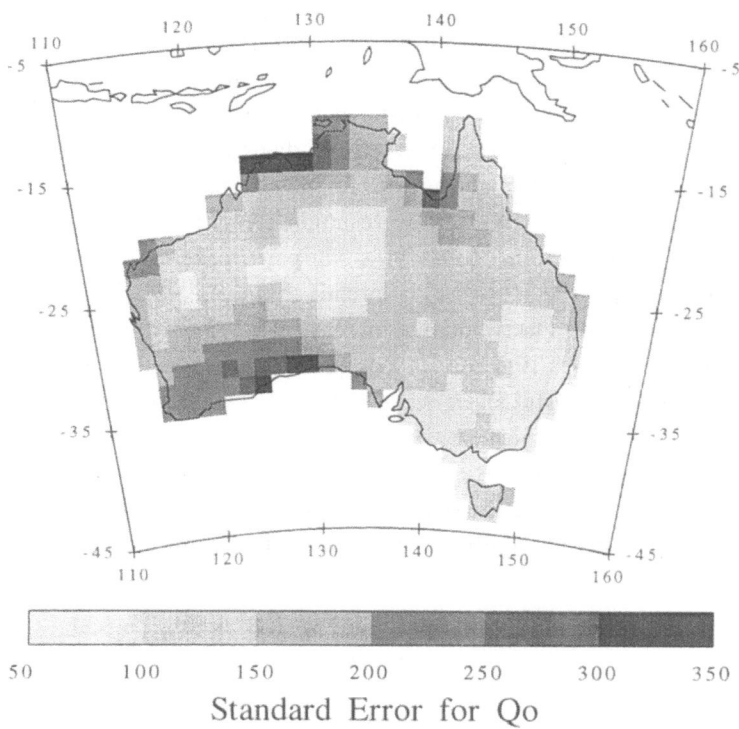

Standard Error for Qo

Figure 6
Standard error distribution for the image of Q_0.

Velocity Gradients at the Crust-mantle Transition

As discussed earlier, our determinations of Q_0 for *Lg* coda *Q* are lower than those obtained in most stable continental regions. BOWMAN and KENNETT (1991) had previously obtained lower than expected values of *Q* for the direct *Lg* wave, and tested the possibility that a velocity gradient rather than a sharp interface at the crust-mantle transition could accelerate *Lg* amplitude decay with distance. Using models with an approximately 25-km thick velocity gradient at the crust-mantle transition, as found by FINLAYSON (1982) from seismic refraction studies, they could explain their Q_{Lg} determinations in north-central Australia.

Although our determinations of Q_{Lg}^c (~ 450) in that region of Australia are not as low as those found for Q_{Lg} (230) by Bowman and Kennett, they are still smaller than expected for a stable continental region. Moreover, except for three small regions (where Q_{Lg}^c may be as high as about 600), all of Australia is characterized by Q_{Lg}^c between 330 and 500. If a velocity gradient at the crust-mantle transition causes a reduction in Q_{Lg}^c, such a gradient must be widespread throughout Australia. Compilation of velocity models obtained from seismic refraction surveys there (DRUMMOND and COLLINS, 1986; COLLINS, 1991) suggest that velocity gradients at the crust-mantle transition occur almost everywhere in Australia. The depth ranges, thicknesses, and severities of gradients in different regions vary widely from province to province and even differ within individual provinces.

We investigated the effect of a velocity gradient on determinations of Q_{Lg} by computing synthetic Lg ground motion for two crust-upper mantle models of north-central Australia (Fig. 8) that were studied earlier by BOWMAN and KENNETT (1991). One model, SJ0, contains a sharp increase in velocity at the crust-mantle boundary while the other, SG3, spreads that same velocity increase over the depth range 37–62 km. The velocities increase from 3.75 to 4.55 km/s for shear waves and from 6.60 to 8.10 km/s for compressional waves for both models.

We tried to find a crustal shear-wave Q model (Q_μ) that, when combined with velocity model SG3 for central Australia, would yield Q_0 and η values that are similar to those observed in this study and shown in Figures 3 and 4. To simplify our search we required both shear-wave Q_μ and the frequency dependence (ζ) of Q_μ to remain constant at all depths. We followed the computational procedure of

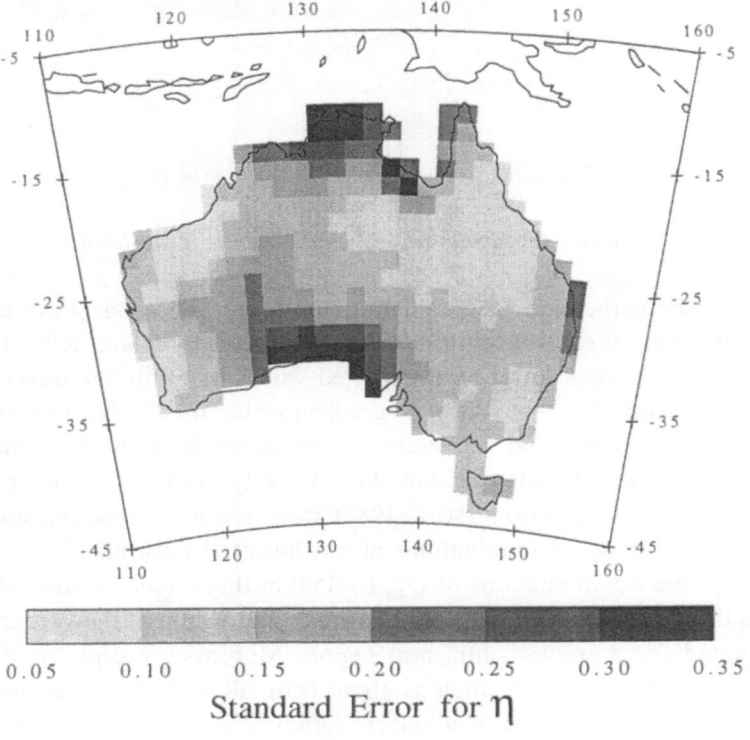

Figure 7
Standard error distribution for the image of η.

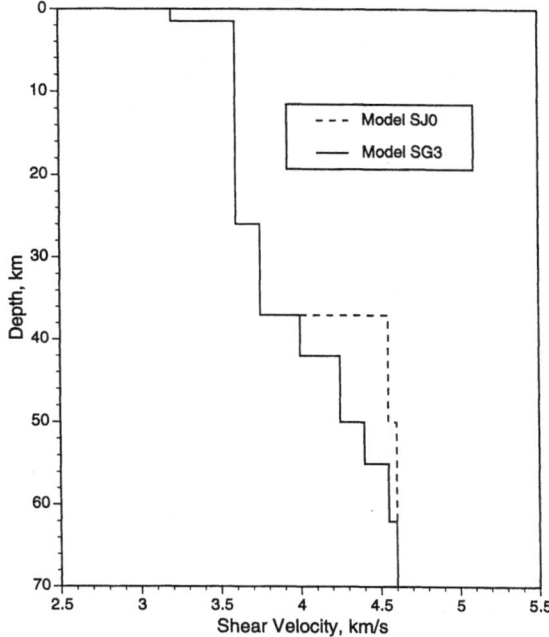

Figure 8
Shear-velocity distributions for a model with a sharp discontinuity at the crust-mantle boundary (SJ0), and model for which velocities grade from crustal to mantle values over a depth range in excess of 20 km (SG3). Both models are taken from BOWMAN and KENNETT (1991).

MITCHELL and XIE (1994) to find the Q_{Lg} value that is predicted by a specified Q_μ model. The best agreement with our *Lg* coda *Q* observations occurred when Q_μ was set at 500 and ζ was 0.70. These model values produced Q_{Lg} and η values of 476 and 0.75, respectively. These compare well with our observed values of 475 and 0.70 in central Australia. The same shear-wave *Q* model, when combined with velocity model SJ0 (sharp crust-mantle boundary), produced Q_{Lg} and η values of 593 and 0.72. The presence of the velocity gradient in model SG3 therefore reduces Q_{Lg} by about 20%.

If we assume that replacing a sharp boundary with a velocity gradient at the crust-mantle boundary will change Q_{Lg}^c by the same amount that it changes Q_{Lg}, and that the same gradient occurs throughout Australia, then values for Q_{Lg}^c would be roughly 20% higher everywhere than those shown in Figure 3. Neglecting the small, very low, values around the fringes of the map, this would produce Q_{Lg}^c that ranges between about 400 and 720. This range is closer to that for typical stable continental regions but still appears to be somewhat low.

We can only speculate on the reason that Australian Q_{Lg}^c is lower than most other continents, even after correcting for the effect of a velocity gradient at the crust-mantle transition. One possibility is that the lateral variability in depth range, thickness and severity of the velocity gradient throughout Australia causes additional energy, above that predicted by a one-dimensional model with a gradient, to leak into the mantle. This possibility cannot presently be studied quantitatively but is consistent with the variability in the velocity gradient found throughout Australia (DRUMMOND and COLLINS, 1986; COLLINS, 1991). It is discussed in additional detail in the present volume by MITCHELL and CONG (1998). Yet another possibility, of course, is that there are additional factors affecting Q_{Lg}^c, or and/or crustal Q_μ, that we have not yet considered.

Conclusions

Q_{Lg}^c obtained from relatively few, but well-distributed digital recordings of *Lg* coda in Australia, varies between 330 and 600. All reliably determined values below 400 lie in the Tasman Fold Belt. The reduced values there may be due to fluids remaining from tectonic or orogenic activity that occurred during Devonian and Carboniferous times or to Jurassic and Cretaceous sediments that may be characterized by low Q, and which are present in that region. A smaller reduction in Q_{Lg}^c in central Australia may be due to fluids produced during Carboniferous deformation or to sediments deposited there over a long period between late Precambrian time and the Carboniferous period.

Q_{Lg}^c through most of Australia is lower than expected for a region that has been stable for a prolonged period. It is 30–60% lower than that observed throughout most of the stable portions in Eurasia, Africa, and South America. About 20% of this reduction can be explained by the presence of a thick velocity gradient, rather than a sharp interface, at the crust-mantle transition that causes some of the shear energy in *Lg* to leak into the upper mantle. Assuming that Q_{Lg}^c and Q_{Lg} are affected to the same degree by that leakage, a velocity gradient such as that found beneath north-central Australia will reduce Q_{Lg}^c by about 20%. Additional reductions in Q_{Lg}^c might be caused by lateral variability in thickness, depth, and severity of the velocity gradient throughout Australia.

Acknowledgment

We are grateful to Dibo Chen who determined many of the Q_0 and η values reported in this paper, to Jiakang Xie, now at Lamont-Doherty Earth Observatory, who provided his programs for determining stacked spectral ratios and for obtaining tomographic images of *Lg* coda Q, and to David Kirschner, of the Department of Earth and Atmospheric Sciences at Saint Louis University, for providing considerable helpful information on the geology and tectonics of Australia.

REFERENCES

BAQER, S., and MITCHELL, B. J. (1998), *Regional Variation of Lg Coda Q in the Continental United States and its Relation to Crustal Structure and Evolution*, Pure appl. geophys., *153*, 613–638.

BOWMAN, J. R., and KENNETT, B. L. N. (1991), *Propagation of Lg Waves in the North Australian Craton: Influence of Crustal Velocity Gradients*, Bull. Seismol. Soc. Am. *81*, 592–610.

COLLINS, C. D. N. (1991), *The Nature of the Crust-mantle Boundary under Australia from Seismic Evidence*, Geol. Soc. Aust. Spec. Publ. 17 (ed. B. J. Drummond) 67–80.

CONG, L., and MITCHELL, B. J. (1998a), *Lg Coda Q and its Relation to the Geology and Tectonics of the Middle East*, Pure appl. Geophys., *153*, 563–585.

CONG, L., and MITCHELL, B. J. (1998b), *Seismic Velocity and Q Structure of the Middle Eastern Crust and Upper Mantle from Surface-wave Dispersion and Attenuation*, Pure. appl. geophys., *153*, 503–538.

DRUMMOND, B. J., and COLLINS, C. D. N. (1986), *Seismic Evidence for Underplanting of the Lower Continental Crust of Australia*, Earth Planet. Sci. Lettrs. *79*, 361–372.

FINLAYSON, D. M. (1982), *Seismic Crustal Structure of the Proterozoic North Australian Craton between Tennant Creek and Mount Isa*, J. Geophys. Res. *87*, 10569–10578.

HUMPHREYS, E., and CLAYTON, R. W. (1988), *Adaptation of Back Projection Tomography to Seismic Travel-time Problems*, J. Geophys. Res. *93*, 1073–1086.

MITCHELL, B. J. (1995), *Anelastic Structure and Evolution of the Continental Crust and Upper Mantle from Seismic Surface Wave Attenuation*, Rev. Geophys. *33*, 441–462.

MITCHELL, B. J., and HWANG, H. J. (1987), *Effect of Low-Q Sediments and Crustal Q on Lg Attenuation in the United States*, Bull. Seismol. Soc. Am. *77*, 1197–1210.

MITCHELL, B. J., and XIE, J. (1994), *Attenuation of Multiphase Surface Waves in the Basin and Range Province—III. Inversion for Crustal Anelasticity*, Geophys. J. Int. *116*, 468–484.

MITCHELL, B. J., PAN, Y., XIE, J., and CONG, L. (1997), *The Variation of Lg Coda Q across Eurasia and its Relation to Crustal Evolution*, J. Geophys. Res. *102*, 22,767–22,779.

MITCHELL, B. J., and CONG, L. (1998), *Lg Coda Q and its Relation to the Structure and Evolution of Continents: A Global Perspective*, Pure appl. geophys., *153*, 655–663.

PLUMB, K. A. (1979), *The Tectonic Evolution of Australia*, Earth-Science Rev. *14*, 205–249.

RUTLAND, R. W. R. (1976), *Orogenic Evolution of Australia*, Earth-Science Rev. *12*, 161–196.

DE SOUZA, J., and MITCHELL, B. J. (1998), *Lg Coda Q Variations across South America and their Relation to Crustal Evolution*, Pure appl. geophys., *153*, 587–612.

XIE, J., and MITCHELL, B. J. (1990), *A Back-projection Method for Imaging Large-scale Lateral Variations of Lg Coda Q with Application to Continental Africa*, Geophys. J. Int. *100*, 161–181.

XIE, J., and NUTTLI, O. W. (1988), *Interpretation of High-frequency Coda at Large Distances: Stochastic Modeling and Method Inversion*, Geophys. J. Int. *95*, 579–595.

WANG, C. Y. (1981), *Wave Theory for Seismogram Synthesis*, Ph.D. Diss., Saint Louis University, 235 pp.

VEEVERS, J. J., *Phanerozoic Earth History of Australia* (Oxford University Press, New York, 1994) 418 pp.

(Received March 5, 1998, revised August 11, 1998, accepted August 25, 1998)

To access this journal online:
http://www.birkhauser.ch

Pure appl. geophys. 153 (1998) 655–663
0033–4553/98/040655–09 $ 1.50 + 0.20/0

❘ Pure and Applied Geophysics

Lg Coda *Q* and its Relation to the Structure and Evolution of Continents: A Global Perspective

B. J. MITCHELL[1] and LIANLI CONG[2]

Abstract—Tomographic maps of *Lg* coda *Q* (Q_{Lg}^c) variation are now available for nearly the entire African, Eurasian, South American, and Australian continents, as well as for the United States. Q_{Lg}^c at 1 Hz (Q_0) varies from less than 200 to more than 1000 and Q_{Lg}^c frequency dependence (η) varies between 0.0 and nearly 1.0. Q_0 appears to increase in proportion to the length of time that has elapsed since the most recent major episode of tectonic or orogenic activity in any region. A plot of Q_0 versus time since that activity indicates that a single Q_0-time relation approximates most mean Q_0 values. Those that deviate most from the trend lay in Australia, the Arabian Peninsula, and the East African rift. The increase in Q_0 with time may be due to a continual increase in crustal shear wave Q (Q_μ) caused by the loss of crustal fluids and reduction of crustal permeability following tectonic or orogenic activity. Extrapolated values of Q_{Lg}^c at 5 Hz (using Q_0 and η values measured at 1 Hz and assuming that η is constant in all regions between 1 and 5 Hz) show a similar percentage-wise increase with times that has elapsed since the most recent activity. Other factors that can reduce Q_0 in continental regions include thick accumulations of sediment (especially sandstone and shale of Mesozoic age and younger), severe velocity gradients at the crust-mantle transition and, possibly, lateral variations in the depth, thickness, and severity of those gradients. Severe and large increases of Q_μ in the mid-crust of some regions can cause relatively large values of η, even if the frequency dependence of Q_μ is small.

Key words: *Lg*, coda, *Q*, continents, crust, structure, evolution.

Introduction

Lg and its coda are often prominent phases on short-period and broadband seismograms when travel paths are restricted to continents. *Lg* can be modeled either as a superposition of numerous higher surface-wave modes or as a summation of supercritically reflected rays that are confined to the crust. *Lg* coda that follows, and is continuous with, direct *Lg*, is known to consist largely of scattered energy. Because *Lg* and its coda are so prominent, they have been extensively studied and their attenuation rates have been determined in many regions.

[1] Department of Earth and Atmospheric Sciences, Saint Louis University, 3507 Laclede Avenue, St. Louis, Missouri 63103, U.S.A. E-mail: mitchell@eas.slu.edu
[2] Now at Department of Earth Sciences, Yunnan University, Kunming, Yunnan, People's Republic of China.

Advances in *Lg* coda analysis (XIE and NUTTLI, 1998) and tomographic inversion (XIE and MITCHELL, 1990) allow stable and relatively precise measurements of *Lg* coda Q (Q_{Lg}^c) and its frequency dependence at 1 Hz (Q_0 and η, respectively). These methods have now been applied to almost the entire continents of Africa (XIE and MITCHELL, 1990), Eurasia (MITCHELL *et al.*, 1997), South America (DE SOUZA and MITCHELL, 1998), and Australia (MITCHELL *et al.*, 1998), as well as to the United States (BAQER and MITCHELL, 1998). Those studies used identical methodologies and attempted to process the *Lg* coda data uniformly. For instance, it is known that the time interval over which coda is analyzed will affect measured values of Q_0, with measurements in the late coda producing higher values than measurements confined to early coda. For that reason, all of the studies referred to above used coda for which the beginning lapse time corresponded to 3.15 km/s and ranged between about 300 and 500 seconds duration. These consistent measurement techniques should allow us to compare results obtained in different continents.

An obvious result of the tomographic studies in each continent is that all regions of current or recent tectonic or orogenic activity exhibit lower Q_0 than do other regions and that most (but not all) old stable regions exhibit high Q_0. Within each continent we have found that the Q_0 distribution is consistent with the proposal (MITCHELL, 1995) that measured Q_0 in any region is directly proportional to the length of time that has elapsed since the most recent major episode of orogenic activity there. In this study we investigate whether or not a single Q_0-time relation can explain the observed Q_0 values in all continents.

Regional Variation of Q_0

The studies cited above present determinations of Q_0 variation across Africa, Eurasia, South America, Australia, and the United States. We would like to know whether or not observed Q_0 values in each continent can be related to those in other continents through some empirical relationship. In order to investigate the possibility that Q_0 in any region increases with time since the most recent episode of tectonic or orogenic activity there, we plotted Q_0 versus time that has elapsed since that activity (e.g., STANLEY, 1986) for the tectonic provinces that we have studied. In most, but not all, stable cratonic regions, we take the time of most recent orogenic activity to be the time of formation of the craton. We take Precambrian ages from GOODWIN (1991).

Figure 1 displays measured $(Q_{Lg}^c)^{-1}$ values for both 1 Hz (Q_0) and 5 Hz for 17 regions where both and Q_0 and η have been well determined. We extrapolated Q_{Lg}^c to 5 Hz using the measured 1-Hz Q_0 and η values for all regions. Like Q_0, η varies greatly from place to place (0.0–1.0). Although low η appears to correlate with low Q_0 in some places, such as throughout Africa, other regions exhibit no systematic relationship between the two parameters. Figure 1 includes values obtained for broad regions across which Q_0 is relatively uniform. The vertical bars for the 1-Hz values approximate the range of Q_0 observed through each region included on the plot. The bars are usually longer for larger regions over which Q_0 is apt to vary more than over smaller regions. For instance, the Tethys region, where bars are relatively long, is a band of plate convergence between the Eurasian and African/ Arabian plates that extends from western Europe into Asia. The vertical bars for the 5-Hz values are extrapolations of the end points of the 1-Hz errors, which use the measured frequency dependence found for the 1-Hz values. We assume that the frequency dependence is constant for frequencies between 1 and 5 Hz.

Q^c_{Lg} at 1 Hz (Q_0) decreases from about 1000 at 2200 My BP to a minimum value of about 250 at 2 My BP or less. The values for the 1-Hz trend can be described by the linear relation $Q^{-1} = -0.25 \times 10^{-3} \ln A + 3.37 \times 10^{-3}$, where A is the time

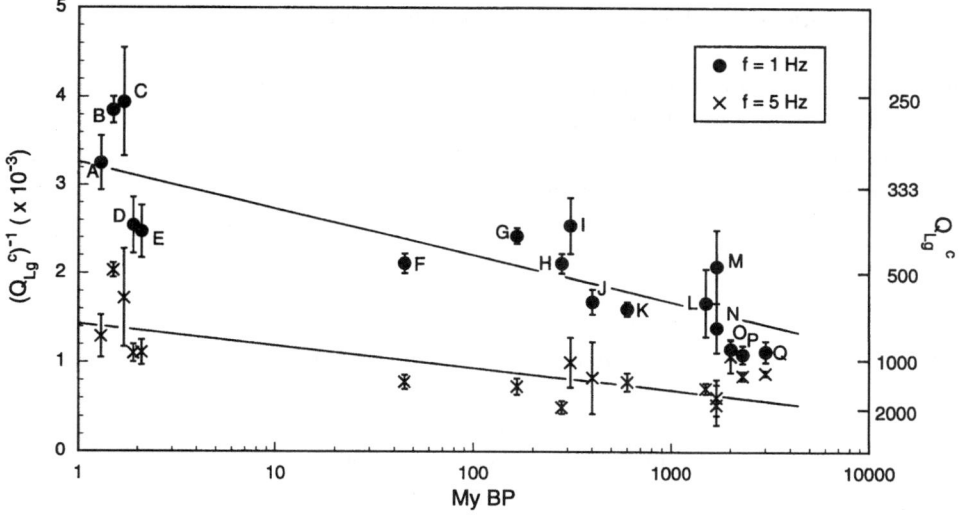

Figure 1

$(Q^c_{Lg})^{-1}$ as a function of time since the most recent tectonic or orogenic activity in a specified region. For cratons and shields that have not been tectonically subjected to orogenic activity, we take reported ages of formation. For regions that have undergone tectonic or orogenic activity, we plot the time since the most recent activity. Points for regions that are still tectonically active are plotted (for clarity) at arbitrary times less than 25 million years before the present. Circles denote measured values of $(Q^c_{Lg})^{-1}$ at 1 Hz and x's denote values of $(Q^c_{Lg})^{-1}$ that have been extrapolated to 5 Hz using measured values for the frequency dependence of (Q^c_{Lg}) in various studies. A—The Andes Mountains; B—Basin and Range province in the western United States; C—Tethys region of convergence of the Eurasian plate with the African, Arabian, and Indian plates; D—the Arabian Peninsula; E—the East African Rift; F—the Rocky Mountains; G—northeastern China; H—the eastern Altaid belt in Eurasia; I—the Tasman province of Australia; J—the Atlantic Shield of South America; K—the African Fold Belts; L—the portion of the North American craton in the United States; M—the Australian craton; N—Eurasian cratons; O—African shields; P—the Brazilian shield; Q—the Indian Shield. The least-squares fit (straight lines) to the plotted values are described by the equations $Q^{-1} = -0.25 \times 10^{-3} \ln A + 3.37 \times 10^{-3}$ at 1 Hz and by $Q^{-1} = 0.98 \times 10^{-4} \ln A + 1.44 \times 10^{-3}$ at 5 Hz, where A is the time that has elapsed since the most recent episode of tectonic or orogenic activity.

that has elapsed since the most recent episode of major tectonic or orogenic activity. There are, however, three perhaps significant deviations from that line. At the young end of the plot, Q_0 is higher (Q^{-1} lower) for both the Arabian Peninsula (*D*) and the East African Rift (*E*) than for the other three regions that are currently active. For the East African rift (XIE and MITCHELL, 1990) it is possible that Q_0 is really considerably lower than measurements indicate, but they occur over a much narrower band than that which appears on the Q_0 map of Africa. The somewhat high values for the rift may therefore only reflect the limited resolution inherent in our method for determining Q_0.

The Arabian Peninsula (*D*) is, however, a broad region over which Q_0 is well resolved and the lower than expected values are likely to be real. In addition, both the cratonic (*M*) and Tasman Fold Belt (*I*) broad regions of Australia that exhibit lower Q_0 than that predicted by the best-fit line. The following section discusses possible reasons for these deviations.

The extrapolated 5-Hz values, because of the greater likelihood of measurement error and because we assume that η is constant in all regions between 1 and 5 Hz, will have greater uncertainties than the 1-Hz values. However if they are realistic they indicate that percentage-wise changes in Q^c_{Lg} at 5 Hz are similar to those at 1 Hz. One interpretation of the difference in trend for 1-Hz and 5-Hz determinations is that whereas at 1-Hz Q^c_{Lg} is largely controlled by intrinsic absorption in the crust, at 5-Hz scattering has become more important.

Reasons for Regional Q^c_{Lg} Variation

Regional variations of Q^c_{Lg} at 1 Hz (Q_0) often correlate with regional variations of shear wave Q (Q_μ) in the upper crust as obtained from the inversion of surface-wave attenuation measurements at periods between a few and several 10's of seconds. For that, and other reasons, MITCHELL (1995) argued that Q^c_{Lg}, in most regions, is largely influenced by intrinsic Q_μ in the upper crust. He proposed that seismic waves expend energy to move fluids through permeable rock during propagation and the energy loss becomes smaller (and Q increases) as the volume of fluids gradually decreases. It is also possible, however, that scattering removes seismic energy and that the observed increase in Q with time reflects a decrease in acoustic impedance due to the loss of fluids.

Figure 2 shows some Q_μ models that have been obtained from the inversion of fundamental-mode Rayleigh-wave attenuation. A model for the Basin and Range province of the western United States that explains both Rayleigh-wave attenuation and Q^c_{Lg} (MITCHELL and XIE, 1994) has very low Q_μ (50–150) at upper crustal depths and high Q_μ (as high as 1000) at lower crustal and uppermost mantle depths. The rapid increase of Q_μ with depth can also cause a high frequency dependence for

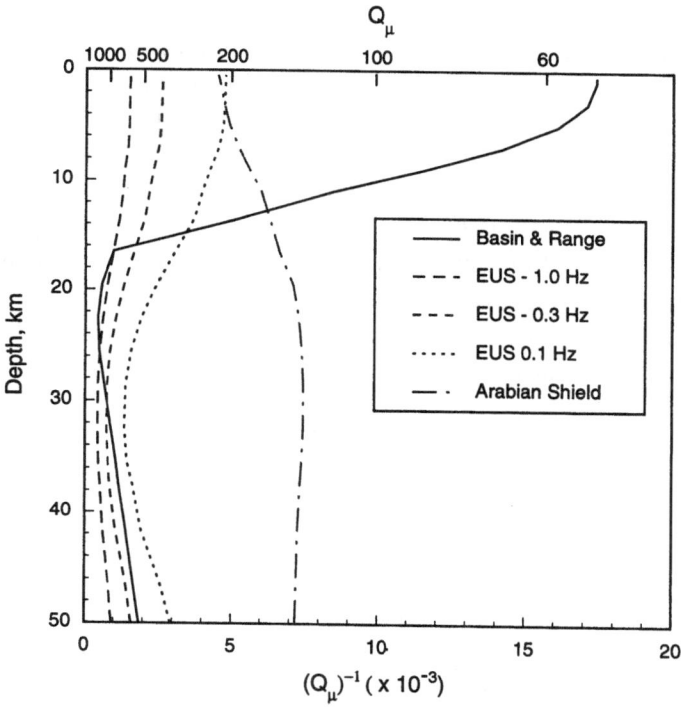

Figure 2

Frequency-independent crustal Q_μ models for the Basin and Range province in the western United States, and the Arabian shield, as well as a freqeuncy-dependent Q_μ model for the eastern United States.

Q_{Lg}^c, even if intrinsic Q_μ is independent of frequency (MITCHELL, 1991). Q_μ values for the eastern United States model (EUS) in Figure 2 (MITCHELL, 1995) are similar to those for the Basin and Range at lower crustal depths but are considerably higher (200–800) at upper crustal depths. In order to simultaneously explain both fundamental-mode Rayleigh-wave and *Lg* attenuation, the EUS Q_μ model must vary with frequency, at least in the upper crust, with the higher frequencies exhibiting higher Q_μ. The differences in upper crustal Q_μ and similarity of lower crustal Q_μ between the eastern United States and the Basin and Range suggest that much of the regional variation of Q_{Lg}^c in continents is due to differences of Q_μ in the upper crust. Figure 2, however, also includes a recently obtained Q_μ model for the crust of the Arabian Peninsula (CONG and MITCHELL, 1998b). Q_μ is lower in the lower crust than it is in the upper crust and is much lower than Q_μ for either the eastern United States or the Basin and Range. This suggests that lateral variations of lower-crustal Q_μ can be responsible for Q_{Lg}^c variations in some regions.

The low Q_μ at depth beneath the Arabian Peninsula causes Q_{Lg}^c to be relatively low there, but not low as in the Andes or Basin and Range province (Fig. 1) that are currently tectonically active. Upper mantle temperatures beneath the Arabian Peninsula are known to be elevated because of current and recent orogenic activity (MCGUIRE and BOHANNON, 1989). Measured heat flow is, however, slightly lower than the continental average (GETTINGS et al., 1986) and the crust is not undergoing significant deformation. CONG and MITCHELL (1998a, b) explained the reduced Q there as resulting from fluids that were generated by upper mantle heating and dehydration and have traveled upward into the crust. These crustal fluids may be less voluminous than those in the crusts of the Andes or Basin and Range which are being actively deformed and where meteoric fluids may also contribute to reduced Q_μ.

Factors other than Q_μ in the deep crust may also affect regional variations of Q_{Lg}^c. MITCHELL and HWANG (1987), for instance, demonstrated that thick accumulations of sediments, especially relatively young (Mesozoic and younger) sandstone and shale, can significantly reduce Lg and Lg coda amplitudes in regions where Q_μ in the deeper crust is high. In regions where Q_μ is low in the deeper crust, low-Q sediments have little effect on measured Q_{Lg}^c.

A factor, unrelated to crustal anelasticity, that can attenuate Lg amplitudes, and by inference Lg coda amplitudes, is the existence of a velocity gradient, rather than a sharp interface at the crust-mantle transition (BOWMAN and KENNETT, 1991). Such gradients have been shown to be prevalent beneath Australia (COLLINS, 1991) and to occur also beneath the southeastern portion of the United States (OWENS et al., 1987). Such gradients reduce Q for direct Lg waves by allowing S-wave energy to leak into the mantle. If the same percentage of Q reduction occurs for Lg coda as for direct Lg, BAQER and MITCHELL (1998) showed that gradients, such as those inferred in Australia and the southeastern United States, can reduce Q_0 by as much as 20%. Correcting for a reduction of this magnitude would place the point for the Tasman Fold Belt (I) very close to the best-fit line. It would also increase Q_0 for the cratonic regions of Australia, but values would still be somewhat lower than those for the best-fit line.

The two gradient models in Figure 3, TCMI-2 and TCMI-3 (FINLAYSON, 1982), exhibit gradients of somewhat different thickness and severity. Lateral variations in the thickness, depth, and severity of velocity gradients, such as these, at the crust-mantle boundary might cause S-wave energy, in excess of that caused by a gradient that is invariant with geographic location, to leak into the mantle. Both of these gradient models were obtained within the North Australian Craton, using shot points separated by only about 450 km. If such variations are pervasive throughout Australia, they may contribute to the reduced Q_0 observed for the continent.

Conclusions

Lg coda *Q* at 1 Hz varies between about 200 or less and more than 1000 throughout the continental regions of the world. Well-determined mean values for selected tectonic regions exhibit a trend that varies between about 1000 for the oldest cratons (2200 My BP) to about 250 for regions that are still tectonically or orogenically active. This variation can be largely attributed to regional variations in intrinsic Q_μ in the crust. Q_{Lg}^c values at 5 Hz can be obtained by extrapolation from Q_0 and η determined at 1 Hz by assuming that η is constant everywhere between 1 and 5 Hz. If these estimates are reliable they indicate that the percentage-wise variation of Q_{Lg}^c at 5 Hz is similar to that at 1 Hz.

Increases in Q_μ can be explained by the loss of crustal fluids that were originally generated by hydrothermal reactions during orogenesis or tectonic activity, and which have subsequently been lost either by migration to the surface or retrograde metamorphism, or have become less mobile because of loss of permeability.

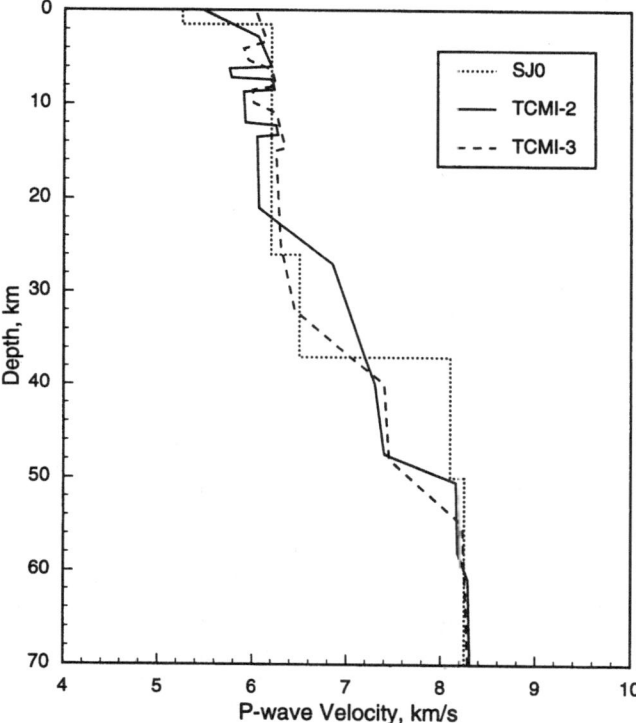

Figure 3

P-wave velocity models for the North Australian Craton. Models TCMI-2 and TCMI-3 are models (with gradients at the crust-mantle transition) obtained from seismic refraction studies along two paths in the craton (FINLAYSON, 1982) and model SJO is a model with a sharp crust-mantle interface (BOWMAN and KENNETT, 1991).

Additional factors that can contribute to reductions in Q_0 include accumulations of sediments, especially if they are young and permeable, and the existence of severe velocity gradients at the crust-mantle boundary. The gradients can cause S-wave energy to leak into the mantle. Lateral variations in the depth, thickness, and severity of those gradients may scatter additional S-wave energy into the mantle and contribute to reductions of observed Q_0.

Acknowledgments

We thank Martha House, John Encarnacion, and David Kirschner for helpful discussions on the geology and tectonics of various regions of the world. Jack Xie, Yu Pan, Dibo Chen, Lianli Cong, and Jorge de Souza compiled various portions of the Lg coda Q information used in this summary. This work was partially supported by the Department of Energy and was monitored by Phillips Laboratory under contract F19628-95-K-0004.

REFERENCES

BAQER, S., and MITCHELL, B. J. (1998), *Regional Variation of Lg Coda Q in the Continental United States and its Relation to Crustal Structure and Evolution*, Pure appl. geophys., *153*, 613–638.

BOWMAN, J. R., and KENNETT, B. L. N. (1991), *Propagation of Lg Waves in the North Australian Craton: Influence of Crustal Velocity Gradients*, Bull. Seismol. Soc. Am. *81*, 592–610.

COLLINS, C. D. N. (1991), *The Nature of the Crust-mantle Boundary under Australia from Seismic Evidence*, Geol. Soc. Aust. Spec. Publ. *17* (ed. B. J. Drummond) 67–80.

CONG, L., and MITCHELL, B. J. (1998a), *Lg Coda Q and its Relation to the Geology and Tectonics of the Middle East*, Pure appl. geophys., *153*, 563–585.

CONG, L., and MITCHELL, B. J. (1998b), *Seismic Velocity and Q Structure of the Middle Eastern Crust and Upper Mantle from Surface-wave Dispersion and Attenuation*, Pure appl. geophys., *153*, 503–538.

FINLAYSON, D. M. (1982), *Seismic Crustal Structure of the Proterozoic North Australian Craton between Tennant Creek and Mount Isa*, J. Geophys. Res. *87*, 10569–10578.

GETTINGS, M. E., BLANK, Jr., H. R., MOONEY, W. D., and HEALY, J. H. (1986), *Crustal Structure of Southwestern Saudi Arabia*, J. Geophys. Res. *91*, 6491–6512.

GOODWIN, A. M., *Precambrian Geology, The Dynamic Evolution of the Continental Crust* (Academic Press, San Diego, CA. 1991) 666 pp.

McGUIRE, A. V., and BOHANNON, R. G. (1989), *Timing of Mantle Upwelling: Evidence for a Passive Origin for the Red Sea Rift*, J. Geophys. Res. *94*, 1677–1682.

MITCHELL, B. J. (1991), *Frequency Dependence of Q_{Lg} and its Relation to Crustal Anelasticity in the Basin and Range Province*, Geophys. Res. Letts. *18*, 621–624.

MITCHELL, B. J. (1995), *Anelastic Structure and Evolution of the Continental Crust and Upper Mantle from Seismic Surface Wave Attenuation*, Rev. Geophys. *33*, 441–462.

MITCHELL, B. J., and HWANG, H. J. (1987), *Effect of Low Q Sediments and Crustal Q on Lg Attenuation in the United States*, Bull. Seismol. Soc. Am. *77*, 1197–1210.

MITCHELL, B. J., and XIE, J. (1994), *Attenuation of Multiphase Surface Waves in the Basin and Range Province, III, Inversion for Crustal Anelasticity*, Geophys. J. Int. *116*, 468–484.

MITCHELL, B. J., PAN, Y., XIE, J., and CONG, L. (1997), *Lg Coda Q Variation across Eurasia and its Relation to Crustal Evolution*, J. Geophys. Res. *102*, 22767–22779.

MITCHELL, B. J., AKINCI, A., and CONG, L. (1998), *Lg coda Q in Australia and its Relation to Crustal Structure and Evolution*, Pure appl. geophys., *153*, 639–653.

OWENS, T. J., TAYLOR, S. R., and ZANDT, G. (1987), *Crustal Structure at Regional Seismic Test Network Stations determined from Inversion of Broadband Teleseismic P Wave Forms*, Bull. Seismol. Soc. Am. *77*, 631–662.

SOUZA, J. L. DE, and MITCHELL, B. J. (1998), *Lg coda Q Variation across South America and their Relation to Crustal Evolution*, Pure appl. geophys. *153*, 587–612.

STANLEY, S. M., *Earth and Life through Time* (W. H. Freeman and Co., New York, NY 1986) pp. 690.

WYLLIE, P. J. (1988), *Magma Genesis, Plate Tectonics, and Chemical Differentiation of the Earth*, Rev. Geophys. *26*, 370–404.

XIE, J., and MITCHELL, B. J. (1990), *A Back-project Method for Imaging Large-scale Lateral Variations of Lg coda Q with Application to Continental Africa*, Geophys. J. Int. *100*, 161–181.

XIE, J., and NUTTLI, O. W. (1988), *Interpretation of High-frequency Coda at Large Distances: Stochastic Modeling and Method of Inversion*, Geophys. J. *95*, 579–595.

(Received March 5, 1998, revised August 17, 1998, accepted August 25, 1988)

To access this journal online:
http://www.birkhauser.ch

Pure appl. geophys. 153 (1998) 665–683
0033–4553/98/040665–19 $ 1.50 + 0.20/0

| Pure and Applied Geophysics

Comparison of Seismic Body Wave and Coda Wave Measures of Q

GOLAM SARKER[1,2] and GEOFFREY A. ABERS[1]

Abstract—Measurements of seismic attenuation (Q^{-1}) can vary considerably when made from different parts of seismograms or using different techniques, particularly at high frequencies. These discrepancies may be methodological, or may reflect earth processes. To investigate this problem, we compare body wave with coda Q^{-1} results utilizing three common techniques: i) parametric fit to spectral decay, ii) coda normalization of S waves, and iii) coda amplitude decay with lapse time. Q^{-1} is measured from both body and coda waves beneath two mountain ranges and one platform, from recordings made at seismic arrays in the Caucasus and Kopet Dagh over paths $\leq 4°$ long. If Q is assumed frequency independent, spectral decay fits show Q_s and Q_{coda} near 700–800 for both mountain paths and near 2100–2200 for platform paths. Similar values are determined with the coda normalization technique. However, frequency-dependent parameterizations fit the data significantly better, with $Q_s(1 \text{ Hz})$ and $Q_{coda}(1 \text{ Hz})$ near 200–300 for mountain paths and near 500–600 for platform paths. Lapse decay measurements are close to the frequency-dependent values, showing that both spectral and lapse decay methods can give similar results when Q has comparable parameterizations. Above 6 Hz, coda measurements suggest some enrichment relative to body waves, perhaps due to scattering, but intrinsic absorption appears to dominate at lower frequencies. All approaches show sharp path differences between the Eurasian platform and adjacent mountains, and all are capable of resolving spatial variations in Q.

Key words: Seismic attenuation, body wave, coda wave, spectral decay technique, lapse time decay technique, Greater Caucasus, Kopet Dagh.

Introduction

Seismic wave amplitudes decay with distance at a rate greater than predicted by geometric spreading. Several different approaches have been used to quantify the attenuation from different parts of seismograms (e.g., AKI, 1969, 1980a; HERRMANN, 1980; MITCHELL, 1995), each of which could be sensitive to different physical processes (e.g., anelastic absorption and scattering). At regional distances ($<10°$) and high frequencies (>1 Hz), the most common methods include (1) those that parameterize the source and fit body wave spectra (e.g., HOUGH *et al.*, 1988;

[1] Department of Geology, University of Kansas, Lawrence, KS 66045, U.S.A.
[2] Now at Department of Earth and Environmental Sciences, Lehigh University, Bethlehem, PA 18015, U.S.A. E-mail: gms5@lehigh.edu

BOATWRIGHT, 1978); (2) methods that cancel the seismic source through spectral ratio of different parts of seismograms (e.g., FRANKEL et al., 1990; AKI, 1980a); (3) methods that measure coda amplitude decay with increasing lapse time (e.g., AKI and CHOUET, 1975); and (4) methods that use a nearby smaller event as an empirical Green's function (e.g., HOUGH, 1997). Some methods assume frequency-dependent Q, while others assume a frequency-independent Q. These different approaches and descriptions of attenuation do not always lead to the same results. For example, in the Caucasus, the Q_s values measured from spectral fall-off (SARKER and ABERS, 1998) are ~3 times larger than Q_{coda} measured from lapse time decay (RAUTIAN et al., 1979). Large differences are also seen in southern California between the coda Q values of ~200 of SINGH and HERRMANN (1983) and body wave Q values of 800 and 1025, determined by FRANKEL et al. (1990) and HOUGH et al. (1988), respectively.

Because these attenuation measurements are generally made by different practitioners, it is unclear whether methodological details or wave propagation processes cause the discrepancies. In this paper, we use seismic data across two moutain belts and one platform in central Eurasia to compare Q measures of S-wave attenuation with coda results, using i) spectral decay, ii) coda normalization, and iii) lapse time decay methods (see Table 1 for Q symbols used in this paper). Both frequency-dependent and frequency-independent parameterizations of Q are considered. Because both body waves and coda waves contain considerable energy at frequencies between 1 and 12 Hz (Fig. 1), we can systematically compare the different methods from different parts of seismograms. A second goal of this study is to better understand the utility of absolute Q_s and Q_{coda} numbers in constraining spatial variations in attenuation. Our results confirm the previously determined low Q through the mountain belts north of the Iranian plateau, and high Q beneath the

Table 1

Q symbols

Methods	Wave types	Q designation
Parameterized source and spectral decay (frequency-independent Q)	S coda	Q_{si} Q_{ci}
Parameterized source and spectral decay (frequency-dependent Q)	S coda	Q_{sf} Q_{cf}
Coda normalized	S	Q_{sn}
Lapse time decay (frequency-dependent Q)	coda	Q_{cfl}

- Q_s and Q_{coda} are general terms for Q of S and coda waves.
- Q_0 is a general term for Q at 1 Hz.
- α_s frequency exponent for S from spectral method.
- α_c frequency exponent for coda from lapse time decay method.

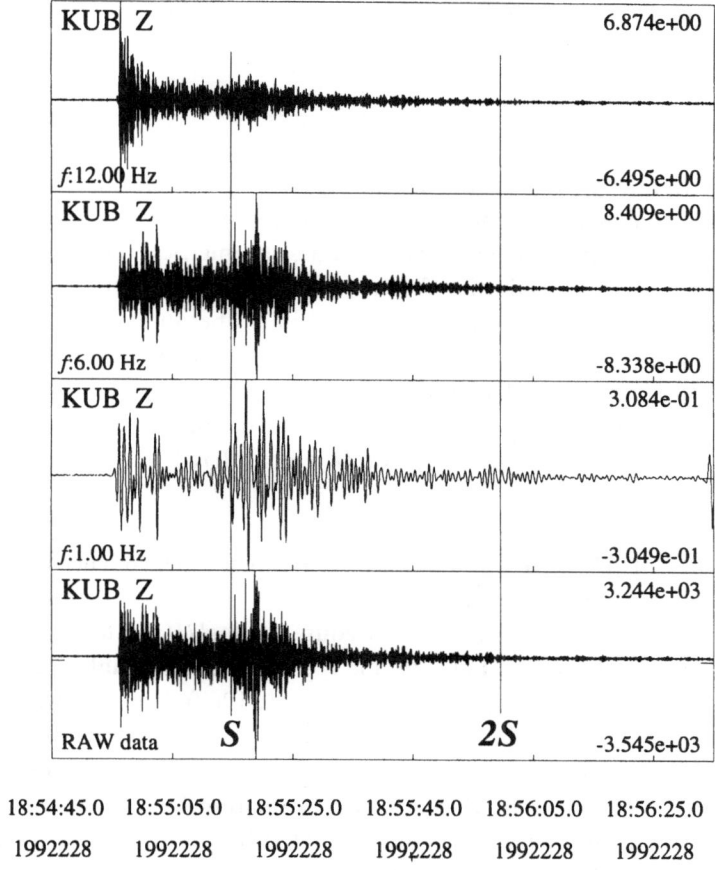

Figure 1

Vertical component seismogram recorded at station KUB of the Caucasus network, comparing energy above noise level for *S* and coda waves, at representative frequency bands (*f*), labeled. Coda windows start at twice the *S* travel time (marked 2*S*). Amplitudes are in nm/s, shown right corners.

adjacent platform. We also find that the data are more consistent with a frequency-dependent description of Q, and that all methods give comparable Q values for both coda and body waves when the frequency dependence is taken into account.

Attenuation Measurement Methods

Parameterized Source and Spectral Decay

For regional paths, spectral fall-off of body waves can be inverted for attenuation along with source parameters in a parametric scheme (e.g., LEES and LINDLEY, 1994; SARKER and ABERS, 1998). The spectral fitting scheme, based on the approach of HOUGH *et al.* (1988) and BOATWRIGHT (1978), fits a three-parameter source model $H(f)$ to observed amplitude spectra, where

$$H(f) = \frac{\Omega_0 I(f)\, e^{-\pi f t^*}}{1 + (f/f_c)^2}.$$ (1)

Here, f is the frequency, $I(f)$ is the instrument response to ground displacement, Ω_0 is the long-period displacement, f_c is the source corner frequency and t^* is the attenuation parameter. The Ω_0 term includes geometrical spreading, radiation pattern, seismic moment, and other frequency-independent effects. These frequency-independent effects are not explicitly included in (1) because spectral fits are done at a single lag time, so they would not affect t^* estimates. The exponent of f_c is assumed to be 2, chosen to facilitate comparison with other studies.

The method allows both frequency-independent and frequency-dependent t^* measurements. Frequency dependence of Q is considered by parameterizing t^* as

$$t^*(f) = t_0^* f^{-\alpha}$$ (2)

where t_0^* represents t^* at 1 Hz and α describes frequency dependence (e.g., LINDLEY and ARCHULETA, 1992; MITCHELL, 1980).

In each inversion, a least-squares procedure fits $\log[H(f)]$ to observed spectra, and determines t_0^*, Ω_0 and f_c (Figs. 2b, c). We measure amplitude spectra in 3–5 s windows using a multitaper algorithm (PARK *et al.*, 1987), from *SH* waves and from coda waves starting at twice the S travel time ($2T_s$). We also measure pre-arrival noise spectra. Evaluations of signal-to-noise ratios determine the usable frequency band for each signal, typically 1 to 12 Hz. When evaluating frequency dependence, the source and propagation parameters are estimated by fitting spectra for a range of α values, and choosing the α value that minimizes least-squares residual for the entire data set.

The observed variation of t_0^* with travel time (T) is used to estimate spatial attenuation by fitting it to

$$t_0^*(T) = t_n^* + \frac{T}{Q}$$ (3)

(HOUGH *et al.*, 1988). The observations of $t_0^*(T)$ are inverted to determine $1/Q$ and t_n^* and associated uncertainties. Here, t_n^* represents a near-surface contribution, and

---

Figure 2

Examples showing fitting procedures to measure attenuation parameters for a Caucasus mountain path record, similar to Figure 1. (a) Transverse component seismogram showing S, coda and lapse-method envelope windows along with pre-event ($n1$) and pre-arrival ($n2$) noise windows. Spectra fitting procedures for S and coda are shown in (b) and (c). The estimated t^* and f_c are in boxes. (d) shows fitting to S-to-coda spectral ratio. The δt^* and corresponding $1/Q_{sn}$ are also shown. (e) Lapse time decay envelope at 1 Hz, along with fitting parameters. Plots are in log-linear scale.

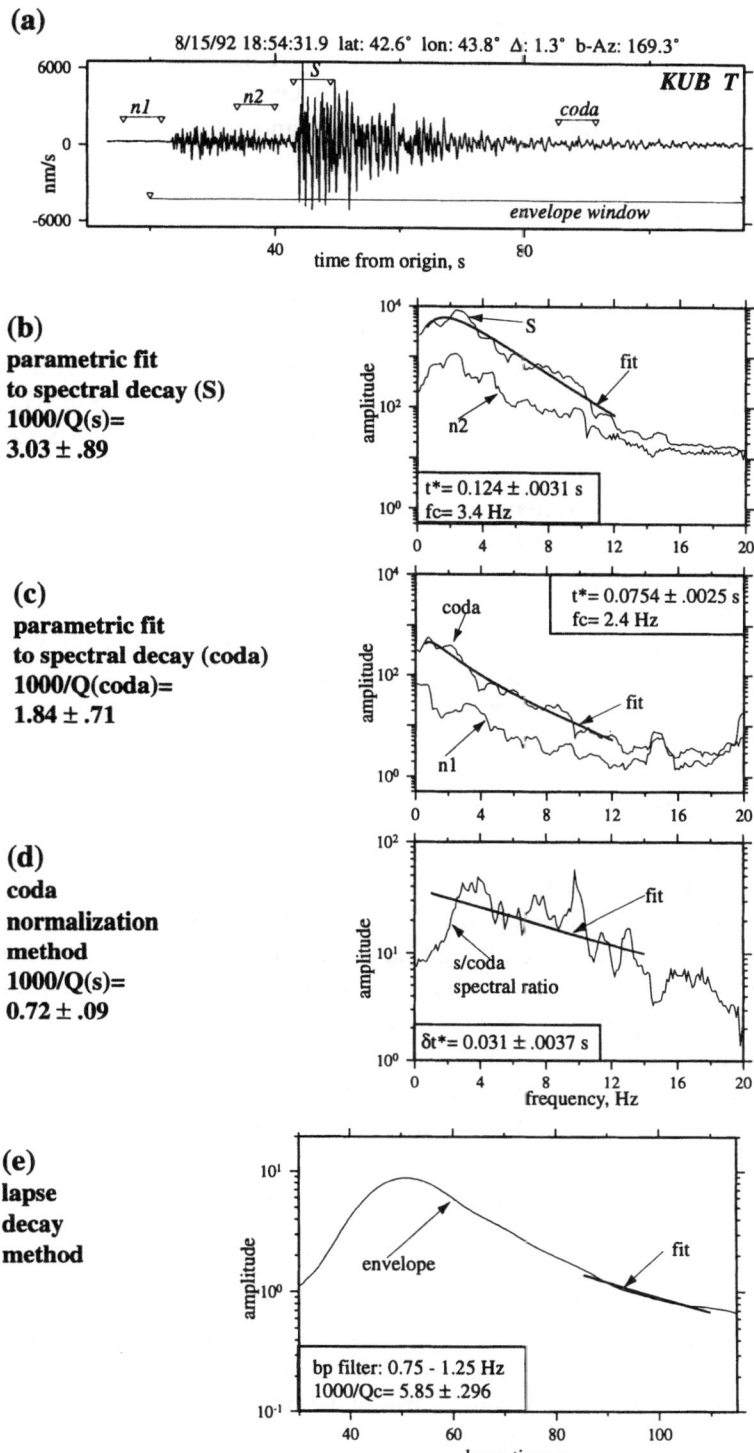

(a)

(b)
parametric fit
to spectral decay (S)
1000/Q(s)=
3.03 ± .89

(c)
parametric fit
to spectral decay (coda)
1000/Q(coda)=
1.84 ± .71

(d)
coda
normalization
method
1000/Q(s)=
0.72 ± .09

(e)
lapse
decay
method

Q is the path-averaged attenuation. Deeper Q variations require more extensive parameterization but are often less significant, so are not explored here. Travel time (T) is calculated from the local velocity models. Other uncertainties and details of this procedure are described elsewhere (SARKER and ABERS, 1998).

Coda Normalization

As an alternative to the parametric inversion of source effects, the S-wave spectra are normalized to coda spectra to estimate $1/Q_{sn}$ following FRANKEL et al. (1990). The underlying assumption of this method is that both S and coda waves exhibit similar source excitation, so that source effects are removed by normalizing S spectra to those of coda. The spectral ratio also cancels common instrument and site responses (e.g., AKI, 1980a; FRANKEL et al., 1990) to give a differential attenuation parameter (δt^*) between S and coda.

For each record, we evaluate spectral ratios at frequencies between 1 and 12 Hz in a manner similar to that described in the previous section. Coda spectra are measured at $2T_s$ in a 3–5 s long window, matching the window length used in evaluating S wave spectra. The logarithm of spectral amplitude ratio is then inverted to calculate δt^* for each record (Fig. 2d), and the attenuation term $1/Q_{sn}$ is determined from δt^* measurements in a manner similar to (3). However, T in (3) now represents the lag time difference between S and the coda window at $2T_s$.

Lapse Time Decay

Temporal decay of coda amplitude also can be inverted for apparent attenuation (e.g., AKI and CHOUET, 1975; ROECKER et al., 1982). Coda amplitude measurements have been compared to a single-scattering model (e.g., AKI and CHOUET, 1975) and to several multiple-scattering models (e.g., FRANKEL and WENNERBERG, 1987; JIN et al., 1994). Because the relationship between coda waves and their sampling path is a complex and controversial subject, it is not clear which of these scattering models is most appropriate. For simplicity, in this study we follow the most common approach, the single-scattering model; our results suggest that this approach sufficiently describes the first-order phenomenology of coda decay.

For single scattering, coda amplitudes decay with lapse time as

$$A(f, T) = A_0(f)T^{-\nu}e^{-\pi fT/Q_{cfl}} \tag{4}$$

(e.g., AKI and CHOUET, 1975; RAUTIAN and KHALTURIN, 1978) where Q_{cfl}^{-1} represents coda attenuation, A_0 is the source factor that includes all lag-independent contributions to attenuation, f is the frequency, ν describes geometric spreading, and T is the elapsed time since earthquake origin time. The approach is valid at lags past twice the S travel time (e.g., AKI, 1980a). Here, Q_{cfl}^{-1} is assumed to vary with frequency in a manner identical to t^* in (2). We estimate Q_{cfl}^{-1} and A_0 at each of a series of frequencies independently, by fitting observed amplitude decay to (4).

We filter each seismogram through a set of 6-pole zero-phase filters centered at 0.5, 1.0, 2.0, 4.0, 6.0 and 8.0 Hz. The narrow-band records are squared in a 25 s window starting at $2T_s$, averaged over three components, and smoothed with a 5 s wide Gaussian operator (Fig. 2e). At each frequency the signal envelopes are compared to similarly filtered pre-event noise, and rejected if signal-to-noise ratio is less than 2. Each envelope gives an estimate of Q_{cfl}^{-1} and A_0 by least-squares regression on $\log[A(f)]$, using (4). Final values represent averages of individual measures of Q_{cfl}^{-1} over all records for each data subset.

The geometrical spreading factor v has been suggested to trade off with $Q(f)$ and α (e.g., FRANKEL, 1991). Tests with our data (not shown) show no consistent correlation between estimates of α and choices of v, and all reasonable choices of v (<2) require frequency-dependent attenuation ($\alpha > 0$ at 95% confidence levels). Estimates of Q_{cfl}^{-1} systematically decrease with increasing v at all frequencies. In the estimates of Q_{cfl}^{-1} below, we choose $v = 1$ to facilitate comparison with the many other studies that use this value (e.g., AKI, 1980a; JIN et al., 1994).

In the spectral fits, near-surface effects are explicitly included as t_n^* in (3). Sampling near-surface layers by coda waves is probably complex and a full treatment is beyond the scope of this paper. However, the Q_{cfl}^{-1} estimates are only sensitive to changes in amplitude with increased lapse time, not time-invariant processes, so any near-receiver amplitude biases such as described by t_n^* in (3) should have no effect in the coda Q estimates. Such effects are implicitly included in the A_0 term of (4) and need not be independently included as parameters. The similarity of body-wave and coda spectra (Fig. 2) also suggests that simple Q models are adequate.

Data Selection and Description

Caucasus Seismic Network

The Caucasus network (CNET) operated on the northern flank of the Greater Caucasus (Fig. 3, top) from 1991 to 1994 as a cooperative program between IRIS's Joint Seismic Program (JSP), Lamont, and OME-Obninsk (Russian) Science group (ABERS, 1994). The network consisted of six stations, of which three northern stations (MIC, NAG, KUB) lied on flat-lying Neogene sediments, while the rest of the stations (KIV, GUM, KNG) lied on tilted, well-lithified Mesozoic limestones and sandstones. The six three-component intermediate-period sensors recorded signals at 60 samples (sps) in an event-triggered mode, giving a flat response to velocity at frequencies between 0.2 and 24 Hz. In this paper, we select wave forms with good signal-to-noise level above 0.25 Hz and with no telemetry gaps within the

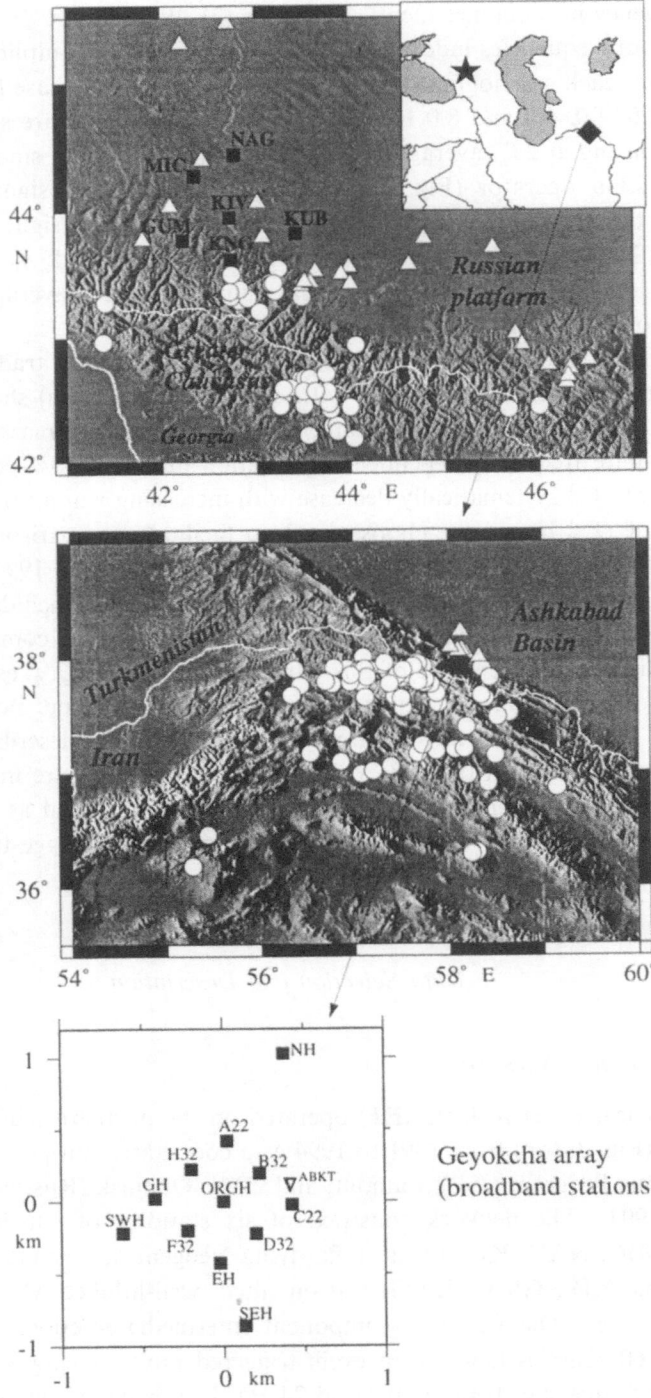

Geyokcha array
(broadband stations)

selected windows. For 1992, these criteria gave a total of 67 events within 4° of the network to determine Q measurements. The data can be subdivided into mountain and platform paths, which exhibit very different Q. The body-wave attenuation results from these data have been reported previously (SARKER and ABERS, 1998) and further details are given therein.

Geyokcha Array and ABKT Station

The tight-aperture (~ 1 km) Geyokcha, Turkmenistan broadband array operated in 1993 and 1994 along the northern foothills of the Kopet Dagh mountains (Fig. 3, bottom) through a cooperative project between the IRIS's JSP, several US institutions and Turkmen Science groups (e.g., PAVLIS et al., 1994; AL-Shukri et al., 1995). The array included 12 digitally telemetered broadband Streckheisen STS-2 sensors as well as numerous short-period sensors and geophone strings. In this study, we make use of the broadband channel only, sampling at 250 sps. Beams formed from these data are used, giving from one year a total of 318 events within 4° of the array (G. PAVLIS, personal communication, 1996). We select 106 of these events for determining attenuation parameters, following the same selection criteria as to the Caucasus data set. Because the triggered windows of Geyokcha records are short and do not include late phases of seismograms, they are supplemented with wave forms from IRIS's Global Seismic Network station ABKT for the 106 events to analyze coda signals. Station ABKT lies within the Geyokcha array (Fig. 3, bottom right). Because few recorded events lie on the adjacent stable Turan platform, we only compare results for mountain paths in the Kopet Dagh.

Hence, in this paper we compare Q_s to Q_{coda} from three sets of data. Two data sets are selected from the Caucasus network, one of which has paths predominantly in the Russian platform and other has paths in the Greater Caucasus. The third data set is selected from the Geyokcha array for paths that sample the Kopet Dagh mountain range.

Results

To better compare techniques for measuring Q from body and coda waves, we present results separately for two different arrays. For each array, we describe and compare the results of each Q measurement method.

Figure 3

The epicenters, the Caucasus network, and the Geyokcha array. Circles denote events for mountain paths; triangles denote events in the platform; squares are network stations. Topographic roughness of Greater Caucasus (top) and Kopet Dagh (center) is illuminated from west. Inset (top) shows location of the two study areas: Caucasus (star) and Kopet Dagh (diamond). (Bottom) shows broadband stations of Geyokcha array (squares) and ABKT station (inverted triangle).

Table 2

1000/Q values of different methods from CNET, GEYOKCHA and ABKT data set

Methods	Symbols	Craton (CNET)		Mountains (CNET)		Mountains (Geyokcha and ABKT)	
		$1000/Q_s$	$1000/Q_{\text{coda}}$	$1000/Q_s$	$1000/Q_{\text{coda}}$	$1000/Q_s$	$1000/Q_{\text{coda}}$
(a) Parameterized source and spectral decay (frequency-independent Q)	Q_{si} Q_{ci}	0.473 ± 0.085	0.450 ± 0.098	1.385 ± 0.172	1.328 ± 0.189	1.340 ± 0.178	1.364 ± 0.305
(b) Parameterized source decay and spectral decay (frequency-dependent Q)	Q_{sf} Q_{cf}	1.660 ± 0.306 @ $\alpha = 0.3$ $(+0.3/-0.1)$		5.030 ± 0.874 @ $\alpha = 0.5$ $(+0.1/-0.2)$		3.230×0.480 @ $\alpha = 0.3$ (± 0.1)	
(c) Coda normalization	Q_{sn}	0.467 ± 0.134		1.286 ± 0.168		1.535 ± 0.374	
(d) Lapse time decay (frequency-dependent Q)	Q_{cfl}		2.030 ± 0.090 @ $\alpha = 0.35 \pm 0.08$		4.500 ± 0.170 @ $\alpha = 0.71 \pm 0.07$		5.110 ± 0.130 @ $\alpha = 0.65 \pm 0.03$

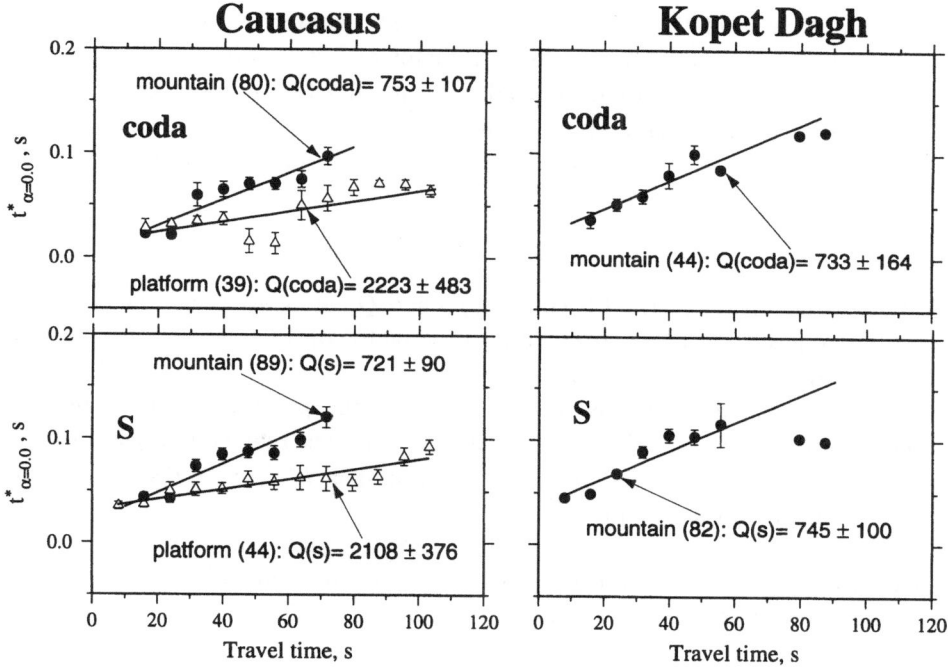

Figure 4

t^* distribution as a function of travel time (T), calculated from spectral fall-off of body and coda waves for frequency-independent parameterization ($\alpha = 0$). Mountain paths: circles; platform paths: triangles. t^* estimates are averaged over $0.5°$ bins. Error bars 1σ standard deviation. Numbers in parenthesis indicate observations. Q values are calculated as described in text.

Caucasus

In each set of spectral measurements, t_0^* gradually increases with distance, giving positive Q. A linear regression on $t_0^*(T)$, following equation (3), gives Q values that are 2–3 times larger in the platform than for mountains, for $\alpha = 0$. For both mountain and platform data sets, Q_{si} is roughly equal to Q_{ci} (Table 2; Fig. 4, left). These regressions also yield a near-surface term, t_n^* in (3), that is variable but always positive.

The fits to the spectra are better when α is positive (Fig. 5). The constraint on α has been evaluated by fitting spectra using (1) and (2) over several assumed α values, allowing t_0^*, Ω_0 and f_c to vary. For the Caucasus, F tests show that the best constrained α values are 0.5 $(+0.1/-0.2)$ for mountain paths and 0.3 $(+0.3/-0.1)$ for platform paths at 95% confidence (Fig. 5). Hence, the data require Q be frequency dependent. In neither Caucasus nor Kopet Dagh data sets could we reliably constrain Q_{cf}, so we only consider Q_{ci}, the result for $\alpha = 0$. Additional tests show that reasonable changes to the geometrical spreading term (v) or the source corner frequency roll off (f_c^2) do not alter this conclusion.

Mountain paths of the Greater Caucasus show greater frequency dependence than adjacent Russian platform paths by a factor of 2 in α (Fig. 6, left). For both frequency-dependent and frequency-independent Q, the attenuation contrasts between platform and mountains are similar, as are fractional uncertainties. Q_{sf} can be compared to Q_{si} at a given frequency by estimating its implied variation of Q with frequency:

$$Q_{sf}(f) = Q_{sf}f^{\alpha} \qquad (5)$$

at the frequency of interest. At 8 Hz, $Q_{sf}(f)$ and Q_{si} are comparable in both data subsets. At one Hz, $Q_{sf}(f)$ and Q_{si} differ by a factor of 3 (Table 2). This suggests that Q_{si} is controlled by signals above 8 Hz.

The coda normalized Q_{sn} is roughly equal to the parameterized Q_{si} in both subsets (Table 2; Fig. 7, left). This similarity indicates that source assumptions made in the source parameter inversion (for example, f^{-2} source decay) have no significant effect on our results.

When Q is determined from lapse time decay, the data require frequency dependence (Fig. 8). For lapse rate measurements, the path differences between mountain and platform are again 2–3 times in Q (Fig. 8, left). These

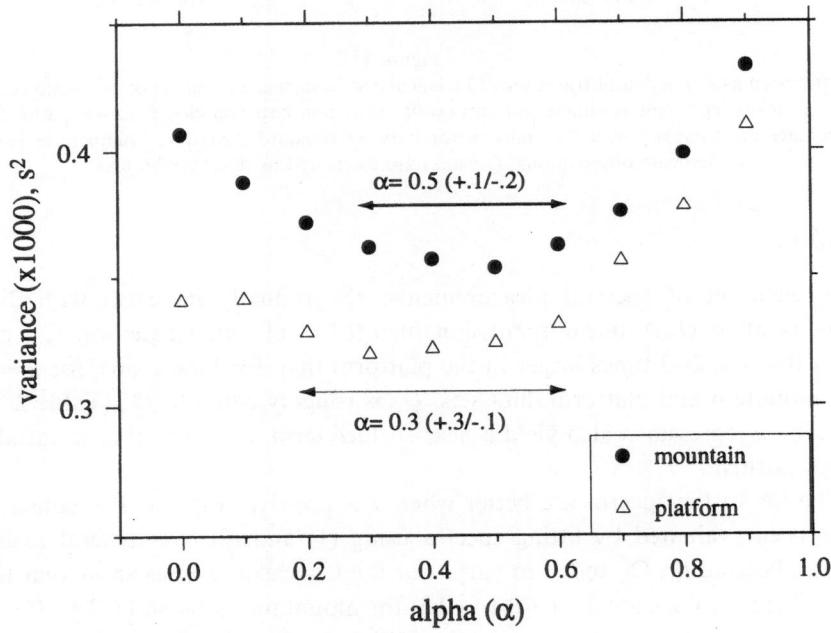

Figure 5

Variances in the t^* estimates are shown as a function of α for both mountain and platform paths. The results are shown from CNET data set. The double arrows show 95% confidence limits calculated from F tests on residual variances. In both data sets, α is significantly greater than zero, and hence frequency dependence of Q is suggested.

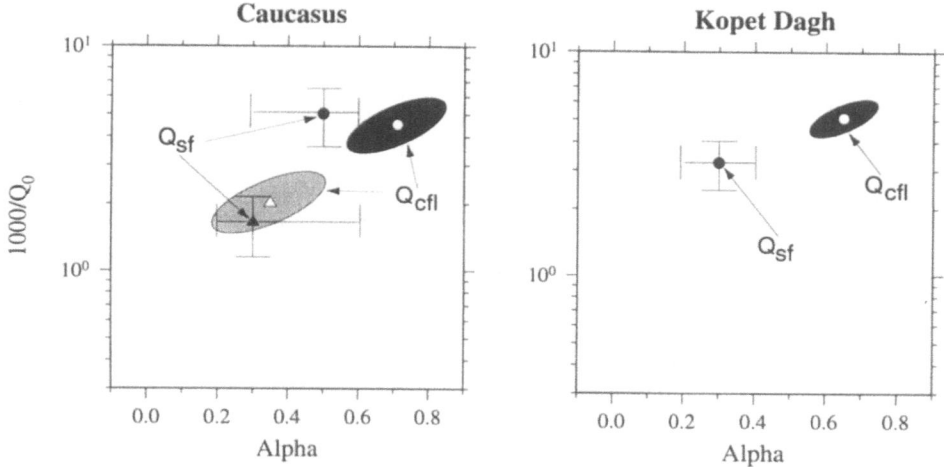

Figure 6

Spectral $1000/Q_0$ estimates for frequency-dependent parameterization (Q_{sf}) (mountain: solid circles; platform: solid triangles). Based on F test of the minimum residual for each α (calculated while determining t^*), the best constrained α vaues are shown at 90% confidence bounds (horizontal error bars). The vertical error bars for Q_{sf} are shown at 1.65σ (equivalent 90% confidence bounds with one degree of freedom). Also shown, error ellipses of Q_{cfl} for mountain paths (dark) and platform paths (light). Error ellipses are calculated at 2.15σ (equivalant 90% confidence bounds with two degrees of freedom). Blank circles and triangles within ellipses are $1000/Q_0$ of coda decay.

frequency-dependent Q_{cfl} values are roughly equal to Q_{sf} values. The frequency exponents (α) for Q_{cfl} and Q_{sf} are also similar. Lapse time decay estimates show $\sim 10-15\%$ smaller fractional uncertainties for both Q_0 and α (Table 2). Hence, lapse decay and spectral methods give similar results when Q has comparable parameterizations.

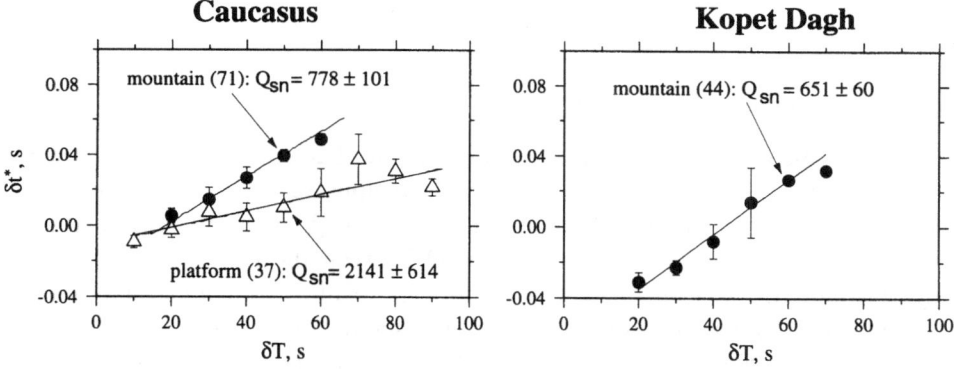

Figure 7

Distribution of differential attenuation parameters (δt^*) as a function of δT from coda normalization method. The δt^* estimates are averaged over $0.5°$ bins. Formats similar to Figure 4.

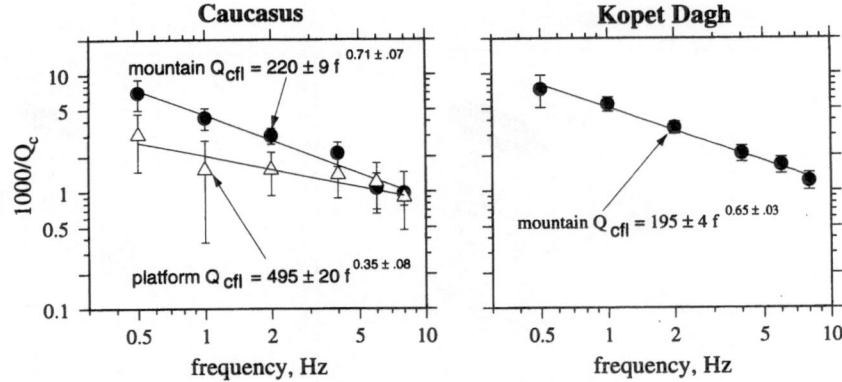

Figure 8

$1000/Q_{coda}$ estimates at different frequency bands, using lapse time decay technique. Error bars are at 1.65σ.

Geyokcha

The same treatment is applied to the Kopet Dagh data. The frequency-independent Q estimates Q_{si}, Q_{ci}, and Q_{sn} are all roughly equal (Figs. 4 and 7, right). The frequency exponents (α) for Q_{sf} and Q_{cfl} do not agree at 90% confidence limits for this data set (Fig. 6, right). However, comparisons of frequency variation in Q show $Q_{sf}(f)$ is similar to $Q_{cfl}(f)$ for most frequencies below 8 Hz (Fig. 9, right). High frequencies appear to be biasing α estimates here. Otherwise, we find similar Q values for both body and coda waves, for comparable parameterizations. For both frequency-independent and frequency-dependent measures of Q, the Q values compare to those from Greater Caucasus, as do fractional uncertainties (Table 2), indicating that the two mountain ranges have similar Q structures.

Discussion

Regional Variations of Q_s and Q_{coda}

Measurements of attenuation variation indicate that attenuation in tectonically active regions greatly exceeds that of stable regions (e.g., NUTTLI et al., 1979; MITCHELL, 1995). Our measurements of Q_{coda} and Q_s similarly show that these variations are roughly a factor of 2–3 in Q between the Eurasian platform and adjacent mountains at all frequencies considered. Values are comparable to those determined in other tectonically active and stable areas (e.g., FLETCHER and BOATWRIGHT, 1991; FRANKEL et al., 1990; SINGH and HERRMANN, 1983), and to our earlier Q estimates from P waves (SARKER and ABERS, 1998). Our frequency-dependent measures of Q_s and Q_{coda} are similar to Lg and Lg-coda Q estimates from central Eurasia described elsewhere at comparable frequencies (e.g.,

MITCHELL et al., 1996; XIE et al., 1996). Hence, for all wave types studied here, attenuation contrasts between mountains and platforms are sharp and similar in to those found in other parts of the world.

Similarity of S and Coda Q Values

Although Q determined from any particular method sometimes gives comparable body and coda Q values (e.g., AKI, 1980a,b), in many cases it is found that different techniques do not always give the same results, and the differences can be large. It is sometimes suspected that differences between coda and body wave attenuation processes are responsible, for example because scattering processes affect the two wave types differently. However, this discrepancy seems attributable to methodological differences here, at least at frequencies <6 Hz.

We illustrate this point by comparing S and coda Q values in two ways. In one approach, we compare best estimates of $1/Q_0$ and α (Fig. 6). Here, α_c (from coda lapse decay) is constrained from linear regression of the coda decay measurements

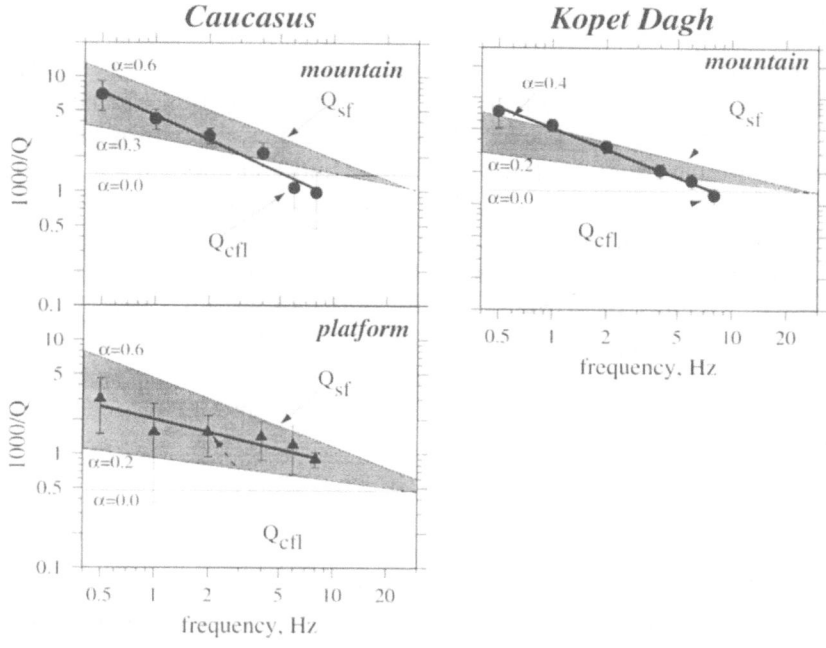

Figure 9

A comparison between $Q(f)$ measured from S spectra and that measured from lapse time decay (coda) methods. Solid circles and triangles are coda results with 1.65σ error bar, from Figure 7. Shaded regions show body-wave spectral measures at 90% confidence bounds in α, calculated according to (5) of text. These show similar rate of attenuation at frequencies below 6 Hz. At frequencies above 6 Hz, some high frequency coda enrichment are seen where $Q_{cfl} > Q_{sf}$.

(Fig. 8), while α_s (from S spectra) is constrained by minimizing the global variance of t_0^* measurements for each α value using (3). In the second approach, we examine variations of $1/Q$ with frequency (Fig. 9). Here, the coda measures are the original $1/Q_{cfl}(f)$ results according to (4), while body wave $1/Q(f)$ is calculated according to (5) at 90% confidence bounds on α described earlier. This second approach compares predictions from S spectra to coda lapse decay observations. In both of these comparisons, we find a close agreement between $Q(f)$ measured from S spectra and that measured from coda decay. Above six Hz, coda measures give slightly lower $1/Q$ estimates (Fig. 9); otherwise values are very close.

When Q is assumed to be frequency independent, our results show that Q_{si} is roughly equal to Q_{ci} for all regions. As well, frequency-dependent Q_{sf} is similar to Q_{cfl}. However, the absolute Q_{sf} and Q_{cfl} values, corresponding to Q at 1 Hz, are three times smaller than the frequency-independent measures of Q_{si} and Q_{ci}. This is true for all regions and all methods used. Hence, the discrepancies between Q_s and Q_{coda} seen in some earlier studies may largely reflect the differences between frequency-dependent and frequency-independent descriptions of Q.

Importance of Frequency Dependence

For each inversion at different α values, the variance in t_0^* is calculated. This "best" (minimum) residual estimate is significantly lower for frequency-dependent parameterizations (Fig. 5). The α values of the Eurasian platform are similar to those made in other stable cratons (e.g., SINGH and HERRMANN, 1983; ATKINSON and MEREU, 1992), while values from the Greater Caucasus are consistent with those from other tectonically active regions (AKI, 1980; RAUTIAN and KHALTURIN, 1978). In Kopet Dagh, the α_c values are similar to those made in other tectonically active areas. Spectral and lapse decay methods also show similar Q values only when frequency dependence is taken into account. Otherwise, Q_{cfl} is similar to Q_{si} at frequencies above 8 Hz. Hence, frequency-dependent parameterization is important in describing Q for all data sets examined, and for reconciling different Q estimates.

High-frequency Scattering

These observations have some implications for the attenuation mechanism. In both Caucasus and Kopet Dagh, our results show that Q_{cfl} for mountain paths is slightly ($<40\%$) larger than Q_{sf} at high frequencies (≥ 6 Hz) (Fig. 9). At lower frequencies, both absolute values and path variations for Q_{cfl} are comparable to Q_{sf} determinations. This high-frequency discrepancy indicates a possible high-frequency coda enrichment, suggesting that scattering attenuation may play an important role for the attenuation of S waves at frequencies ≥ 6 Hz. High-frequency coda enrichment has been reported from several other regions (FRANKEL,

1991). However, at frequencies <6 Hz, S and coda show similar Q. For scattering processes such as that described by FRANKEL and WENNERBERG (1987), this implies that intrinsic absorption dominates and scattering is unimportant. This is true for all three data subsets examined.

Conclusions

Spectral fall-off and lapse decay methods give similar attenuation values at frequencies $1-10$ Hz through the active tectonic belts and adjacent platforms of Greater Caucasus and Kopet Dagh. For all methods, regional variations of attenuation are consistent for both body and coda waves, and mountain paths are about three times more attenuating than the adjacent platform paths. Frequency dependence is also observed, with α for platform paths is roughly half that for mountain paths. Measurements of Q_s from spectral parameterization are similar to Q_{coda} from lapse-time decay, but only if frequency dependence is considered. Hence, differing Q_s and Q_{coda} values seen in earlier studies may largely reflect the differences in assumed frequency dependence. Although Q_{sf} values are consistent with Q_{cfl} at most of the frequencies studied, the 40% larger Q_{cfl} values at 6 and 8 Hz indicate that scattering may be significant at these frequencies. Importantly, our results indicate that the body and coda wave measures give comparable Q values at the frequencies studied here, and both frequency-dependent and frequency-independent methods are similarly capable of resolving spatial variations in Q.

Acknowledgments

This work would not have been possible without the effort of those who established and operated the Caucasus network and Geyokcha array: A. Lerner-Lam, D. Lentrichia, G. Pavlis, F. Vernon, and many others. Initial data processing from Caucasus for 1992 was largely done by D. Harvey at the JSP center, and all data were made available through IRIS DMC. G. Pavlis and H. Al-Shukri kindly made available beam-formed data and associated hypocenters from the Geyokcha array. Significant improvement to the manuscript resulted from the comments of Dr. Susan Hough and an anonymous reviewer. This work was supported by AFOSR award F49620-95-1-0002.

REFERENCES

ABERS, G. A. (1994), *The Caucasus Seismic Network*, IRIS Newsletter *13*, 16–17.
AL-SHUKRI, H., PAVLIS, G., and VERNON, F. (1995), *Site Effect Observations from Broadband Arrays*, Bull. Seismol. Soc. Am. *85*, 1758–1769.

AKI, K. (1980a), *Attenuation of Shear-waves in the Lithosphere for Frequencies from 0.05 to 25 Hz*, Phys. Earth Planet. Inter. *21*, 50–60.

AKI, K. (1980b), *Scattering and Attenuation of Shear Waves in the Lithosphere*, J. Geophys. Res. *85*, 6496–6504.

AKI, K., and CHOUET, B. (1975), *Origin of Coda Waves: Source, Attenuation and Scattering Effects*, J. Geophys. Res *23*, 3322–3342.

AKI, K. (1969), *Analysis of the Seismic Coda of Local Earthquakes as Scattered Waves*, J. Geophys. Res. *74*, 615–631.

ATKINSON, G. M., and MEREU, R. F. (1992), *The Shape of Ground Motion Attenuation Curves in Southeast Canada*, Bull. Seismol. Soc. Am. *82*, 2014–2031.

BOATWRIGHT, J. (1978), *Detailed Spectral Analysis of Two Small New York State Earthquakes*, Bull. Seismol. Soc. Am. *68*, 1117–1131.

FLETCHER, J., and BOATWRIGHT, J. (1991), *Source Parameters of Loma Prieta Aftershocks and Wave Propagation Characteristics along the San Francisco Peninsula from a Joint Inversion of Digital Seismograms*, Bull. Seismol. Soc. Am. *81*, 1783–1812.

FRANKEL, A., MCGARR, A., BICKNELL, J., MORI, J., SEEBER, L., and CRANSWICK, E. (1990), *Attenuation of High-frequency Shear Waves in the Crust: Measurements from New York State, South Africa, and Southern California*, J. Geophys. Res. *95*, 17441–17457.

FRANKEL, A. (1991), *Mechanisms of Seismic Attenuation in the Crust: Scattering and Anelasticity in New York State, South Africa, and Southern California*, J. Geophys. Res. *96*, 6269–6289.

FRANKEL, A., and WENNERBERG, L. (1987), *Energy-flux Model for Seismic Coda: Separation of Scattering and Intrinsic Attenuation*, Bull. Sesimol. Soc. Am. *77*, 1223–1251.

HARVEY, D. and others (1994), *Caucasus Network Information Product Triggered Events from January 1, 1992 to November 9, 1992*, version 1.0, IRIS-JSP.

HERRMANN, R. (1980), *Q Estimates Using the Coda of Local Earthquakes*, Bull. Seismol. Soc. Am. *70*, 447–468.

HOUGH, S., ANDERSON, J. G., BRUNE, J., VERNON, F., BERGER, J., FLETCHER, J., HAAR, L., HANKS, T., and BAKER, L. (1988), *Attenuation near Anza, California*, Bull. Seismol. Soc. Am. *78*, 672–691.

HOUGH, S. E. (1997), *Empirical Green's Function Analysis: Taking the Next Step*, J. Geophys. Res. *102*, 5369–5384.

JIN, A., MAYEDA, K., ADAMS, D., and AKI, K. (1994), *Separation of Intrinsic and Scattering Attenuation in Southern California Using TERRAscope Data*, J. Geophys. Res. *99*, 17835–17848.

LEES, J. M., and LINDLEY, G. T. (1994), *Three-dimensional Attenuation Tomography at Loma Prieta: Inversion of t* for Q*, J. Geophys. Res *99*, 6843–6863.

LINDLEY, G., and ARCHULETA, R. (1992), *Earthquake Source Parameters and the Frequency Dependence of Attenuation at Coalinga, Mammoth Lakes, and the Santa Cruz Mountains, California*, J. Geophys. Res. *97*, 14137–14154.

MITCHELL, B., XIE, J., and PAN, Y. (1996), *Attenuation and blockage of Lg in Eurasia*. In *Monitoring a Comprehensive Test Ban Treaty* (eds. E. S. Husebye and A. M. Dainty) pp. 645–654.

MITCHELL, B. (1995), *Anelastic Structure and Evolution of the Continental Crust and Upper Mantle from Seismic Surface Wave Attenuation*, Rev. Geophys. *33*, 441–462.

MITCHELL, B. J. (1980), *Frequency Dependence of Shear Wave Internal Friction in the Crust of Eastern North America*, J. Geophys. Res. *85*, 5212–5218.

NUTTLI, O. W., BOLLINGER, G. A., and GRIFFITHS, D. W. (1979), *On the Relation between Modified Mercalli Intensity and Body Wave Magnitude*, Bull. Seismol. Soc. Am. *69*, 893–910.

PARK, J., LINDBERG, C. R., and VERNON, F. L. (1987), *Multitaper Spectral Analysis of High-frequency Seismograms*, J. Geophys. Res. *92*, 12675–12684.

PAVLIS, G., AL-SHUKRI, H., MAHDI, H., REPKIN, D., and VERNON, F. (1994), *JSP Arrays and Networks in Central Asia*, IRIS Newsletter *13*, 9–12.

RAUTIAN, T. G., KHALTURIN, V. I., and SHENGELIYA, I. S. (1979), *Seismic Coda Envelopes and Assessment of Earthquake Magnitudes in the Caucasus*, Phys. Solid Earth *15*, 393–398.

RAUTIAN, T. G., and KHALTURIN, V. I. (1978), *The Use of the Coda for Determination of the Earthquake Source Spectrum*, Bull. Seismol. Soc. Am *68*, 923–948.

ROECKER, S. W., TUCKER, B., KING, J., and HATZFELD, D. (1982), *Estimates of Q in Central Asia as a Function of Frequency and Depth Using the Coda of Locally Recorded Earthquakes*, Bull. Seismol. Soc. Am *72*, 129–149.

SARKER, G., and ABERS, G. A. (1998), *Deep Structures along the Boundary of a Collisional Belt: Attenuation Tomography of P and S Waves in the Greater Caucasus*, Geophys. J. Int. *133*, 326–340.

SINGH, S., and HERRMANN, R. (1983), *Regionalization of Crustal Coda Q in the Continental United States*, J. Geophys. Res. *88*, 527–538.

XIE, J., CONG, L., and MITCHELL, B. J. (1996), *Spectral Characteristics of the Excitation and Propagation of Lg from Underground Nuclear Explosions in Central Asia*, J. Geophys. Res. *101*, 5813–5822.

(Received October 14 1997, revised May 6, 1998, accepted May 20, 1998)

 To access this journal online:
http://www.birkhauser.ch

Pure appl. geophys. 153 (1998) 685–702
0033–4553/98/040685–18 $ 1.50 + 0.20/0

Pure and Applied Geophysics

Estimation of the Intrinsic Absorption and Scattering Attenuation in Northeastern Venezuela (Southeastern Caribbean) Using Coda Waves

ARANTZA UGALDE,[1] LUIS G. PUJADES,[2] JOSÉ ANTONIO CANAS,[3] and ANTONIO VILLASEÑOR[4]

Abstract—Northeastern Venezuela has been studied in terms of coda wave attenuation using seismograms from local earthquakes recorded by a temporary short-period seismic network. The studied area has been separated into two subregions in order to investigate lateral variations in the attenuation parameters. Coda-Q^{-1} (Q_c^{-1}) has been obtained using the single-scattering theory. The contribution of the intrinsic absorption (Q_I^{-1}) and scattering (Q_s^{-1}) to total attenuation (Q_T^{-1}) has been estimated by means of a multiple lapse time window method, based on the hypothesis of multiple isotropic scattering with uniform distribution of scatterers. Results show significant spatial variations of attenuation: the estimates for intermediate depth events and for shallow events present major differences. This fact may be related to different tectonic characteristics that may be due to the presence of the Lesser Antilles subduction zone, because the intermediate depth seismic zone may be coincident with the southern continuation of the subducting slab under the arc.

Key words: Coda waves, coda-Q^{-1}, attenuation, scattering, Venezuela, Caribbean.

1. Introduction

Coda waves have been widely used to estimate seismic wave attenuation in the earth's lithosphere. Total anelastic attenuation of seismic waves can be characterized by the inverse quality factor, Q_I^{-1}, defined as the fraction of energy lost during

[1] Universitat Politècnica de Catalunya. Departament d'Enginyeria del Terreny i Cartogràfica. UPC-Jordi Girona Salgado, s/n, D-2, 08034-Barcelona, Spain. Tel.: 34-3-4017407, Fax: 34-3-4016504, E-mail: dugalde@etseccpb.upc.es

[2] Universitat Politècnica de Catalunya. Departament d'Enginyeria del Terreny i Cartogràfica. UPC-Jordi Girona Salgado, s/n, D-2, 08034-Barcelona, Spain. Tel.: 34-3-4017258, Fax: 34-3-4016504, E-mail: pujades@etseccpb.upc.es

[3] Instituto Geográfico Nacional, c/General Ibáñez de Ibero, 3, 28003-Madrid, Spain. Tel.: 34-1-5979410, E-mail: canas@ign.es

[4] US Geological Survey, Box 25046, MS966, Denver, CO-80225, U.S.A. Tel.: 303-273-8590, Fax: 303-273-8600, E-mail: antonio@gldvisitor.cr.usgs.gov

a wave cycle. Two effects contribute to the observed total attenuation, as expressed by the equation: $Q_t^{-1} = Q_i^{-1} + Q_s^{-1}$ (DAINTY, 1981). Q_i^{-1} represents the intrinsic absorption, caused by the conversion of seismic energy into heat, and Q_s^{-1} is the scattering attenuation, due to the redistribution of energy that occurs when seismic waves interact with the heterogeneities of the medium.

Many theoretical studies have been developed to model the coda shape. AKI and CHOUET (1975) proposed the single-backscattering model, under the assumption of weak scattering. The inverse quality factors obtained from their formulation have been called Q_c^{-1} or coda-Q^{-1}. Within the framework of the single-scattering theory Q_c^{-1} represents an effective attenuation, including the contributions of both intrinsic absorption and scattering loss. However, numerical experiments (FRANKEL and CLAYTON, 1986), laboratory experiments (MATSUNAMI, 1991) and theoretical studies (FRANKEL and WENNERBERG, 1987; HOSHIBA, 1991) concluded that coda-Q^{-1} measured from the time window later than the mean-free time (mean-free path divided by the wave velocity, where the mean-free path is the parameter which controls the energy transferred from the primary to the scattered waves throughout the traveled path, HERRAIZ and ESPINOSA, 1987) should correspond only to the intrinsic absorption, not including the scattering effect. On the other hand, observational work does not always support this conclusion: empirically Q_c^{-1} is bounded between Q_i^{-1} and Q_t^{-1}.

In order to separately estimate the contribution of intrinsic absorption and scattering to total attenuation, multiple scattering models are needed. A multiple scattering model, under the assumptions of isotropic scattering and uniform distribution of scatterers in a background elastic medium which is uniform and unbounded, was considered by WU (1985). He modeled the spatial distribution of the seismic energy integrated over infinite time, using the radiative energy transfer approach. Under the same assumptions of WU (1985), HOSHIBA (1991) modeled the spatial and temporal distribution of the multiscattered seismic wave energy using a Monte-Carlo numerical simulation. ZENG *et al.* (1991) found the numerical solution of the same problem from an integral equation.

In the present study, the Multiple Lapse Time Window Analysis (MLTWA) (HOSHIBA *et al.*, 1991), will be applied to seismic data from northeastern Venezuela. The following assumptions are made: scattering is multiple and isotropic, the distribution of scatterers is uniform, and the coda is only composed of S-to-S scattered waves. Two attenuation parameters are calculated (WU, 1985): the seismic albedo (B_0), defined as the dimensionless ratio of the scattering loss to total attenuation, and the inverse of the extinction length (L_e^{-1}) that is the inverse of the distance (in km) over which the primary S-wave energy is decreased by e^{-1}. The multiple lapse time window method allows one to estimate B_0 and L_e^{-1} by comparing the energy density predicted by the multiple isotropic scattering theory, in space and time, with the observations.

We think this methodology is appropriate for the present study because the MLTWA has been widely and successfully applied to different regions in the world: FEHLER *et al.* (1992) in the Kanto-Tokai region (Japan); MAYEDA *et al.* (1992) in Hawaii, Long Valley and central California; HOSHIBA (1993) in Japan; JIN *et al.* (1994) in southern California; AKINCI *et al.* (1995) in southern Spain and western Anatolia (Turkey); PUJADES *et al.* (1997) in the southeastern Iberian peninsula; and CANAS *et al.* (1998) in the Canary Islands. In the present study, the Caribbean-South America plate boundary in northeastern Venezuela is studied in terms of the anelastic attenuation by means of the MLTWA. To the best of our knowledge, no quantitative studies of seismic wave attenuation in this region have been made to date. The only work in existence is the paper by AMBEH and LYNCH (1993), who estimated Q_c^{-1} in the eastern Caribbean (Lesser Antilles arc) using a different methodology than the one in this paper, by means of the SATO's (1977) extension of the S-S single-scattering coda model of AKI and CHOUET (1975).

A discussion of the Q parameters obtained in this study will be performed and a correlation of the observed attenuation parameters with the tectonics of the region will be made.

2. Method

Firstly, the decay rate of the coda amplitudes Q_c^{-1} is estimated by applying the following equation based on the single-scattering approximation (AKI and CHOUET, 1975):

$$\ln[t^2 A_{obs}(f|r, t)] = c - Q_c^{-1} 2\pi f t, \tag{1}$$

where c is a constant and $A_{obs}(f|r, t)$ represent the mean-squared amplitudes of the bandpass-filtered wave forms, as a function of hypocentral distance r and lapse time t measured from the origin time of the earthquake. Q_c^{-1} is easily obtained as the slope of the straight line fitting the measured $\ln[t^2 A_{obs}(f|r, t)]$ versus ft. The geometrical spreading is considered to be proportional to r^{-1}, which only applies to body waves in a uniform medium.

Equation (1) has been validated in many studies, as summarized by HERRÁIZ and ESPINOSA (1987). Empirically, it holds for $t > 2t_S$, where t_S is the S-wave travel time (RAUTIAN and KHALTURIN, 1978). Sometimes (e.g., GAGNEPAIN-BEYNEIX, 1987) there is a significant dependence of the slope (Q_c^{-1}) on the time window for which the fit of equation (1) is made. Therefore, it has become necessary to specify the time window used.

Secondly, in order to determine the contribution of intrinsic absorption and scattering to total attenuation, the Multiple Lapse Time Window Method (HOSHIBA *et al.*, 1991) is applied. It consists of comparing the predicted seismic energy with observations in several time windows following the S-wave onset, varying with hypocentral distance. The comparisons give values of the seismic albedo (B_0) and the inverse of the extinction length (L_e^{-1}) in a specified frequency band. The observed seismic energy as a function of hypocentral distance, is calculated following HOSHIBA (1993).

The model curves of the energy density are calculated at a given lapse time and hypocentral distance, under the assumptions of multiple isotropic scattering and uniform distribution of scatterers, by means of the following integral equation (ZENG *et al.*, 1991):

$$E(\vec{r}, t) = E_0\left(t - \frac{|\vec{r} - \vec{r}_0|}{\beta}\right)\frac{e^{-L_e^{-1}|\vec{r} - \vec{r}_0|}}{4\pi|\vec{r} - \vec{r}_0|^2} + \int_V gE\left(\vec{r}_1, t - \frac{|\vec{r}_1 - \vec{r}|}{\beta}\right) \cdot \frac{e^{-L_e^{-1}|\vec{r}_1 - \vec{r}|}}{4\pi|\vec{r}_1 - \vec{r}|^2} dV_1, \quad (2)$$

where $E(\vec{r}, t)$ is the seismic energy density per unit volume at 3-D position $\vec{r}$ and time t, for a point source at $t = 0$ located at $\vec{r}_0$. The first term on the right-hand side of (2) represents the direct wave energy traveling from $\vec{r}_0$ to the receiver located at $\vec{r}$, and the second term is the contribution of scattered energy of all orders. β is the S-wave velocity and $g = L_e^{-1}B_0$ is the scattering coefficient.

ZENG *et al.* (1991) solved equation (2) in the integral transform domain, and found the following solution:

$$E(r, t) = \delta\left(t - \frac{r}{\beta}\right)\frac{e^{-L_e^{-1}\beta t}}{4\pi\beta r^2} + \sum_{n=1}^{2} E_n(r, t) + \int_{-\infty}^{\infty} \frac{e^{i\Omega t}}{2\pi} d\Omega$$

$$\cdot \int_0^{\infty} \frac{g^3}{2\pi^2\beta r} \frac{\left[\tan^{-1}\left(\dfrac{k}{L_e^{-1} + \dfrac{i\Omega}{\beta}}\right)\right]^4}{\left[1 - \dfrac{g}{k}\tan^{-1}\left(\dfrac{k}{L_e^{-1} + \dfrac{i\Omega}{\beta}}\right)\right]} \frac{\sin(kr)}{k^3} dk, \quad (3)$$

where

$$E_1(r, t) = \frac{g}{4\pi r\beta t} H\left(t - \frac{r}{\beta}\right) e^{-L_e^{-1}\beta t} \ln\frac{1 + \dfrac{r}{\beta t}}{1 - \dfrac{r}{\beta t}}, \quad (4)$$

$$E_2(r, t) = \frac{g^2}{16\pi} H\left(t - \frac{r}{\beta}\right) e^{-L_e^{-1}\beta t}\left[\frac{\pi^2}{\beta t} - \frac{3}{r}\int_0^{r/\beta t}\left(\ln\frac{1 + \alpha}{1 - \alpha}\right)^2 d\alpha\right]. \quad (5)$$

The integral with respect to the wave number k in (3) can be computed using the method of discrete wave number sum (BOUCHON, 1979). Its convergence is guaranteed by the presence of $\sin(kr)/k^3$ in the integrand (ZENG *et al.*, 1991).

Starting from the estimated parameters L_e^{-1} and B_0, the inverse quality factors Q_t^{-1}, Q_s^{-1} and Q_i^{-1} are then calculated through the expressions $Q_t^{-1} = L_e^{-1}(\beta/\omega)$;

$Q_s^{-1} = B_0 Q_t^{-1}$ and $Q_i^{-1} = (1 - B_0)Q_t^{-1}$ (HOSHIBA et al., 1991), where ω represents the angular frequency.

3. Tectonic Framework

The Caribbean-South America plate boundary in northeastern Venezuela is characterized by a broad deformation zone, some 300 km in width, that extends from the Venezuela basin (part of the Caribbean plate) in the north to the undeformed Maturin basin and Precambrian Guayana shield in the south (Fig. 1). The tectonic history of the elements that we observe in the Caribbean-South America plate boundary zone can be explained by a right-oblique collision of the Lesser Antilles arc (the leading edge of the Caribbean plate) with the passive margin of continental South America during the Cenozoic (SPEED, 1985). This collision began at approximately 65°W in Late Eocene or Oligocene time and, as a result, allochthonous terranes transported on the leading edge of the magmatic arc were emplaced above the South American margin on north-dipping thrusts. This collision has migrated from west to east and is presumably ongoing north of Trinidad (SPEED, 1985; RUSSO and SPEED, 1992).

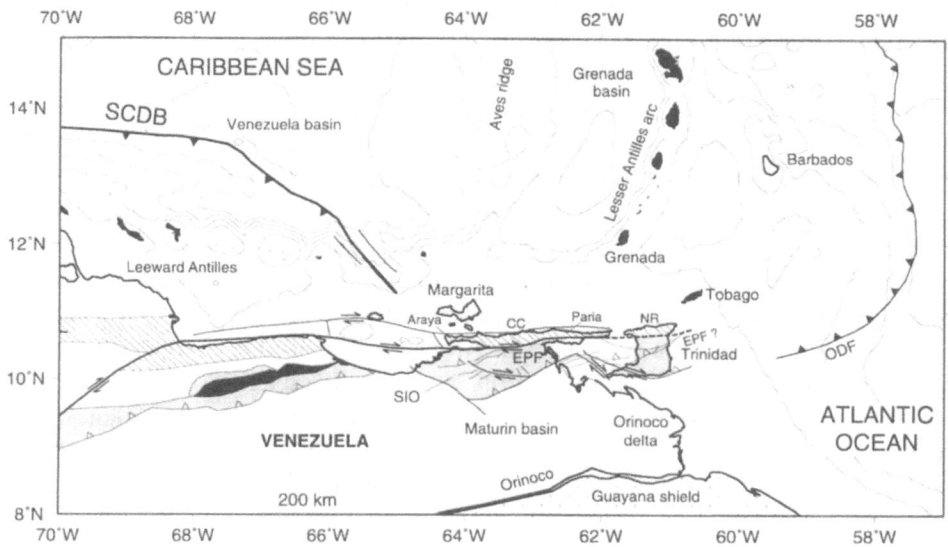

Figure 1

Structural map of northern South America and southeastern Caribbean, showing the main elements of the plate boundary zone (after BELLIZIA et al., 1976 and AVÉ LALLEMANT, 1997). Arc volcanic rocks are shown in black. CC: Cordillera de la Costa, EPF: El Pilar Fault, ODF: outer deformation front of the Barbados accretionary prism, SCDB: southern Caribbean deformed belt, SIO: Serranía del Interior Oriental.

The main elements of the plate boundary zone exposed on land in northeastern Venezuela are: an allochthonous metamorphic terrane (Cordillera de la Costa Oriental, formed by the Araya and Paria peninsulas, and the northern range in Trinidad) thrust over the South American former passive margin; a foreland fold and thrust belt (Serranía del Interior Oriental) formed by the imbrication of the metamorphic terrane; and a foreland basin (Maturin basin), interpreted as the southern limit of the plate boundary zone.

The Caribbean plate moves generally to the east relative to the North and South American plates. On the eastern part, old oceanic lithosphere parts of the North American and/or South American plates are subducted under the Lesser Antilles arc. To the south, the Caribbean-South America relative movement is slow (1 to 1.5 cm/year) and poorly constrained. Lacking direct measurements (e.g., geodetic studies) it is difficult to resolve whether movement on the Caribbean-South America plate boundary is pure dextral strike-slip or instead there exists a significant component of convergence (DEMETS *et al.*, 1990). Various models have been proposed for the transition from the Lesser Antilles subduction zone to northeastern Venezuela and Trinidad, including hinge faulting (MOLNAR and SYKES, 1969) and tectonic wedging as a result of an ongoing oblique collision (RUSSO and SPEED, 1992).

Seismicity in the southeastern Caribbean is separated into two distinct zones, characterized by different depth ranges and tectonic regimes. The most remarkable seismic feature in northeastern Venezuela is the northeast-trending cluster of intermediate-depth (70–150 km) earthquakes north of the Paria peninsula (Paria cluster). The cluster defines a northwest-dipping seismic zone, interpreted as the continuation of the slab of oceanic lithosphere subducted under the Lesser Antilles arc. Focal mechanisms of moderate-to-large earthquakes in the Paria cluster are consistent with down-dip extension which indicates that the slab is sinking in the surrounding mantle. On the other hand, shallow seismicity in northeastern Venezuela concentrates on the central Araya-Paria region, following two east-west lineations. One of the lineations is coincident with the surface trace of the El Pilar fault and the second one parallels the Araya-Paria coastline. Mechanisms of all large, shallow earthquakes on the El Pilar fault are characterized by pure right-lateral strike-slip movement on near-vertical east-west striking fault planes. This pattern is confirmed by microearthquake seismicity (VILLASEÑOR *et al.*, 1997).

4. Data

Wave-form data used in this study were acquired during a 3-month microseismicity experiment consisting of a 57-station short-period network (VILLASEÑOR *et al.*, 1997). The covered area was of about 32,000 km². The instruments (with a flat velocity response from 2.5 Hz to 39.2 Hz) continuously recorded the three

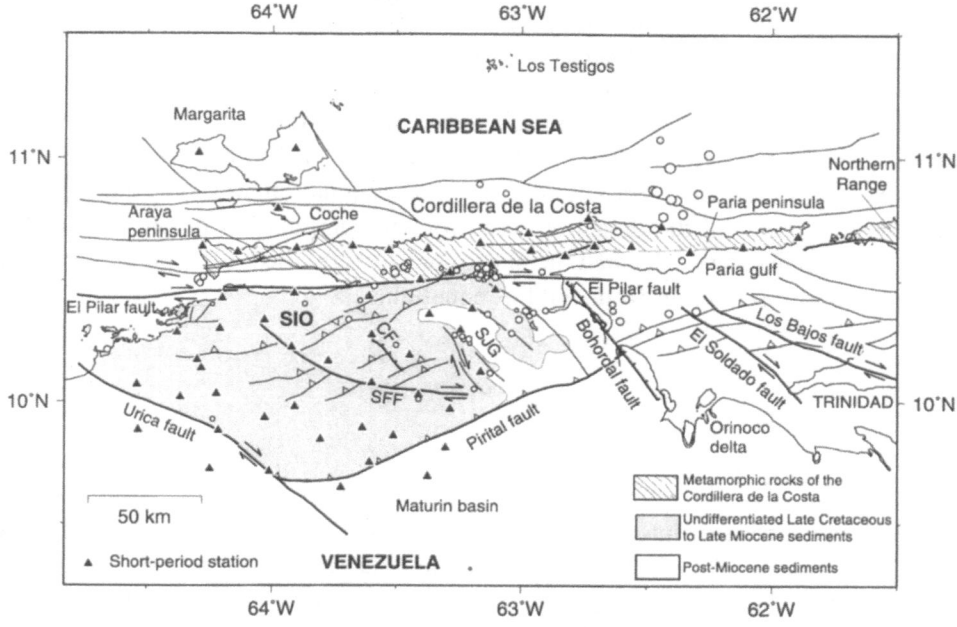

Figure 2
Simplified geological map of northeastern Venezuela, showing the station distribution of the local network, and the location of the 96 microearthquakes discussed in the text. CF: Caripe fault, SFF: San Francisco fault, SIO: Serranía del Interior Oriental, SJG: San Juan graben.

components of the movement and a coded time signal on analog audio tapes (IFGHH, 1987). The analog tapes were processed off-line to obtain the occurrence times of possible seismic events, and the selected time windows were digitized at a sampling interval of 8 ms. Station locations are shown in Figure 2 on a simplified geologic map of northeastern Venezuela.

Earthquake locations were determined using HYPO71 (LEE and LAHR, 1975) and a layered velocity model (2 layers over an infinite half-space), obtained by minimization of travel-time residuals (VILLASEÑOR et al., 1997). A total of 245 events were located during the three months of network operation. For this study only 96 well recorded (8 or more phase readings), well located (RMS < 0.5 s, ERH < 5 km, ERZ < 10 km) earthquakes were used. The magnitudes of the events are mostly small ($M < 3$) and focal depths range between 1 km and 90 km.

All the high quality remaining seismograms after a visual inspection were processed. However, only those giving reliable attenuation values with high correlation coefficients were kept. As can be seen in Figure 3, the hypocenters used are clearly separated into two regions which will be called subregion 1 and subregion 2. Shallow earthquakes occur following the lineation of the main faults in the area,

part of the Caribbean-South America plate boundary (see Fig. 2). The intermediate
depth events seem to be associated with the presence of a subducted oceanic
lithosphere under the Lesser Antilles (VILLASEÑOR *et al.*, 1997) (see Figs. 1 and 3).

In addition, when processing the available seismograms, only amplitudes with a
signal-to-noise ratio greater than 2 were considered. This last requirement revealed
that the separation of intrinsic and scattering attenuation was only possible for
subregion 1. Thus all the data were used for the computation of Q_c^{-1} values but
only those hypocenters located in subregion 1 were used to separate intrinsic
absorption and scattering from total attenuation.

5. Results

5.1 The Coda-Q^{-1}

Each seismogram time series was first bandpass-filtered over the following
frequency bands: 2.5–4.5 Hz; 4.5–8.5 Hz; and 8.5–16.5 Hz. From the bandpass-
filtered data, the function $A_{obs}(f|r, t)$ was calculated by averaging the squared
amplitudes in a time window of $t \pm 1$ s for each frequency band. An example of a
function $A_{obs}(f|r, t)$ for an actual seismogram used is shown in Figure 4.

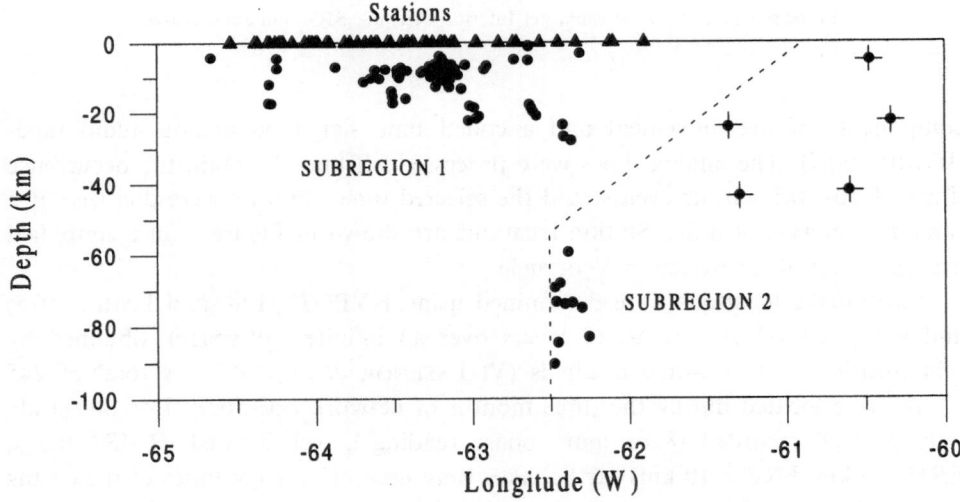

Figure 3

Separation of the studied area into two distinct subregions. Subregion 1 is characterized by shallow
seismicity and subregion 2 presents intermediate depth events. The shallowest events of subregion 2,
marked with a plus symbol are not significant, because the hypocentral distances to all the stations are
extensive.

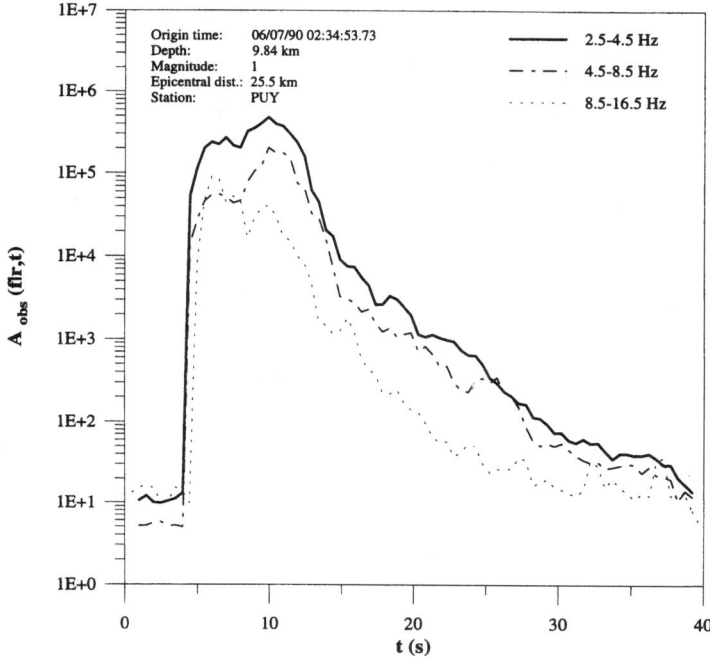

Figure 4
Mean-squared amplitudes for an actual bandpass-filtered seismogram.

Then, for determining Q_c^{-1}, a time window of $2(r/\beta)$ to $2(r/\beta) + 20$ s from the origin time of the earthquake was adopted to perform a least-squares regression of equation (1) for each frequency band.

Two estimates of coda-Q^{-1} are obtained for each frequency band. One corresponds to the so-called subregion 1, characterized by shallow events (focal depths less than 20 km). The other group of earthquakes belongs to the subregion 2 and exhibits intermediate focal depths. These two separated groups of events are shown in Figure 3.

The resulting Q_c^{-1} values are plotted in Figure 5, where a dot represents an estimate from a single seismogram. Only the values obtained from least-squares fits with correlation coefficients corr ≥ 0.9 were kept. Average numerical values are shown in Table 1. From these results it can be seen that there is a vast difference between both zones: subregion 1 exhibits a stronger coda attenuation.

5.2 Q_i^{-1}, Q_s^{-1} and Q_t^{-1}

The estimation of L_e^{-1} and B_0 was performed by means of the MLTWA, which compares the predicted and observed energy density.

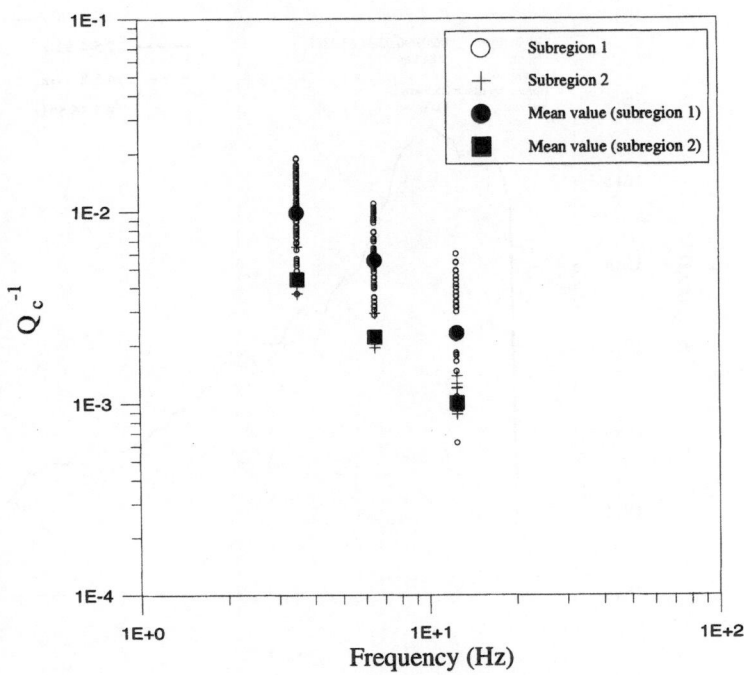

Figure 5

Coda-Q^{-1} as a function of frequency for the three analyzed frequency bands and the two subregions considered. A dot represents an estimate from a single seismogram.

The observed energy density data were obtained by integrating the function $A_{obs}(f|r, t)$ over the three consecutive time windows: 0–15 s; 15–30 s and 30–45 s from the S-wave onset. They were then normalized to a common source by the

Table 1

The following attenuation parameters: inverse of the extinction length (L_e^{-1}), seismic albedo (B_0), intrinsic absorption (Q_i^{-1}), scattering attenuation (Q_s^{-1}), total attenuation (Q_t^{-1}) and observed coda attenuation for subregions 1 (Q_{c1}^{-1}) and 2 (Q_{c2}^{-1}) are shown. Three frequency bands have been studied: 2.5–4.5 Hz, 4.5–8.5 Hz, and 8.5–16.5 Hz.

Frequency (Hz)	L_e^{-1} (km^{-1})	B_0	Q_i^{-1} ($\times 10^3$)	Q_s^{-1} ($\times 10^3$)	Q_t^{-1} ($\times 10^3$)	Q_{c1}^{-1} ($\times 10^3$)	Q_{c2}^{-1} ($\times 10^3$)
2.5–4.5	0.020 + 0.006 − 0.004	0.075 + 0.125 − 0.075	2.78	0.23	3.01	9.85 ± 0.12	4.37 ± 0.13
4.5–8.5	0.018 + 0.002 − 0.004	0.050 + 0.025 − 0.050	1.38	0.07	1.45	5.48 ± 0.11	2.24 ± 0.12
8.5–16.5	0.012 + 0.004 − 0.006	0.050 + 0.050 − 0.050	0.48	0.03	0.51	2.28 ± 0.07	1.04 ± 0.04

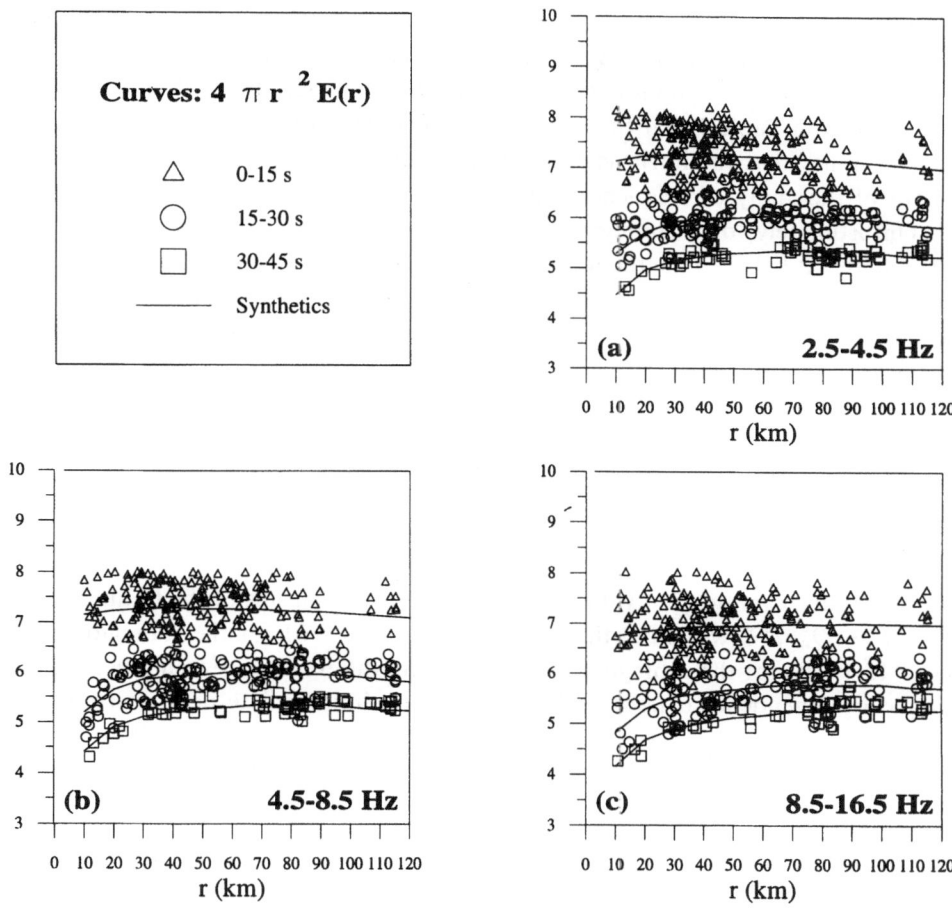

Figure 6

The observed integrated seismic energies for the 0–15 s lapse time window (triangles), 15–30 s (circles) and 30–45 s (squares) are plotted. The continuous line represents the synthetic energy best fitting the observed data for (a) 2.5–4.5 Hz, (b) 4.5–8.5 Hz, and (c) 8.5–16.5 Hz.

coda normalization method of AKI (1980), for which a 5 s long reference time window centered at 70 s past the origin time of the earthquake was used.

The best fit model parameters L_e^{-1} and B_0 were calculated by means of a multiple least-squares regression between the observed data and the theoretical curves, where the synthetics were obtained by solving equation (3) numerically. The S-wave velocity is considered to be $\beta = 3.3$ km·s^{-1} for shallow events in this region (VILLASEÑOR et al., 1997).

In Figure 6 the theoretical curves best fitting the observed data are plotted for the three studied frequency bands. Table 1 contains the estimated attenuation parameters L_e^{-1} and B_0 with their respective associated errors, which are not symmetric due to the asymmetry of the confidence areas (Fig. 7). These uncertainties have been calculated using F-distribution tests (DRAPER and SMITH, 1981). Table 1 also shows the corresponding Q_i^{-1}, Q_s^{-1} and Q_t^{-1} for each studied frequency band. It is seen from Table 1 that the seismic albedo is lower than 0.5 for all frequencies, which implies that the intrinsic absorption dominates over scattering for the scale length of the analyzed frequencies in the region.

Unfortunately, only subregion 1 has been studied by means of the MLTWA, because the fit between the theoretical and observed data was poor, due to the small magnitude of most of the events, which makes the signal to be buried in the noise at long travel times. This fact gives few observed points for the fit of the energy curves in the case of subregion 2.

6. Discussion

6.1 Frequency Dependence

Fitting the average Q_c^{-1} values shown in Table 1 by means of the frequency law: $Q^{-1}(f) = Q_0^{-1}f^{-\nu}$, we found the values of $\nu = 1.17 \pm 0.01$ for the subregion 1 and $\nu = 1.12 \pm 0.03$ for the subregion 2. No significant differences are found between the frequency-dependence exponents for shallow and intermediate depth earthquakes.

In the case of Q_i^{-1}, Q_s^{-1} and Q_s^{-1} and Q_t^{-1} (only subregion 1), the following frequency dependences have been found: $\nu = 1.41 \pm 0.13$ for the intrinsic absorption;

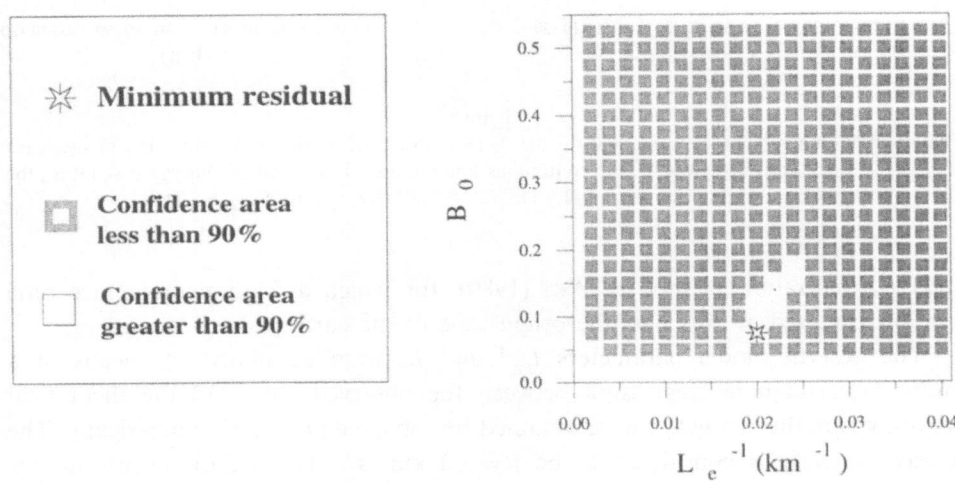

Figure 7
Example of an actual residual plot and its associated error area calculated by means of the F-test at the 90% confidence level.

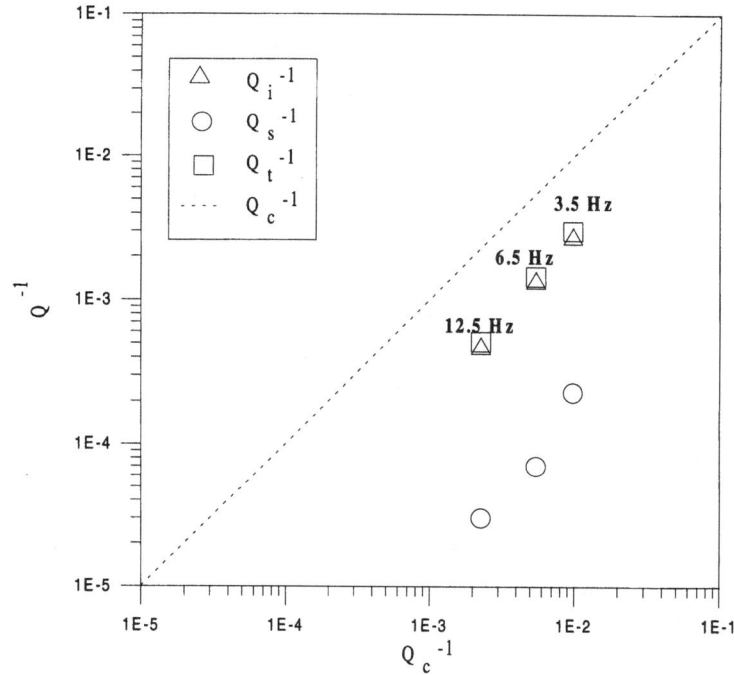

Figure 8
Plots of Q_i^{-1}, Q_s^{-1} and Q_t^{-1} as a function of Q_c^{-1} for: (a) subregion 1 and (b) subregion 2.

$v = 1.60 \pm 0.18$ for the scattering attenuation; and $v = 1.38 \pm 0.14$ for total attenuation. The frequency-dependent exponents are comprised in the range 1.3 to 1.6. A decrease of Q_s^{-1} faster than f^{-1} with increasing frequency implies that the medium may be characterized by a Gaussian rather than exponential autocorrelation function (WU, 1985).

6.2 Comparison among Q_i^{-1}, Q_t^{-1} and Q_c^{-1}

A comparative plot among the attenuation parameters obtained in this study is presented in Figure 8, where Q_s^{-1} (triangles), Q_t^{-1} (squares), Q_i^{-1} (circles) and Q_c^{-1} (dashed line) are plotted as a function of Q_c^{-1}. It can be seen that Q_c^{-1} is very close to but greater than Q_t^{-1} in the case of subregion 1.

6.3 Comparison with Other Attenuation Studies in the Region

AMBEH and LYNCH (1993) estimated coda-Q^{-1} for the frequency range from 1 Hz to 30 Hz in the eastern Caribbean (Lesser Antilles arc), using the single isotropic scattering method of SATO (1977). They subdivided the study area into four

Table 2

Volumes sampled by coda waves. $\bar{\Delta}$ is the average epicentral distance, h_{av} is the average hypocentral depth, a_1 and a_2 are the semi-axes of the ellipsoid, h is the sampled depth, and t is the average lapse time.

Subregion	Δ (km)	h_{av} (km)	a_1 (km)	a_2 (km)	h (km)	t_{av} (s)
1	55	10	66	60	70	40
2	114	72	157	146	218	70

subregions, one of them the Trinidad-Tobago region, which is very close to the area analyzed in the present work (Fig. 1).

According to PULLI (1984), the ellipsoidal volume sampled by coda waves at a lapse time t has an average depth of $h = h_{av} + a_2$, where h_{av} is the average hypocentral depth and a_2 is the small semi-axis of the surface projection of the ellipsoid, given by $a_2 = \sqrt{a_1^2 - r^2/4}$. r is the source-receiver distance and a_1 is the large semi-axis, where $a_1 = \beta t/2$. The average sampled volume can be assumed to be represented by $t_{av} = t_{start} + \Delta t_{win}/2$, where t_{av} is the average lapse time, t_{start} is the starting lapse time, and Δt_{win} is the window length.

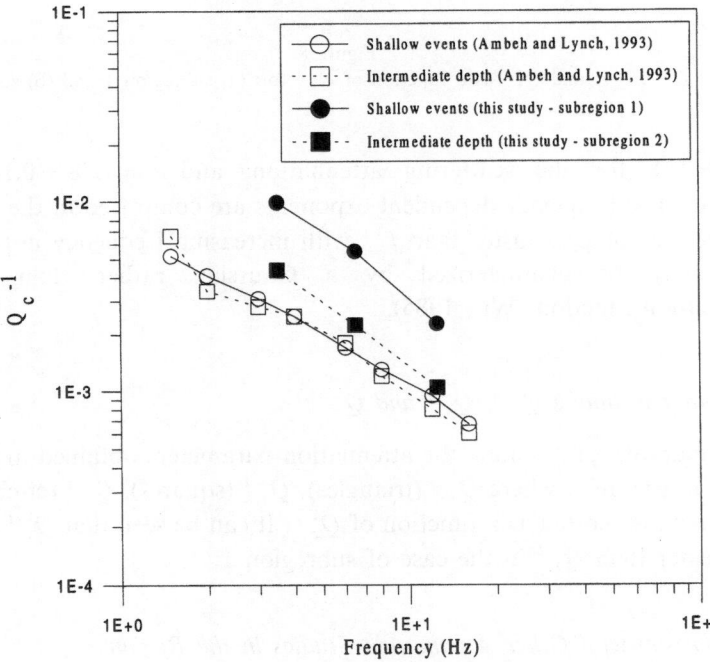

Figure 9
Qualitative comparison between the Q_c^{-1} estimates in this study and those obtained by AMBEH and LYNCH (1993) in the Trinidad-Tobago region.

In this study, S-wave velocities of 3.3 km·s^{-1} and 4.5 km·s^{-1} have been considered for shallow (focal depth between 0 and 30 km) and intermediate depth (focal depth between 60 and 90 km) events, respectively. Table 2 shows the estimated sampled volumes for the two subregions. The average lapse time is 40 s for shallow events, which corresponds to sampling volumes of about 70 km in depth. For intermediate depth events the average lapse time is 70 s, and corresponds to 218 km in depth.

AMBEH and LYNCH (1993) reported sampling volumes of about 208 km in lateral extent and 137 km in depth, indicating that the coda waves of intermediate depth events ($h > 70$ km) sample larger volumes. Therefore, a significant part of the volume sampled by coda waves for our subregion 2 overlaps the volume sampled by the coda waves for the intermediate depth events in the Trinidad-Tobago region.

Figure 9 exhibits a qualitative comparison between the Q_c^{-1} estimates by AMBEH and LYNCH (1993) and those obtained in this study. Taking into account the volumes sampled by coda waves, it seems that our results for intermediate depth events are not in disagreement with the values obtained in the near Trinidad-Tobago region. However, significant differences are observed for shallow events.

6.4 Correlation with the Geotectonics of the Region

The strong observed total attenuation correlates well with the seismicity pattern of the region. Seismically active regions, or areas with a high intensity of tectonic activity seem to be associated with high Q^{-1} values (CANAS and MITCHELL, 1981; ROECKER et al., 1982; SINGH and HERRMANN, 1983; JIN et al., 1988).

On the other hand, seismicity in northeastern Venezuela is separated into two distinct zones (the so-called subregions 1 and 2). Whereas the shallow earthquakes occur following the lineations of the main faults in the region, part of the Caribbean-South America plate boundary (Fig. 2), the intermediate depth events seem to be associated with the presence of a subducted oceanic lithosphere under the Lesser Antilles arc (VILLASEÑOR et al., 1997) (Figs. 1 and 3). The different Q_c^{-1} values obtained in this study support different tectonic behavior between subregions 1 and 2 that may be due to the presence of the subducted slab under the region. Although the methodology in this work is different than the one used in AMBEH and LYNCH (1993), coda-Q^{-1} parameters estimated for subregion 2 seem to agree qualitatively with those for the neighboring Lesser Antilles.

The great differences between the Q_c^{-1} values in subregions 1 and 2 for shallow events may be an indication of the different tectonic behavior which causes the observed seismic wave attenuation in both regions.

7. Conclusions

Northeastern Venezuela has been studied in terms of coda wave attenuation. The single-scattering model of AKI and CHOUET (1975) has been applied to estimate Q_c^{-1}, and the Multiple Lapse Time Window Analysis (HOSHIBA *et al.*, 1991), which assumes multiple isotropic scattering and uniform distribution of scatterers, has been used to separate the contributions of the intrinsic absorption (Q_i^{-1}) and scattering (Q_s^{-1}) from total attenuation (Q_t^{-1}). The following conclusions were reached:

(1) Seismicity in northeastern Venezuela is distributed in two zones with distinct tectonic regimes: a shallow domain and an intermediate depth region (VILLASEÑOR *et al.*, 1997). Coda-Q^{-1} estimates present considerable differences between both regions: the coda attenuation is stronger in the area characterized by shallow events.

(2) The seismic albedo is lower than 0.5 for all the studied frequencies from 2.5 Hz to 16.5 Hz. This result indicates a low degree of heterogeneity for the scale length of the analyzed frequencies in the region, and that the intrinsic absorption dominates in the attenuation process.

(3) The frequency dependent exponents for the attenuation parameters range between 1.1 to 1.6. The lowest values corresponds to Q_c^{-1} whereas Q_s^{-1} appears to be strongly frequency dependent. This fact may be consistent with a medium characterized by a Gaussian autocorrelation function.

(4) The attenuation parameters obtained in the shallow and intermediate depth subregions present significant differences. VILLASEÑOR *et al.* (1997) suggested that the intermediate depth seismic zone was coincident with the southern continuation of the subducting slab under the Lesser Antilles arc. The differing results obtained for both subregions may indicate different tectonic behavior supporting the idea of VILLASEÑOR *et al.* (1997).

Acknowledgements

We thank B. J. Mitchell, G. Abers, and an anonymous reviewer for their suggestions offered to improve the manuscript. This work has been partially financed by DGICYT and CICYT projects: PB93–0972 and MAR95–1916, respectively.

REFERENCES

AKI, K. (1980), *Attenuation of Shear-waves in the Lithosphere for Frequencies from 0.05 to 25 Hz*, Phys. Earth Planet. Inter. *21*, 50–60.
AKI, K., and CHOUET, B. (1975), *Origin of Coda Waves: Source, Attenuation and Scattering Effects*, J. Geophys. Res. *80*, 3322–3342.

AKINCI, A., DEL PEZZO, E., and IBÁÑEZ, J. M. (1995), *Separation of Scattering and Intrinsic Attenuation in Southern Spain and Western Anatolia (Turkey)*, Geophys. J. Int. *121*, 337–353.

AMBEH, W. B., and LYNCH, L. L. (1993), *Coda-Q_i Eastern Caribbean*, Geophys. J. Int. *112*, 507–516.

AVÉ LALLEMANT, H. G. (1997), *Transpréssion, Displacement Partitioning, and Exhumation in the Eastern Caribbean/South America Plate Boundary Zone*, Tectonics *16*, 272–289.

BELLIZIA, A., PIMENTEL, N., and BAJO, R., *Mapa geológico estructural de Venezuela, Ministerio de Minas e Hidrocarburos* (Foninves, Caracas, 1976).

BOUCHON, M. (1979), *Discrete Wave number Representation of Elastic Wave Fields in Three Space Dimensions*, J. Geophys. Res. *84*, 3609–3614.

CANAS, J. A., and MITCHELL, B. J. (1981), *Rayleigh Wave Attenuation and its Variation across the Atlantic Ocean*, Geophys. J. R. Astron. Soc. *67*, 159–176.

CANAS, J. A., UGALDE, A., PUJADES, L. G., CARRACEDO, J. C., BLANCO, M. J., and SOLER, V. (1998), *Intrinsic and Scattering Seismic Wave Attenuation in the Canary Islands*, J. Geophys. Res. *103*, 15037–15049.

DAINTY, A. M. (1981), *A Scattering Model to Explain Seismic Q Observations in the Lithosphere between 1 and 30 Hz*, Geophys. Res. Lett. *8*, 1126–1128.

DEMETS, C. R., GORDON, G., ARGUS, D. F., and STEIN, S. (1990), *Current Plate Motions*, Geophys. J. Int. *101*, 425–478.

DRAPER, N. R., and SMITH, H., *Applied Regression Analysis* (John Wiley and Sons 1981).

GAGNEPAIN-BEYNEIX, J. (1987), *Evidence of Spatial Variations of Attenuation in the Western Pyrenean Range*, Geophys. J. R. Astron. Soc. *89*, 681–704.

FEHLER, M., HOSHIBA, M., SATO, H., and OBARA, K. (1992), *Separation of Scattering and Intrinsic Attenuation for the Kanto-Tokai Region, Japan, Using Measurements of S-wave Energy vs. Hypocentral Distance*, Geophys. J. Int. *108*, 787–800.

FRANKEL, A., and CLAYTON, R. W. (1986), *Finite Differences Simulation of Seismic Scattering: Implications for the Propagation of Short-period Seismic Waves in the Crust and Models of Crustal Heterogeneity*, J. Geophys. Res. *91*, 6465–6489.

FRANKEL, A., and WENNERBERG, L. (1987), *Energy-flux Model of the Seismic Coda: Separation of Scattering and Intrinsic Attenuation*, Bull. Seismol. Soc. Am. *77*, 1223–1251.

HERRÁIZ, M., and ESPINOSA, A. F. (1987), *Coda Waves: A Review*, Pure and appl. geophys. *125*, 499–577.

HOSHIBA, M. (1991), *Simulation of Multiple Scattered Coda Wave Excitation Based on the Energy Conservation Law*, Phys. Earth Planet. Inter. *67*, 123–136.

HOSHIBA, M. (1993), *Separation of Scattering Attenuation and Intrinsic Absorption in Japan Using the Multiple Lapse Time Window Analysis of Full Seismogram Envelope*, J. Geophys. Res. *98*, 15809–15824.

HOSHIBA, M., SATO, H., and FEHLER, M. (1991), *Numerical Basis of the Separation of Scattering and Intrinsic Absorption from Full Seismogram Envelope—A Monte-Carlo Simulation of Multiple Isotropic Scattering*, Pap. Geophys. Meteorol. *42*, 65–91, Meteorol. Res. Inst. of Jpn.

IFGHH, *The 3C-LOBS: A Portable Seismic Station for Refraction and Seismicity Studies* (Institute of Geophysics, Hamburg University, unpublished report, 1987).

JIN, A., and AKI, K. (1988), *Spatial and Temporal Correlation between Coda Q and Seismicity in China*, Bull. Seismol. Soc. Am. *78*, 741–769.

JIN, A., MAYEDA, K., ADAMS, D., and AKI, K. (1994), *Separation of Intrinsic and Southern California Using TERRAscope Data*, J. Geophys. Res. *99*, 17835–17848.

LEE, W. H. K., and LAHR, J. C., *HYPO71: A Computer Program for Determining Hypocenter, Magnitude, and First Motion Pattern of Local Earthquakes* (U.S. Geol. Surv. Open-File Report, 1975).

MATSUNAMI, K. (1991), *Laboratory Tests of Excitation and Attenuation of Coda Waves Using 2-D Models of Scattering Media*, Phys. Earth Planet. Inter. *67*, 104–114.

MAYEDA, K., KOYANAGI, S., HOSHIBA, M., AKI, K., and ZENG, Y. (1992), *A Comparative Study of Scattering, Intrinsic and Coda Q^{-1} for Hawaii, Long Valley and Central California between 1.5 and 15.0 Hz*, Geophys. Res. *97*, 6643–6659.

MOLNAR, P., and SYKES, L. R. (1969), *Tectonics of the Caribbean and Middle America Regions from Focal Mechanisms and Seismicity*, Geol. Soc. Am. Bull. *80*, 1639–1648.

PULLI, J. J. (1984), *Attenuation of Coda Waves in New England*, Bull. Seismol. Soc. Am. *74*, 1149–1166.

PUJADES, L. G., UGALDE, A., CANAS, J. A., NAVARRO, M., BADAL, F. J., and CORCHETE, V. (1997), *Intrinsic and Scattering Attenuation from Observed Coda Q Frequency Dependence. Application to the Almeria Basin (Southeastern Iberian Peninsula)*, Geophys. J. Int. *129*, 281–291.

RAUTIAN, T. J., and KHALTURIN, V. I. (1978), *The Use of the Coda for the Determination of the Earthquake Source Spectrum*, Bull. Seismol. Soc. Am. *68*, 923–948.

ROECKER, S. W., TUCKER, B., KING, J., and HATZFELD, D. (1982), *Estimates of Q in Central Asia as a Function of Frequency and Depth Using the Coda of Locally Recorded Earthquakes*, Bull. Seismol. Soc. Am. *72*, 129–149.

RUSSO, R. M., and SPEED, R. C. (1992), *Oblique Collision and Tectonic Wedging of the South American Continent and Caribbean Terranes*, Geology *20*, 447–450.

SATO, H. (1977), *Energy Propagation Including Scattering Effects. Single Isotropic Scattering Approximation*, J. Phys. Earth *25*, 27–41.

SINGH, S., and HERRMANN, R. B. (1983), *Regionalization of Crustal Coda Q in the Continental United States*, J. Geophys. Res. *88*, 527–538.

SPEED, R. C. (1985), *Cenozoic Collision of the Lesser Antilles Arc and Continental South America and the Origin of the El Pilar Fault*, Tectonics *4*, 41–69.

VILLASEÑOR, A., BANDA, E., GAJARDO, E., FRANKE, M., and MAKRIS, J. (1998), *The El Pilar Fault Zone and the Seismotectonics of the Caribbean-South America Plate Boundary in Northeastern Venezuela*, Geophys. J. Int. (in press).

WU, R. S. (1985), *Multiple Scattering and Energy Transfer of Seismic Waves: Separation of Scattering Effect from Intrinsic Attenuation, I, Theoretical Modelling*, Geophys. J. R. Astron. Soc. *82*, 57–80.

ZENG, Y., SU, F., and AKI, K. (1991), *Scattered Wave Energy Propagation in a Random Isotropic Scattering Medium, I, Theory*, J. Geophys. Res. *96*, 607–619.

(Received November 10, 1997, revised April 30, 1998, accepted May 20, 1998)

To access this journal online:
http://www.birkhauser.ch

Pure appl. geophys. 153 (1998) 703–712
0033–4553/98/040703–10 $ 1.50 + 0.20/0

▌Pure and Applied Geophysics

Intrinsic and Scattering Seismic Attenuation in W. Greece

G-Akis Tselentis[1]

Abstract—Intrinsic (Q_i^{-1}) and scattering (Q_s^{-1}) attenuation parameters have been determined in the seismically active region of W. Greece, which is continuously monitored by the University of Patras microearthquake network. One hundred and twenty-three local and shallow earthquakes close to the recording stations have been used and Wennerberg's (1993) approach has been adopted. Results for 1 to 12 Hz range show that Q_i^{-1} is higher than Q_s^{-1} and coda Q values are close to Q_i, indicating that coda Q can be a reasonable estimate of intrinsic Q.

Key words: Intrinsic, scattering, attenuation, Greece.

1. Introduction

The attenuation of short-period S waves, expressed by the inverse of the quality factor (Q^{-1}), is of particular importance in understanding the physical laws according to which the elastic energy of an earthquake propagates through the earth's lithosphere.

Among the many influential factors on the propagation path are attenuation due to scattering, i.e., deviation of energy from the general propagation direction, and attenuation due to intrinsic losses or anelasticity, i.e., conversion of energy into heat.

Intrinsic attenuation is conveniently described by the quality factor Q_i of the medium which depends, among others, on viscous processes between the rock matrix and liquid inclusions (Gorich and Muller, 1987), such as pore fluids and on movements of dislocations through the mineral grains.

Since scattering has the same effect as anelasticity, namely a reduction of amplitudes, it is also described by a quality factor Q_s which depends on the spatial structure of the scattering heterogeneities in the medium, on the size of the velocity and density fluctuations.

[1] University of Patras, Seismological Laboratory, Rio 261 10, Greece. Fax: 30 619 90639, E-mail: tselenti@upatras.gr

Attenuation inferred from the decay rate of the coda (AKI and CHOUET, 1975; SINGH and HERRMANN, 1983; SATO, 1988) is also a combination of scattering and intrinsic attenuation.

During recent years, many researchers have tried to separate the contribution of scattering and intrinsic attenuation. TSUJIURA (1978) and AKI (1980) stated that scattering attenuation plays a more significant role than intrinsic attenuation, while FRANKEL and WENNERBERG (1987) stated the opposite.

WU (1985) developed a method, based on the radiative transfer theory, allowing an estimation of the relative amounts of scattering and intrinsic attenuation from the dependence of the entire S-wave energy on hypocentral distance (TÖKSÖZ et al., 1988; MAYEDA et al., 1992; HATZIDIMITRIOU, 1994). HOSHIBA (1991) proposed a multiple lapse time window analysis method to estimate scattering and intrinsic attenuation by considering energy for three consecutive time windows as a function of hypocentral distance (FEHLER et al., 1992).

GUSEV and ABUBAKIROV (1987) and HOSHIBA (1991) found that for short lapse times the single-scattering model (AKI and CHOUET, 1975) is adequate and for long lapse times the diffusion model is appropriate. FEHLER et al. (1992) and MAYEDA et al. (1992) applied HOSHIBA's (1991) method, based on Monte Carlo simulations, to estimate both intrinsic and scattering quality factors.

ZENG (1991) extended WU's work and developed a coda model based on multiple scattering. Obviously, neither the single nor the multiple scattering hypothesis can completely describe the complexities of the earth's crust, however the latter is probably the most relevant to the physical processes which control the shape of the coda envelope.

WENNERBERG (1993) proposed a methodology to estimate Q_i and Q_s from measurements of the direct S-wave Q (Q_d) and coda Q (Q_c), based on the approximation given by ABUBAKIROV and GUSEV (1990) to the model developed by ZENG (1991).

According to WENNERBERG (1993) the lapse-time dependence of Q_c cannot be fully explained with a deficiency of the single-scattering model but it is more reasonably interpreted in terms of a non-uniform medium. A possible explanation of the observed lapse-time dependence of Q_c could be attributed to depth decreasing intrinsic attenuation in the lithosphere (DEL PEZZO et al., 1995) and could be plausibly related to depth-decreasing intrinsic attenuation in the lithosphere.

The goal of this study is to separate the contributions to the apparent decay, namely, the intrinsic and scattering attenuation, Q_i^{-1} and Q_s^{-1} in the crust of W. Greece by applying WENNERBERG's (1993) methodology.

2. Method of Analysis

Comparing single-backscattering (AKI and CHOUET, 1975) and multiple-scattering (ZENG, 1991) attenuation models and assuming that the intrinsic attenuation is increased as a common overall exponential factor $(\exp - \omega t/2Q_i)$, WENNERBERG (1993) indicated that it is possible to express the observed value of Q_c in terms of the intrinsic and scattering Q as follows

$$\frac{1}{Q_c} = \frac{1}{Q_i} + \frac{1 - 2\delta t}{Q_s} \tag{1}$$

where

$$\delta(\tau) = \frac{0.72}{4.44 + 0.738\tau} - 0.5 \tag{2}$$

$$\tau = \frac{\omega t}{Q_s} \tag{3}$$

and ω, t are the angular frequency and the lapse time, respectively.

Let Q_d be the quality factor for direct S waves, corresponding to an earth volume equivalent to the volume sampled by coda waves and assuming that it describes the total attenuation, we can write

$$\frac{1}{Q_d} = \frac{1}{Q_i} + \frac{1}{Q_s}. \tag{4}$$

Following DEL PEZZO et al. (1995) we express Q_s and Q_i as

$$\frac{1}{Q_s} = \frac{1}{2\delta(\tau)}\left(\frac{1}{Q_d} - \frac{1}{Q_c(\tau)}\right) \tag{5}$$

$$\frac{1}{Q_i} = \frac{1}{2\delta(\tau)}\left(\frac{1}{Q_c(\tau)} + \frac{2\delta(\tau) - 1}{Q_d}\right). \tag{6}$$

Considering (1), (2) and (3) Q_s is given as the positive root of the following equation

$$4.44\left(\frac{1}{Q_d} - \frac{1}{Q_c}\right)Q_s^2 + \left[0.738\left(\frac{1}{Q_d} - \frac{1}{Q_c}\right)\omega t - 5.88\right]Q_s - 0.738\omega t = 0. \tag{7}$$

The above relations indicate that if we measure Q_c as a function of lapse time t, then we can estimate Q_i and Q_s as a function of lapse time for different frequencies.

In the present investigation we use short-period seismic data and calculate Q_c as a function of lapse time for short epicentral distances. To estimate Q_c, we use the well-known method of AKI and CHOUET (1975), a brief outline of which follows.

Assuming that coda waves are composed of single-scattered wavelets, the coda envelope can be approximately expressed by the following formula

$$A(\omega|t) = A_0(\omega)t^{-1}\exp[-\omega t/(2Q_c)] \tag{8}$$

where $A(\omega|t)$ is the moving Fourier spectrum of the coda, depending on lapse time

t, $A_0(\omega)$ is the so-called coda source and ω is the angular frequency. Rewriting (8) as

$$\ln[A(\omega|t)t] = \ln[A_0(\omega)] - \omega t/(2Q_c) \qquad (9)$$

we can calculate Q_c by applying a linear regression analysis between $\ln[A(\omega|t)t]$ and lapse time t for each frequency.

Q_d is estimated following the spectral ratio method (TSUJIURA, 1966; FERRUCCI and HIRN, 1985). In this method, the observed amplitude $A(f)$ of body waves is expressed as

$$A(f) \propto \frac{A_0(f)R(f)\exp(-\pi f t/Q_d)}{r} \qquad (10)$$

where $A_0(f)$ is the spectral amplitude at the source and $R(f)$ the site response. By assigning two frequencies f_1 and f_2 and taking the logarithm of the ratio of the corresponding amplitude, we can write

$$\ln\frac{A(f_1)}{A(f_2)} = \ln\frac{A_0(f_1)}{A_0(f_2)} + \ln\frac{R(f_1)}{R(f_2)} - \frac{\pi(f_1-f_2)}{Q_d}t. \qquad (11)$$

We can estimate Q_d by applying a linear regression between the function $\ln[A(f_1)/A(f_2)]$ and time t. All the above are based on the assumption that Q_d can be considered constant over the frequency band studied and that source spectrum is similar for the events analyzed. The later assumption may hold if we use shocks relating to small focal volumes and lying in a small magnitude range. Next, by substituting in (5) and (6) equation (3) we can estimate Q_i and Q_s.

3. Data Used

The data used in the present investigation are 136 digitally recorded seismograms corresponding to local earthquakes at epicentral distances less than 50 km, recorded by the University of Patras seismological network. All records chosen exhibit excellent signal to noise and were free from glitches and spurious signals. Figure 1 shows the distribution of events and stations used.

The network commenced operation in 1990 and the data used cover the period 1993–1996. The network consists of 17 stations equipped with (Teledyne S-13 1 Hz) seismometers operating at 90 dB dynamic range in a low-noise environment. The signals are radio-telemetered via FM subcarriers to the central recording site at Patras seismological center in real time. There, each channel signal is filtered for aliasing with a 30-Hz Butterworth low-pass filter, sampled at 100 sps and converted to digital form with a resolution of 16 bit.

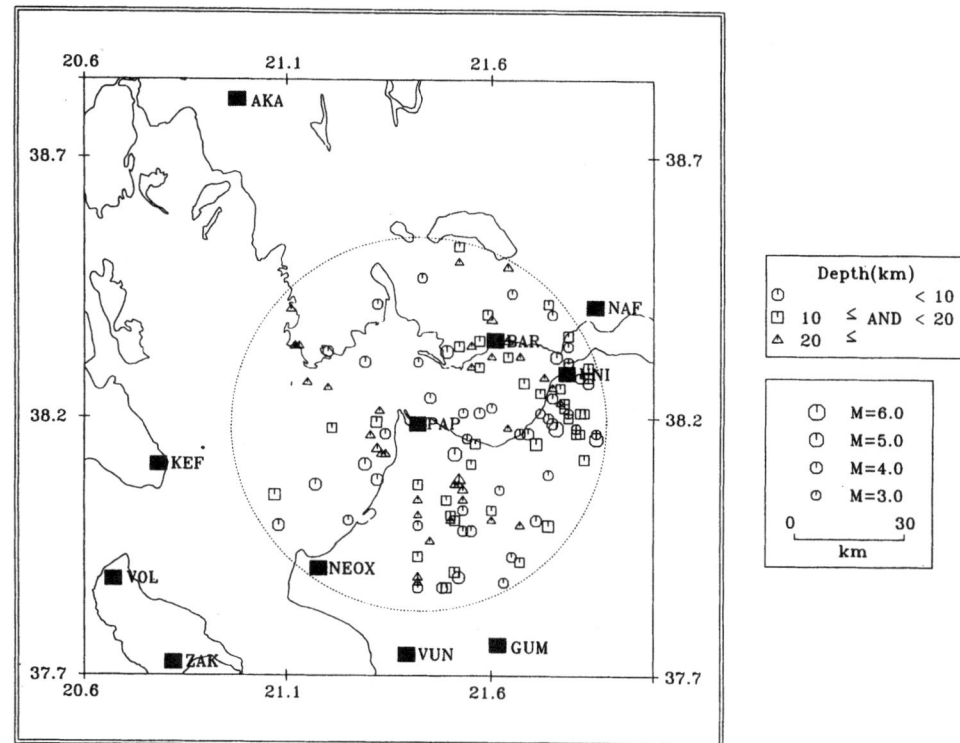

Figure 1
Map showing the location of the events used in the present analysis. Filled rectangles show the seismological stations.

For estimating Q_c, only the vertical components of the seismograms have been used. Beginning at times $t > 2rV_s$, where r and V_s are the hypocentral distance and shear-wave velocity respectively, values of $A(\omega|t)$ were calculated using successive overlapping time windows, corrected for instrument response and averaged over octave frequency bands centered at 1.0, 2.0, 4.0, 8.0 and 12.0 Hz. The effect of noise

Table 1

Measures of Q_c, Q_d, and estimates of Q_i and Q_s.

Freq. (Hz)	Q_d	Q_c	Q_i	Q_s
1	119	175	157	487
2	130	223	190	407
4	178	337	278	494
8	246	511	406	623
12	337	752	580	803

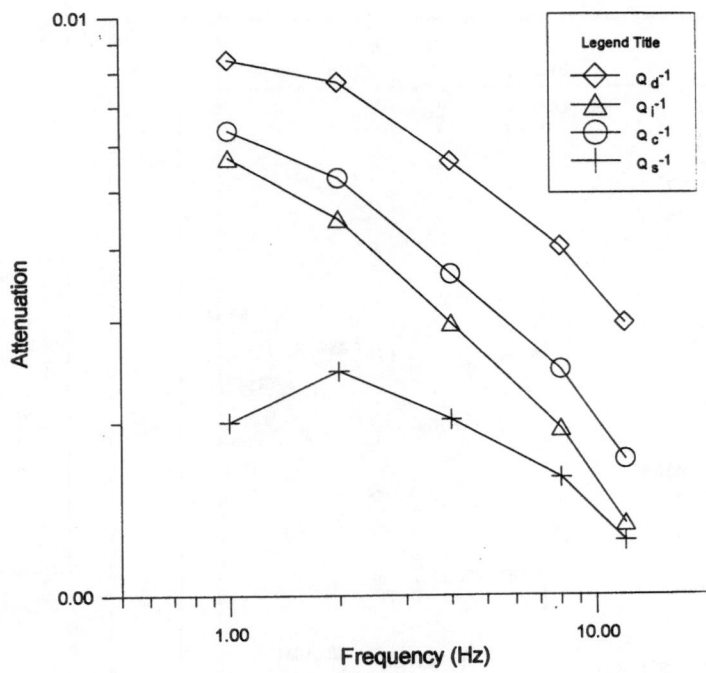

Figure 2

Plot showing the total, scattering and intrinsic attenuation determined in the present study, versus frequency together with coda attenuation.

was removed from each seismogram $S(t)$ in the following way (assuming the general case of uncorrelated signals $s(t)$ and noise $n(t)$)

$$\langle s(t) \rangle_T^2 = \langle S(t) \rangle_T^2 - \langle n(t) \rangle_T^2 \qquad (12)$$

where the quantity between brackets refers to the mean over the time interval T. Thus, the rms amplitude of the signal is derived as:

$$A(r, \omega | t) = [\langle s(t) \rangle_T^2]^{1/2}. \qquad (13)$$

4. Results and Discussion

By applying eq. (9), Q_c values were calculated in all the frequency bands. Values corresponding to regression coefficients less than 0.80 were not considered. Averaged Q_c values are depicted in Table 1.

For estimating Q_d, we follow the method of TSUJIURA (1966) as described above. Spectral amplitudes were estimated over 2 Hz wide bands centered at f_1 and f_2. With f_2 set equal to 1 Hz and allowing f_1 to vary in 2 Hz steps between 2 and 12 Hz. Averaged Q_d values over all the measuring stations are depicted in Table 1.

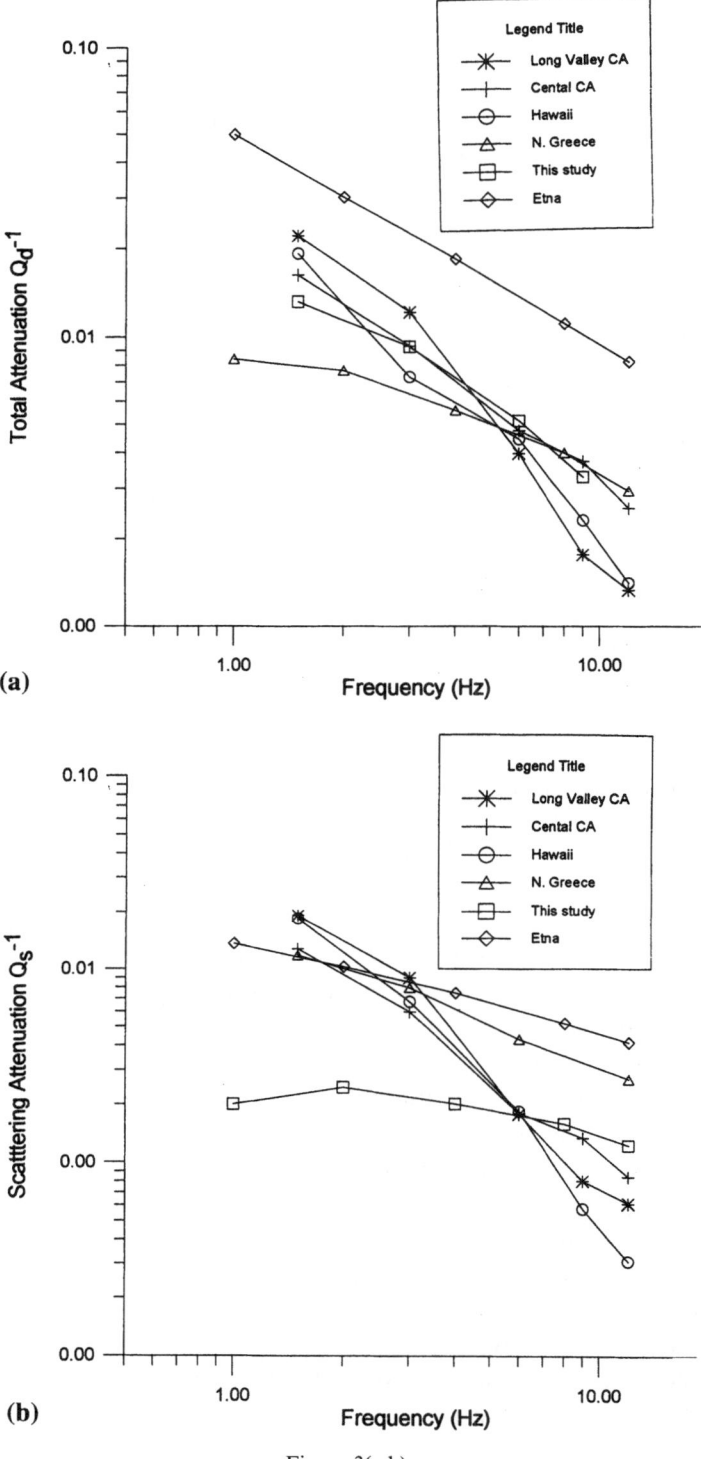

Figure 3(a,b)

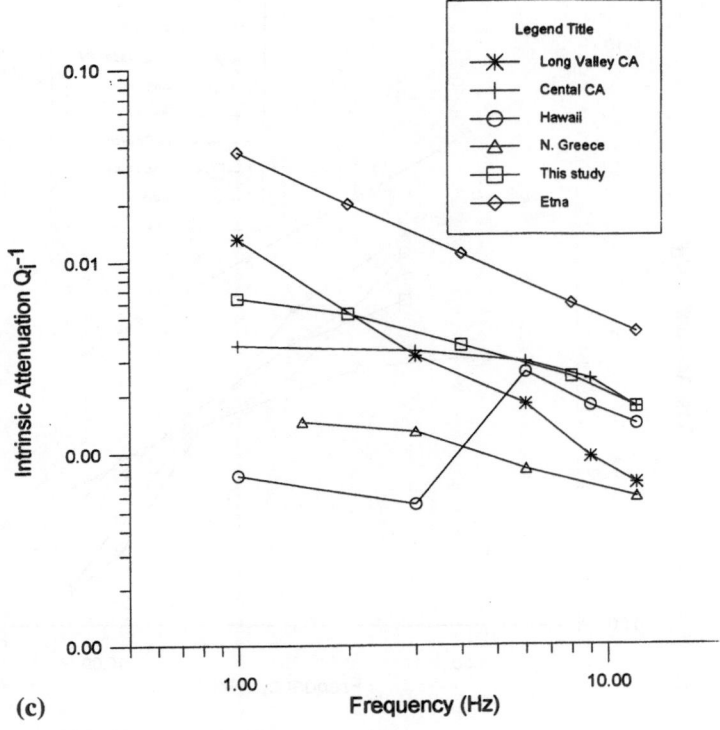

Figure 3

Plot of the total (a), scattering (b) and intrinsic (c) attenuation calculated in the present study, versus frequency. The values of the same parameters calculated for Hawaii, Long Valley and Central California (MAYEDA et al., 1992), N. Greece (HATZIDIMITRIOU, 1994), and Etna (DEL PEZZO et al., 1995) are also shown for comparison.

The next step was to evaluate Q_s and Q_i by solving eq. (7) and eq. (6), respectively and using the previously computed values of Q_d and Q_c. Table 1 lists the corresponding values for scattering and intrinsic attenuation for the five frequency bands examined.

In order to compare the scattering and the intrinsic attenuation with that inferred from the coda waves, Figure 2 depicts the Q_d^{-1}, Q_s^{-1}, and Q_i^{-1} values versus frequency, together with the obtained Q_c^{-1} values. Judging from this diagram we clearly see that the observed Q_c^{-1} is close to Q_i^{-1}, in agreement with finite-difference simulation results of FRANKEL and WENNERBERG (1987) and laboratory experiments of MATSUNAMI (1991). The analytical results of SHANG and GAO (1988) point that in a highly scattering medium, coda decay is identical to Q_i^{-1}.

Similar observations have been published by TÖKSÖZ et al. (1988) for North America, WU and AKI (1988) for Hindu-Kush, while MAYEDA et al. (1991) for South California and MCSWEENEY et al. (1991) for Alaska have reported Q_c^{-1}

values close to Q_s^{-1}. HATZIDIMITRIOU (1994) reported Q_c^{-1} values very close to Q_s^{-1} for short lapse times and intermediate between Q_s^{-1} and Q_i^{-1} for longer lapse times for North Greece.

In Figure 3 we plot for comparison the total scattering and intrinsic attenuation estimated in the present study versus frequency together with the results obtained from other regions. Judging from this figure we can see that both the total and intrinsic attenuation estimates for W. Greece agree well with those obtained for Central California, while the scattering attenuation estimate in this study displays a different behavior. The difference between the results from N. Greece and the present work can possibly be attributed to the higher tectonic activity and consequently higher heterogeneity present in the region of W. Greece. The coda Q values obtained are in good agreement with those obtained for the same region by TSELENTIS (1993) using a different data set.

From Table 1 we can obtain the frequency dependence of Q_i^{-1}, Q_s^{-1} and Q_c^{-1} for the region of W. Greece. The results show that Q_i^{-1} has a frequency dependence of $f^{-0.52}$, Q_s^{-1} has a frequency dependence of $f^{-0.36}$ and Q_c has a frequency dependence of $f^{0.58}$ which is similar to that of Q_i. This might imply that the frequency dependence of the coda wave attenuation in this region is due to the frequency dependence of intrinsic attenuation.

5. Conclusions

WENNERBERG's (1993) methodology has been applied to seismological data from Central Greece, in order to study the relative contribution of scattering and intrinsic attenuation to the total S-wave attenuation for frequencies 1, 2, 4, 8 and 12 Hz. Results show that both Q_i^{-1} and Q_s^{-1} have a frequency dependence with higher variations for Q_i^{-1}. Also that coda wave attenuation is very close to the intrinsic attenuation. The patterns of Q_i^{-1} and Q_s^{-1} with frequency are comparable to other estimates obtained from other areas.

Acknowledgments

The author wishes to thank Edoardo Del Pezzo and an anonymous reviewer for their most useful comments.

REFERENCES

ABUBAKIROV, I. R., and GUSEV, A. A. (1990), *Estimation of Scattering Properties of Lithosphere of Kamchatka Based on Monte-Carlo Simulation of Record Envelope of a Near Earthquake*, Phys. Earth. Planet. Int. *64*, 52–67.

AKI, K. (1980), *Scattering and Attenuation of Shear Waves in the Lithosphere*, J. Geophys. Res. *85*, 6496–6504.

AKI, K., and CHOUET, B. (1975), *Origin of Coda Waves: Source, Attenuation and Scattering Effects*, J. Geophys. Res. *80*, 3322–3342.

DEL PEZZO, E., IBANEZ, J., MORALES, J., AKINCI, A., and MARESCA, R. (1995), *Measurements of Intrinsic and Scattering Seismic Attenuation in the Crust*, Bull. Seismol. Soc. Am. *85* (5), 1373–1380.

FEHLER, M., HOSHIBA, M., SATO, H., and OBARA, K. (1992), *Separation of Scattering and Intrinsic Attenuation for the Kanto-Tokai Region, Japan Using Measurements of S-wave Energy Versus Hypocentral Distance*, Geophys. J. Int. *108*, 787–800.

FERRUCCI, F., and HIRN, A. (1985), *Differential Attenuation of Seismic Waves at a Dense Array: A Marker of the Travale Field from Sources at Regional Distances*, Geothermics *14*, 723–730.

FRANKEL, A., and WENNERBERG, L. (1987), *Energy-flux Model of Seismic Coda: Separation of Scattering and Intrinsic Attenuation*, Bull. Seismol. Soc. Am. *77*, 1223–1251.

GORICH, M., and MULLER, G. (1987), *Apparent and Intrinsic Q. The One-dimensional Case*, J. Geophys. *61*, 46–54.

GUSEV, A. A., and ABUBAKIROV, I. R. (1987), *Monte-Carlo Simulation of Record Envelope of a Near Earthquake*, Phys. Earth Planet. Int. *49*, 30–67.

HATZIDIMITRIOU, P. M. (1994), *Scattering and Anelastic Attenuation of Seismic Energy in N. Greece*, Pure and appl. geophys. *143* (4), 587–601.

HOSHIBA, M. (1991), *Simulation of Multiple Scattered Coda Wave Excitation Based on the Energy Conservation Law*, Phys. Earth Planet. Int. *67*, 123–136.

MATSUNAMI, K. (1991), *Laboratory Tests of Excitation and Attenuation of Coda Waves Using 2-D Models of Scattering Media*, Phys. Earth Planet. Int. *67*, 36–47.

MAYEDA, K., SU, F., and AKI, K. (1991), *Seismic Albedo from the Total Seismic Energy Dependence on Hypocentral Distance in Southern California*, Phys. Earth Planet. Int. *67*, 104–114.

MAYEDA, K., KOYNAGI, S., HOSHIBA, M., AKI, K., and ZENG, Y. (1992), *A Comparative Study of Scattering, Intrinsic and Coda Q^{-1} for Hawaii, Long Valley, and Central California between 1.5 and 15 Hz*, J. Geophys. Res. *97*, 6643–6659.

McSWEENEY, T., BISWAS, N., MAYEDA, K., and AKI, K. (1991), *Scattering and Anelastic Attenuation of Seismic Energy in Central and South Central Alaska*, Phys. Earth Planet. Int. *67*, 115–122.

SATO, H. (1988), *Temporal Change in Scattering and Attenuation Associated with the Earthquake Occurrence*, Pure and appl. geophys. *126*, 465–498.

SHANG, T., and GAO, L. (1988), *Transportation Theory of Multiple Scattering and its Application to Seismic Coda Waves of Impulsive Source*, Sci. Sin. Ser. *B31*, 1503–1514.

SINGH, S., and HERRMANN, R. B. (1983), *Regionalization of Crustal Coda Q in the Continental United States*, J. Geophys. Res. *88*, 527–538.

TÖKSÖZ, M. N., DAINTY, A. M., REITER, E., and WU, R. S. (1988), *A Model for Attenuation and Scattering in the Earth's Crust*, Pure and appl. geophys. *128*, 81–100.

TSELENTIS, G.-A. (1993), *Depth Dependent Seismic Attenuation in Western Greece*, Tectonophysics *225*, 523–528.

TSUJIURA, M. (1966), *Frequency Analysis of the Seismic Waves*, Bull. Earth. Res. Inst. Tokio Univ. *44*, 873–891.

TSUJIURA, M. (1978), *Spectral Analysis of Coda Waves from Local Earthquakes*, Bull. Earthquake Res. Inst. Univ. Tokyo *53*, 1–48.

WENNERBERG, L. (1993), *Multiple-scattering Interpretation of coda-Q Measurements*, Bull. Seismol. Soc Am. *83*, 279–290.

WU, R. S. (1985), *Multiple Scattering and Energy Transfer of Seismic Waves, Separation of Scattering Effect from Intrinsic Attenuation, I. Theoretical Modeling*, Geophys, J. R. Astron. Soc. *82*, 57–80.

WU, R. S., and AKI, K. (1988), *Multiple Scattering and Energy Transfer of Seismic Waves—Separation of Scattering Effect from Intrinsic Attenuation, II. Application of the Theory to Hindu-Kush Region*, Pure and appl. geophys. *128*, 49–80.

ZENG, Y. (1991), *Compact Solutions for Multiple Scattered Wave Energy in Time Domain*, Bull. Seismol. Soc. Am. *81*, 1022–1029.

(Received April 14, 1997, revised December 15, 1998, accepted April 5, 1998)

Pure appl. geophys. 153 (1998) 713–731
0033–4553/98/040713–19 $ 1.50 + 0.20/0

Pure and Applied Geophysics

Coda Q Estimates in the Koyna Region, India

S. C. Gupta,[1] S. S. Teotia,[2] S. S. Rai[3] and Navneet Gautam[4]

Abstract—The coda Q, Q_c, have been estimated for the Koyna region of India. The coda waves of 76 seismograms from thirteen local earthquakes, recorded digitally in the region during July–August, 1996, have been analyzed for this purpose at nine central frequencies viz., 1.5, 2.0, 3.0, 4.0, 6.0, 8.0, 12.0, 16.0 and 24.0 Hz using a single backscattering model. All events with magnitude less than 3 fall in the epicentral distances less than 60 km and have focal depths which range from 0.86 to 9.43 km. For the 30 sec coda window length the estimated Q_c values vary from 81 to 261 at 1.5 Hz and 2088 to 3234 at 24 Hz, whereas the mean values of Q_c with the standard error vary from 148 ± 13.5 at 1.5 Hz to 2703 ± 38.8 at 24 Hz. Both the estimated Q_c values and their mean values exhibit the clear dependence on frequency in the region and a frequency dependence average attenuation relationship, $Q_c = 96f^{1.09}$, has been obtained for the region, covering an approximate area of 11 500 km² with the surfacial extent of about 120 km and depth of 60 km.

Lapse time dependence of Q_c has also been studied for the region, with the coda waves analyzed at five lapse time windows from 20 to 60 sec duration with the difference of 10 sec. The frequency dependence average Q_c relationships obtained at these window lengths $Q_c = 66f^{1.16}$ (20 sec), $Q_c = 96f^{1.09}$ (30 sec), $Q_c = 131f^{1.04}$ (40 sec), $Q_c = 148f^{1.04}$ (50 sec), $Q_c = 182f^{1.02}$ (60 sec) show that the frequency dependence (exponent n) remains mostly stationary at all the lapse time window lengths, while the change in Q_0 value is significant. Lapse time dependence of Q_c in the region is also interpreted as the function of depth.

Key words: Q_c, coda waves, Koyna region, single backscattering and lapse time window.

1. Introduction

The knowledge of both seismic wave attenuation and velocity, in a given region, is necessary as these two physical parameters effect the propagation of seismic waves and therefore are required for the exact determination of earthquake source parameters and also for the assessment of seismic hazards in a region.

[1] Department of Earthquake Engineering, University of Roorkee, Roorkee 247 667, India. Fax: 01332-76899, E-mail: quake@rurkiu.ernet.in
[2] Department of Geophysics, Kurukshetra University, Kurukshetra 136 119, India.
[3] National Geophysical Research Institute, Hyderabad 500 006, India.
[4] CGG Pan India Ltd., Gurgaon 122015, India.

Attenuation of seismic waves is described by the dimensionless quantity called quality factor 'Q' (KNOPOFF, 1964), which expresses the decay of wave amplitude during its propagation in the medium. It is a combination of both intrinsic attenuation, the loss of elastic energy to heat or other form of energy, and scattering, which is deflection and/or mode conversion of elastic energy due to heterogeneities in the transmitting medium.

High frequency coda waves recorded during the occurrence of local earthquakes, which arrive at station after the arrival of all direct phases, are assumed to be the superposition of backscattered primary S waves generated by the numerous heterogeneities distributed randomly in the earth's crust and upper mantle. These waves arrive at the station on different time intervals and form a coda. Therefore, the decay of these waves with time, in a seismogram, provides the average attenuation characteristics of the medium instead of the property of a single path connecting from a source to the station. As these waves are the result of numerous heterogeneities distributed randomly, they cannot be explained by any deterministic approach in which a number of parameters is required to describe a small portion of seismogram. However, they can be solved by applying the statistical method in which a small number of parameters is sufficient to describe the properties of coda waves. AKI (1969) and AKI and CHOUET (1975) are pioneers in this field and they proposed a single backscattering model to use the coda waves of local earthquakes for the estimation of quality factor (Q_c) of coda waves in a region.

Numerous studies have been carried out in different parts of the world to determine the seismic wave attenuation properties of the medium. These studies analyze the coda waves of local earthquakes for the estimation of Q_c values, using the single backscattering model (e.g., ROECKER *et al.*, 1982; PULLI, 1984; REHA, 1984; IBANEZ *et al.*, 1990; WOODGOLD, 1994; AKINCI *et al.*, 1994; GUPTA *et al.*, 1995, 1996). A frequency-independent Q study has been done in the Koyna region applying the strong motion acceleration data (GUPTA and RAMBABU, 1994). However, a frequency-dependent Q study has not been conducted in the region.

In the present study, the Q_c values have been estimated for the Koyna region by analyzing the high frequency coda waves of local earthquakes at various central frequency bands. The single backscattering model is used for the estimation of Q_c values and to provide average attenuation properties of the medium around the Koyna region. The coda waves amplitudes have also been analyzed, at five different lapse time window lengths, to study the variation of estimated Q_c values with the lapse time windows.

2. Study Area

The Koyna region (Fig. 1) of the Indian Peninsula is a seismically active region. Seismicity of this zone started after the impoundment of the reservoir in 1963 for the construction of the Koyna dam. The Koyna earthquakes' sequence is considered as one of the most outstanding examples of the reservoir-induced seismicity (e.g., GUPTA *et al.*, 1972; GUPTA, 1985). Intense seismicity, with a 6.3-magnitude earthquake, occurred during the period 1967–1969, subsequent to reaching the appropriate reservoir filling level which again increased during 1973–1980 when earthquakes of magnitude 5 occurred (RASTOGI *et al.*, 1992). High seismic flow rate continues in the vicinity of Koyna, and on average 2 to 3 local earthquakes occur per day in the region.

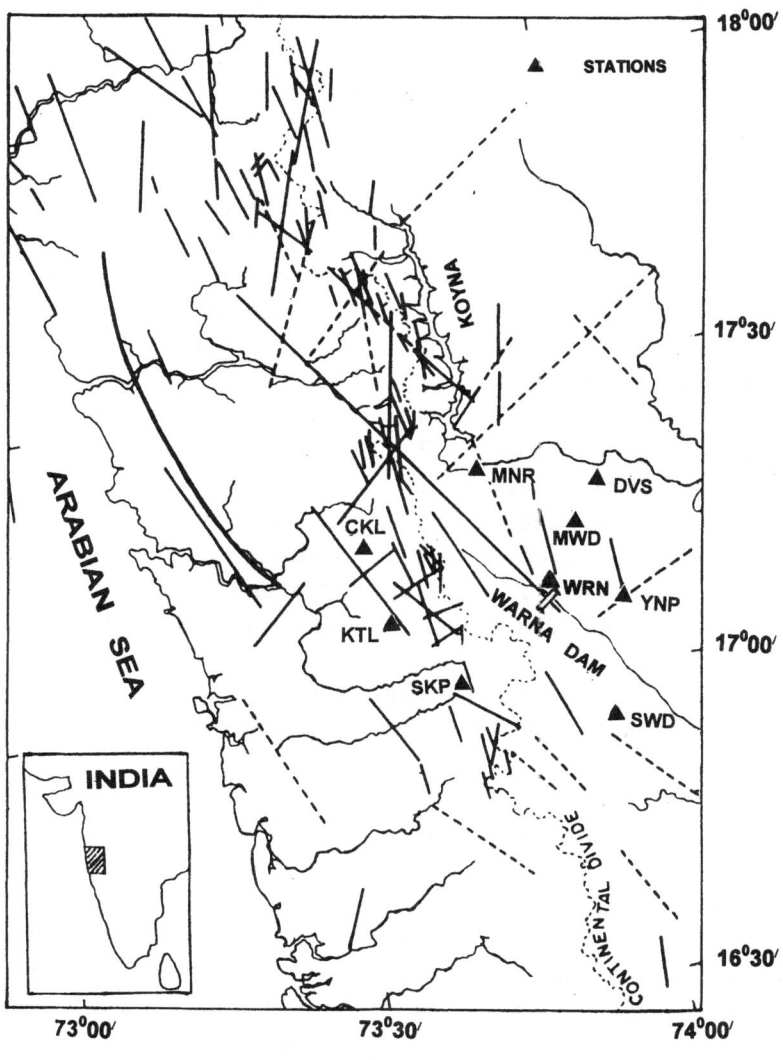

Figure 1

Map showing lineaments in the Koyna–Warna region along with the stations location of the seismological network installed in the region (originally prepared by LANGSTON, 1981 and modified by RASTOGI *et al.*, 1992).

Table 1

Parameters of seismological stations (e.g., name, codes and their locations)

Sl. No.	Station name	Code	Latitude (°N)	Longitude (°E)
1	Warna	WRN	17–07.41	73–53.06
2	Sakharpa	SKP	16–59.77	73–42.81
3	Shahuwadi	SWD	16–54.84	73–57.21
4	Katwali	KTL	17–07.23	73–37.25
5	Chikhali	CKL	17–14.83	73–35.15
6	Marathwadi	MWD	17–14.27	73–56.26
7	Divasi	DVS	17–17.84	73–59.07
8	Yenpe	YNP	17–06.45	74–02.26
9	Maneri	MNR	17–20.51	73–47.73

From a different view the seismicity of this region is related with the two lineaments, NW and NE are present in the area (LANGSTON, 1981). A major lineament is in the NW direction, coinciding with the Warna river in its southeastern part. This is probably a deep fault which is currently active. A deep NNW-trending fault has also been detected by Deep Seismic Soundings just west of the Koyna reservoir (KAILA *et al.*, 1981a,b). This trending fault, with an 8-m throw on the western side, has been observed along the Warna river course for a few meters length in the excavation done for the Warna Dam (RAWAT, 1982). In this region, where the geological observation of faults is difficult, there is a cover of volcanic basalt about 2 km thick. Faults are occasionally seen in fresh excavations.

The geology of the Koyna region is fundamental. For miles around to great depths the rocks are all Deccan traps. The basement under the trap consists of meta sedimentary gneisses, schists and granite of Dharwarian and Cuddapah age. The Deccan rocks include a variety of textures, compact, fine-grained, occasionally porphyritic basalt. Crude columnar joints and extensive curved fractures are common to these homogeneous hard, brittle rocks (KRISHNAN, 1960).

3. The Data Set and Analysis

The data consisting of 76 seismograms of thirteen local earthquakes recorded during July–August, 1996, have been used in the study. These earthquakes were obtained from the operation of the Seismological Network (Fig. 1) deployed by the National Geophysical Research Institute, Hyderabad in the Koyna–Warna region. All events were recorded digitally on 4 to 7 stations, using short-period (1 Hz) three-component seismometers, employing Ref-Teck recording instruments at the sampling rate of 100 samples/sec. However, in the present study, the events recorded only on vertical components have been used. The outstation parameters (e.g., names, codes and locations) are given in Table 1 and their locations are marked on the map of the region (Fig. 1). At all stations the time signals were electronically impinged using GPS, which ensures the accuracy of the time for an event recorded at these stations.

The hypocentral parameters, namely origin time, latitude, longitude and focal depth of the events, have been computed using the HYPO71PC computer program (LEE and LAHR, 1975) and are listed in the Table 2. All events are of local origin with magnitudes less than 3 and are recorded in the epicentral distances of 11 to 55 km with shallow depths from 0.86 to 9.43 km. The epicentral locations of these events are shown in Figure 2.

The coda waves of 76 seismograms related to thirteen local earthquakes have been analyzed to estimate the quality factor (Q_c) for the Koyna region employing the single backscattering model. According to this model, the RMS coda wave amplitudes, $A(f, t)$ in a seismogram, for a central frequency f over a narrow bandwidth signal and lapse time (t), measured from the origin time of the seismic event, can be expressed as

$$A(f, t) = C(f)t^{-a} \exp(-\pi ft / Q_c) \tag{1}$$

where, $C(f)$ is the coda source factor which is considered as constant; a is the geometrical spreading factor ($a = 1$ for body waves), and Q_c is the quality factor of coda waves representing the average attenuation properties of the medium for a given region.

Taking the natural logarithm of equation (1) and rearranging the terms we get,

$$\ln[A(f, t) \cdot t] = c - bt \tag{2}$$

Table 2

Hypocentral parameters of the events considered in this study

Sl. No.	Date	Origin time (IST)			Latitude (°N)	Longitude (°E)	Focal depth (km)
		hr	min	sec			
1	960711	05	47	02.44	17–15.98	73–43.86	6.13
2	960713	15	12	48.82	17–15.85	73–41.99	2.26
3	960715	20	47	55.83	17–16.00	73–41.52	4.68
4	960717	21	55	59.58	17–16.03	73–42.92	8.48
5	960718	09	05	44.95	17–15.98	73–43.00	7.98
6	960721	07	44	48.82	17–13.86	73–45.24	5.76
7	960801	09	39	44.37	17–09.92	73–42.98	1.43
8	960804	04	20	29.23	17–16.21	73–42.84	9.43
9	960812	07	27	16.52	17–06.68	73–46.10	0.86
10	960815	17	51	21.88	17–09.77	73–42.98	5.57
11	960816	01	07	13.37	17–09.66	73–43.29	5.66
12	960816	08	27	46.90	17–09.68	73–43.18	5.13
13	960817	06	40	28.55	17–09.40	73–43.22	1.29

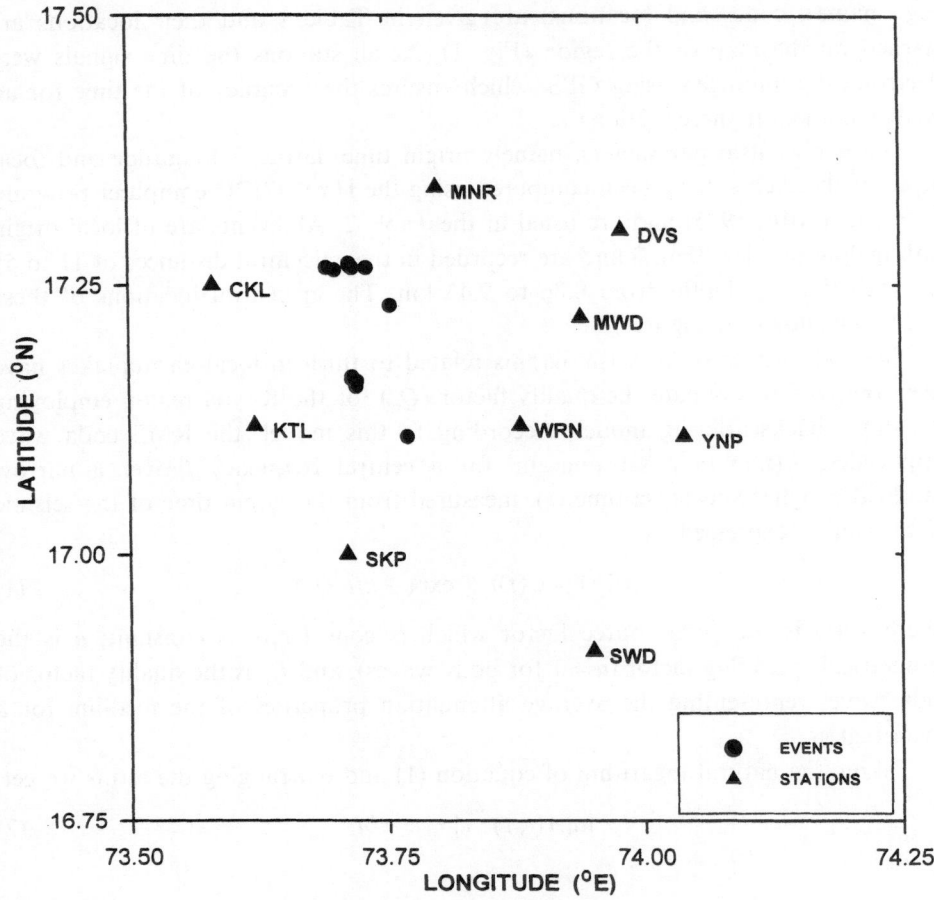

Figure 2
Epicenter map of events located around the Koyna region along with the stations' locations.

where, $b = \pi f / Q_c$ and $c = \ln C(f)$.

Equation (2) represents a straight line and slope of which $b(= \pi f / Q_c)$ provides the Q_c values at a central frequency f. This equation (2) has been used for the estimation of Q_c for the Koyna region.

Each seismogram is filtered using the Butterworth bandpass filter of eight poles (STEARNS and DAVID, 1988). Nine frequency bands (bandwidth $0.67f_c$, where f_c is the central frequency) are used for this purpose, low cut-off and high cut-off of these bands are given in Table 3. The response of this filter, for a bandwidth of 10 to 20 Hz, is shown in Figure 3.

Seismograms thus filtered are used for the detailed study of the decay of coda wave amplitudes with time to estimate Q_c values at nine central frequencies. Figure 4a shows an example of a seismogram recorded on 15.08.96 at station SWD and

Table 3

Central frequency components of bandpass filter with low and high cut-off frequencies

Low cut-off	Central frequency (*f*) (Hz)	High cut-off
1.00	1.5	2.00
1.33	2.0	2.67
2.00	3.0	4.00
2.67	4.0	5.33
4.00	6.0	8.00
5.33	8.0	10.67
8.00	12.0	16.00
10.67	16.0	21.33
16.00	24.0	32.00

filtered at five central frequencies. The coda waves of a fixed time window for each filtered seismogram, which starts from the lapse time $t > 2t_s$ where t_s is the direct *S*-wave travel time (RAUTIAN and KHALTURIN, 1978), are chosen for the analysis. The coda waves of five window lengths of 20, 30, 40, 50 and 60 sec lapse time for each seismogram are used for the study. The arrows (↑) in Figure 4a show the coda window length of 30 sec. The coda wave amplitudes of this window length are smoothed, using a root-mean-square technique which calculates the RMS value of amplitudes of the filtered seismograms in a window of 512 samples (for 1.5 Hz) and 256 samples (for 2, 3, 4, 6, 8, 12, 16 and 24 Hz) with a sliding window along the coda in steps of half of the window, i.e., 256 and 128 samples respectively, and

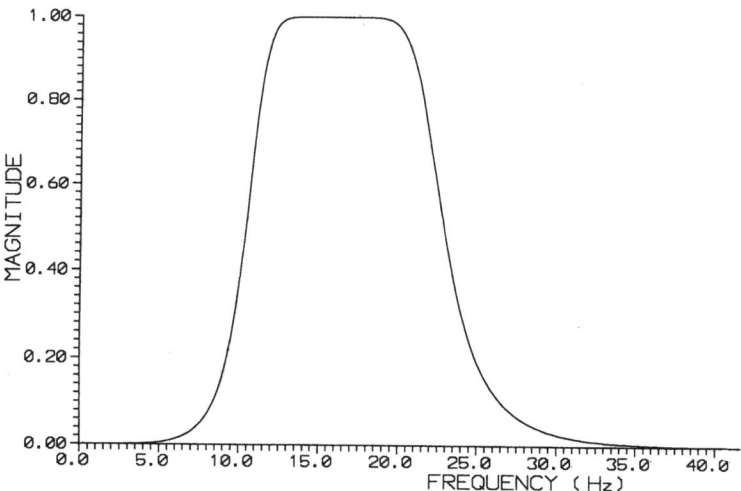

Figure 3

Response of the Butterworth bandpass filter for frequency band 10 to 20 Hz.

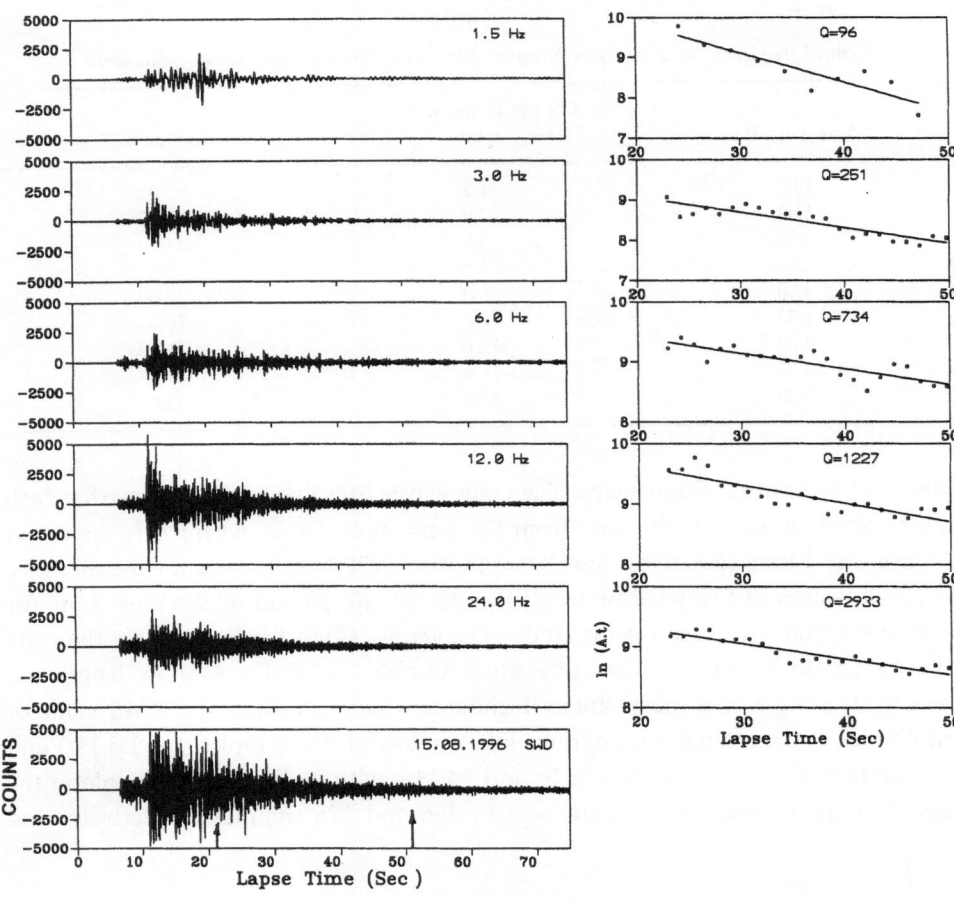

Figure 4a
An example of original and bandpass-filtered seismograms from station SWD recorded on 15.08.96 (no. 10 in Table 2). The coda waves portion of 30 sec window length is indicated by arrows (↑).

Figure 4b
Plot of logarithmic of geometrical spreading corrected and smoothed coda amplitudes as a function of lapse time for the window selected in Figure 4a. The best fitted linear line and estimated Q_c value for each central frequency are also shown in the figure.

which evaluates the RMS value at each step in the same frequency band. These RMS values constitute a smoother envelope of the coda which multiply by the lapse time (t) for applying the geometrical spreading correction. Figure 4b displays the logarithmic smoothed geometrical spreading corrected coda amplitudes of the coda part of filtered seismograms as shown in Figure 4a versus lapse time. The slope of this least-squares straight line, corresponding to each plot, provides the Q_c value for each central frequency using equation (2).

4. Results and Discussion

The quality factor of coda waves (Q_c) has been estimated for the Koyna region, using the digital data of thirteen local earthquakes. The events considered for the analysis are of local origin, with magnitudes less than 3, epicentral distances less than 55 km and focal depths are shallow and lie between 0.86 to 9.43 km. The single backscattering model has been applied on the coda waves of five lapse time windows of 20, 30, 40, 50 and 60 sec duration for 76 seismograms of these local earthquakes. Nine frequency bands of central frequency e.g., 1.5, 2.0, 3.0, 4.0, 6.0, 8.0, 12.0, 16.0 and 24.0 Hz have been used for this purpose. First, the results obtained on Q_c estimation using a 30 sec window length have been discussed, and later the Q_c values estimated as a function of lapse time windows are studied.

From the analysis of coda waves of 30 sec window length, a total 684 Q_c measurements were made in the nine frequency bands however, due to the constraint of high background noise level at some events/stations/frequencies, only 321 Q_c measurements of those which fulfill the criteria of having correlation coefficients (≥ 0.75), are considered. Values of Q_c measurements vary from 81 to 261 at frequency 1.5 Hz and from 2088 to 3234 at frequency 24.0 Hz, and as a

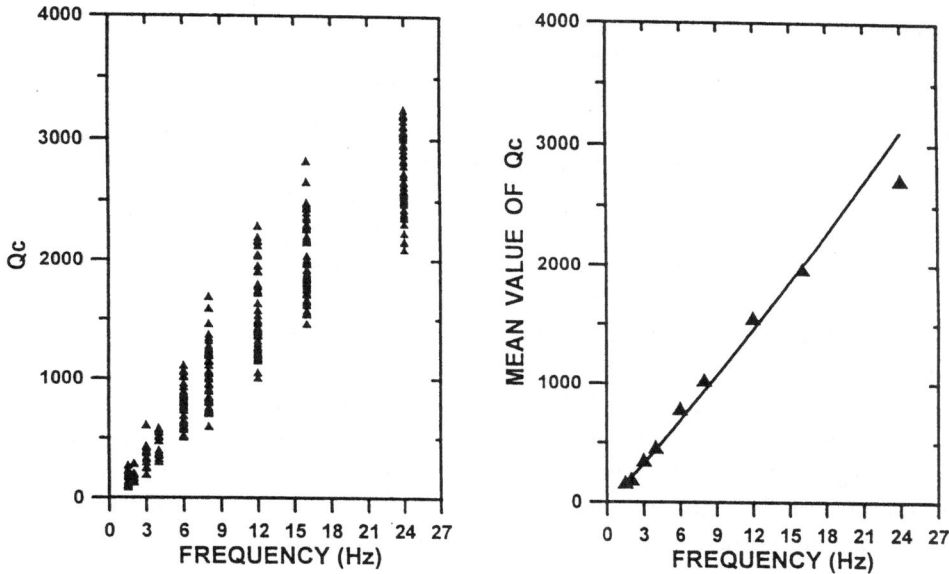

Figure 5a
Plot of all Q_c values as a function of frequency obtained for the Koyna region using 30 sec coda window length.

Figure 5b
Plot of mean values of Q_c as a function of frequency. A power law of the form $Q_c = Q_0 f^n$ has also been fitted, using all mean values and a relationship $Q_c = 96 f^{1.09}$, is obtained for the Koyna region.

Table 4

Mean value of Q_c and estimated error at each lapse time window length. N is the total number of observations made for each central frequency

f (Hz)	Mean value of Q_c and error at lapse time window of									
	20 sec		30 sec		40 sec		50 sec		60 sec	
	Q_c	N	Q_c	N	Q_c	N	Q_c	N	Q_c	N
1.5	105 ± 11.8	14	148 ± 13.5	17	187 ± 14.6	21	206 ± 13.2	18	233 ± 13.7	15
2.0	116 ± 16.4	6	178 ± 17.3	8	228 ± 13.1	14	250 ± 11.2	12	322 ± 18.8	13
3.0	247 ± 27.1	10	335 ± 26.3	15	421 ± 19.8	23	480 ± 22.3	20	560 ± 25.7	13
4.0	354 ± 13.1	8	443 ± 23.9	16	588 ± 29.5	24	714 ± 36.5	23	845 ± 40.7	14
6.0	640 ± 31.3	27	768 ± 29.8	37	1004 ± 37.2	43	1139 ± 44.8	36	1429 ± 55.9	32
8.0	768 ± 25.8	39	1014 ± 39.2	44	1325 ± 46.6	52	1524 ± 51.6	43	1839 ± 66.8	41
12.0	1231 ± 40.2	54	1539 ± 44.7	60	1784 ± 36.8	61	2144 ± 47.7	57	2493 ± 58.2	54
16.0	1604 ± 40.6	61	1956 ± 41.8	63	2209 ± 37.8	61	2535 ± 41.2	60	2849 ± 60.5	57
24.0	2229 ± 44.3	63	2703 ± 38.8	61	3002 ± 41.0	58	3252 ± 48.2	53	3531 ± 52.9	50

function of frequency are plotted in Figure 5a. The variation in the Q_c values observed at various central frequencies may be due (i) to the difference in local site geological conditions of various recording stations, (ii) to the difference in focal depths of the events (from 0.86 to 9.43 km), and (iii) to the difference in the epicentral distances of various events/stations pairs from 11 to 55 km. However, we tried to minimize this variation by considering a mostly similar coda analysis window length (e.g., 20–50 sec in the case of 30 sec window) for all the seismograms. It can be further minimized by reducing above differences in the data set. Therefore, to provide an average picture of the region, averaging of all the Q_c has been done by computing mean values of Q_c from all estimated Q_c values at each central frequency. For the 30 sec coda window length, these mean values of Q_c along with their errors from mean values estimated from the standard deviation are listed in Table 4, which vary from 148 ± 13.5 at 1.5 Hz to 2703 ± 38.8 at 24 Hz. These mean values of Q_c are plotted as a function of frequency in Figure 5b and a power law of the form ($Q_c = Q_0 f^n$) has been obtained as $Q_c = 96 f^{1.09}$ for the Koyna region, where $Q_0 = 96$ at $f = 1$ Hz and $n = 1.09$. Figures 5a and 5b and their relationship, $Q_c = 96 f^{1.09}$, illustrate that the Q_c is a function of frequency for the Koyna region and the Q_c value increases as frequency increases. The frequency-dependent Q_c relationship, $Q_c = 96 f^{1.09}$, provides an average attenuation characteristic of the medium properties of a localized zone around the Koyna region, covering an approximate surface area of about 11 500 km² with the lateral extent of about 120 km and depth of approximately 60 km.

To study any lateral changes in Q_c estimates in the region, all Q_c value measurements have been divided into two ellipsoids, A1 and A2. The area of A1 ellipsoid falls in the northern part of the region while the area of A2 ellipsoid falls in the southern part of the region (Fig. 6). The attenuation characteristics for A1 ellipsoid are obtained on the basis of 189 Q_c values estimated from the analysis of coda waves recorded at four stations namely, MNR, DVS, MWD and CKL, while the attenuation characteristics of A2 ellipsoid obtained from the 201 Q_c values estimated from the coda waves of six stations namely, MWD, YNP, WRN, SWD, SKP and KTL. Some of the area in both ellipsoids A1 and A2 is found to be common and some paths of events-stations-pairs for both these ellipsoids are also shown in Figure 6. The frequency-dependent Q_c relationships obtained for both these ellipsoids are $Q_c = 110f^{1.03}$ (for A1) and $Q_c = 92f^{1.10}$ (for A2). The study shows a small lateral variation in Q_c values from north to south. The variation in Q_0 and n indicates that the area of the southern part of the region seems to be more heterogeneous/active compared to the northern part of the region.

GUPTA and RAMBABU (1994) obtained Q in the region by using the data of thirty-one accelerograms with strong motion record of local earthquakes which have a hypocentral distance extending to 30 km. They found that Q value obtained in the region corresponds to the frequency-independent anelastic attenuation. A

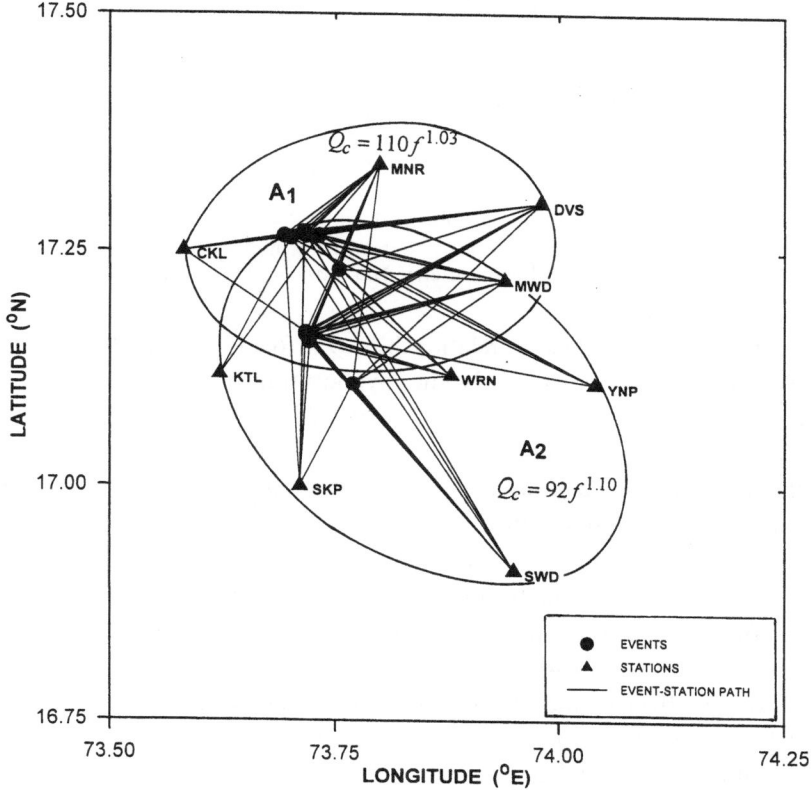

Figure 6

Map showing two zones of ellipsoids A1 and A2 located in north and south directions. Frequency dependence Q_c relationships obtained for both these ellipsoids are also given in the figure.

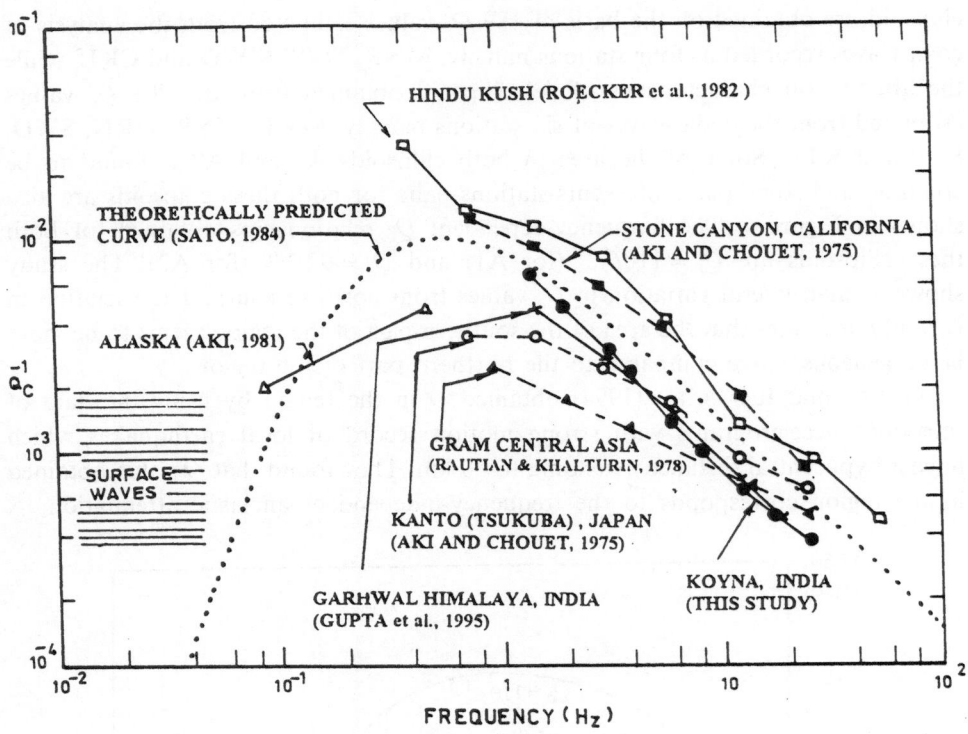

Figure 7
A comparison of Q_c^{-1} as a function of frequency obtained for the Koyna region with other tectonic regions in the world (modified after HERRAIZ and ESPINOSA, 1987).

source-site distance-dependent Q relationship, $Q = (6.456 \pm 1.093)D$, has been obtained for the region, where D is the hypocentral distance. They have also computed an empirical relationship as $Q = 19.22D\, e^{-0.214M_L}$ for the estimation of magnitude and distance-dependent Q for the region.

A comparison of Q_c^{-1} measurements, as a function of frequency, for the Koyna region, has been made in Figure 7 with the Q_c^{-1} observations of other tectonic regions in the world obtained by numerous investigators. Figure 7 represents that the Q_c^{-1} values of the Koyna region follow a substantially similar trend of Q_c^{-1} decay with frequency as the other tectonic regions and also a theoretically predicted curve is given by SATO (1984).

To study the effect of increasing time window length on the estimation of Q_c values in the region, the coda waves of all 76 seismograms have been analyzed at five lapse time window lengths of 20, 30, 40, 50 and 60 sec duration. The distribution of Q_c observations, made for each lapse time window, which fulfill the criteria of having correlation coefficient (≥ 0.75), are represented in Figure 8 and plots of these Q_c values as a function of frequency are shown in Figure 9.

Averaging of Q_c values has been accomplished at each frequency and their mean values of Q_c, along with the error estimated using the standard deviation, are listed in Table 4. The mean values of Q_c as a function of frequency show an increasing trend with the increasing window length (Fig. 10). This figure shows the empirical relationships, in the form of power law, obtained from the fitting of all the mean values as a function of frequency for each window length and these relationships along with the errors in Q_0 and n. Additionally the correlation coefficients for each window length are listed in Table 5. The increase in Q_c values with lapse time is attributed to an increase of Q_c with depth since the longer the analysis window, the larger will be the sampled area of the earth's crust and upper mantle (Table 5). The increase in Q_c with lapse time has also been interpreted as a function of depth by many investigators (e.g., ROECKER *et al.*, 1982; KVAMME and HAVSKOV, 1989; IBANEZ *et al.*, 1990; WOODGOLD, 1994; AKINCI, 1994; GUPTA *et al.*, 1996).

According to WOODGOLD (1994), the increase in Q_c with lapse time can be caused by any of several factors such as (i) consideration of non-zero source-receiver distance with the nonisotropic scattering, (ii) use of a 2-D model instead of a 3-D model, (iii) use of a single scattering model where multiple scattering is significant. How these factors are taken into consideration in the present study are briefly discussed here.

The consideration of source-receiver distance becomes significant in the case in which the lapse time window length is small enough and close to the direct S-wave arrival. However, in the present analysis, the starting time of all five lapse time

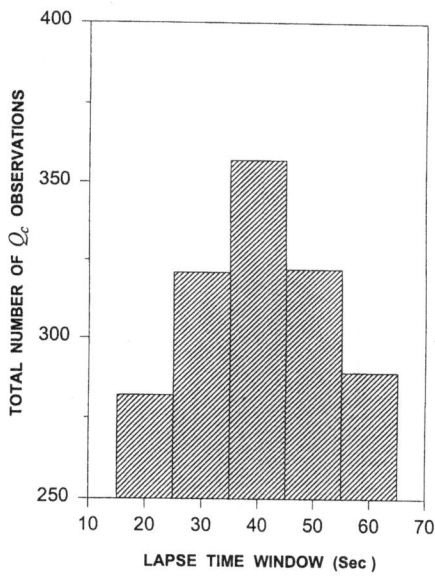

Figure 8

Distribution of Q_c observations made for five lapse time window lengths.

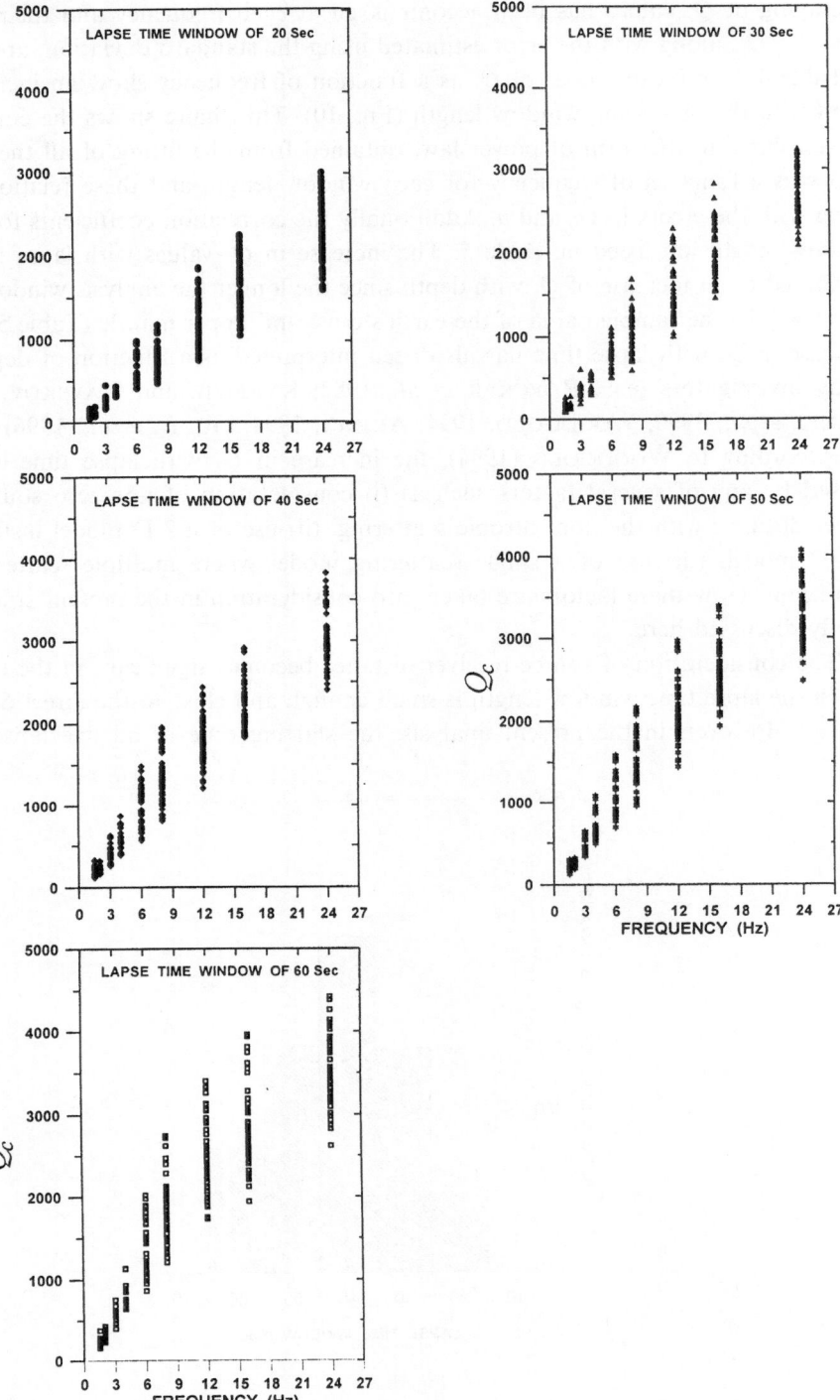

Figure 9
Plot of all Q_c values as a function of frequency obtained at different lapse time windows.

windows has been taken twice of the *S*-wave travel time (RAUTIAN and KHAL-TURIN, 1978). This time, in the seismogram, falls long after the arrival of all direct waves and where only backscattered waves arrive (AKI and CHOUET, 1975). Therefore, the consideration of the source receiver as coincident with isotropic scattering will have a similar effect on the results obtained from all the coda window lengths.

The consideration of the single scattering model for all the time windows will also effect the results in a similar manner as KOPNICHEV (1977) and GAO *et al.* (1983) demonstrated that the effect of multiple scattering becomes insignificant for local events with a lapse time less than 100 sec duration, and in the present analysis the lapse time for all the window lengths falls less below than that of the 100 sec duration.

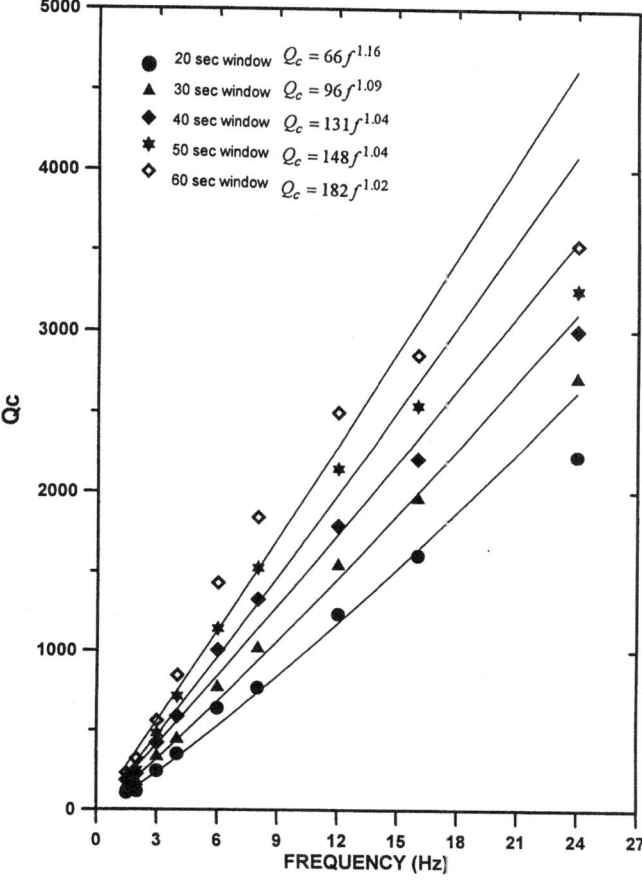

Figure 10

A comparison of mean values of Q_c as a function of frequency obtained at five lapse time windows. Power law fitted for each window is also shown in the figure.

Table 5

Empirical relationships for five lapse time windows are obtained from the mean values of Q_c (Table 4). The ranges given in Q_0 and n are computed at 90 percent confidence interval. The area coverage is computed using the formulation given by PULLI (1984)

Lapse time window (sec)	Empirical relationship	Correlation coefficient	Coverage of area (km²)	depth (km)
20	$Q_c = (66 \pm 11) f^{(1.16 \pm 0.10)}$	0.99	8500	50
30	$Q_c = (96 \pm 12) f^{(1.09 \pm 0.07)}$	0.99	11 500	60
40	$Q_c = (131 \pm 22) f^{(1.04 \pm 0.09)}$	0.98	15 400	70
50	$Q_c = (148 \pm 30) f^{(1.04 \pm 0.11)}$	0.97	20 000	80
60	$Q_c = (182 \pm 42) f^{(1.02 \pm 0.13)}$	0.97	25 500	90

The choice of the 2-D model (single scattering model) instead of the 3-D model seems to be appropriate as the analysis window is greater than 20 sec in which considerable energy is transferred into the mantle (WOODGOLD, 1994). However, the choice of the 2-D or 3-D model only has a small effect on the resultant Q_c value (JIN and AKI, 1988; WOODGOLD, 1994).

Variation in the degree of frequency dependence (n) with increasing lapse time window lengths has been observed in the study area. This value varies from 1.16 at the 20 sec window to 1.02 at the 60 sec window length. A strong correlation between the degree of frequency dependence and the level of tectonic activity in the area of measurement has also been made by several authors for a number of tectonic regions (e.g., AKI, 1980; PULLI and AKI, 1981; ROECKER *et al.*, 1982; VAN ECK, 1988; AKINCI *et al.*, 1994). They ascertained higher n value for tectonically active regions compared to the tectonically stable region. However, in an identical region of study, significant variation in the value of n with increasing lapse time as well as increasing depth, has also been observed by e.g., PULLI (1984) as $f^{0.95}$ (for < 100 sec lapse time) and $f^{0.40}$ (for > 100 sec lapse time), for New England from the data set with a depth range of 0 to 500 km, while ROECKER *et al.* (1982) obtained $f^{1.0}$ (100 km depth) to $f^{0.75}$ (400 km depth) and $f^{0.50}$ (1000 km depth) for the Hindukush region. Results from both these regions supply information for a very wide area with extensive depth. While in a region of study with smaller lapse time window lengths and shallow focal depths of events, no significant variation in n with lapse time has been observed, which can also be considered as constant (e.g., IBANEZ *et al.*, 1990 and AKINCI *et al.*, 1994 for the Granada basin, South Spain and the western Antolia region, Turkey, respectively).

In the present study, from the analysis of coda waves at different lapse time windows, in the frequency range from 1.5 to 24 Hz, the variation in the value of n as a function of lapse time window is small, which can also be considered as stationary (Table 5). This minor variation in the value of n can be so interpreted that the scattering effect, which is the dominant contributor for the degree of frequency dependence (n), shows a decreasing trend with depth in the depth range from 50–90 km. This may be due to a decrease in heterogeneities (density variation and fractures, etc.) of the medium with depth beneath the Koyna region, while Q_0 value (Q_c at 1 Hz) increases significantly in the region, with a lapse time from 66 to 182. This demonstrates that the effect due to the intrinsic attenuation dominates in the region in the depth range from 50–90 km.

5. Conclusions

In the present study, the Q_c values have been estimated for the Koyna region. The coda waves of 76 seismograms from thirteen local earthquakes recorded digitally around the Koyna region have been analyzed for five lapse time window lengths (e.g., 20, 30, 40, 50 and 60 sec) at nine frequency bands with a central frequency in the range of 1.5 Hz to 24.0 Hz. The focal depths and epicentral distances of these events, with magnitudes less than 3, vary from 0.86 to 9.43 km and 11 to 55 km, respectively.

The estimated Q_c values, for the lapse time window of 30 sec, vary from 81 to 261 at 1.5 Hz and from 2088 to 3234 at 24.0 Hz, while the average value of Q_c along with the standard error from mean vary from 148 ± 13.5 at 1.5 Hz to 2703 ± 38.8 at 24 Hz. Both from the Q_c and their mean values it is clear that the Q_c is a function of frequency in the Koyna region. The Q_c values increase as frequency increases. A frequency-dependent relationship, $Q_c = 96 f^{1.09}$, has also been obtained for the region, which represents the average attenuation properties of the medium of a localized zone around Koyna sampling an approximate surface area of about 11 500 km^2, with the surface extent of about 120 km and depth coverage of about 60 km by assuming the single scattering.

An analysis of coda waves at five lapse time window lengths indicates that Q_c is lapse time dependent in the region. The Q_c value increases as the time window increases. The increase in Q_c value with the time window is attributed to the increase in Q_c with depth. The frequency-dependent average Q_c relationships obtained at these window lengths are as $Q_c = 66 f^{1.16}$ (20 sec), $Q_c = 96 f^{1.09}$ (30 sec), $Q_c = 131 f^{1.04}$ (40 sec), $Q_c = 148 f^{1.04}$ (50 sec), $Q_c = 182 f^{1.02}$ (60 sec) show that there is a significant increase in Q_0 value (Q_c at 1 Hz) with increasing window length, while there is a nominal decrease in the degree of frequency dependence (n) with increasing window length, which can be considered almost stationary. This can also be interpreted that the scattering effect in the region exhibits a decreasing trend, with increasing depth from 50 to 90 km, which may be due to a decrease in the heterogeneities level of the medium while the intrinsic attenuation manifests a dominant role in the region in this depth range.

Acknowledgements

The authors are grateful to the Department of Science and Technology, Earth Sciences (ESS), Govt. of India, New Delhi for providing funds to operate the Seismological Network around the Koyna–Warna region, deployed by the National Geophysical Research Institute, Hyderabad. The Head of Department Earthquake Engineering is acknowledged for providing facilities to carry out this study. Suggestions and comments from Aybige Akinci for enhancement of the manuscripts are highly appreciated.

REFERENCES

AKI, K. (1969), *Analysis of the Seismic Coda of Local Earthquakes as Scattered Waves*, J. Geophys. Res. *74*, 615–631.

AKI, K. (1980), *Attenuation of Shear Waves in the Lithosphere for Frequencies from 0.05 to 25 Hz*, Phys. Earth Planet Inter. *21*, 50–60.

AKI, K., and CHOUET, B. (1975), *Origin of the Coda Waves: Source Attenuation and Scattering Effects*, J. Geophys. Res. *80*, 3322–3342.

AKINCI, A., TAKTAK, A. G., and ERGINTAV, S. (1994), *Attenuation of Coda Waves in Western Anatolia*, Phys. Earth Planet Inter. *87*, 155–165.

GAO, L. S., LEE, L. C., BISWAS, N. N., and AKI, K. (1983), *Comparison of the Single and Multiple Scattering on Coda Waves for Local Earthquakes*, Bull. Seismol. Soc. Am. *73*, 377–389.

GUPTA, H. K. (1985), *The Present Status of the Reservoir-induced Seismicity with the Special Emphasis on Koyna Earthquake*, Tectonophysics *118*, 257–507.

GUPTA, H. K., RASTOGI, B. K., and NARAIN, H. (1972), *Some Discriminatory Characteristics of Earthquakes Near the Kariba, Kremasta and Koyna Artificial Lakes*, Bull. Seismol. Soc. Am. *62*, 493–507.

GUPTA, I. D., and RAMBABU, V. (1994), *High-frequency Q Values from Strong Motion Acceleration Data for the Koyna Dam Region, India*, Proc. of Tenth Symp. on Earthq. Engin., Nov. 16–18, 1994, held at University of Roorkee, 83–92.

GUPTA, S. C., SINGH, V. N., and KUMAR, A. (1995), *Attenuation of Coda Waves in the Garhwal Himalaya, India*, Phys. Earth and Planet, Inter. *87*, 247–253.

GUPTA, S. C., KUMAR, A., SINGH, V. N., and BASU, S. (1996), *Lapse-time Dependence of Q_c in the Garhwal Himalaya*, Bull. Ind. Soc. Earth. Tech. *33*, 147–159.

HERRAIZ, M., and ESPINOSA, A. F. (1987), *Coda Waves: A Review*, Pure appl. geophys. *125*, 499–577.

IBANEZ, J. M., PEZZO, E. D., DE MIGUEL, F., HERRAIZ, M., ALGUACIE, G., and MORALES, J. (1990), *Depth-dependent Seismic Attenuation in the Granada Zone (Southern Spain)*, Bull. Seismol. Soc. Am. *80*, 1232–1244.

JIN, A., and AKI, K. (1988), *Spatial and Temporal Correlation between Coda Q and Seismicity in China*, Bull. Seismol. Soc. Am. *78*, 741–769.

KAILA, K. L., REDDY, P. R., DIXIT, M. M., and LAZARENKO, M. A. (1981a), *Deep Crustal Structure at Koyna, Maharashtra Indicated by Deep Seismic Sounding*, J. Geol. Soc. India *22*, 1–16.

KAILA, K. L., MURTHY, P. R. K., RAO, V. K., and KHARETCHKO, G. E. (1981b), *Crustal Structure from Deep Seismic Sounding along the Koyna 11 (Kelsi-Loni) Profile in the Deccan Trap Area, India*, Tectonophysics *73*, 365–384.

KNOPOFF, L. (1964), *Q*, Rev. Geophys. *2*, 625–660.

KOPNICHEV, Y. F. (1977), *The Role of Multiple Scattering in the Formation of Seismogram's Tail*, Izv. Akad. Nauk. SSSR, Fiz. Zemli *13*, 394–398 (in Russian).

KRISHNAN, M. S., *Geology of India and Burma*, 4th ed. (Higginbothams, Madras, 1960).

KVAMME, L. B. and HAVSKOV, J. (1989), *Q in Southern Norway*, Bull. Seismol. Soc. Am. *79*, 1575–1588.

LANGSTON, C. A. (1981), *Source Inversion of Seismic Waveforms: The Koyna India, Earthquake of September 13, 1967*, Bull. Seismol. Soc. Am. *75*, 1–24.

LEE, W. H. K., and LAHR, J. C. (1975), *HYPO71 (revised): A Computer Program for Determining the Hypocenter, Magnitude and First-motion Pattern of Local Earthquakes*, USGS Open File Report *75–311*, 1–116.

PULLI, J. J. (1984), *Attenuation of Coda Waves in New England*, Bull. Seismol. Soc. Am. *74*, 1149–1166.

PULLI, J. J., and AKI, K., *Attenuation of seismic waves in the lithosphere: comparison of active and stable areas. In Earthquakes and Earthquake Engineering: The Eastern United States* (ed. Beavers, J. E.) (Ann Arbor Science Publishers Inc., Ann Arbor, Michigan 1981) pp. 129–141.

RASTOGI, B. K., SHARMA, C. P. S., CHADDHA, R. K., and KUMAR, N., *Current Seismicity at the Koyna Reservoir, Maharashtra, India, Induced Seismicity* (Peter Knoll, ed., 1992) pp. 321–329.

RAUTIAN, T. G., and KHALTURIN, V. I. (1978), *The Use of the Coda for the Determination of the Earthquake Source Spectrum*, Bull. Seismol. Soc. Am. *68*, 923–948.

RAWAT, J. S. (1982), *Engineering Geological Studies at Warna Dam, Maharashtra, India*, Proc. 4th Congress Inter. Assoc. Eng. Geol. III, Theme I., 59–66.

ROECKER, S. W., TUCKER, B., KING, J., and HATZFIELD, D. (1982), *Estimates of Q in Central Asia as a Function of Frequency and Depth Using the Coda of Locally Recorded Earthquakes*, Bull. Seismol. Soc. Am. *72*, 129–149.

REHA, S. (1984), *Q Determined from Local Earthquakes in the South Carolina Coastal Plain*, Bull. Seismol. Soc. Am. *74*, 2257–2268.

SATO, H. (1984), *Attenuation of Envelope Formation of Three-component Seismograms of Small Local Earthquakes in Randomly Inhomogeneous Lithosphere*, J. Geophys. Res. *89*, 1221–1241.

STEARNS, D. S., and DAVID, R. A., *Signal Processing Algorithms* (Prentice-Hall, Inc., Englewood Cliffs, New Jersey 07632, 1988).

VAN ECK, T. (1988), *Attenuation of Coda Waves in the Dead Sea Region*, Bull. Seismol. Soc. Am. *78*, 770–779.

WOODGOLD, C. R. D. (1994), *Coda Q in the Charlevoix, Quebec, Region: Lapse-time Dependence and Spatial and Temporal Comparison*, Bull. Seismol. Soc. Am. *84*, 1123–1131.

(Received December 2, 1997, received June 16, 1998, accepted July 25, 1998)

 To access this journal online:
http://www.birkhauser.ch